Video Solutions

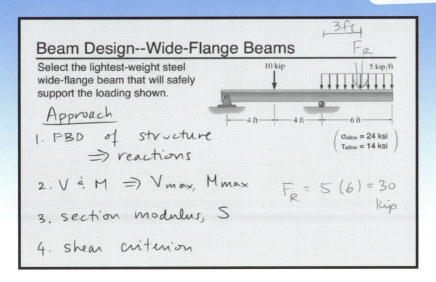

Video **Solutions** are step-by-step solution walkthroughs of representative homework problems.

Redeem access to:
Video **Solutions**, Pearson eText, and more at:
www.pearsonhighered.com/hibbeler

Use a coin to scratch off the coating and reveal your student access code.
Do not use a knife or other sharp object as it may damage the code.

The access code on this page can only be used once to establish a subscription. If the access code has already been scratched off, it may no longer be valid. If this is the case, visit the website above and select "Get Access" to purchase a new subscription. Please note that the access code does not include access to MasteringEngineering.

Technical Support is available at www.247pearsoned.com.

 PEARSON Pearson Education | One Lake Street, Upper Saddle River, NJ 07458

Beam Deflections and Slopes (continued)

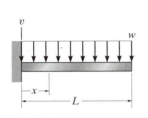

$$v_{max} = -\frac{wL^4}{8EI}$$
at $x = L$

$$\theta_{max} = -\frac{wL^3}{6EI}$$
at $x = L$

$$v = -\frac{w}{24EI}(x^4 - 4Lx^3 + 6L^2x^2)$$

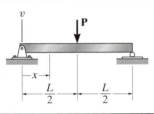

$$v_{max} = -\frac{PL^3}{48EI}$$
at $x = L/2$

$$\theta_{max} = \pm\frac{PL^2}{16EI}$$
at $x = 0$ or $x = L$

$$v = \frac{P}{48EI}(4x^3 - 3L^2x),$$
$$0 \le x \le L/2$$

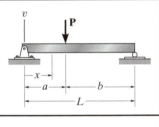

$$\theta_L = -\frac{Pab(L + b)}{6LEI}$$

$$\theta_R = \frac{Pab(L + a)}{6LEI}$$

$$v = -\frac{Pbx}{6LEI}(L^2 - b^2 - x^2)$$
$$0 \le x \le a$$

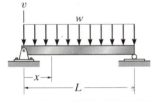

$$v_{max} = -\frac{5wL^4}{384EI}$$
at $x = \frac{L}{2}$

$$\theta_{max} = \pm\frac{wL^3}{24EI}$$

$$v = -\frac{wx}{24EI}(x^3 - 2Lx^2 + L^3)$$

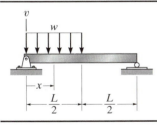

$$\theta_L = -\frac{3wL^3}{128EI}$$

$$\theta_R = \frac{7wL^3}{384EI}$$

$$v = -\frac{wx}{384EI}(16x^3 - 24Lx^2 + 9L^3)$$
$$0 \le x \le L/2$$

$$v = -\frac{wL}{384EI}(8x^3 - 24Lx^2 + 17L^2x - L^3)$$
$$L/2 \le x \le L$$

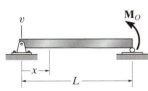

$$v_{max} = -\frac{M_OL^2}{9\sqrt{3}EI}$$

$$\theta_L = -\frac{M_OL}{6EI}$$

$$\theta_R = \frac{M_OL}{3EI}$$

$$v = -\frac{M_Ox}{6EIL}(L^2 - x^2)$$

STRUCTURAL ANALYSIS

NINTH EDITION

R. C. HIBBELER

PEARSON

Boston Columbus Indianapolis New York San Francisco Upper Saddle River
Amsterdam Cape Town Dubai London Madrid Milan Munich Paris
Montréal Toronto Delhi Mexico City São Paulo Sydney Hong Kong
Seoul Singapore Taipei Tokyo

Library of Congress Cataloging-in-Publication Data on File

Vice President and Editorial Director, ECS: *Marcia J. Horton*
Acquisitions Editor: *Norrin Dias*
Editorial Assistant: *Sandra Rodriguez*
Managing Editor: *Scott Disanno*
Production Editor: *Rose Keman*
Art Director: *Marta Samsel*
Art Editor: *Gregory Dulles*
Manufacturing Manager: *Mary Fischer*
Manufacturing Buyer: *Maura Zaldivar-Garcia*
Product Marketing Manager: *Bram Van Kempen*
Field Marketing Manager: *Demetrius Hall*
Marketing Assistant: *Jon Bryant*
Cover Designer: *Black Horse Designs*
Cover Image: © *Acnalesky/Fotolia*

Photos not otherwise credited are © R.C. Hibbeler.

Pearson Education Ltd., *London*
Pearson Education Australia Pty. Ltd., *Sydney*
Pearson Education Singapore, Pte. Ltd.
Pearson Eduction North Asia Ltd., *Hong Kong*
Pearson Education Canada, Inc., *Toronto*
Pearson Educación de Mexico, S.A. de C.V.
Pearson Education--Japan, *Tokyo*
Pearson Education Malaysia, Pte. Ltd.
Pearson Education, *Upper Saddle River, New Jersey*

Printed in the United States of America.

10 9 8 7 6 5 4 3 2

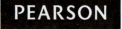

www.pearsonhighered.com

ISBN-10: 0-13-394284-8
ISBN-13: 978-0-13-394284-2

To The Student

With the hope that this work will stimulate
an interest in Structural Analysis
and provide an acceptable guide to its understanding.

PREFACE

This book is intended to provide the student with a clear and thorough presentation of the theory and application of structural analysis as it applies to trusses, beams, and frames. Emphasis is placed on developing the student's ability to both model and analyze a structure and to provide realistic applications encountered in professional practice.

For many years now, engineers have been using matrix methods to analyze structures. Although these methods are most efficient for a structural analysis, it is the author's opinion that students taking a first course in this subject should also be well versed in some of the more important classicial methods. Practice in applying these methods will develop a deeper understanding of the basic engineering sciences of statics and mechanics of materials. Also, problem-solving skills are further developed when the various techniques are thought out and applied in a clear and orderly way. By solving problems in this way one can better grasp the way loads are transmitted through a structure and obtain a more complete understanding of the way the structure deforms under load. Finally, the classical methods provide a means of checking computer results rather than simply relying on the generated output.

New Material and Content Revision. This edition now includes examples of the causes of structural failures, the concept of a load path, and an enhanced discussion for drawing shear and moment diagrams and the deflection of beams and frames. Chapter 17 has been added, which now provides a discussion of structural modeling concepts and a general description of how computer software is applied. Included are some structural modeling projects, along with a set of problems that require a computer analysis.

Structural Terminology. There are several places throughout the text where illustrations and discussion of additional terminology has been added, so that the student becomes familiar with the basic forms of fundamental structures and the names of their members.

Problem Arrangement. Different from the previous edition, the problems in each chapter are now placed at the end of the chapter. They are grouped with section headings for the convenience of assigning problems for homework.

New Problems. There are approximately 70% new problems in this edition. They retain a balance of easy, medium, and difficult applications. In addition, some new fundamental problems have been added that stress the importance of drawing frame moment diagrams and drawing deflected structures. Apart from the author, the problems have been checked by four other parties, namely Scott Hendricks, Karim Nora, Kurt Norlin, and Kai Beng Yap.

Additional Photos. The relevance of knowing the subject matter is reflected by the realistic applications depicted in many new and updated photos along with captions that are placed throughout the book.

Organization and Approach

The contents of each chapter are arranged into sections with specific topics categorized by title headings. Discussions relevant to a particular theory are succinct, yet thorough. In most cases, this is followed by a "procedure for analysis" guide, which provides the student with a summary of the important concepts and a systematic approach for applying the theory. The example problems are solved using this outlined method in order to clarify its numerical application. Problems are given at the end of each chapter, and are arranged to cover the material in sequential order. Moreover, for any topic they are arranged in approximate order of increasing difficulty.

Hallmark Elements

- **Photographs.** Many photographs are used throughout the book to explain how the principles of structural analysis apply to real-world situations.

- **Problems.** Most of the problems in the book depict realistic situations encountered in practice. It is hoped that this realism will both stimulate the student's interest in structural analysis and develop the skill to reduce any such problem from its physical description to a model or symbolic representation to which the appropriate theory can be applied. This modeling process is further discussed in Chapter 17. Throughout the book there is an approximate balance of problems using either SI or FPS units. The intent has been to develop problems that test the student's ability to apply the theory, keeping in mind that those problems requiring tedious calculations can be relegated to computer analysis.

- **Answers to Selected Problems.** The answers to selected problems are listed in the back of the book. Extra care has been taken in the presentation and solution of the problems, and all the problem sets have been reviewed and the solutions checked and rechecked to ensure both their clarity and numerical accuracy.

- **Example Problems.** All the example problems are presented in a concise manner and in a style that is easy to understand.

- **Illustrations.** Throughout the book, an increase in two-color art has been added, including many photorealistic illustrations that provide a strong connection to the 3-D nature of structural engineering.

- **Triple Accuracy Checking.** The edition has undergone rigorous accuracy checking and proofing of pages. Besides the author's review of all art pieces and pages, Scott Hendricks of Virginia Polytechnic Institute, Karim Nohra of the University of South Florida, and Kurt Norlin of Laurel Technical Services rechecked the page proofs and together reviewed the Solutions Manual.

- **Fundamental Problems.** These problem sets are selectively located at the end of most chapters. They offer students simple applications of the concepts and, therefore, provide them with the chance to develop their problem-solving skills before attempting to solve any of the standard problems that follow. You may consider these problems as extended examples since they *all have solutions and answers* that are given in the back of the book. Additionally, the fundamental problems offer students an excellent means of studying for exams, and they can be used at a later time to prepare for the exam necessary to obtain a professional engineering license.

Contents

This book is divided into three parts. The first part consists of seven chapters that cover the classical methods of analysis for statically determinate structures. Chapter 1 provides a discussion of the various types of structural forms and loads. Chapter 2 discusses the determination of forces at the supports and connections of statically determinate beams and frames. The analysis of various types of statically determinate trusses is given in Chapter 3, and shear and bending-moment functions and diagrams for beams and frames are presented in Chapter 4. In Chapter 5, the analysis of simple cable and arch systems is presented, and in Chapter 6 influence lines for beams, girders, and trusses are discussed. Finally, in Chapter 7 several common techniques for the approximate analysis of statically indeterminate structures are considered.

In the second part of the book, the analysis of statically indeterminate structures is covered in six chapters. Geometrical methods for calculating deflections are discussed in Chapter 8. Energy methods for finding deflections are covered in Chapter 9. Chapter 10 covers the analysis of statically indeterminate structures using the force method of analysis, in addition to a discussion of influence lines for beams. Then the displacement methods consisting of the slope-deflection method in Chapter 11 and moment distribution in Chapter 12 are discussed. Finally, beams and frames having nonprismatic members are considered in Chapter 13.

The third part of the book treats the matrix analysis of structures using the stiffness method. Trusses are discussed in Chapter 14, beams in Chapter 15, and frames in Chapter 16. Finally, Chapter 17 provides some basic ideas as to how to model a structure, and for using available software for solving problem in structural analysis. A review of matrix algebra is given in Appendix A.

Resources for Instructors

- **MasteringEngineering.** This online Tutorial Homework program allows you to integrate dynamic homework with automatic grading and adaptive tutoring. MasteringEngineering allows you to easily track the performance of your entire class on an assignment-by-assignment basis, or the detailed work of an individual student.

- **Instructor's Solutions Manual.** An instructor's solutions manual was prepared by the author. The manual was also checked as part of the Triple Accuracy Checking program.

- **Presentation Resources.** All art from the text is available in PowerPoint slide and JPEG format. These files are available for download from the Instructor Resource Center at www.pearsonhighered. com. If you are in need of a login and password for this site, please contact your local Pearson Prentice Hall representative.

- **Video Solutions.** Located on the Companion Website, Video Solutions offer step-by-step solution walkthroughs of representative homework problems from each chapter of the text. Make efficient use of class time and office hours by showing students the complete and concise problem solving approaches that they can access anytime and view at their own pace. The videos are designed to be a flexible resource to be used however each instructor and student prefers. A valuable tutorial resource, the videos are also helpful for student self-evaluation as students can pause the videos to check their understanding and work alongside the video. Access the videos at www.pearsonhighered.com/hibbeler and follow the links for the *Structural Analysis* text.

- **STRAN.** Developed by the author and Barry Nolan, a practicing engineer, STRAN is a downloadable program for use with Structural Analysis problems. Access STRAN on the Companion Website, www. pearsonhighered.com/hibbeler and follow the links for the *Structural Analysis* text. Complete instructions for how to use the software are included on the Companion Website.

Resources for Students

- **MasteringEngineering.** Tutorial homework problems emulate the instrutor's office-hour environment.

- **Companion Website.** The Companion Website provides practice and review materials including:
  - **Video Solutions**—Complete, step-by-step solution walkthroughs of representative homework problems from each chapter.

Videos offer:

- **Fully worked Solutions**—Showing every step of representative homework problems, to help students make vital connections between concepts.

- **Self-paced Instruction**—Students can navigate each problem and select, play, rewind, fast-forward, stop, and jump-to-sections within each problem's solution.

- **24/7 Access**—Help whenever students need it with over 20 hours of helpful review.

○ **STRAN**—A program you can use to solve two and three dimensional trusses and beams, and two dimensional frames. Instructions for downloading and how to use the program are available on the Companion Website.

An access code for the *Structural Analysis*, Ninth Edition Companion Website is included with this text. To redeem the code and gain access to the site, go to www.pearsonhighered.com/hibbeler and follow the directions on the access code card. Access can also be purchased directly from the site.

Acknowledgments

Through the years, over one hundred of my colleagues in the teaching profession and many of my students have made valuable suggestions that have helped in the development of this book, and I would like to hereby acknowledge all of their comments. I personally would like to thank the reviewers contracted by my editor for this new edition, namely:

Delong Zuo, *Texas Tech University*
Husam Najm, *Rutgers University*
Tomasz Arciszewski, *University of Colorado—Boulder*
Brian Swartz, *University of Hartford*
Vicki May, *Dartmouth College*
Thomas Boothby, *Penn State University*
Leroy Hulsey, *University of Alaska—Fairbanks*
Reagan Herman, *University of Houston*
Des Penny, *Southern Utah University*
Ahmet Pamuk, *Flordia State University*

Also, the constructive comments from Kai Beng Yap, and Barry Nolan, both practicing engineers are greatly appreciated. Finally, I would like to acknowledge the support I received from my wife Conny, who has always been very helpful in preparing the manuscript for publication.

I would greatly appreciate hearing from you if at any time you have any comments or suggestions regarding the contents of this edition.

Russell Charles Hibbeler
hibbeler@bellsouth.net

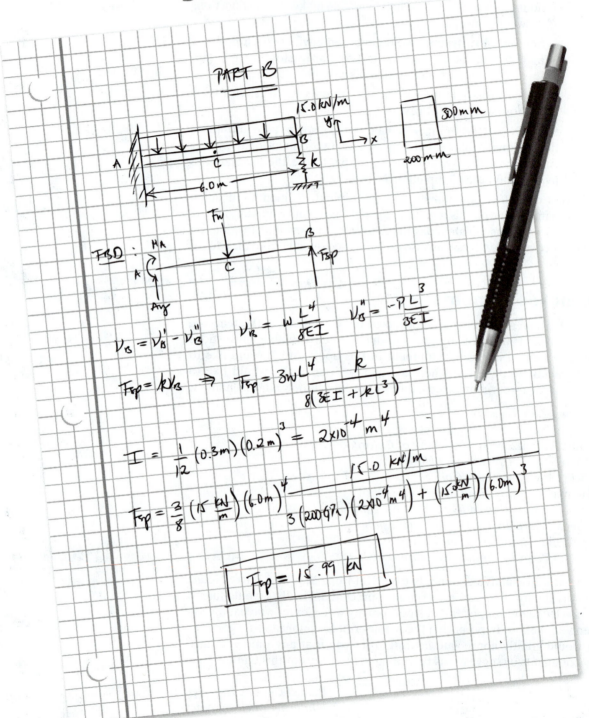

PART B

FBD:

$$V_B = V_B' - V_B'' \qquad V_B' = \frac{wL^4}{8EI} \qquad V_B'' = \frac{-PL^3}{3EI}$$

$$F_{sp} = kV_B \quad \Rightarrow \quad F_{sp} = \frac{3wL^4}{8} \cdot \frac{k}{(3EI + kL^3)}$$

$$I = \frac{1}{12}(0.3m)(0.2m)^3 = 2\times10^{-4}\, m^4$$

$$F_{sp} = \frac{3}{8}\left(15\,\tfrac{kN}{m}\right)(6.0m)^4 \cdot \frac{15.0\,\tfrac{kN}{m}}{3(200\,GPa)(2\times10^{-4}\,m^4) + \left(15\,\tfrac{kN}{m}\right)(6.0m)^3}$$

$$\boxed{F_{sp} = 15.99\ kN}$$

your answer specific feedback

Part B - Spring force at *B*

Using the method of superposition, determine the force $\mathbf{F}_{sp}$ that the spring at *B* exerts on the bar. Assume that this force acts in the positive *y* direction.

Express your answer to three significant figures and include the appropriate units.

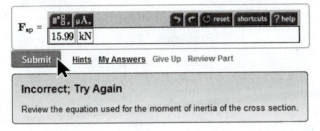

$\mathbf{F}_{sp} = $ | 15.99 | kN

Submit   Hints   My Answers   Give Up   Review Part

Incorrect; Try Again

Review the equation used for the moment of inertia of the cross section.

www.MasteringEngineering.com

CONTENTS

3
Analysis of Statically Determinate Trusses 83

4
Internal Loadings Developed in Structural Members 139

5
Cables and Arches 187

6
Influence Lines for Statically Determinate Structures 211

7
Approximate Analysis of Statically Indeterminate Structures 269

8
Deflections 305

9
Deflections Using Energy Methods 349

10
Analysis of Statically Indeterminate Structures by the Force Method 405

11

Displacement Method of Analysis: Slope-Deflection Equations   461

14
Truss Analysis Using the Stiffness Method 549

15
Beam Analysis Using the Stiffness Method 585

16
Plane Frame Analysis Using the Stiffness Method 605

17
Structural Modeling and Computer Analysis 623

Appendices

STRUCTURAL ANALYSIS

Chapter 1

© Joel Stahl/AP Images

Severe wind loadings caused by a hurricane have caused noticeable damage to the windows of the high-rise building.

Types of Structures and Loads

This chapter provides a discussion of some of the preliminary aspects of structural analysis. The phases of activity necessary to produce a structure are presented first, followed by an introduction to the basic types of structures, their components, and supports. Finally, a brief explanation is given of the various types of loads that must be considered for an appropriate analysis and design.

1.1   Introduction

In this text we will present many of the different ways engineers model and then analyze the loadings and deflections of various types of structures. Important examples related to civil engineering include buildings, bridges, and towers; and in other branches of engineering, ship and aircraft frames, and mechanical, and electrical supporting structures are important.

A *structure* refers to a system of connected parts used to support a load. When designing a structure to serve a specified function for public use, the engineer must account for its safety, esthetics, and serviceability, while taking into consideration economic and environmental constraints. Often this requires several independent studies of different solutions before final judgment can be made as to which structural form is most appropriate. This design process is both creative and technical and requires a fundamental knowledge of material properties and the laws of mechanics which govern material response. Once a preliminary design of a structure is proposed, the structure must then be *analyzed* to ensure that it has its required stiffness and strength. To analyze a structure properly, certain idealizations must be made as to how the members are supported and connected together. The loadings are determined from codes and local specifications, and the forces in the members and their displacements are found using the theory of structural analysis, which is

1

the subject matter of this text. The results of this analysis then can be used to redesign the structure, accounting for a more accurate determination of the weight of the members and their size. Structural design, therefore, follows a series of successive approximations in which every cycle requires a structural analysis. In this book, the structural analysis is applied to civil engineering structures; however, the method of analysis described can also be used for structures related to other fields of engineering.

1.2   Classification of Structures

It is important for a structural engineer to recognize the various types of elements composing a structure and to be able to classify structures as to their form and function. We will introduce some of these aspects now and discuss others throughout the text.

Structural Elements.   Some of the more common elements from which structures are composed are as follows.

Tie Rods. Structural members subjected to a *tensile force* are often referred to as **tie rods** or **bracing struts.** Due to the nature of this load, these members are rather slender, and are often chosen from rods, bars, angles, or channels, Fig. 1–1.

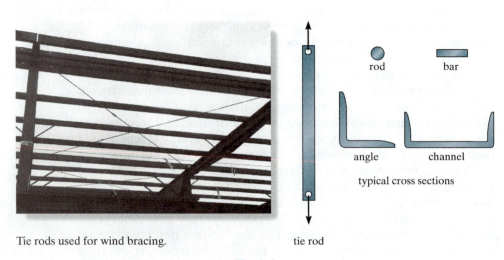

Tie rods used for wind bracing.

tie rod

rod

bar

angle

channel

typical cross sections

Fig. 1–1

Beams. **Beams** are usually straight horizontal members used primarily to carry vertical loads. Quite often they are classified according to the way they are supported, as indicated in Fig. 1–2. In particular, when the cross section varies the beam is referred to as tapered or haunched. Beam cross sections may also be "built up" by adding plates to their top and bottom.

Beams are primarily designed to resist bending moment; however, if they are short and carry large loads, the internal shear force may become quite large and this force may govern their design. When the material used for a beam is a metal such as steel or aluminum, the cross section is most efficient when it is shaped as shown in Fig. 1–3. Here the forces developed in the top and bottom *flanges* of the beam form the necessary couple used to resist the applied moment **M**, whereas the **web** is effective in resisting the applied shear **V**. This cross section is commonly referred to as a "wide flange," and it is normally formed as a single unit in a rolling mill in lengths up to 75 ft (23 m). If shorter lengths are needed, a cross section having tapered flanges is sometimes selected. When the beam is required to have a very large span and the loads applied are rather large, the cross section may take the form of a *plate girder*. This member is fabricated by using a large plate for the web and welding or bolting plates to its ends for flanges. The girder is often transported to the field in segments, and the segments are designed to be spliced or joined together at points where the girder carries a small internal moment.

Concrete beams generally have rectangular cross sections, since it is easy to construct this form directly in the field. Because concrete is rather weak in resisting tension, steel "reinforcing rods" are cast into the beam within regions of the cross section subjected to tension. Precast concrete beams or girders are fabricated at a shop or yard in the same manner and then transported to the job site.

Beams made from timber may be sawn from a solid piece of wood or laminated. **Laminated beams** are constructed from solid sections of wood, which are fastened together using high-strength glues.

Shown are typical splice plate joints used to connect the steel girders of a highway bridge.

The prestressed concrete girders are simply supported and are used for this highway bridge.

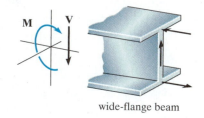

wide-flange beam

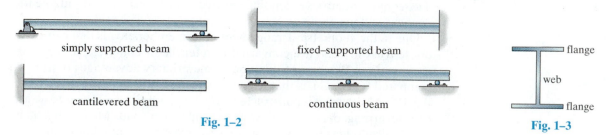

simply supported beam

fixed–supported beam

cantilevered beam

continuous beam

Fig. 1–2

flange

web

flange

Fig. 1–3

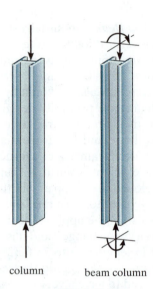

Wide-flange members are often used for columns. Here is an example of a beam column.

column   beam column

Fig. 1–4

Columns. Members that are generally vertical and resist axial compressive loads are referred to as *columns,* Fig. 1–4. Tubes and wide-flange cross sections are often used for metal columns, and circular and square cross sections with reinforcing rods are used for those made of concrete. Occasionally, columns are subjected to both an axial load and a bending moment as shown in the figure. These members are referred to as *beam columns.*

Types of Structures.

The combination of structural elements and the materials from which they are composed is referred to as a *structural system.* Each system is constructed of one or more of four basic types of structures. Ranked in order of complexity of their force analysis, they are as follows.

Trusses. When the span of a structure is required to be large and its depth is not an important criterion for design, a truss may be selected. *Trusses* consist of slender elements, usually arranged in triangular fashion. *Planar trusses* are composed of members that lie in the same plane and are frequently used for bridge and roof support, whereas *space trusses* have members extending in three dimensions and are suitable for derricks and towers.

Due to the geometric arrangement of its members, loads that cause the entire truss to bend are converted into tensile or compressive forces in the members. Because of this, one of the primary advantages of a truss, compared to a beam, is that it uses less material to support a given load, Fig. 1–5. Also, a truss is constructed from *long and slender elements,* which can be arranged in various ways to support a load. Most often it is economically feasible to use a truss to cover spans ranging from 30 ft (9 m) to 400 ft (122 m), although trusses have been used on occasion for spans of greater lengths.

Loading causes bending of the truss, which develops compression in the top members, and tension in the bottom members.

Fig. 1–5

Cables and Arches. Two other forms of structures used to span long distances are the cable and the arch. *Cables* are usually flexible and carry their loads in tension. They are commonly used to support bridges, Fig. 1–6a, and building roofs. When used for these purposes, the cable has an advantage over the beam and the truss, especially for spans that are greater than 150 ft (46 m). Because they are always in tension, cables will not become unstable and suddenly collapse, as may happen with beams or trusses. Furthermore, the truss will require added costs for construction and increased depth as the span increases. Use of cables, on the other hand, is limited only by their sag, weight, and methods of anchorage.

The *arch* achieves its strength in compression, since it has a reverse curvature to that of the cable. The arch must be rigid, however, in order to maintain its shape, and this results in secondary loadings involving shear and moment, which must be considered in its design. Arches are frequently used in bridge structures, Fig. 1–6b, dome roofs, and for openings in masonry walls.

Cables support their loads in tension.

(a)

Arches support their loads in compression.

(b)

Fig. 1–6

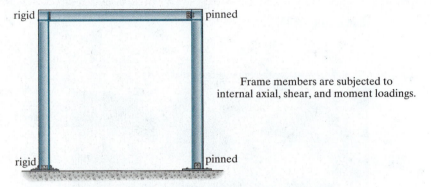

rigid   pinned

Frame members are subjected to
internal axial, shear, and moment loadings.

rigid   pinned

Fig. 1–7

Typical steel framework.

Frames. *Frames* are often used in buildings and are composed of beams and columns that are either pin or fixed connected, Fig. 1–7. Like trusses, frames extend in two or three dimensions. The loading on a frame causes bending of its members, and if it has rigid joint connections, this structure is generally "indeterminate" from a standpoint of analysis. The strength of such a frame is derived from the moment interactions between the beams and the columns at the rigid joints.

Surface Structures. A *surface structure* is made from a material having a very small thickness compared to its other dimensions. Sometimes this material is very flexible and can take the form of a tent or air-inflated structure. In both cases the material acts as a membrane that is subjected to pure tension.

Surface structures may also be made of rigid material such as reinforced concrete. As such they may be shaped as folded plates, cylinders, or hyperbolic paraboloids, and are referred to as *thin plates* or *shells.* These structures act like cables or arches since they support loads primarily in tension or compression, with very little bending. In spite of this, plate or shell structures are generally very difficult to analyze, due to the three-dimensional geometry of their surface. Such an analysis is beyond the scope of this text and is instead covered in texts devoted entirely to this subject.

© Bob Krist/Documentary Value/Corbis

The roof of the "Georgia Dome" in Atlanta, Georgia can be considered as a thin membrane.

1.3 Loads

Once the dimensional requirements for a structure have been defined, it becomes necessary to determine the loads the structure must support. Often, it is the anticipation of the various loads that will be imposed on the structure that provides the basic type of structure that will be chosen for design. For example, high-rise structures must endure large lateral loadings caused by wind, and so shear walls and tubular frame systems are selected, whereas buildings located in areas prone to earthquakes must be designed having ductile frames and connections.

Once the structural form has been determined, the actual design begins with those elements that are subjected to the primary loads the structure is intended to carry, and proceeds in sequence to the various supporting members until the foundation is reached. Thus, a building floor slab would be designed first, followed by the supporting beams, columns, and last, the foundation footings. In order to design a structure, it is therefore necessary to first specify the loads that act on it.

The design loading for a structure is often specified in codes. In general, the structural engineer works with two types of codes: general building codes and design codes. *General building codes* specify the requirements of governmental bodies for minimum design loads on structures and minimum standards for construction. *Design codes* provide detailed technical standards and are used to establish the requirements for the actual structural design. Table 1.1 lists some of the important codes used in practice. It should be realized, however, that codes provide only a general guide for design. *The ultimate responsibility for the design lies with the structural engineer.*

Since a structure is generally subjected to several types of loads, a brief discussion of these loadings will now be presented to illustrate how one must consider their effects in practice.

TABLE 1.1   Codes

General Building Codes

Minimum Design Loads for Buildings and Other Structures, ASCE/SEI 7-10, American Society of Civil Engineers
International Building Code

Design Codes

Building Code Requirements for Reinforced Concrete, Am. Conc. Inst. (ACI)
Manual of Steel Construction, American Institute of Steel Construction (AISC)
Standard Specifications for Highway Bridges, American Association of State Highway and Transportation Officials (AASHTO)
National Design Specification for Wood Construction, American Forest and Paper Association (AFPA)
Manual for Railway Engineering, American Railway Engineering Association (AREA)

Dead Loads. *Dead loads* consist of the weights of the various structural members and the weights of any objects that are permanently attached to the structure. Hence, for a building, the dead loads include the weights of the columns, beams, and girders, the floor slab, roofing, walls, windows, plumbing, electrical fixtures, and other miscellaneous attachments.

In some cases, a structural dead load can be estimated satisfactorily from simple formulas based on the weights and sizes of similar structures. Through experience one can also derive a "feeling" for the magnitude of these loadings. For example, the average weight for timber buildings is 40–50 lb/ft² (1.9–2.4 kN/m²), for steel framed buildings it is 60–75 lb/ft² (2.9–3.6 kN/m²), and for reinforced concrete buildings it is 110–130 lb/ft² (5.3–6.2 kN/m²). Ordinarily, though, once the materials and sizes of the various components of the structure are determined, their weights can be found from tables that list their densities.

TABLE 1.2 Minimum Densities for Design Loads from Materials*

	lb/ft³	kN/m³
Aluminum	170	26.7
Concrete, plain cinder	108	17.0
Concrete, plain stone	144	22.6
Concrete, reinforced cinder	111	17.4
Concrete, reinforced stone	150	23.6
Clay, dry	63	9.9
Clay, damp	110	17.3
Sand and gravel, dry, loose	100	15.7
Sand and gravel, wet	120	18.9
Masonry, lightweight solid concrete	105	16.5
Masonry, normal weight	135	21.2
Plywood	36	5.7
Steel, cold-drawn	492	77.3
Wood, Douglas Fir	34	5.3
Wood, Southern Pine	37	5.8
Wood, spruce	29	4.5

*Minimum Densities for Design Loads from Materials, Reproduced with permission from American Society of Civil Engineers *Minimum Design Loads for Buildings and Other Structures*, ASCE/SEI 7-10. Copies of this standard may be purchaed from ASCE at www.pubs.asce.org, American Society of Civil Engineers.

TABLE 1.3 Minimum Design Dead Loads*

Walls	psf	kN/m²
4-in. (102 mm) clay brick	39	1.87
8-in. (203 mm) clay brick	79	3.78
12-in. (305 mm) clay brick	115	5.51
Frame Partitions and Walls		
Exterior stud walls with brick veneer	48	2.30
Windows, glass, frame and sash	8	0.38
Wood studs 2 × 4 in. (51 × 102 mm), unplastered	4	0.19
Wood studs 2 × 4 in. (51 × 102 mm), plastered one side	12	0.57
Wood studs 2 × 4 in. (51 × 102 mm), plastered two sides	20	0.96
Floor Fill		
Cinder concrete, per inch (mm)	9	0.017
Lightweight concrete, plain, per inch (mm)	8	0.015
Stone concrete, per inch (mm)	12	0.023
Ceilings		
Acoustical fiberboard	1	0.05
Plaster on tile or concrete	5	0.24
Suspended metal lath and gypsum plaster	10	0.48
Asphalt shingles	2	0.10
Fiberboard, ½-in. (13 mm)	0.75	0.04

*Minimum Design Dead Loads, Reproduced with permission from American Society of Civil Engineers *Minimum Design Loads for Buildings and Other Structures*, ASCE/SEI 7-10, American Society of Civil Engineers.

The densities of typical materials used in construction are listed in Table 1.2, and a portion of a table listing the weights of typical building components is given in Table 1.3. Although calculation of dead loads based on the use of tabulated data is rather straightforward, it should be realized that in many respects these loads will have to be estimated in the initial phase of design. These estimates include nonstructural materials such as prefabricated facade panels, electrical and plumbing systems, etc. Furthermore, even if the material is specified, the unit weights of elements reported in codes may vary from those given by manufacturers, and later use of the building may include some changes in dead loading. As a result, estimates of dead loadings can be in error by 15% to 20% or more.

Normally, the dead load is not large compared to the design load for simple structures such as a beam or a single-story frame; however, for multistory buildings it is important to have an accurate accounting of all the dead loads in order to properly design the columns, especially for the lower floors.

EXAMPLE 1.1

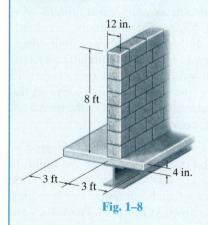

12 in.

8 ft

3 ft

3 ft

4 in.

Fig. 1–8

The floor beam in Fig. 1–8 is used to support the 6-ft width of a lightweight plain concrete slab having a thickness of 4 in. The slab serves as a portion of the ceiling for the floor below, and therefore its bottom is coated with plaster. Furthermore, an 8-ft-high, 12-in.-thick lightweight solid concrete block wall is directly over the top flange of the beam. Determine the loading on the beam measured per foot of length of the beam.

SOLUTION

Using the data in Tables 1.2 and 1.3, we have

Concrete slab: $\left[8\text{ lb}/(\text{ft}^2 \cdot \text{in.})\right](4\text{ in.})(6\text{ ft}) = $ 192 lb/ft

Plaster ceiling: $(5\text{ lb}/\text{ft}^2)(6\text{ ft}) = $ 30 lb/ft

Block wall: $(105\text{ lb}/\text{ft}^3)(8\text{ ft})(1\text{ ft}) = $ 840 lb/ft

Total load   1062 lb/ft = 1.06 k/ft

Ans.

Here the unit k stands for "kip," which symbolizes kilopounds. Hence, 1 k = 1000 lb.

1

It is important to find the position of this moving load where it causes the largest compression in this bridge pier.

Live Loads. *Live loads* can vary both in their magnitude and location. They may be caused by the weights of objects temporarily placed on a structure, moving vehicles, or natural forces. The minimum live loads specified in codes are determined from studying the history of their effects on existing structures. Usually, these loads include additional protection against excessive deflection or sudden overload. In Chapter 6 we will develop techniques for specifying the proper location of live loads on the structure so that they cause the greatest stress or deflection of the members. Various types of live loads will now be discussed.

Building Loads. The floors of buildings are assumed to be subjected to *uniform live loads,* which depend on the purpose for which the building is designed. These loadings are generally tabulated in local, state, or national codes. A representative sample of such *minimum live loadings,* taken from the ASCE 7-10 Standard, is shown in Table 1.4. The values are determined from a history of loading various buildings. They include some protection against the possibility of overload due to emergency situations, construction loads, and serviceability requirements due to vibration. In addition to uniform loads, some codes specify *minimum concentrated live loads,* caused by hand carts, automobiles, etc., which must also be applied anywhere to the floor system. For example, both uniform and concentrated live loads must be considered in the design of an automobile parking deck.

TABLE 1.4 Minimum Live Loads*

Occupancy or Use	Live Load psf	Live Load kN/m²	Occupancy or Use	Live Load psf	Live Load kN/m²
Assembly areas and theaters			Residential		
Fixed seats	60	2.87	Dwellings (one- and two-family)	40	1.92
Movable seats	100	4.79	Hotels and multifamily houses		
Garages (passenger cars only)	40	1.92	Private rooms and corridors	40	1.92
Office buildings			Public rooms and corridors	100	4.79
Lobbies	100	4.79	Schools		
Offices	50	2.40	Classrooms	40	1.92
Storage warehouse			First-floor corridors	100	4.79
Light	125	6.00	Corridors above first floor	80	3.83
Heavy	250	11.97			

*Minimum Live Loads, Reproduced with permission from American Society of Civil Engineers *Minimum Design Loads for Buildings and Other Structures*, ASCE/SEI 7-10, American Society of Civil Engineers.

For some types of buildings having very large floor areas, many codes will allow a *reduction* in the uniform live load for a *floor,* since it is unlikely that the prescribed live load will occur simultaneously throughout the entire structure at any one time. For example, ASCE 7-10 allows a reduction of live load on a member having an *influence area* $(K_{LL} A_T)$ of 400 ft² (37.2 m²) or more. This reduced live load is calculated using the following equation:

$$L = L_o \left(0.25 + \frac{15}{\sqrt{K_{LL} A_T}} \right) \quad \text{(FPS units)}$$

$$L = L_o \left(0.25 + \frac{4.57}{\sqrt{K_{LL} A_T}} \right) \quad \text{(SI units)}$$

$$(1\text{--}1)$$

where

L = reduced design live load per square foot or square meter of area supported by the member.

L_o = unreduced design live load per square foot or square meter of area supported by the member (see Table 1.4).

K_{LL} = live load element factor. For interior columns $K_{LL} = 4$.

A_T = tributary area in square feet or square meters.*

The reduced live load defined by Eq. 1–1 is limited to not less than 50% of L_o for members supporting one floor, or not less than 40% of L_o for members supporting more than one floor. No reduction is allowed for loads exceeding 100 lb/ft² (4.79 kN/m²), or for structures used for public assembly, garages, or roofs. Example 1.2 illustrates Eq. 1–1's application.

*Specific examples of the determination of tributary areas for beams and columns are given in Sec. 2.1.

EXAMPLE   **1.2**

A two-story office building shown in the photo has interior columns that are spaced 22 ft apart in two perpendicular directions. If the (flat) roof loading is 20 lb/ft², determine the reduced live load supported by a typical interior column located at ground level.

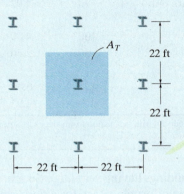

Fig. 1–9

SOLUTION

As shown in Fig. 1–9, each interior column has a tributary area or effective loaded area of $A_T = (22 \text{ ft})(22 \text{ ft}) = 484 \text{ ft}^2$. A ground-floor column therefore supports a roof live load of

$$F_R = (20 \text{ lb/ft}^2)(484 \text{ ft}^2) = 9680 \text{ lb} = 9.68 \text{ k}$$

This load cannot be reduced, since it is not a floor load. For the second floor, the live load is taken from Table 1.4: $L_o = 50 \text{ lb/ft}^2$. Since $K_{LL} = 4$, then $4A_T = 4(484 \text{ ft}^2) = 1936 \text{ ft}^2$ and $1936 \text{ ft}^2 > 400 \text{ ft}^2$, the live load can be reduced using Eq. 1–1. Thus,

$$L = 50\left(0.25 + \frac{15}{\sqrt{1936}}\right) = 29.55 \text{ lb/ft}^2$$

The load reduction here is $(29.55/50)100\% = 59.1\% > 50\%$. O.K. Therefore,

$$F_F = (29.55 \text{ lb/ft}^2)(484 \text{ ft}^2) = 14\,300 \text{ lb} = 14.3 \text{ k}$$

The total live load supported by the ground-floor column is thus

$$F = F_R + F_F = 9.68 \text{ k} + 14.3 \text{ k} = 24.0 \text{ k} \qquad \textit{Ans.}$$

Highway Bridge Loads. The primary live loads on bridge spans are those due to traffic, and the heaviest vehicle loading encountered is that caused by a series of trucks. Specifications for truck loadings on highway bridges are reported in the *LRFD Bridge Design Specifications* of the American Association of State and Highway Transportation Officials (AASHTO). For two-axle trucks, these loads are designated with an H, followed by the weight of the truck in tons and another number which gives the year of the specifications in which the load was reported. H-series truck weights vary from 10 to 20 tons. However, bridges located on major highways, which carry a great deal of traffic, are often designed for two-axle trucks plus a one-axle semitrailer as in Fig. 1–10. These are designated as HS loadings. In general, a truck loading selected for design depends upon the type of bridge, its location, and the type of traffic anticipated.

Fig. 1–10

The size of the "standard truck" and the distribution of its weight is also reported in the specifications. Although trucks are assumed to be on the road, all lanes on the bridge need not be fully loaded with a row of trucks to obtain the critical load, since such a loading would be highly improbable. The details are discussed in Chapter 6.

Railroad Bridge Loads. The loadings on railroad bridges, as in Fig. 1–11, are specified in the *Specifications for Steel Railway Bridges* published by the American Railroad Engineers Association (AREA). Normally, E loads, as originally devised by Theodore Cooper in 1894, were used for design. B. Steinmann has since updated Cooper's load distribution and has devised a series of M loadings, which are currently acceptable for design. Since train loadings involve a complicated series of concentrated forces, to simplify hand calculations, tables and graphs are sometimes used in conjunction with influence lines to obtain the critical load. Also, computer programs are used for this purpose.

Fig. 1–11

1

Impact Loads. Moving vehicles may bounce or sidesway as they move over a bridge, and therefore they impart an *impact* to the deck. The percentage increase of the live loads due to impact is called the *impact factor, I.* This factor is generally obtained from formulas developed from experimental evidence. For example, for highway bridges the AASHTO specifications require that

$$I = \frac{50}{L + 125} \qquad \text{but not larger than 0.3}$$

where L is the length of the span in feet that is subjected to the live load.

In some cases provisions for impact loading on the structure of a building must also be taken into account. For example, the ASCE 7-10 Standard requires the weight of elevator machinery to be increased by 100%, and the loads on any hangers used to support floors and balconies to be increased by 33%.

Wind Loads. When the speed of the wind is very high, it can cause massive damage to a structure. The reason is that the pressure created by the wind is proportional to the *square* of the wind speed. In large *hurricanes* this speed can reach over 100 mi/h (161 km/h); however, by comparison, an F5 *tornado* (Fujita scale) has wind speeds over 300 mi/h (483 km/h)!

To understand the effect of a horizontal wind blowing over and around a building, consider the simple structure shown in Fig. 1–12. Here the positive pressure (pushing) on the front of the building is intensified, because the front will arrest the flow and redirect it over the roof and along the sides. Because air flows faster around the building, by the Bernoulli effect this higher velocity will cause a lower pressure (suction). This is especially true at the corners, under the eaves, and at the ridge of the roof. Here the wind is redirected and the damage is the greatest. Behind the building there is also a suction, which produces a wake within the air stream.

The destruction due to the wind is increased if the building has an opening, If the opening is at the front, then the pressure within the building is increased and this intensifies the external suction on the back, side walls, and the roof. If the opening is on a side wall, then the opposite effect occurs. Air will be sucked out of the building, lowering its inside pressure, and intensifying the pressure acting externally on the front of the building.

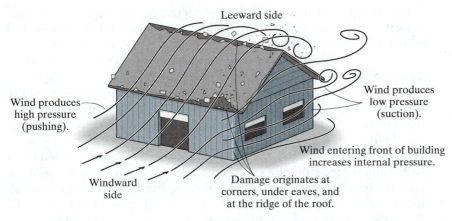

Leeward side

Wind produces high pressure (pushing).

Wind produces low pressure (suction).

Wind entering front of building increases internal pressure.

Windward side

Damage originates at corners, under eaves, and at the ridge of the roof.

Fig. 1–12

For a high-rise building, the wind loading can be quite complex, and so these structures are often designed based on the behavior of a model of the building, tested in a wind tunnel. When doing so, it is important to consider the wind striking the structure from *any direction*.*

The effects of lateral loadings developed by wind, can cause ***racking***, or leaning of a building frame. To resist this effect, engineers often use cross bracing, knee or diagonal bracing, or shear walls. Examples that show the use of these members are indicated in the photos.

wind

Shear walls.

Cross bracing.

Knee bracing.

Diagonal bracing.

*You may want to investigate the case of the initial design of Citigroup Center. Construction of this skyscraper was completed in New York City in 1977, and only *afterwards* was it realized that the *quartering winds,* that is, those directed at the corners of the building, would produce enough force to actually collapse the building. Retrofits had to be made to the connections to provide the necessary strength to stiffen the structure. See http://failures.wikispaces.com/Citicorp+Center.

1

The effect of wind on a structure depends upon the density and velocity of the air, the angle of incidence of the wind, the shape and stiffness of the structure, and the roughness of its surface. For design purposes, wind loadings can be treated using either a static or a dynamic approach.

For the static approach, the fluctuating pressure caused by a constantly blowing wind is approximated by a mean velocity pressure that acts on the structure. This pressure q is defined by the air's kinetic energy per unit volume, $q = \frac{1}{2}\rho V^2$, where ρ is the density of the air and V is its velocity. According to the ASCE 7-10 Standard, this equation is modified to account for the structure's height, and the terrain in which it is located. Also the importance of the structure is considered, as it relates to the risk to human life or the public welfare if it is damaged or loses its functionality. This modified equation is represented by the following equation.

Hurricane winds caused this damage to a condominium in Miami, Florida.

© Jeff Greenberg/Alamy

Some high-rise buildings must be able to resist hurricane winds having speeds of over 120 mi / h.

$$q_z = 0.00256K_z K_{zt} K_d V^2 \ (\text{lb/ft}^2)$$

$$q_z = 0.613K_z K_{zt} K_d V^2 \ (\text{N/m}^2)$$

(1–2)

where

V = the velocity in mi/h (m/s) of a 3-second gust of wind measured 33 ft (10 m) above the ground. Specific values depend upon the "category" of the structure obtained from a specified wind map. For example, the interior portion of the continental United States reports a wind speed of 105 mi/h (47 m/s) if the structure is an agricultural or storage building, since it is of low risk to human life in the event of a failure. The wind speed is 120 mi/h (54 m/s) for cases where the structure is a hospital, since its failure would cause substantial loss of human life.

K_z = the velocity pressure exposure coefficient, which is a function of height and depends upon the ground terrain. Table 1.5 lists values for a structure which is located in open terrain with scattered low-lying obstructions.

K_{zt} = a factor that accounts for wind speed increases due to hills and escarpments. For flat ground K_{zt} = 1.0.

K_d = a factor that accounts for the direction of the wind. It is used only when the structure is subjected to combinations of loads (see Sec. 1.4). For wind acting alone, K_d = 1.0.

TABLE 1.5 Velocity Pressure Exposure Coefficient for Terrain with Low-Lying Obstructions

z (ft)	z (m)	$\dfrac{K_z}{z}$
0–15	0–4.6	0.85
20	6.1	0.90
25	7.6	0.94
30	9.1	0.98
40	12.2	1.04
50	15.2	1.09

1

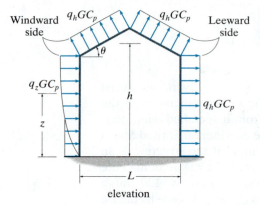

		Windward angle θ		Leeward angle
Wind direction	h/L	10°		$\theta = 10°$
Normal to ridge	≤0.25	−0.7		−0.3
	0.5	−0.9		−0.5
	>1.0	−1.3		−0.7

Maximum negative roof pressure
coefficients, C_p, for use with q_h

elevation

Fig. 1–13

Application of Eq. 1–3 will involve calculations of wind pressures from each side of the building, with due considerations for the possibility of either positive or negative pressures acting on the building's interior.*

For high-rise buildings or those having a shape or location that makes them wind sensitive, it is recommended that a *dynamic approach* be used to determine the wind loadings. The methodology for doing this is also outlined in the ASCE 7-10 Standard. It requires wind-tunnel tests to be performed on a scale model of the building and those surrounding it, in order to simulate the natural environment. The pressure effects of the wind on the building can be determined from pressure transducers attached to the model. Also, if the model has stiffness characteristics that are in proper scale to the building, then the dynamic deflections of the building can be determined.

*As with using any code, application of the requirements of a ASCE 7-10 demands careful attention to details related to the use of formulas and graphs within the code. The recent failure of a fabric-covered steel truss structure, used by the Dallas Cowboys for football practice, was due to high winds. A review of the engineer's calculations, as recorded in *Civil Engineering*, April 2013, indicated a simple arithmetic error was made in calculating the slope angle θ of the roof (see Fig. 1–13). Also, the internal pressure within the structure was not considered, along with other careless mistakes in modeling the structure for analysis. All this led to an underdesigned structure, which failed at a wind speed lower than the anticipated design speed. *The importance of a careful, accurate, and complete analysis cannot be overemphasized.*

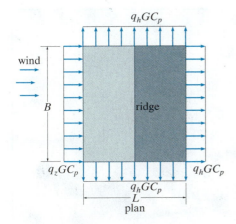

Surface	L/B	C_p	Use with
Windward wall	All values	0.8	q_z
Leeward wall	0–1 2 ≥4	−0.5 −0.3 −0.2	q_h
Side walls	All values	−0.7	q_h

Wall pressure coefficients, C_p

Design Wind Pressure for Enclosed Buildings. Once the value for q_z is obtained, the design pressure can be determined from a list of relevant equations listed in the ASCE 7-10 Standard. The choice depends upon the flexibility and height of the structure, and whether the design is for the main wind-force resisting system, or for the building's components and cladding. For example, using a "directional procedure" the *wind-pressure* on an enclosed building of any height is determined using a two-termed equation resulting from both external and internal pressures, namely,

$$p = qGC_p - q_h(GC_{pi}) \qquad (1\text{–}3)$$

Here

$q = q_z$ for the windward wall at height z above the ground (Eq. 1–2), and $q = q_h$ for the leeward walls, side walls, and roof, where $z = h$, the mean height of the roof.

G = a wind-gust effect factor, which depends upon the exposure. For example, for a rigid structure, $G = 0.85$.

C_p = a wall or roof pressure coefficient determined from a table. These tabular values for the walls and a roof pitch of $\theta = 10°$ are given in Fig. 1–13. Note in the elevation view that the pressure will vary with height on the windward side of the building, whereas on the remaining sides and on the roof the pressure is assumed to be constant. Negative values indicate pressures acting away from the surface.

(GC_{pi}) = the internal pressure coefficient, which depends upon the type of openings in the building. For fully enclosed buildings $(GC_{pi}) = \pm 0.18$. Here the signs indicate that either positive or negative (suction) pressure can occur within the building.

1

EXAMPLE   **1.3**

The enclosed building shown in the photo and in Fig. 1–14a is used for storage purposes and is located outside of Chicago, Illinois on open flat terrain. When the wind is directed as shown, determine the design wind pressure acting on the roof and sides of the building using the ASCE 7-10 Specifications.

SOLUTION

First the wind pressure will be determined using Eq. 1–2. The basic wind speed is $V = 105$ mi/h, since the building is used for storage. Also, for flat terrain, $K_{zt} = 1.0$. Since only wind loading is being considered, $K_d = 1.0$. Therefore,

$$q_z = 0.00256 \, K_z K_{zt} K_d V^2$$
$$= 0.00256 \, K_z (1.0)(1.0)(105)^2$$
$$= 28.22 \, K_z$$

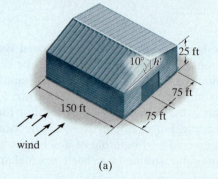

(a)

Fig. 1–14

From Fig. 1–14a, $h' = 75 \tan 10° = 13.22$ ft so that the mean height of the roof is $h = 25 + 13.22/2 = 31.6$ ft. Using the values of K_z in Table 1.5, calculated values of the pressure profile are listed in the table in Fig. 1–14b. Note the value of K_z was determined by linear interpolation for $z = h$, i.e., $(1.04 - 0.98)/(40 - 30) = (1.04 - K_z)/(40 - 31.6)$, $K_z = 0.990$, and so $q_h = 28.22(0.990) = 27.9$ psf.

In order to apply Eq. 1–3 the gust factor is $G = 0.85$, and $(GC_{pi}) = \pm 0.18$. Thus,

$$p = qGC_p - q_h(GC_{pi})$$
$$= q(0.85)C_p - 27.9(\pm 0.18)$$
$$= 0.85qC_p \mp 5.03 \qquad (1)$$

The pressure loadings are obtained from this equation using the calculated values for q_z listed in Fig. 1–14b in accordance with the wind-pressure profile in Fig. 1–13.

z (ft)	K_z	q_z (psf)
0–15	0.85	24.0
20	0.90	25.4
25	0.94	26.5
$h = 31.6$	0.990	27.9

(b)

Windward Wall. Here the pressure varies with height z since $q_z G C_p$ must be used. For all values of L/B, $C_p = 0.8$, so that from Eq. (1),

$$p_{0-15} = 11.3 \text{ psf} \quad \text{or} \quad 21.3 \text{ psf}$$
$$p_{20} = 12.2 \text{ psf} \quad \text{or} \quad 22.3 \text{ psf}$$
$$p_{25} = 13.0 \text{ psf} \quad \text{or} \quad 23.1 \text{ psf}$$

Leeward Wall. Here $L/B = 2(75)/150 = 1$, so that $C_p = -0.5$. Also, $q = q_h$ and so from Eq. (1),

$$p = -16.9 \text{ psf} \quad \text{or} \quad -6.84 \text{ psf}$$

Side Walls. For all values of L/B, $C_p = -0.7$, and therefore since we must use $q = q_h$ in Eq. (1), we have

$$p = -21.6 \text{ psf} \quad \text{or} \quad -11.6 \text{ psf}$$

Windward Roof. Here $h/L = 31.6/2(75) = 0.211 < 0.25$, so that $C_p = -0.7$ and $q = q_h$. Thus,

$$p = -21.6 \text{ psf} \quad \text{or} \quad -11.6 \text{ psf}$$

Leeward Roof. In this case $C_p = -0.3$; therefore with $q = q_h$, we get

$$p = -12.2 \text{ psf} \quad \text{or} \quad -2.09 \text{ psf}$$

These two sets of loadings are shown on the elevation of the building, representing either positive or negative (suction) internal building pressure, Fig. 1–14c. The main framing structure of the building must resist these loadings as well as for separate loadings calculated from wind blowing on the front or rear of the building.

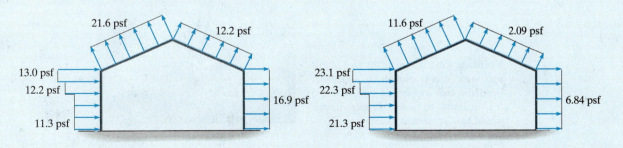

(c)

Design Wind Pressure for Signs. If the structure represents a sign, the wind will produce a *resultant force* acting on the face of the sign which is determined from

$$F = q_h G C_f A_s \tag{1–4}$$

Here

q_h = the wind pressure evaluated at the height h, measured from the ground to the top of the sign.

G = the wind-gust coefficient factor defined previously.

C_f = a force coefficient which depends upon the aspect ratio (width B of the sign to height s of the sign), and the clear area ratio (sign height s to the elevation h, measured from the ground to the top of the sign). For cases of wind directed normal to the sign and through its center, for $B/s = 4$, values are listed in Table 1.6.

A_s = the area of the face of the sign in ft² (m²).

TABLE 1.6 Force Coefficients for Above-Ground Solid Signs, C_f	
s/h	C_f
1	1.35
0.9	1.45
0.5	1.70
0.2	1.80
≤0.16	1.85

To allow for normal and oblique wind directions, the calculated resultant force is assumed to act either through the geometric center of the face of the sign or at other specified locations on the face of the sign which depend upon the ratios s/h and B/s.

Hurricane winds acting on the face of this sign were strong enough to noticeably bend the two supporting arms causing the material to yield. Proper design would have prevented this.

Snow Loads. In some parts of the country, roof loading due to snow can be quite severe, and therefore protection against possible failure is of primary concern. Design loadings typically depend on the building's general shape and roof geometry, wind exposure, location, its importance, and whether or not it is heated. Like wind, snow loads in the ASCE 7-10 Standard are generally determined from a zone map reporting 50-year recurrence intervals of an extreme snow depth. For example, on the relatively flat elevation throughout the mid-section of Illinois and Indiana, the ground snow loading is 20 lb/ft^2 (0.96 kN/m^2). However, for areas of Montana, specific case studies of ground snow loadings are needed due to the variable elevations throughout the state. Specifications for snow loads are covered in the ASCE 7-10 Standard, although no single code can cover all the implications of this type of loading.

If a roof is flat, defined as having a slope of less than 5%, then the pressure loading on the roof can be obtained by modifying the ground snow loading, p_g, by the following empirical formula

$$p_f = 0.7C_eC_tI_sp_g \qquad (1\text{--}5)$$

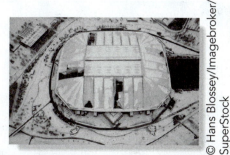

Excessive snow and ice loadings act on this roof.

Here

C_e = an exposure factor which depends upon the terrain. For example, for a fully exposed roof in an unobstructed area, $C_e = 0.8$, whereas if the roof is sheltered and located in the center of a large city, then $C_e = 1.2$.

C_t = a thermal factor which refers to the average temperature within the building. For unheated structures kept below freezing $C_t = 1.2$, whereas if the roof is supporting a normally heated structure, then $C_t = 1.0$.

I_s = the importance factor as it relates to occupancy. For example, $I_s = 0.80$ for agriculture and storage facilities, and $I_s = 1.20$ for schools and hospitals.

If $p_g \leq 20$ lb/ft^2 (0.96 kN/m^2), then use the *largest value* for p_f, either computed from the above equation or from $p_f = I_sp_g$. If $p_g > 20$ lb/ft^2 (0.96 kN/m^2), then use $p_f = I_s(20$ lb/ft$^2)$.

EXAMPLE | 1.4

Fig. 1–15

The unheated storage facility shown in Fig. 1–15 is located on flat open terrain in southern Illinois, where the specified ground snow load is 15 lb/ft². Determine the design snow load on the roof which has a slope of 4%.

SOLUTION

Since the roof slope is < 5%, we will use Eq. 1–5. Here, $C_e = 0.8$ due to the open area, $C_t = 1.2$ and $I_s = 0.8$. Thus,

$$p_f = 0.7C_eC_tI_sp_g$$

$$= 0.7(0.8)(1.2)(0.8)(15 \text{ lb/ft}^2) = 8.06 \text{ lb/ft}^2$$

Since $p_g = 15 \text{ lb/ft}^2 < 20 \text{ lb/ft}^2$, then also

$$p_f = Ip_g = 1.2(15 \text{ lb/ft}^2) = 18 \text{ lb/ft}^2$$

By comparison, choose

$$p_f = 18 \text{ lb/ft}^2 \qquad \textit{Ans.}$$

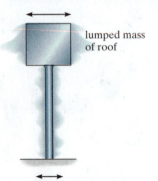

lumped mass
of roof

Fig. 1–16

Earthquake Loads. Earthquakes produce lateral loadings on a structure through the structure's interaction with the ground. The magnitude of an earthquake load depends on the amount and type of ground accelerations and the mass and stiffness of the structure. In order to show how earthquake loads occur, consider the simple structural model in Fig. 1–16. This model may represent a single-story building, where the block is the "lumped" mass of the roof, and the column has a total stiffness representing all the building's columns. During an earthquake, the ground vibrates both horizontally and vertically. The horizontal accelerations create shear forces in the column that put the block in sequential motion with the ground. If the column is *stiff* and the block has a *small* mass, the period of vibration of the block will be *short* and the block will accelerate with the same motion as the ground and undergo only slight relative displacements. For an actual structure that is designed to have large amounts of bracing and stiff connections, this can be beneficial, since less stress is developed in the members. On the other hand, if the column in Fig. 1–16 is very flexible and the block has a large mass, then earthquake-induced motion will cause small accelerations of the block and large relative displacements.

Some codes require that specific attention be given to earthquake design, especially in areas of the country where strong earthquakes predominate. To find this out, one can check the seismic ground-acceleration maps published in the ASCE 7-10 Standard. These maps provide the peak ground accelerations caused by an earthquake along with risk coefficients. Regions vary from low risk, such as parts of Texas, to very high risk, such as along the west coast of California.

For high-rise structures, or, say, nuclear power plants, an earthquake analysis can be quite elaborate. It requires attaining an **acceleration response spectrum,** then using a computer to calculate the earthquake loadings based on the theory of structural dynamics.

For small structures, a *static analysis* for earthquake design may be satisfactory. This case approximates the dynamic loads by a set of externally applied *static forces* that are applied laterally to the structure. One such method for doing this is reported in the ASCE 7-10 Standard. It is based upon finding a seismic response coefficient, C_s, determined from the soil properties, the ground accelerations, and the vibrational response of the structure. For most structures, this coefficient is then multiplied by the structure's total dead load W to obtain the "base shear" in the structure. The value of C_s is actually determined from

$$C_s = \frac{S_{DS}}{R/I_e}$$

where

S_{DS} = the spectral response acceleration for short periods of vibration.

R = a response modification factor that depends upon the ductility of the structure. Steel frame members which are highly ductile can have a value as high as 8, whereas reinforced concrete frames can have a value as low as 3.

I_e = the importance factor that depends upon the use of the building. For example, $I_e = 1$ for agriculture and storage facilities, and $I_e = 1.5$ for hospitals and other essential facilities.

With each new publication of the Standard, values of these coefficients are updated as more accurate data about earthquake response become available.

Hydrostatic and Soil Pressure. When structures are used to retain water, soil, or granular materials, the pressure developed by these loadings becomes an important criterion for their design. Examples of such types of structures include tanks, dams, ships, bulkheads, and retaining walls. Here the laws of hydrostatics and soil mechanics are applied to define the intensity of the loadings on the structure.

Other Natural Loads. Several other types of live loads may also have to be considered in the design of a structure, depending on its location or use. These include the effect of blast, temperature changes, and differential settlement of the foundation.

1.4 Structural Design

Whenever a structure is designed, it is important to give consideration to both material and load uncertainties. These uncertainties include a possible variability in material properties, residual stress in materials, intended measurements being different from fabricated sizes, loadings due to vibration or impact, and material corrosion or decay.

ASD. Allowable-stress design (ASD) methods include *both* the material and load uncertainties into a single factor of safety. The many types of loads discussed previously can occur simultaneously on a structure, but it is very unlikely that the maximum of all these loads will occur at the same time. For example, both maximum wind and earthquake loads normally do not act simultaneously on a structure. For *allowable-stress design* the computed elastic stress in the material must not exceed the allowable stress for each of various load combinations. Typical load combinations as specified by the ASCE 7-10 Standard include

- dead load
- 0.6 (dead load) + 0.6 (wind load)
- 0.6 (dead load) + 0.7 (earthquake load)

LRFD. Since uncertainty can be considered using probability theory, there has been an increasing trend to *separate* material uncertainty from load uncertainty. This method is called *strength design* or LRFD (load and resistance factor design). For example, to account for the uncertainty of loads, this method uses load factors applied to the loads or combinations of loads. According to the ASCE 7-10 Standard, some of the load factors and combinations are

- 1.4 (dead load)
- 1.2 (dead load) + 1.6 (live load) + 0.5 (snow load)
- 0.9 (dead load) + 1.0 (wind load)
- 0.9 (dead load) + 1.0 (earthquake load)

In all these cases, the combination of loads is thought to provide a maximum, yet realistic loading on the structure.

PROBLEMS

1–1. The floor of a heavy storage warehouse building is made of 6-in.-thick stone concrete. If the floor is a slab having a length of 15 ft and width of 10 ft, determine the resultant force caused by the dead load and the live load.

1–2. The wall is 12-ft high and consists of 2×4 studs. On each side is acoustical fiberboard and 4-in. clay brick. Determine the average load in lb/ft of length of wall that the wall exerts on the floor.

Prob. 1–2

1–3. A building wall consists of 12-in. clay brick and $\frac{1}{2}$-in. fiberboard on one side. If the wall is 10 ft high, determine the load in pounds per foot that it exerts on the floor.

*1–4. The "New Jersey" barrier is commonly used during highway construction. Determine its weight per foot of length if it is made from plain stone concrete.

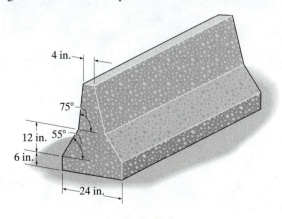

Prob. 1–4

1–5. The precast floor beam is made from concrete having a specific weight of 150 lb/ft³. If it is to be used for a floor in an office of an office building, calculate its dead and live loadings per foot length of beam.

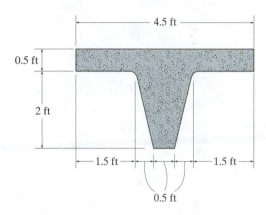

Prob. 1–5

1–6. The floor of a light storage warehouse is made of 150-mm-thick lightweight plain concrete. If the floor is a slab having a length of 7 m and width of 3 m, determine the resultant force caused by the dead load and the live load.

1–7. The pre-cast T-beam has the cross-section shown. Determine its weight per foot of length if it is made from reinforced stone concrete and eight $\frac{3}{4}$-in. cold-formed steel reinforcing rods.

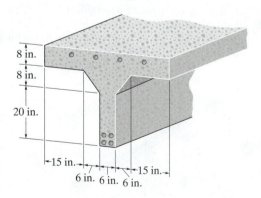

Prob. 1–7

*1–8. The building wall consists of 8-in. clay brick. In the interior, the wall is made from 2×4 wood studs, plastered on one side. If the wall is 10 ft high, determine the load in pounds per foot of length of wall that the wall exerts on the floor.

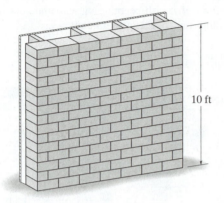

10 ft

Prob. 1–8

1–9. A building wall consists of exterior stud walls with brick veneer and 13 mm fiberboard on one side. If the wall is 4 m high, determine the load in kN/m that it exerts on the floor.

1–10. The interior wall of a building is made from 2×4 wood studs, plastered on two sides. If the wall is 12 ft high, determine the load in lb/ft of length of wall that it exerts on the floor.

1–11. The beam supports the roof made from asphalt shingles and wood sheathing boards. If the boards have a thickness of $1\frac{1}{2}$ in. and a specific weight of 50 lb/ft³, and the roof's angle of slope is 30°, determine the dead load of the roofing—per square foot—that is supported in the x and y directions by the purlins.

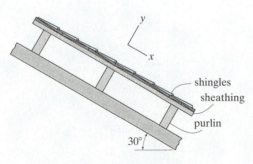

y

x

shingles
sheathing

purlin

30°

Prob. 1–11

*1–12. A three-story hotel has interior columns that are spaced 20 ft apart in two perpendicular directions. If the loading on the flat roof is estimated to be 30 lb/ft², determine the live load supported by a typical interior column at (a) the ground-floor level, and (b) the second-floor level.

1–13. A hospital is to be built on open flat terrain in central Texas. If the building is 9.1 meters high, determine the internal pressure within the building if it is fully enclosed. Also, what is the external wind pressure acting on the side walls of the building? Each wall of the building is 25 meters long.

1–14. The office building has interior columns spaced 5 m apart in perpendicular directions. Determine the reduced live load supported by a typical interior column located on the first floor under the offices.

Prob. 1–14

1–15. A hospital located in Chicago, Illinois, has a flat roof, where the ground snow load is 25 lb/ft². Determine the design snow load on the roof of the hospital.

*1–16.   Wind blows on the side of a fully enclosed hospital located on open flat terrain in Arizona. Determine the external pressure acting over the windward wall, which has a height of 30 ft. The roof is flat.

1–18.   Determine the resultant force acting on the face of the sign if $q_h = 3.70$ kPa. The sign has a width of 12 m and a height of 3 m as indicated.

Prob. 1–16

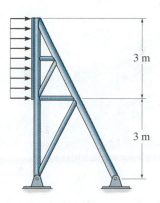

Prob. 1–18

1–17.   Wind blows on the side of the fully enclosed hospital located on open flat terrain in Arizona. Determine the external pressure acting on the leeward wall, which has a length of 200 ft and a height of 30 ft.

1–19.   The light metal storage building is on open flat terrain in central Oklahoma. If the side wall of the building is 14 ft high, what are the two values of the external wind pressure acting on this wall when the wind blows on the back of the building? The roof is essentially flat and the building is fully enclosed.

Prob. 1–17

Prob. 1–19

*1–20. The horse stall has a flat roof with a slope of 80 mm/m. It is located in an open field where the ground snow load is 1.20 kN/m². Determine the snow load that is required to design the roof of the stall.

Prob. 1–20

1–21. The horse stall has a flat roof with a slope of 80 mm/m. It is located in an open field where the ground snow load is 0.72 kN/m². Determine the snow load that is required to design the roof of the stall.

Prob. 1–21

1–22. A hospital located in central Illinois has a flat roof. Determine the snow load in kN/m² that is required to design the roof.

1–23. The school building has a flat roof. It is located in an open area where the ground snow load is 0.68 kN/m². Determine the snow load that is required to design the roof.

Prob. 1–23

*1–24. Wind blows on the side of the fully enclosed agriculture building located on open flat terrain in Oklahoma. Determine the external pressure acting over the windward wall, the leeward wall, and the side walls. Also, what is the internal pressure in the building which acts on the walls? Use linear interpolation to determine q_h.

1–25. Wind blows on the side of the fully enclosed agriculture building located on open flat terrain in Oklahoma. Determine the external pressure acting on the roof. Also, what is the internal pressure in the building which acts on the roof? Use linear interpolation to determine q_h and C_p in Fig. 1–13.

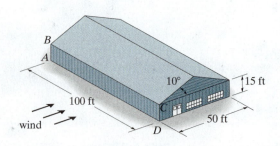

Probs. 1–24/25

CHAPTER REVIEW

The basic structural elements are:

Tie Rods—Slender members subjected to tension. Often used for bracing.

Beams—Members designed to resist bending moment. They are often fixed or pin supported and can be in the form of a steel plate girder, reinforced concrete, or laminated wood.

Columns—Members that resist axial compressive force. If the column also resists bending, it is called a *beam column*.

The types of structures considered in this book consist of *trusses* made from slender pin-connected members forming a series of triangles, *cables and arches,* which carry tensile and compressive loads, respectively, and *frames* composed of pin- or fixed-connected beams and columns.

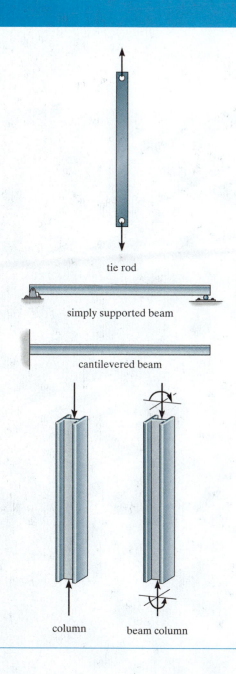

tie rod

simply supported beam

cantilevered beam

column   beam column

Loads are specified in codes such as the ASCE 7-10 code. *Dead loads* are fixed and refer to the weights of members and materials. *Live loads* are movable and consist of uniform building floor loads, traffic and train loads on bridges, impact loads due to vehicles and machines, wind loads, snow loads, earthquake loads, and hydrostatic and soil pressure.

Chapter 2

© SergeyIT/Shutterstock

Oftentimes the elements of a structure, like the beams and girders of this building frame, are connected together in a manner whereby the analysis can be considered statically determinate.

Analysis of Statically Determinate Structures

In this chapter we will direct our attention to the most common form of structure that the engineer will have to analyze, and that is one that lies in a plane and is subjected to a force system that lies in the same plane. We begin by discussing the importance of choosing an appropriate analytical model for a structure so that the forces in the structure may be determined with reasonable accuracy. Then the criteria necessary for structural stability are discussed. Finally, the analysis of statically determinate, planar, pin-connected structures is presented.

2.1   Idealized Structure

An exact analysis of a structure can never be carried out, since estimates always have to be made of the loadings and the strength of the materials composing the structure. Furthermore, points of application for the loadings must also be estimated. It is important, therefore, that the structural engineer develop the ability to model or idealize a structure so that he or she can perform a practical force analysis of the members. In this section we will develop the basic techniques necessary to do this.

Notice that the deck of this concrete bridge is made so that one section can be considered roller supported on the other section.

2

Support Connections.

Structural members are joined together in various ways depending on the intent of the designer. The three types of joints most often specified are the pin connection, the roller support, and the fixed joint. A pin-connected joint and a roller support allow some freedom for slight rotation, whereas a fixed joint allows no relative rotation between the connected members and is consequently more expensive to fabricate. Examples of these joints, fashioned in metal and concrete, are shown in Figs. 2–1 and 2–2, respectively. For most timber structures, the members are assumed to be pin connected, since bolting or nailing them will not sufficiently restrain them from rotating with respect to each other.

Idealized models used in structural analysis that represent pinned and fixed supports and pin-connected and fixed-connected joints are shown in Figs. 2–3a and 2–3b. In reality, however, all connections exhibit some stiffness toward joint rotations, owing to friction and material behavior. In this case a more appropriate model for a support or joint might be that shown in Fig. 2–3c. If the torsional spring constant $k = 0$, the joint is a pin, and if $k \rightarrow \infty$, the joint is fixed.

Pin-connected steel members.

typical "pin-supported" connection (metal)
(a)

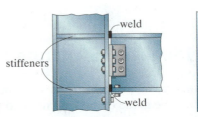

typical "fixed-supported" connection (metal)
(b)

Fig. 2–1

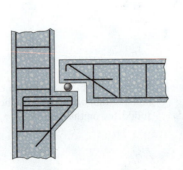

typical "roller-supported" connection (concrete)
(a)

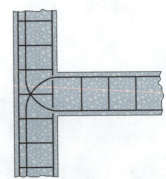

typical "fixed-supported" connection (concrete)
(b)

Fig. 2–2

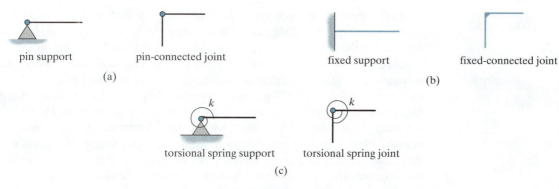

pin support   pin-connected joint   fixed support   fixed-connected joint

(a)   (b)

torsional spring support   torsional spring joint

(c)

Fig. 2–3

When selecting a particular model for each support or joint, the engineer must be aware of how the assumptions will affect the actual performance of the member and whether the assumptions are reasonable for the structural design. For example, consider the beam shown in Fig. 2–4a, which is used to support a concentrated load **P**. The angle connection at support A is like that in Fig. 2–1a and can therefore be idealized as a typical pin support. Furthermore, the support at B provides an approximate point of smooth contact and so it can be idealized as a roller. The beam's thickness can be neglected since it is small in comparison to the beam's length, and therefore the idealized model of the beam is as shown in Fig. 2–4b. The analysis of the loadings in this beam should give results that closely approximate the loadings in the actual beam. To show that the model is appropriate, consider a specific case of a beam made of steel with $P = 8$ k (8000 lb) and $L = 20$ ft. One of the major simplifications made here was assuming the support at A to be a pin. Design of the beam using standard code procedures* indicates that a W10 × 19 would be adequate for supporting the load. Using one of the deflection methods of Chapter 8, the rotation at the "pin" support can be calculated as $\theta = 0.0103$ rad $= 0.59°$. From Fig. 2–4c, such a rotation only moves the top or bottom flange a distance of $\Delta = \theta r = (0.0103 \text{ rad})(5.12 \text{ in.}) = 0.0528$ in. This *small amount* would certainly be accommodated by the connection fabricated as shown in Fig. 2–1a, and therefore the pin serves as an appropriate model.

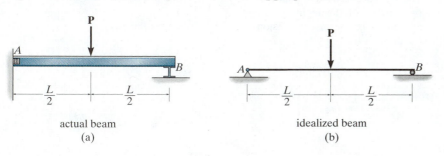

actual beam
(a)

idealized beam
(b)

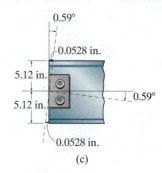

(c)

Fig. 2–4

*Codes such as the *Manual of Steel Construction*, American Institute of Steel Construction.

2

A typical rocker support used for a bridge girder.

Rollers and associated bearing pads are used to support the prestressed concrete girders of a highway bridge.

Other types of connections most commonly encountered on coplanar structures are given in Table 2.1. It is important to be able to recognize the symbols for these connections and the kinds of reactions they exert on their attached members. This can easily be done by noting how the connection *prevents* any degree of freedom or displacement of the member. In particular, the support will develop a *force* on the member if it *prevents translation* of the member, and it will develop a *moment* if it *prevents rotation* of the member. For example, a member in contact with a smooth surface (3) is prevented from translating only in one direction, which is perpendicular or normal to the surface. Hence, the surface exerts only a *normal force* **F** on the member in this direction. The magnitude of this force represents *one unknown*. Also note that the member is free to rotate on the surface, so that a moment cannot be developed by the surface on the member. As another example, the fixed support (7) prevents *both* translation and rotation of a member at the point of connection. Therefore, this type of support exerts two force components and a moment on the member. The "curl" of the moment lies in the plane of the page, since rotation is prevented in that plane. Hence, there are *three unknowns* at a fixed support.

In reality, all supports actually exert *distributed surface loads* on their contacting members. The concentrated forces and moments shown in Table 2.1 represent the *resultants* of these load distributions. This representation is, of course, an idealization; however, it is used here since the surface area over which the distributed load acts is considerably *smaller* than the *total* surface area of the connecting members.

Concrete smooth or "roller" support.

The short link is used to connect the two girders of the highway bridge and allow for thermal expansion of the deck.

Steel pin support.

2

TABLE 2.1 Supports for Coplanar Structures

Type of Connection	Idealized Symbol	Reaction	Number of Unknowns
(1) light cable / weightless link		θ $\mathbf{F}$	One unknown. The reaction is a force that acts in the direction of the cable or link.
(2) rollers / rocker		$\mathbf{F}$	One unknown. The reaction is a force that acts perpendicular to the surface at the point of contact.
(3) smooth contacting surface		$\mathbf{F}$	One unknown. The reaction is a force that acts perpendicular to the surface at the point of contact.
(4) smooth pin-connected collar		$\mathbf{F}$	One unknown. The reaction is a force that acts perpendicular to the surface at the point of contact.
(5) smooth pin or hinge	θ	$\mathbf{F}_y$ $\mathbf{F}_x$	Two unknowns. The reactions are two force components.
(6) slider / fixed-connected collar		$\mathbf{M}$ $\mathbf{F}$	Two unknowns. The reactions are a force and a moment.
(7) fixed support		$\mathbf{F}_y$ $\mathbf{M}$ $\mathbf{F}_x$	Three unknowns. The reactions are the moment and the two force components.

2

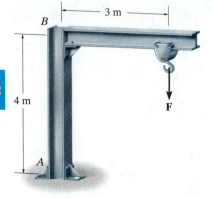

actual structure

(a)

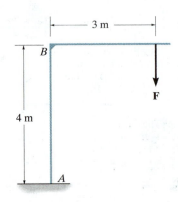

idealized structure

(b)

Fig. 2–5

Idealized Structure.

Having stated the various ways in which the connections on a structure can be idealized, we are now ready to discuss some of the techniques used to represent various structural systems by idealized models.

As a first example, consider the jib crane and trolley in Fig. 2–5a. For the structural analysis we can neglect the thickness of the two main members and will assume that the joint at B is fabricated to be rigid. Furthermore, the support connection at A can be modeled as a fixed support and the details of the trolley excluded. Thus, the members of the idealized structure are represented by two connected lines, and the load on the hook is represented by a single concentrated force $\mathbf{F}$, Fig. 2–5b. This idealized structure shown here as a *line drawing* can now be used for applying the principles of structural analysis, which will eventually lead to the design of its two main members.

Beams and girders are often used to support building floors. In particular, a *girder* is the main load-carrying element of the floor, whereas the smaller elements having a shorter span and connected to the girders are called *beams*. Often the loads that are applied to a beam or girder are transmitted to it by the floor that is supported by the beam or girder. Again, it is important to be able to appropriately idealize the system as a series of models, which can be used to determine, to a close approximation, the forces acting in the members. Consider, for example, the framing used to support a typical floor slab in a building, Fig. 2–6a. Here the slab is supported by *floor joists* located at even intervals, and these in turn are supported by the two side girders AB and CD. For analysis it is reasonable to assume that the joints are pin and/or roller connected to the girders and that the girders are pin and/or roller connected to the columns. The top view of the structural framing plan for this system is shown in Fig. 2–6b. In this "graphic" scheme, notice that the "lines" representing the joists do not touch the girders and the lines for the girders do not touch the columns. This symbolizes pin- and/ or roller-supported connections. On the other hand, if the framing plan is intended to represent fixed-connected members, such as those that are welded

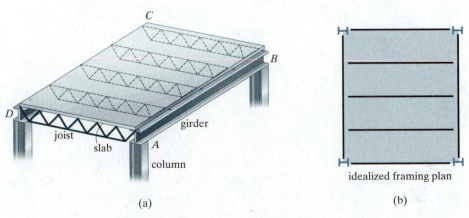

(a)

idealized framing plan

(b)

Fig. 2–6

instead of simple bolted connections, then the lines for the beams or girders would touch the columns as in Fig. 2–7. Similarly, a fixed-connected overhanging beam would be represented in top view as shown in Fig. 2–8. If reinforced concrete construction is used, the beams and girders are represented by double lines. These systems are generally all fixed connected and therefore the members are drawn to the supports. For example, the structural graphic for the cast-in-place reinforced concrete system in Fig. 2–9a is shown in top view in Fig. 2–9b. The lines for the beams are dashed because they are below the slab.

Structural graphics and idealizations for timber structures are similar to those made of metal. For example, the structural system shown in Fig. 2–10a represents beam-wall construction, whereby the roof deck is supported by wood joists, which deliver the load to a masonry wall. The joists can be assumed to be simply supported on the wall, so that the idealized framing plan would be like that shown in Fig. 2–10b.

fixed-connected beam

idealized beam

Fig. 2–7

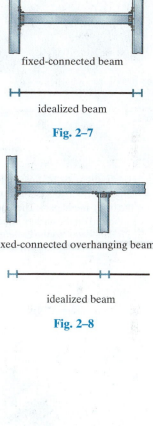

fixed-connected overhanging beam

idealized beam

Fig. 2–8

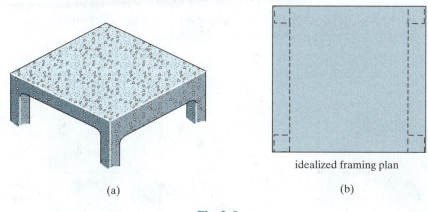

idealized framing plan

(a)   (b)

Fig. 2–9

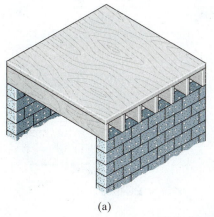

idealized framing plan

(a)   (b)

Fig. 2–10

2

The structural framework of this building consists of concrete floor joists, which were formed on site using metal pans. These joists are simply supported on the girders, which in turn are simply supported on the columns.

Tributary Loadings.
When flat surfaces such as walls, floors, or roofs are supported by a structural frame, it is necessary to determine how the load on these surfaces is transmitted to the various structural elements used for their support. There are generally two ways in which this can be done. The choice depends on the geometry of the structural system, the material from which it is made, and the method of its construction.

One-Way System.
A slab or deck that is supported such that it delivers its load to the supporting members by one-way action, is often referred to as a *one-way slab*. To illustrate the method of load transmission, consider the framing system shown in Fig. 2–11a where the beams AB, CD, and EF rest on the girders AE and BF. If a uniform load of 100 lb/ft^2 is placed on the slab, then the center beam CD is assumed to support the load acting on the **tributary area** shown dark shaded on the structural framing plan in Fig. 2–11b. Member CD is therefore subjected to a *linear* distribution of load of $(100 \text{ lb/ft}^2)(5 \text{ ft}) = 500 \text{ lb/ft}$, shown on the idealized beam in Fig. 2–11c. The reactions on this beam (2500 lb) would then be applied to the center of the girders AE (and BF), shown idealized in Fig. 2–11d. Using this same concept, do you see how the remaining portion of the slab loading is transmitted to the ends of the girder as 1250 lb?

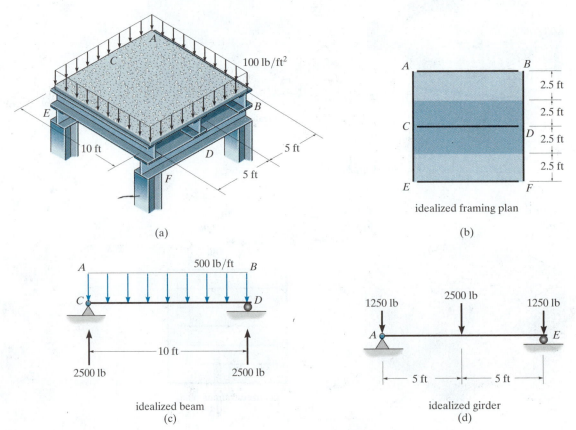

(a)

idealized framing plan

(b)

idealized beam
(c)

idealized girder
(d)

Fig. 2–11

An example of one-way slab construction of a steel frame building having a poured concrete floor on a corrugated metal deck. The load on the floor is considered to be transmitted to the beams, not the girders.

For some floor systems the beams and girders are connected to the columns at the *same elevation*, as in Fig. 2–12a. If this is the case, the slab can in some cases also be considered a ***one-way slab***. For example, if the slab is reinforced concrete with reinforcement in *only one direction*, or the concrete is poured on a corrugated metal deck, as in the above photo, then one-way action of load transmission can be assumed. On the other hand, if the slab is reinforced in *two directions*, then consideration must be given to the *possibility* of the load being transmitted to the supporting members from either one or two directions. For example, consider the slab and framing plan in Fig. 2–12b. According to the American Concrete Institute, ACI 318 code, if $L_2 > L_1$ and if the span ratio $(L_2/L_1) > 2$, *the slab will behave as a one-way slab*, since as L_1 becomes smaller, the beams AB, CD, and EF provide the greater stiffness to carry the load.

Floor beams are often coped, that is, the top flange is cut back, so that the beam is at the same level as the girder.

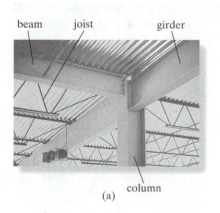

beam   joist   girder

column

(a)

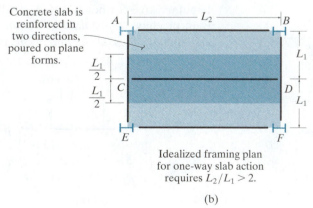

Concrete slab is reinforced in two directions, poured on plane forms.

A ⊢————— L_2 —————⊣ B

$\dfrac{L_1}{2}$

$\dfrac{L_1}{2}$ C D

L_1

L_1

L_1

E F

Idealized framing plan for one-way slab action requires $L_2/L_1 > 2$.

(b)

Fig. 2–12

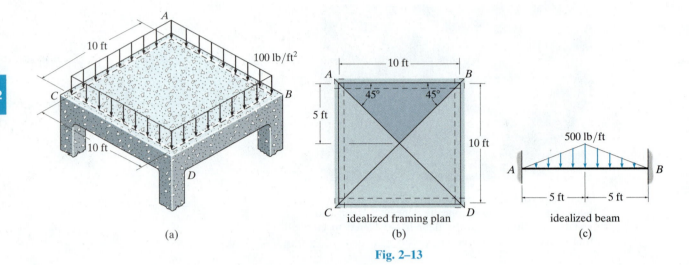

(a)   idealized framing plan   idealized beam
                              (b)   (c)

Fig. 2–13

Two-Way System.

According to the ACI 318 concrete code, if $L_2 > L_1$ and the support ratio $(L_2/L_1) \le 2$, then the load is assumed to be transferred to the supporting beams and girders in *two directions.* When this is the case the slab is referred to as a ***two-way slab.*** To show how to treat this case, consider the square reinforced concrete slab in Fig. 2–13a, which is supported by four 10-ft-long edge beams, AB, BD, DC, and CA. Here $L_2/L_1 = 1$. As the load on the slab intensifies, numerous experiments have shown that 45° cracks form at the corners of the slab. As a result, the tributary area is constructed using diagonal 45° lines as shown in Fig. 2–13b. This produces the dark shaded tributary area for beam AB. Hence if a uniform load of 100 lb/ft² is applied to the slab, a peak intensity of $(100 \text{ lb/ft}^2)(5 \text{ ft}) = 500 \text{ lb/ft}$ will be applied to the center of beam AB, resulting in the *triangular* load distribution shown in Fig. 2–13c. For other geometries that cause two-way action, a similar For example, if $L_2/L_1 = 1.5$ it is then necessary to construct 45° lines that intersect as shown in Fig. 2–14a. This produce the dark shaded tributary area for beam AB. A 100-lb/ft² loading placed on the slab will then produce *trapezoidal* and *triangular* distributed loads on members AB and AC, Figs. 2–14b and 2–14c, respectively.

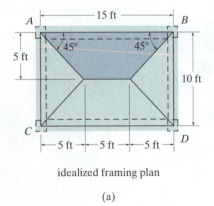

idealized framing plan

(a)

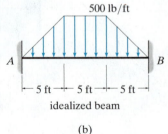

idealized beam

(b)

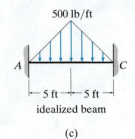

idealized beam

(c)

Fig. 2–14

The ability to reduce an actual structure to an idealized form, as shown by these examples, can only be gained by experience. To provide practice at doing this, the example problems and the problems for solution throughout this book are presented in somewhat realistic form, and the associated problem statements aid in explaining how the connections and supports can be modeled by those listed in Table 2.1. In engineering practice, if it becomes doubtful as to how to model a structure or transfer the loads to the members, it is best to consider *several* idealized structures and loadings and then design the actual structure so that it can resist the loadings in all the idealized models.

EXAMPLE | 2.1

The floor of a classroom is to be supported by the bar joists shown in Fig. 2–15a. Each joist is 15 ft long and they are spaced 2.5 ft on centers. The floor itself is to be made from lightweight concrete that is 4 in. thick. Neglect the weight of the joists and the corrugated metal deck, and determine the load that acts along each joist.

(a)

SOLUTION

The dead load on the floor is due to the weight of the concrete slab. From Table 1.3 for 4 in. of lightweight concrete it is $(4)(8 \text{ lb/ft}^2) = 32 \text{ lb/ft}^2$. From Table 1.4, the live load for a classroom is 40 lb/ft^2. Thus the total floor load is $32 \text{ lb/ft}^2 + 40 \text{ lb/ft}^2 = 72 \text{ lb/ft}^2$. For the floor system, $L_1 = 2.5$ ft and $L_2 = 15$ ft. Since $L_2/L_1 > 2$ the concrete slab is treated as a one-way slab. The tributary area for each joist is shown in Fig. 2–15b. Therefore the uniform load along its length is

$$w = 72 \text{ lb/ft}^2(2.5 \text{ ft}) = 180 \text{ lb/ft}$$

This loading and the end reactions on each joist are shown in Fig. 2–15c.

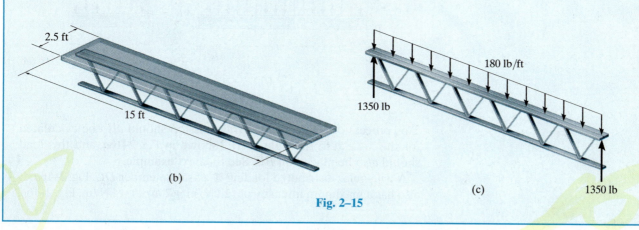

2.5 ft

15 ft

(b)

180 lb/ft

1350 lb

(c)

1350 lb

Fig. 2–15

EXAMPLE 2.2

The flat roof of the steel-frame building shown in the photo is intended to support a total load of 2 kN/m^2 over its surface. Determine the roof load within region $ABCD$ that is transmitted to beams BC and DC. The dimensions are shown in Fig. 2–16a.

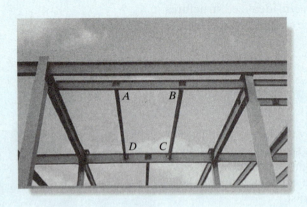

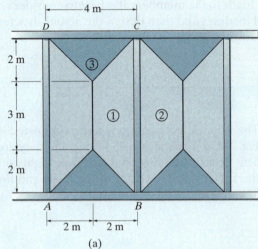

(a)

SOLUTION

In this case $L_2 = 7$ m and $L_1 = 4$ m. Since $L_2/L_1 = 1.75 < 2$, we have two-way slab action. The tributary loading along each edge beam is shown in Fig. 2–16a, where the lighter shaded trapezoidal area ① of loading is transmitted to member BC. The peak intensity of this loading is $(2 \text{ kN/m}^2)(2 \text{ m}) = 4 \text{ kN/m}$. As a result, the distribution of the load along BC is as shown in Fig. 2–16b.

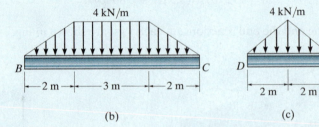

(b)   (c)

Fig. 2–16

This process of tributary load transmission should *also* be calculated for the area ② to the right of BC shown in Fig. 2–16a, and this load should *also* be placed on BC. See the next example.

A triangular distributed loading ③ acts on member DC, Fig. 2–16a. It also has a maximum intensity of $(2 \text{ kN/m}^2)(2 \text{ m}) = 4 \text{ kN/m}$, Fig. 2–16c.

EXAMPLE 2.3

The concrete girders shown in the photo of the passenger car parking garage span 30 ft and are 15 ft on center. If the floor slab is 5 in. thick and made of reinforced stone concrete, and the specified live load is 50 lb/ft^2 (see Table 1.4), determine the distributed load the floor system transmits to each interior girder.

SOLUTION

Here, $L_2 = 30$ ft and $L_1 = 15$ ft, so that $L_2/L_1 = 2$. We have a two-way slab. From Table 1.2, for reinforced stone concrete, the specific weight of the concrete is 150 lb/ft^3. Thus the design floor loading is

$$p = 150 \text{ lb/ft}^3\left(\frac{5}{12}\text{ ft}\right) + 50 \text{ lb/ft}^2 = 112.5 \text{ lb/ft}^2$$

A trapezoidal distributed loading is transmitted to each interior girder AB from each of its two sides ① and ②. The maximum intensity of each of these distributed loadings is $(112.5 \text{ lb/ft}^2)(7.5 \text{ ft}) = 843.75 \text{ lb/ft}$, so that on the girder this intensity becomes $2(843.75 \text{ lb/ft}) = 1687.5 \text{ lb/ft}$, Fig. 2–17b. *Note:* For design, consideration should also be given to the weight of the girder.

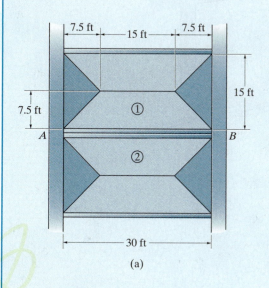

(a)

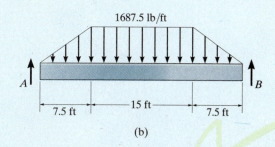

(b)

Fig. 2–17

2

2.2   Load Path

The various elements that make up a structure should be designed in such a way that they transmit the primary load acting on the structure to its foundation in the most efficient way possible. Hence, as a first step in any design or analysis of a structure, it is very important to understand how the loads are transmitted through it, if damage or collapse of the structure is to be avoided. This description is called a ***load path***, and by visualizing how the loads are transmitted the engineer can sometimes eliminate unnecessary members, strengthen others, or identify where there may be potential problems. Like a chain, which is "as strong as its weakest link", so a structure is only as strong as the weakest part along its load path. To show how to construct a load path, let us consider a few examples. In Fig. 2–18*a*, the loading acting on the floor of the building is transmitted from the slab to the floor joists then to the spandrel and interior girder, and finally to the columns and foundation footings. The loading on the deck of the suspension bridge in Fig. 2–18*b* is transmitted to the hangers or suspenders, then the cables, and finally the towers and piers.

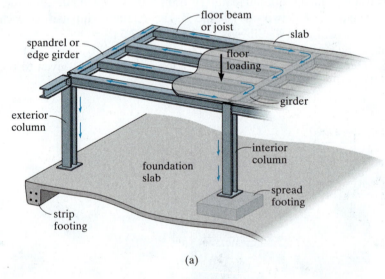

(a)

Fig. 2–18

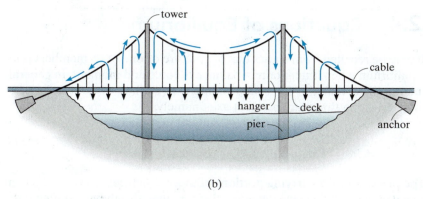

(b)

Fig. 2–18

2.3   Principle of Superposition

The principle of superposition forms the basis for much of the theory of structural analysis. It may be stated as follows: *The total displacement or internal loadings (stress) at a point in a structure subjected to several external loadings can be determined by adding together the displacements or internal loadings (stress) caused by each of the external loads acting separately.* For this statement to be valid it is necessary that a *linear relationship* exist among the loads, stresses, and displacements.

Two requirements must be imposed for the principle of superposition to apply:

1. The material must behave in a linear-elastic manner, so that Hooke's law is valid, and therefore the load will be proportional to displacement.

2. The geometry of the structure must not undergo significant change when the loads are applied, i.e., small displacement theory applies. Large displacements will significantly change the position and orientation of the loads. An example would be a cantilevered thin rod subjected to a force at its end, causing it to bend.

Throughout this text, these two requirements will be satisfied. Here only linear-elastic material behavior occurs; and the displacements produced by the loads will not significantly change the directions of applied loadings nor the dimensions used to compute the moments of forces.

2.4   Equations of Equilibrium

It may be recalled from statics that a structure or one of its members is in equilibrium when it maintains a balance of force and moment. In general this requires that the force and moment equations of equilibrium be satisfied along three independent axes, namely,

$$\Sigma F_x = 0 \quad \Sigma F_y = 0 \quad \Sigma F_z = 0$$
$$\Sigma M_x = 0 \quad \Sigma M_y = 0 \quad \Sigma M_z = 0 \tag{2-1}$$

The principal load-carrying portions of most structures, however, lie in a single plane, and since the loads are also coplanar, the above requirements for equilibrium reduce to

$$\Sigma F_x = 0$$
$$\Sigma F_y = 0 \tag{2-2}$$
$$\Sigma M_O = 0$$

Here ΣF_x and ΣF_y represent, respectively, the algebraic sums of the x and y components of all the forces acting on the structure or one of its members, and ΣM_O represents the algebraic sum of the moments of these force components about an axis perpendicular to the x–y plane (the z axis) and passing through point O.

Whenever these equations are applied, *it is first necessary to draw a free-body diagram of the structure or its members*. If a member is selected, it must be *isolated* from its supports and surroundings and its outlined shape drawn. All the forces and couple moments must be shown that act *on the member*. In this regard, the types of reactions at the supports can be determined using Table 2.1. Also, recall that forces common to two members act with equal magnitudes but opposite directions on the respective free-body diagrams of the members.

If the *internal loadings* at a specified point in a member are to be determined, the *method of sections* must be used. This requires that a "cut" or section be made perpendicular to the axis of the member at the point where the internal loading is to be determined. A free-body diagram of either segment of the "cut" member is isolated and the internal loads are then determined from the equations of equilibrium applied to the segment. In general, the internal loadings acting at the section will consist of a normal force **N**, shear force **V**, and bending moment **M**, as shown in Fig. 2–19.

We will cover the principles of statics that are used to determine the external reactions on structures in Sec. 2.5. Internal loadings in structural members will be discussed in Chapter 4.

internal loadings

Fig. 2–19

2.5 Determinacy and Stability

Before starting the force analysis of a structure, it is necessary to establish the determinacy and stability of the structure.

Determinacy. The equilibrium equations provide both the *necessary and sufficient* conditions for equilibrium. When all the forces in a structure can be determined strictly from these equations, the structure is referred to as *statically determinate*. Structures having more unknown forces than available equilibrium equations are called *statically indeterminate*. As a general rule, a stable structure can be identified as being either statically determinate or statically indeterminate by drawing free-body diagrams of all its members, or selective parts of its members, and then comparing the total number of unknown reactive force and moment components with the total number of available equilibrium equations.* For a coplanar structure there are at most *three* equilibrium equations for each part, so that if there is a total of n parts and r force and moment reaction components, we have

$$\boxed{\begin{aligned} r &= 3n, \text{ statically determinate} \\ r &> 3n, \text{ statically indeterminate} \end{aligned}} \qquad (2\text{--}3)$$

In particular, if a structure is *statically indeterminate,* the additional equations needed to solve for the unknown reactions are obtained by relating the applied loads and reactions to the displacement or slope at different points on the structure. These equations, which are referred to as *compatibility equations,* must be equal in number to the *degree of indeterminacy* of the structure. Compatibility equations involve the geometric and physical properties of the structure and will be discussed further in Chapter 10.

We will now consider some examples to show how to classify the determinacy of a structure. The first example considers beams; the second example, pin-connected structures; and in the third we will discuss frame structures. Classification of trusses will be considered in Chapter 3.

---

*Drawing the free-body diagrams is not strictly necessary, since a "mental count" of the number of unknowns can also be made and compared with the number of equilibrium equations.

EXAMPLE | **2.4**

Classify each of the beams shown in Figs. 2–20a through 2–20d as statically determinate or statically indeterminate. If statically indeterminate, report the number of degrees of indeterminacy. The beams are subjected to external loadings that are assumed to be known and can act anywhere on the beams.

SOLUTION

Compound beams, i.e., those in Figs. 2–20c and 2–20d, which are composed of pin-connected members must be disassembled. Note that in these cases, the unknown reactive forces acting between each member must be shown in equal but opposite pairs. The free-body diagrams of each member are shown. Applying $r = 3n$ or $r > 3n$, the resulting classifications are indicated.

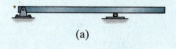
(a)

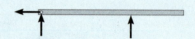

$r = 3, n = 1, 3 = 3(1)$ Statically determinate.   *Ans.*

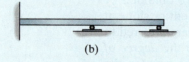
(b)

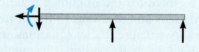

$r = 5, n = 1, 5 > 3(1)$ Statically indeterminate to the second degree.   *Ans.*

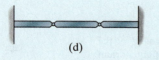
(c)

$r = 6, n = 2, 6 = 3(2)$ Statically determinate.   *Ans.*

(d)

$r = 10, n = 3, 10 > 3(3)$ Statically indeterminate to the first degree.   *Ans.*

Fig. 2–20

EXAMPLE 2.5

Classify each of the pin-connected structures shown in Figs. 2–21*a* through 2–21*d* as statically determinate or statically indeterminate. If statically indeterminate, report the number of degrees of indeterminacy. The structures are subjected to arbitrary external loadings that are assumed to be known and can act anywhere on the structures.

SOLUTION

Classification of pin-connected structures is similar to that of beams. The free-body diagrams of the members are shown. Applying $r = 3n$ or $r > 3n$, the resulting classifications are indicated.

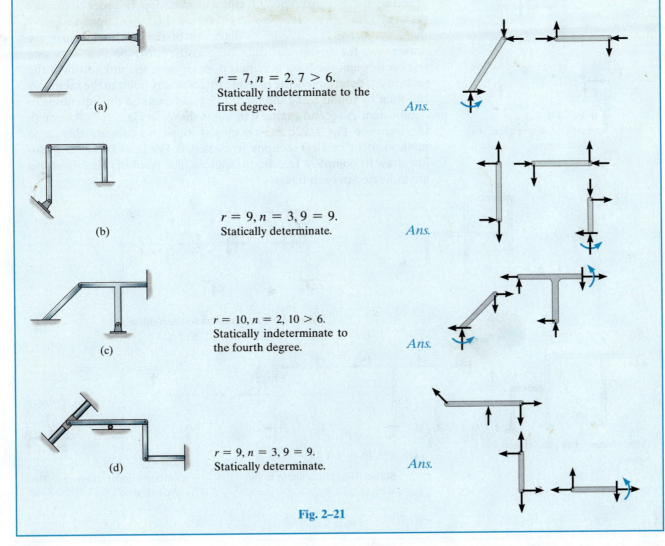

(a)

$r = 7, n = 2, 7 > 6$.
Statically indeterminate to the first degree. *Ans.*

(b)

$r = 9, n = 3, 9 = 9$.
Statically determinate. *Ans.*

(c)

$r = 10, n = 2, 10 > 6$.
Statically indeterminate to the fourth degree. *Ans.*

(d)

$r = 9, n = 3, 9 = 9$.
Statically determinate. *Ans.*

Fig. 2–21

EXAMPLE   2.6

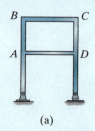

(a)

Classify each of the frames shown in Figs. 2–22a through 2–22c as statically determinate or statically indeterminate. If statically indeterminate, report the number of degrees of indeterminacy. The frames are subjected to external loadings that are assumed to be known and can act anywhere on the frames.

SOLUTION

Unlike the beams and pin-connected structures of the previous examples, frame structures consist of members that are connected together by rigid joints. Sometimes the members form internal loops as in Fig. 2–22a. Here *ABCD* forms a closed loop. In order to classify these structures, it is necessary to use the method of sections and "cut" the loop apart. The free-body diagrams of the sectioned parts are drawn and the frame can then be classified. Notice that only *one section* through the loop is required, since once the unknowns at the section are determined, the internal forces at any point in the members can then be found using the method of sections and the equations of equilibrium. A second example of this is shown in Fig. 2–22b. Although the frame in Fig. 2–22c has no closed loops, we can use this same method, using vertical sections, to classify it. For this case we can *also* just draw its complete free-body diagram. The resulting classifications are indicated in each figure.

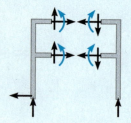

$r = 9, n = 2, 9 > 6,$
Statically indeterminate to the third degree.   *Ans.*

(a)

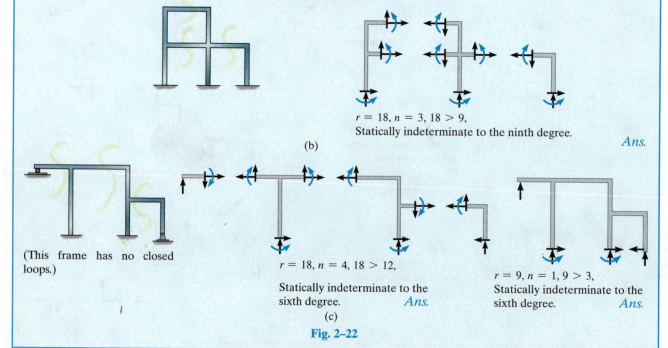

$r = 18, n = 3, 18 > 9,$
Statically indeterminate to the ninth degree.
Ans.

(b)

(This frame has no closed loops.)

$r = 18, n = 4, 18 > 12,$
Statically indeterminate to the sixth degree.   *Ans.*

(c)

$r = 9, n = 1, 9 > 3,$
Statically indeterminate to the sixth degree.   *Ans.*

Fig. 2–22

Stability.

To ensure the equilibrium of a structure or its members, it is not only necessary to satisfy the equations of equilibrium, but the members must also be properly held or constrained by their supports regardless of how the structure is loaded. Two situations may occur where the conditions for proper constraint have not been met.

Partial Constraints.

Instability can occur if a structure or one of its members has *fewer* reactive forces than equations of equilibrium that must be satisfied. The structure then becomes only *partially constrained*. For example, consider the member shown in Fig. 2–23 with its corresponding free-body diagram. Here the equation $\Sigma F_x = 0$ will not be satisfied for the loading conditions, and therefore the member will be unstable.

Improper Constraints.

In some cases there may be as many unknown forces as there are equations of equilibrium; however, *instability or movement of a structure or its members can develop because of improper constraining by the supports.* This can occur if all the *support reactions are concurrent* at a point. An example of this is shown in Fig. 2–24. From the free-body diagram of the beam it is seen that the summation of moments about point O will *not* be equal to zero ($Pd \neq 0$); thus rotation about point O will take place.

Another way in which improper constraining leads to instability occurs when the *reactive forces* are all *parallel*. An example of this case is shown in Fig. 2–25. Here when an inclined force **P** is applied, the summation of forces in the horizontal direction will not equal zero.

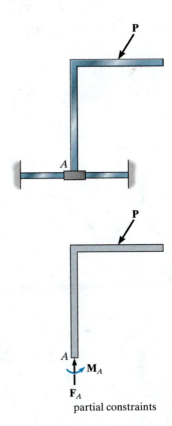

partial constraints

Fig. 2–23

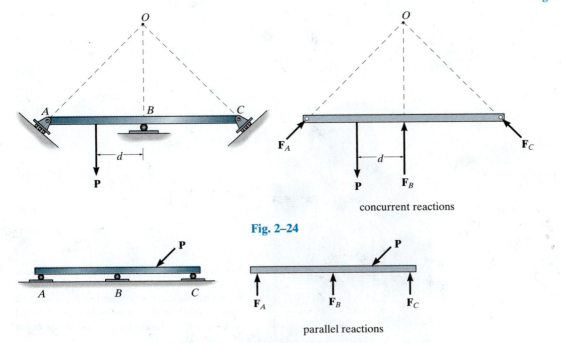

concurrent reactions

Fig. 2–24

parallel reactions

Fig. 2–25

In general, then, a structure will be geometrically unstable—that is, it will move slightly or collapse—if there are fewer reactive forces than equations of equilibrium; or if there are enough reactions, instability will occur if the lines of action of the reactive forces intersect at a common point or are parallel to one another. If the structure consists of several members or components, local instability of one or several of these members can generally be determined *by inspection*. If the members form a collapsible mechanism, the structure will be unstable. We will now formalize these statements for a *coplanar structure* having n members or components with r unknown reactions. Since three equilibrium equations are available for each member or component, we have

$$
\begin{array}{ll}
r < 3n & \text{unstable} \\
r \geq 3n & \text{unstable if member reactions are} \\
& \text{concurrent or parallel or some of the} \\
& \text{components form a collapsible mechanism}
\end{array}
\qquad (2\text{–}4)
$$

If the structure is unstable, *it does not matter* if it is statically determinate or indeterminate. In all cases such types of structures must be avoided in practice.

The following examples illustrate how structures or their members can be classified as stable or unstable. Structures in the form of a truss will be discussed in Chapter 3.

The K-bracing on this frame provides lateral stability from wind and vertical support of the floor girders. The framework has been sprayed with concrete grout in order to insulate the steel to keep it from losing its stiffness in the event of a fire.

EXAMPLE 2.7

Classify each of the structures in Figs. 2–26a through 2–26d as stable or unstable. The structures are subjected to arbitrary external loads that are assumed to be known.

SOLUTION

The structures are classified as indicated.

(a)

Fig. 2–26

The member is *stable* since the reactions are nonconcurrent and nonparallel. It is also statically determinate. *Ans.*

(b)

The member is statically indeterminate, but *unstable* since the three reactions are concurrent at *B*. *Ans.*

(c)

The beam is statically indeterminate, but *unstable* since the three reactions are all parallel. *Ans.*

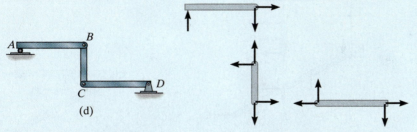

(d)

The structure is *unstable* since $r = 7$, $n = 3$, so that, by Eq. 2–4, $r < 3n$, $7 < 9$. Also, this can be seen by inspection, since AB can move horizontally without restraint. *Ans.*

2.6   Application of the Equations of Equilibrium

Occasionally, the members of a structure are connected together in such a way that the joints can be assumed as pins. Building frames and trusses are typical examples that are often constructed in this manner. Provided a pin-connected coplanar structure is properly constrained and contains no more supports or members than are necessary to prevent collapse, the forces acting at the joints and supports can be determined by applying the three equations of equilibrium ($\Sigma F_x = 0$, $\Sigma F_y = 0$, $\Sigma M_O = 0$) to each member. Understandably, once the forces at the joints are obtained, the size of the members, connections, and supports can then be determined on the basis of design code specifications.

To illustrate the method of force analysis, consider the three-member frame shown in Fig. 2–27a, which is subjected to loads $\mathbf{P}_1$ and $\mathbf{P}_2$. The free-body diagrams of each member are shown in Fig. 2–27b. In total there are nine unknowns; however, nine equations of equilibrium can be written, three for each member, so the problem is *statically determinate*. For the actual solution it is *also* possible, and sometimes convenient, to consider a portion of the frame or its entirety when applying some of these nine equations. For example, a free-body diagram of the entire frame is shown in Fig. 2–27c. One could determine the three reactions $\mathbf{A}_x$, $\mathbf{A}_y$, and $\mathbf{C}_x$ on this "rigid" pin-connected system, then analyze *any two* of its members, Fig. 2–27b, to obtain the other six unknowns. Furthermore, the answers can be checked in part by applying the three equations of equilibrium to the remaining "third" member. To summarize, this problem can be solved by writing *at most* nine equilibrium equations using free-body diagrams of any members and/or combinations of connected members. Any more than nine equations written would *not* be unique from the original nine and would only serve to check the results.

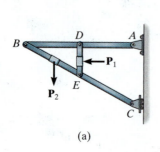

(a)

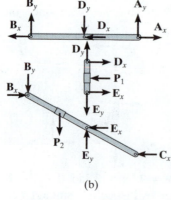

(b)

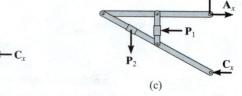

(c)

Fig. 2–27

Consider now the two-member frame shown in Fig. 2–28a. Here the free-body diagrams of the members reveal six unknowns, Fig. 2–28b; however, six equilibrium equations, three for each member, can be written, so again the problem is statically determinate. As in the previous case, a free-body diagram of the entire frame can also be used for part of the analysis, Fig. 2–28c. Although, as shown, the frame has a tendency to collapse without its supports, by rotating about the pin at B, this will not happen since the force system acting on it must still hold it in equilibrium. Hence, if so desired, all six unknowns can be determined by applying the three equilibrium equations to the entire frame, Fig. 2–28c, and also to either one of its members.

The above two examples illustrate that if a structure is properly supported and contains no more supports or members than are necessary to prevent collapse, the frame becomes statically determinate, and so the unknown forces at the supports and connections can be determined from the equations of equilibrium applied to each member. Also, if the structure remains *rigid* (noncollapsible) when the supports are removed (Fig. 2–27c), all three support reactions can be determined by applying the three equilibrium equations to the entire structure. However, if the structure appears to be nonrigid (collapsible) after removing the supports (Fig. 2–28c), it must be dismembered and equilibrium of the individual members must be considered in order to obtain enough equations to determine *all* the support reactions.

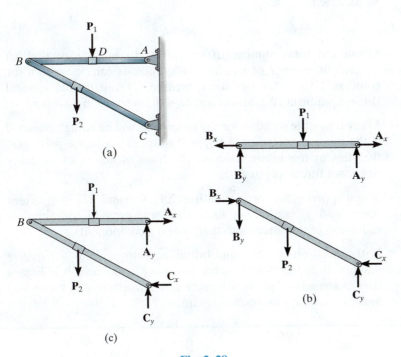

Fig. 2–28

2

Procedure for Analysis

The following procedure provides a method for determining the *joint reactions* for structures composed of pin-connected members.

Free-Body Diagrams

- Disassemble the structure and draw a free-body diagram of each member. Also, it may be convenient to supplement a member free-body diagram with a free-body diagram of the *entire structure*. Some or all of the support reactions can then be determined using this diagram.

- Recall that reactive forces common to two members act with equal magnitudes but opposite directions on the respective free-body diagrams of the members.

- All two-force members should be identified. These members, regardless of their shape, have no external loads on them, and therefore their free-body diagrams are represented with equal but opposite collinear forces acting on their ends.

- In many cases it is possible to tell by inspection the proper arrowhead sense of direction of an unknown force or couple moment; however, if this seems difficult, the directional sense can be assumed.

Equations of Equilibrium

- Count the total number of unknowns to make sure that an equivalent number of equilibrium equations can be written for solution. Except for two-force members, recall that in general three equilibrium equations can be written for each member.

- Many times, the solution for the unknowns will be straightforward if the moment equation $\Sigma M_O = 0$ is applied about a point (O) that lies at the intersection of the lines of action of as many unknown forces as possible.

- When applying the force equations $\Sigma F_x = 0$ and $\Sigma F_y = 0$, orient the x and y axes along lines that will provide the simplest reduction of the forces into their x and y components.

- If the solution of the equilibrium equations yields a *negative* magnitude for an unknown force or couple moment, it indicates that its arrowhead sense of direction is *opposite* to that which was assumed on the free-body diagram.

How Important is the Free-Body Diagram? For any structural analysis it is very important! Not only does it greatly reduce the chance for errors, by accounting for all the forces and the geometry used in the equilibrium equations, but it is also a source of communication to other engineers, who may check or use your calculations.

To emphase the importance of the free-body diagram, consider the case of the collapse of the second and fourth story walkways that crossed an open atrium in the Kansas City Hyatt Regency Hotel. This event occurred in July 1981, when about 2000 people assembled in the atrium for a dance contest. Some of the observers gathered on the walkways, then when the walkways suddenly collapsed, 114 people died, and more than 200 were injured.

The main reason for the failure was the faulty design of the connections used to support the walkways. The original design, shown in Fig. 2–29a, called for single tie or hanger rods that were attached to the roof and passed through holes in fabricated box beams, each made from two channels welded together, Fig. 2–29c. The connection of the rods to the beams required a nut and washer. A dispute arose between the contractor and the engineering firm over this design since it required unconventional fully threaded rods. As a result the design was changed to end threaded rods as shown in Fig. 2–29b.

To show why failure occurred, study the free-body diagrams of each design, along with the equilibrium analysis of the anticipated 180 kN loading shown in Figs. 2–29a, and 2–29b. Notice that the altered design requires the load supported by the top washer to be *twice* that of the original design. Since this new design was not properly reviewed, it resulted in the box beam splitting open at the weld seam and the rod pulling the washer through, resulting in the collapse, Fig. 2–29c.

Since the design of the box beam and tie rod assembly were substandard, and did not meet code, the engineers at the firm were found to have performed with gross negligence and unprofessional conduct. Consequently, they lost their license and the firm went bankrupt. It seems needless to say, but *engineers are responsible for their design and will be held accountable if anything goes wrong. So be neat and accurate in your work, and draw your free-body diagrams!*

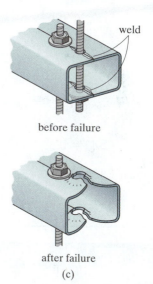

weld

before failure

after failure

(c)

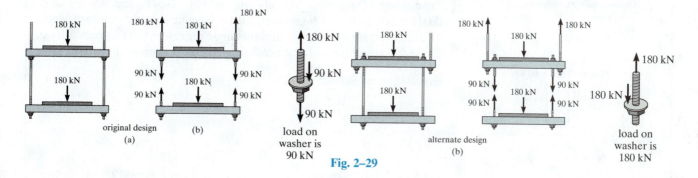

original design
(a)

(b)

load on
washer is
90 kN

alternate design
(b)

load on
washer is
180 kN

Fig. 2–29

2

EXAMPLE 2.8

Determine the reactions on the beam shown in Fig. 2–30a.

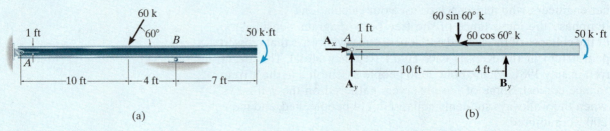

(a)   (b)

Fig. 2–30

SOLUTION

Free-Body Diagram. As shown in Fig. 2–30b, the 60-k force is resolved into x and y components. Furthermore, the 7-ft dimension line is not needed since a couple moment is a *free vector* and can therefore act anywhere on the beam for the purpose of computing the external reactions.

Equations of Equilibrium. Applying Eqs. 2–2 in a sequence, using previously calculated results, we have

$$\xrightarrow{+}\Sigma F_x = 0; \qquad A_x - 60 \cos 60° = 0 \qquad\qquad A_x = 30.0 \text{ k} \qquad Ans.$$

$$\zeta + \Sigma M_A = 0; \qquad -60 \sin 60°(10) + 60 \cos 60°(1) + B_y(14) - 50 = 0 \qquad B_y = 38.5 \text{ k} \qquad Ans.$$

$$+\uparrow\Sigma F_y = 0; \qquad -60 \sin 60° + 38.5 + A_y = 0 \qquad\qquad A_y = 13.4 \text{ k} \qquad Ans.$$

EXAMPLE 2.9

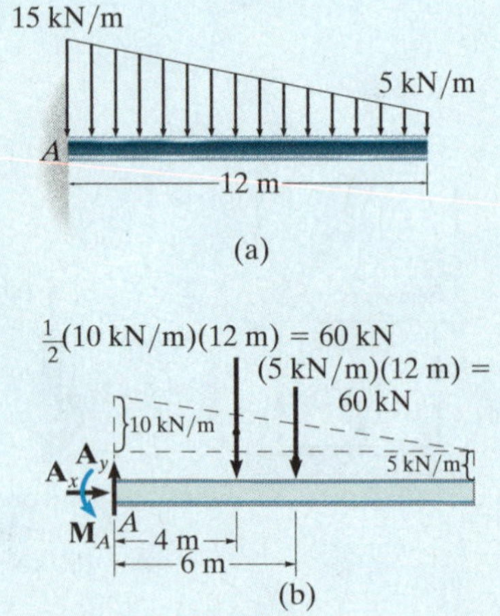

Determine the reactions on the beam in Fig. 2–31a.

SOLUTION

Free-Body Diagram. As shown in Fig. 2–31b, the trapezoidal distributed loading is segmented into a triangular and a uniform load. The *areas* under the triangle and rectangle represent the *resultant* forces. These forces act through the centroid of their corresponding areas.

Equations of Equilibrium.

$$\xrightarrow{+}\Sigma F_x = 0; \quad A_x = 0 \qquad\qquad\qquad\qquad\qquad Ans.$$

$$+\uparrow\Sigma F_y = 0; \quad A_y - 60 - 60 = 0 \qquad A_y = 120 \text{ kN} \qquad Ans.$$

$$\zeta + \Sigma M_A = 0; \quad -60(4) - 60(6) + M_A = 0 \quad M_A = 600 \text{ kN} \cdot \text{m} \qquad Ans.$$

Fig. 2–31

EXAMPLE 2.10

Determine the reactions on the beam in Fig. 2–32a. Assume A is a pin and the support at B is a roller (smooth surface).

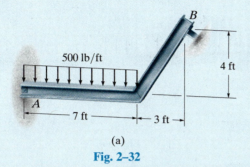

500 lb/ft

B

4 ft

A

— 7 ft — — 3 ft —

(a)

Fig. 2–32

SOLUTION

Free-Body Diagram.   As shown in Fig. 2–32b, the support ("roller") at B exerts a *normal force* on the beam at its point of contact. The line of action of this force is defined by the 3–4–5 triangle.

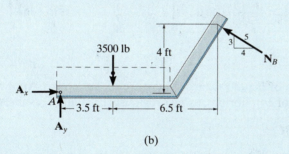

3500 lb

4 ft

3 5
4

N_B

A_x

A — 3.5 ft — — 6.5 ft —

A_y

(b)

Equations of Equilibrium.   Resolving N_B into x and y components and summing moments about A yields a direct solution for N_B. Why? Using this result, we can then obtain A_x and A_y.

$$\zeta + \Sigma M_A = 0; \quad -3500(3.5) + \left(\tfrac{4}{5}\right)N_B(4) + \left(\tfrac{3}{5}\right)N_B(10) = 0 \qquad \text{\textit{Ans.}}$$

$$N_B = 1331.5 \text{ lb} = 1.33 \text{ k}$$

$$\xrightarrow{+} \Sigma F_x = 0; \quad A_x - \tfrac{4}{5}(1331.5) = 0 \qquad A_x = 1.07 \text{ k} \qquad \text{\textit{Ans.}}$$

$$+\uparrow \Sigma F_y = 0; \quad A_y - 3500 + \tfrac{3}{5}(1331.5) = 0 \qquad A_y = 2.70 \text{ k} \qquad \text{\textit{Ans.}}$$

EXAMPLE   2.11

The compound beam in Fig. 2–33a is fixed at *A*. Determine the reactions at *A*, *B*, and *C*. Assume that the connection at *B* is a pin and *C* is a roller.

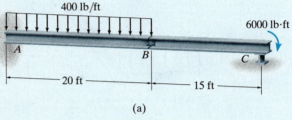

(a)

Fig. 2–33

SOLUTION

Free-Body Diagrams. The free-body diagram of each segment is shown in Fig. 2–33b. Why is this problem statically determinate?

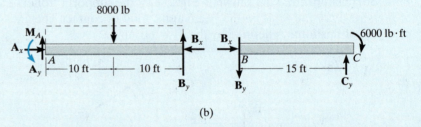

(b)

Equations of Equilibrium. There are six unknowns. Applying the six equations of equilibrium, using previously calculated results, we have

Segment *BC*:

$\zeta + \Sigma M_C = 0;$	$-6000 + B_y(15) = 0$	$B_y = 400 \text{ lb}$	*Ans.*
$+\uparrow \Sigma F_y = 0;$	$-400 + C_y = 0$		*Ans.*
		$C_y = 400 \text{ lb}$	
$\xrightarrow{+} \Sigma F_x = 0;$	$B_x = 0$		*Ans.*

Segment *AB*:

$\zeta + \Sigma M_A = 0;$	$M_A - 8000(10) + 400(20) = 0$		
	$M_A = 72.0 \text{ k} \cdot \text{ft}$		*Ans.*
$+\uparrow \Sigma F_y = 0;$	$A_y - 8000 + 400 = 0$	$A_y = 7.60 \text{ k}$	*Ans.*
$\xrightarrow{+} \Sigma F_x = 0;$	$A_x - 0 = 0$	$A_x = 0$	*Ans.*

EXAMPLE 2.12

Determine the horizontal and vertical components of reaction at the pins A, B, and C of the two-member frame shown in Fig. 2–34a.

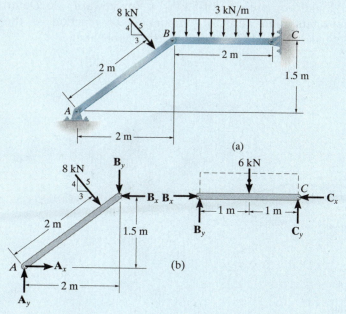

(a)

(b)

Fig. 2–34

SOLUTION

Free-Body Diagrams.   The free-body diagram of each member is shown in Fig. 2–34b.

Equations of Equilibrium.   Applying the six equations of equilibrium in the following sequence allows a direct solution for each of the six unknowns.

Member BC:

$$\zeta + \Sigma M_C = 0; \quad -B_y(2) + 6(1) = 0 \qquad B_y = 3 \text{ kN} \qquad \textit{Ans.}$$

Member AB:

$$\zeta + \Sigma M_A = 0; \quad -8(2) - 3(2) + B_x(1.5) = 0 \quad B_x = 14.7 \text{ kN} \qquad \textit{Ans.}$$

$$\xrightarrow{+} \Sigma F_x = 0; \quad A_x + \tfrac{3}{5}(8) - 14.7 = 0 \qquad A_x = 9.87 \text{ kN} \qquad \textit{Ans.}$$

$$+\uparrow \Sigma F_y = 0; \quad A_y - \tfrac{4}{5}(8) - 3 = 0 \qquad A_y = 9.40 \text{ kN} \qquad \textit{Ans.}$$

Member BC:

$$\xrightarrow{+} \Sigma F_x = 0; \quad 14.7 - C_x = 0 \qquad C_x = 14.7 \text{ kN} \qquad \textit{Ans.}$$

$$+\uparrow \Sigma F_y = 0; \quad 3 - 6 + C_y = 0 \qquad C_y = 3 \text{ kN} \qquad \textit{Ans.}$$

EXAMPLE   2.13

The side of the building in Fig. 2–35a is subjected to a wind loading that creates a uniform *normal* pressure of 15 kPa on the windward side and a suction pressure of 5 kPa on the leeward side. Determine the horizontal and vertical components of reaction at the pin connections A, B, and C of the supporting gable arch.

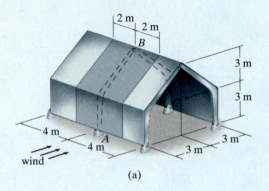

(a)

Fig. 2–35

SOLUTION

Since the loading is evenly distributed, the central gable arch supports a loading acting on the walls and roof of the dark-shaded tributary area. This represents a uniform distributed load of $(15 \text{ kN/m}^2)(4 \text{ m}) = 60 \text{ kN/m}$ on the windward side and $(5 \text{ kN/m}^2)(4 \text{ m}) = 20 \text{ kN/m}$ on the leeward side, Fig. 2–35b.

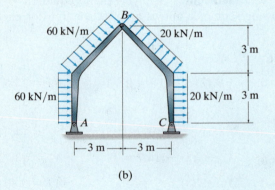

(b)

Free-Body Diagrams. Simplifying the distributed loadings, the free-body diagrams of the entire frame and each of its parts are shown in Fig. 2–35c.

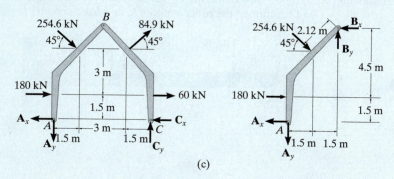
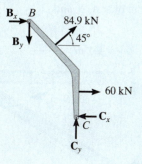

(c)

Equations of Equilibrium. Simultaneous solution of equations is avoided by applying the equilibrium equations in the following sequence using previously computed results.*

Entire Frame:

$\zeta + \Sigma M_A = 0;$ $-(180 + 60)(1.5) - (254.6 + 84.9) \cos 45°(4.5)$

$\qquad - (254.6 \sin 45°)(1.5) + (84.9 \sin 45°)(4.5) + C_y(6) = 0$

$$C_y = 240.0 \text{ kN} \qquad \textit{Ans.}$$

$+\uparrow \Sigma F_y = 0;$ $-A_y - 254.6 \sin 45° + 84.9 \sin 45° + 240.0 = 0$

$$A_y = 120.0 \text{ kN} \qquad \textit{Ans.}$$

Member AB:

$\zeta + \Sigma M_B = 0;$ $A_x(6) + 120.0(3) + 180(4.5) - 254.6(2.12) = 0$

$$A_x = 285.0 \text{ kN} \qquad \textit{Ans.}$$

$\overset{+}{\rightarrow} \Sigma F_x = 0;$ $-285.0 + 180 + 254.6 \cos 45° - B_x = 0$

$$B_x = 75.0 \text{ kN} \qquad \textit{Ans.}$$

$+\uparrow \Sigma F_y = 0;$ $-120.0 - 254.6 \sin 45° + B_y = 0$

$$B_y = 300.0 \text{ kN} \qquad \textit{Ans.}$$

Member CB:

$\overset{+}{\rightarrow} \Sigma F_x = 0;$ $-C_x + 60 + 84.9 \cos 45° + 75.0 = 0$

$$C_x = 195.0 \text{ kN} \qquad \textit{Ans.}$$

*The problem can also be solved by applying the six equations of equilibrium only to the two members. If this is done, it is best to first sum moments about point A on member AB, then point C on member CB. By doing this, one obtains two equations to be solved simultaneously for B_x and B_y.

FUNDAMENTAL PROBLEMS

F2–1. Determine the horizontal and vertical components of reaction at the pins A, B, and C.

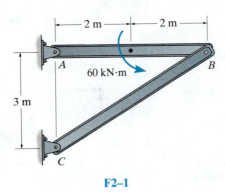

F2–1

F2–2. Determine the horizontal and vertical components of reaction at the pins A, B, and C.

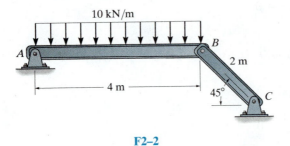

F2–2

F2–3. Determine the horizontal and vertical components of reaction at the pins A, B, and C.

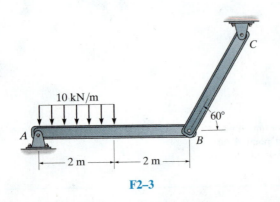

F2–3

F2–4. Determine the horizontal and vertical components of reaction at the roller support A, and fixed support B.

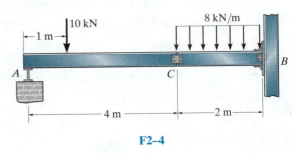

F2–4

F2–5. Determine the horizontal and vertical components of reaction at pins A, B, and C of the two-member frame.

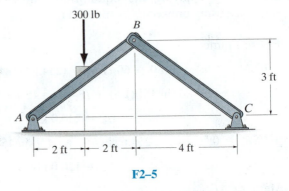

F2–5

F2–6. Determine the components of reaction at the roller support A and pin support C. Joint B is fixed connected.

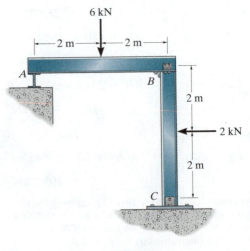

F2–6

F2–7. Determine the horizontal and vertical components of reaction at the pins *A*, *B*, and *D* of the three-member frame. The joint at *C* is fixed connected.

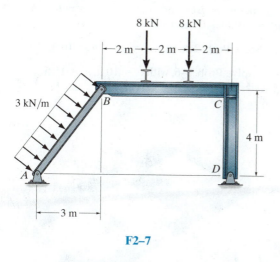

F2–7

F2–9. Determine the components of reaction at the fixed support *D* and the pins *A*, *B*, and *C* of the three-member frame. Neglect the thickness of the members.

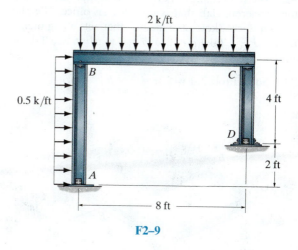

F2–9

F2–8. Determine the components of reaction at the fixed support *D* and the pins *A*, *B*, and *C* of the three-member frame. Neglect the thickness of the members.

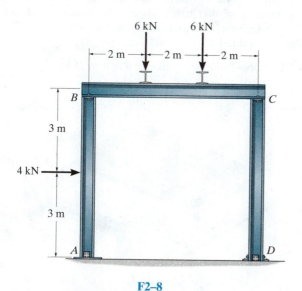

F2–8

F2–10. Determine the components of reaction at the fixed support *D* and the pins *A*, *B*, and *C* of the three-member frame. Neglect the thickness of the members.

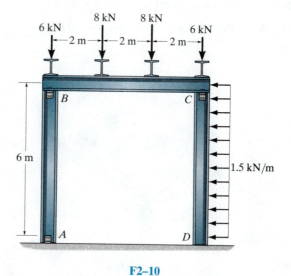

F2–10

PROBLEMS

Sec. 2.1

2–1. The steel framework is used to support the reinforced stone concrete slab that is used for an office. The slab is 200 mm thick. Sketch the loading that acts along members *BE* and *FED*. Take $a = 2$ m, $b = 5$ m. *Hint:* See Tables 1.2 and 1.4.

2–2. Solve Prob. 2–1 with $a = 3$ m, $b = 4$ m.

2–6. The frame is used to support a 2-in.-thick plywood floor of a residential dwelling. Sketch the loading that acts along members *BG* and *ABCD*. Set $a = 6$ ft, $b = 18$ ft. *Hint:* See Tables 1.2 and 1.4.

2–7. Solve Prob. 2–6, with $a = 10$ ft, $b = 10$ ft.

*2–8. Solve Prob. 2–6, with $a = 10$ ft, $b = 15$ ft.

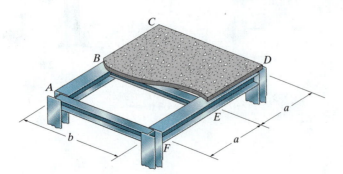

Probs. 2–1/2

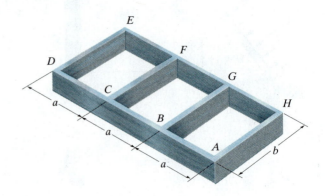

Probs. 2–6/7/8

2–3. The floor system used in a school classroom consists of a 4-in. reinforced stone concrete slab. Sketch the loading that acts along the joist *BF* and side girder *ABCDE*. Set $a = 10$ ft, $b = 30$ ft. *Hint:* See Tables 1.2 and 1.4.

*2–4. Solve Prob. 2–3 with $a = 10$ ft, $b = 15$ ft.

2–5. Solve Prob. 2–3 with $a = 7.5$ ft, $b = 20$ ft.

2–9. The steel framework is used to support the 4-in. reinforced stone concrete slab that carries a uniform live loading of 400 lb/ft^2. Sketch the loading that acts along members *BE* and *FED*. Set $a = 9$ ft, $b = 12$ ft. *Hint:* See Table 1.2.

2–10. Solve Prob. 2–9, with $a = 6$ ft, $b = 18$ ft.

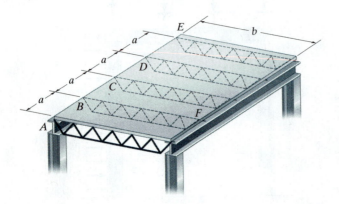

Probs. 2–3/4/5

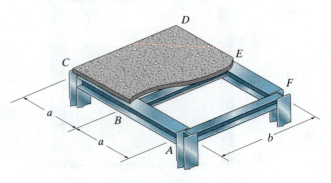

Probs. 2–9/10

Sec. 2.2–2.5

2–11. Classify each of the structures as statically determinate or indeterminate. If indeterminate, specify the degree of indeterminacy.

*2–12. Classify each of the frames as statically determinate or indeterminate. If indeterminate, specify the degree of indeterminacy. All internal joints are fixed connected.

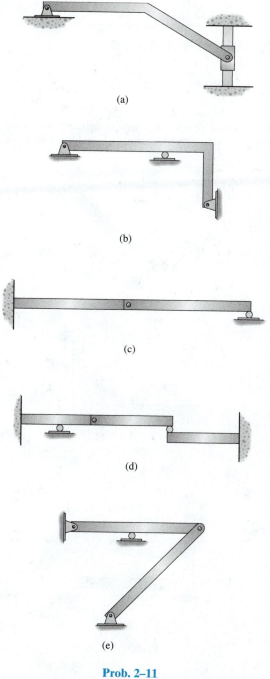

(a)

(b)

(c)

(d)

(e)

Prob. 2–11

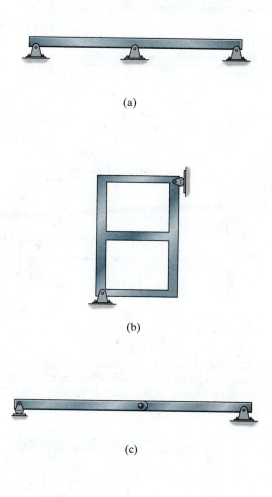

(a)

(b)

(c)

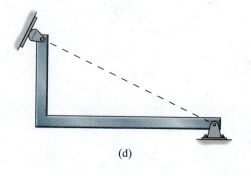

(d)

Prob. 2–12

2–13. Classify each of the structures as statically determinate, statically indeterminate, stable, or unstable. If indeterminate, specify the degree of indeterminacy. The supports or connections are to be assumed as stated.

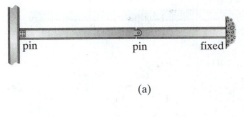

pin   pin   fixed

(a)

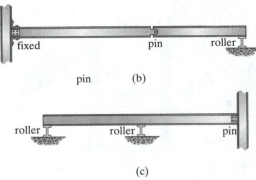

fixed   pin   roller

(b)

pin

roller   roller   pin

(c)

Prob. 2–13

2–14. Classify each of the structures as statically determinate, statically indeterminate, stable, or unstable. If indeterminate, specify the degree of indeterminacy. The supports or connections are to be assumed as stated.

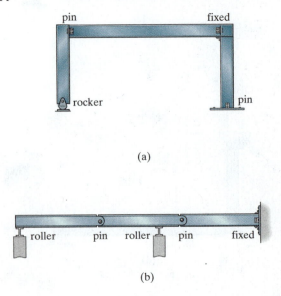

pin   fixed

rocker   pin

(a)

roller   pin   roller   pin   fixed

(b)

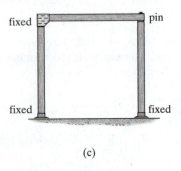

fixed   pin

fixed   fixed

(c)

Prob. 2–14

2–15. Classify each of the structures as statically determinate, statically indeterminate, stable, or unstable.

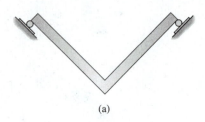

(a)

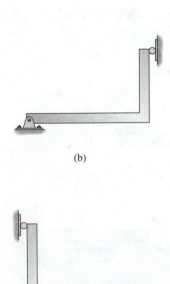

(b)

(c)

Prob. 2–15

*2–16. Classify each of the structures as statically determinate, statically indeterminate, or unstable. If indeterminate, specify the degree of indeterminacy.

2–17. Classify each of the structures as statically determinate, statically indeterminate, or unstable. If indeterminate, specify the degree of indeterminacy.

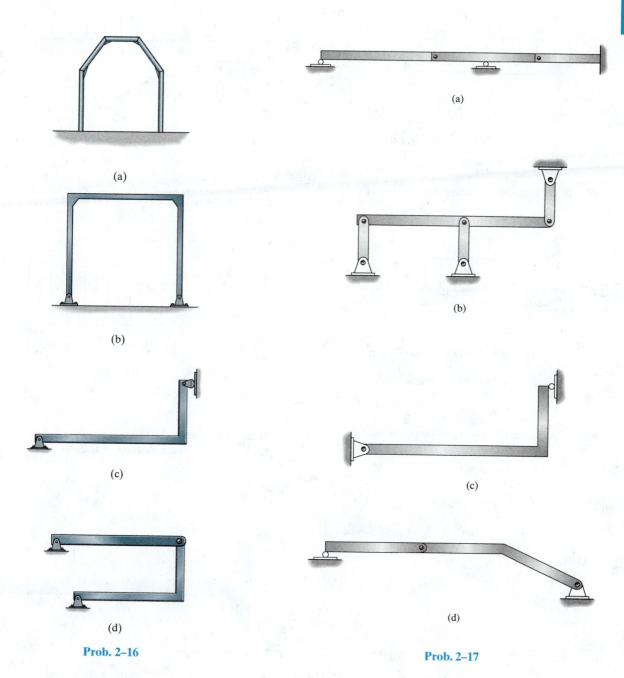

(a)

(a)

(b)

(b)

(c)

(c)

(d)

(d)

Prob. 2–16

Prob. 2–17

Sec. 2.6

*2–18. Determine the reactions on the beam.

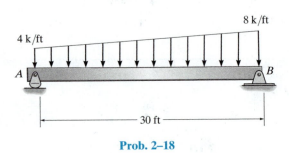

Prob. 2–18

2–19. Determine the reactions at the supports.

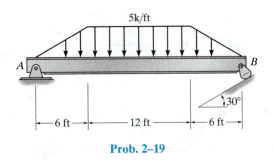

Prob. 2–19

*2–20. Determine the reactions on the beam.

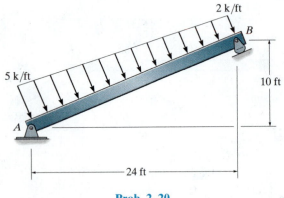

Prob. 2–20

2–21. Determine the reactions at the supports A and B of the compound beam. There is a pin at C.

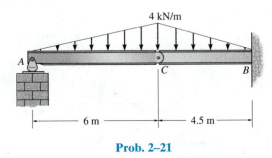

Prob. 2–21

2–22. Determine the reactions at the supports.

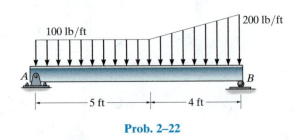

Prob. 2–22

2–23. Determine the reactions at the supports A and C of the compound beam. Assume A is fixed, B is a pin, and C is a roller.

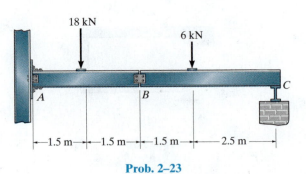

Prob. 2–23

*2–24. Determine the reactions on the beam. The support at *B* can be assumed to be a roller.

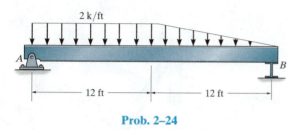

2 k/ft

12 ft   12 ft

Prob. 2–24

2–25. Determine the horizontal and vertical components of reaction at the pins *A* and *C*.

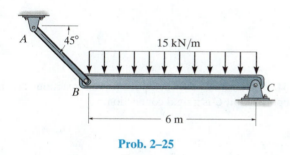

A
45°
15 kN/m
B
6 m
C

Prob. 2–25

2–26. Determine the reactions at the truss supports *A* and *B*. The distributed loading is caused by wind.

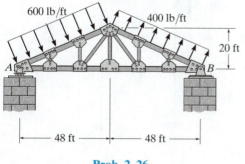

600 lb/ft
400 lb/ft
20 ft
A   *B*
48 ft   48 ft

Prob. 2–26

2–27. The compound beam is fixed at *A* and supported by a rocker at *E* and *C*. There are hinges (pins) at *D* and *B*. Determine the reactions at the supports.

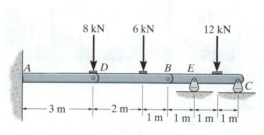

8 kN   6 kN   12 kN
A   *D*   *B* *E*   *C*
3 m   2 m
1 m 1 m 1 m 1 m

Prob. 2–27

*2–28. Determine the reactions on the beam. The support at *B* can be assumed as a roller.

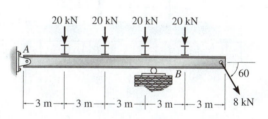

20 kN   20 kN   20 kN   20 kN
A
B
60
3 m   3 m   3 m   3 m   3 m   8 kN

Prob. 2–28

2–29. Determine the reactions at the supports *A*, *B*, *C*, and *D*.

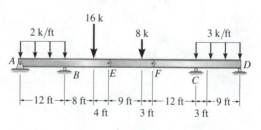

16 k
2 k/ft   8 k   3 k/ft
A   *D*
B   *E*   *F*   *C*
12 ft   8 ft   9 ft   12 ft   9 ft
4 ft   3 ft   3 ft

Prob. 2–29

2-30. Determine the reactions at the supports A and B of the compound beam. Assume A is a roller, C is a pin, and B is fixed.

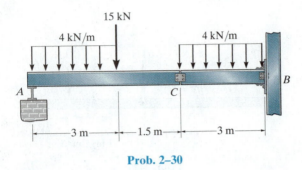

Prob. 2–30

2-31. The beam is subjected to the two concentrated loads as shown. Assuming that the foundation exerts a linearly varying load distribution on its bottom, determine the load intensities w_1 and w_2 for equilibrium (a) in terms of the parameters shown; (b) set $P = 500$ lb, $L = 12$ ft.

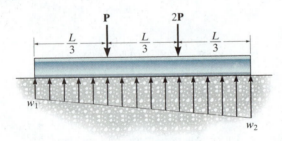

Prob. 2–31

*2-32. Determine the horizontal and vertical components of reaction at the supports A and C. Assume the members are pin connected at A, B, and C.

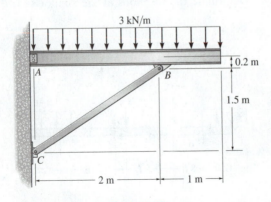

Prob. 2–32

2-33. Determine the horizontal and vertical components of reaction at the supports A and C.

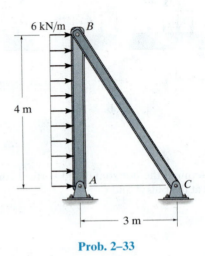

Prob. 2–33

2-34. Determine the components of reaction at the supports. Joint C is a rigid connection.

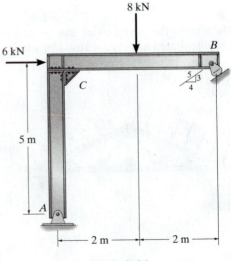

Prob. 2–34

2–35. The bulkhead AD is subjected to both water and soil-backfill pressures. Assuming AD is "pinned" to the ground at A, determine the horizontal and vertical reactions there and also the required tension in the ground anchor BC necessary for equilibrium. The bulkhead has a mass of 800 kg.

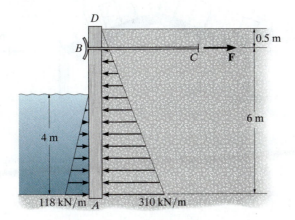

Prob. 2–35

2–37. Determine the horizontal and vertical reactions at A and C of the two-member frame.

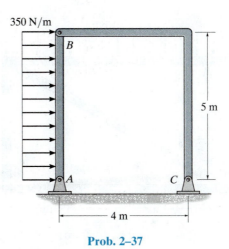

Prob. 2–37

***2–36.** Determine the reactions at the supports A and B. Assume the support at B is a roller. C is a fixed-connected joint.

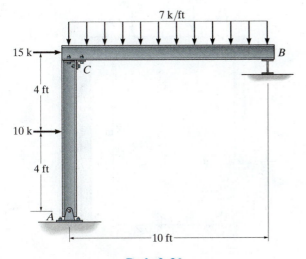

Prob. 2–36

2–38. The wall crane supports a load of 700 lb. Determine the horizontal and vertical components of reaction at the pins A and D. Also, what is the force in the cable at the winch W?

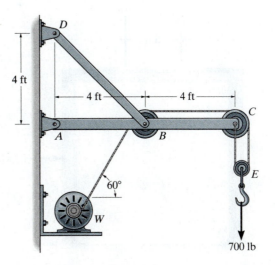

Prob. 2–38

2–39. Determine the horizontal and vertical force components that the pins support at *A* and *D* exert on the four-member frame.

2–41. Determine the components of reaction at the pinned supports *A* and *C* of the two-member frame. Neglect the thickness of the members. Assume *B* is a pin.

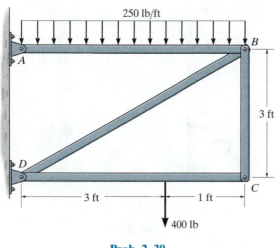

Prob. 2–39

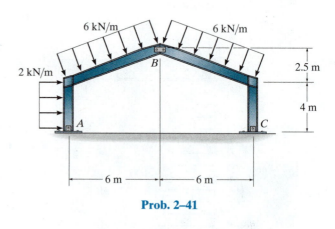

Prob. 2–41

***2–40.** Determine the reactions at the supports *A* and *D*. Assume *A* is fixed and *B*, *C* and *D* are pins.

2–42. Determine the horizontal and vertical components of reaction at *A*, *C*, and *D*. Assume the frame is pin connected at *A*, *C*, and *D*, and there is a fixed-connected joint at *B*.

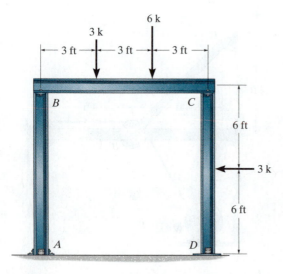

Prob. 2–40

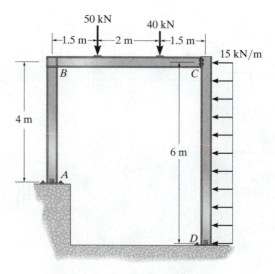

Prob. 2–42

2–43. The bridge frame consists of three segments which can be considered pinned at A, D, and E, rocker supported at C and F, and roller supported at B. Determine the horizontal and vertical components of reaction at all these supports due to the loading shown.

*2–44. Determine the horizontal and vertical reactions at the connections A and C of the gable frame. Assume that A, B, and C are pin connections. The purlin loads such as D and E are applied perpendicular to the center line of each girder.

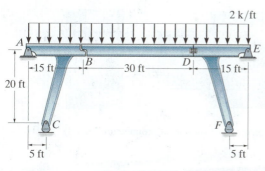

Prob. 2–43

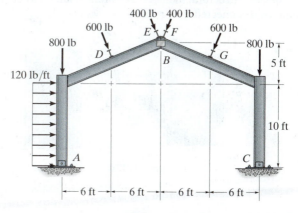

Prob. 2–44

PROJECT PROBLEM

2–1P. The railroad trestle bridge shown in the photo is supported by reinforced concrete bents. Assume the two simply supported side girders, track bed, and two rails have a weight of 0.5 k/ft and the load imposed by a train is 7.2 k/ft. Each girder is 20 ft long. Apply the load over the entire bridge and determine the compressive force in the columns of each bent. For the analysis assume all joints are pin connected and neglect the weight of the bent. Are these realistic assumptions?

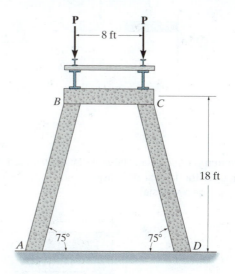

Project Prob. 2–1P

CHAPTER REVIEW

Supports—Structural members are often assumed to be pin connected if slight relative rotation can occur between them, and fixed connected if no rotation is possible.

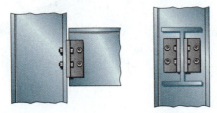

typical "pin-supported" connection (metal)

typical "fixed-supported" connection (metal)

Idealized Structures—By making assumptions about the supports and connections as being either roller supported, pinned, or fixed, the members can then be represented as lines, so that we can establish an idealized model that can be used for analysis.

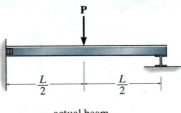
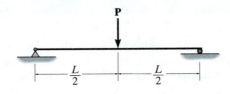

actual beam

idealized beam

The tributary loadings on slabs can be determined by first classifying the slab as a one-way or two-way slab. As a general rule, if L_2 is the largest dimension, and $L_2/L_1 > 2$, the slab will behave as a one-way slab. If $L_2/L_1 \leq 2$, the slab will behave as a two-way slab.

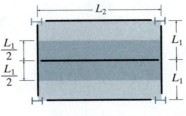
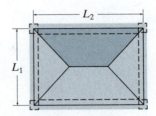

One-way slab action
requires $L_2/L_1 > 2$.

Two-way slab action
requires $L_2/L_1 \leq 2$.

Principle of Superposition—Either the loads or displacements can be added together provided the material is linear elastic and only small displacements of the structure occur.

Equilibrium—Statically determinate structures can be analyzed by disassembling them and applying the equations of equilibrium to each member. The analysis of a statically determinate structure requires first drawing the free-body diagrams of all the members, and then applying the equations of equilibrium to each member.

$$\Sigma F_x = 0$$
$$\Sigma F_y = 0$$
$$\Sigma M_O = 0$$

The number of equations of equilibrium for all n members of a structure is $3n$. If the structure has r reactions, then the structure is *statically determinate* if

$$r = 3n$$

and *statically indeterminate* if

$$r > 3n$$

The additional number of equations required for the solution refers to the degree of indeterminacy.

Stability—If there are fewer reactions than equations of equilibrium, then the structure will be unstable because it is partially constrained. Instability due to improper constraints can also occur if the lines of action of the reactions are concurrent at a point or parallel to one another.

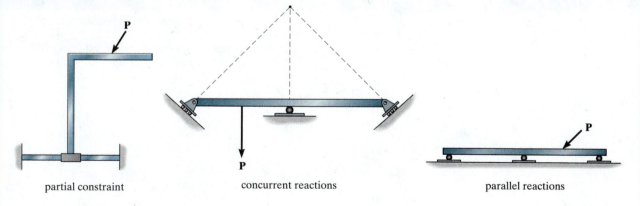

partial constraint   concurrent reactions   parallel reactions

Chapter 3

© Jeff Greenberg/AGE Fotostock/SuperStock

The forces in the members of this bridge can be analyzed using either the method of joints or the method of sections.

Analysis of Statically Determinate Trusses

In this chapter we will develop the procedures for analyzing statically determinate trusses using the method of joints and the method of sections. First, however, the determinacy and stability of a truss will be discussed. Then the analysis of three forms of planar trusses will be considered: simple, compound, and complex. Finally, at the end of the chapter we will consider the analysis of a space truss.

3.1 Common Types of Trusses

A **truss** is a structure composed of slender members joined together at their end points. The members commonly used in construction consist of wooden struts, metal bars, angles, or channels. The joint connections are usually formed by bolting or welding the ends of the members to a common plate, called a **gusset plate**, as shown in Fig. 3–1, or by simply passing a large bolt or pin through each of the members. Planar trusses lie in a single plane and are often used to support roofs and bridges.

This gusset plate is used to connect eight members of the truss supporting structure for a water tank.

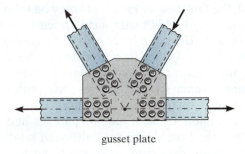

gusset plate

Fig. 3–1

Gusset plates used for a wood truss.

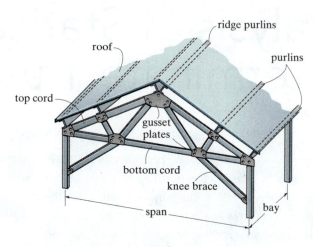

Fig. 3–2

Roof Trusses.   Roof trusses are often used as part of an industrial building frame, such as the one shown in Fig. 3–2. Here, the roof load is transmitted to the truss at the joints by means of a series of *purlins*. The roof truss along with its supporting columns is termed a *bent*. Ordinarily, roof trusses are supported either by columns of wood, steel, or reinforced concrete, or by masonry walls. To keep the bent rigid, and thereby capable of resisting horizontal wind forces, *knee braces* are sometimes used at the supporting columns. The space between adjacent bents is called a *bay*. Bays are economically spaced at about 15 ft (4.6 m) for spans around 60 ft (18 m) and about 20 ft (6.1 m) for spans of 100 ft (30 m). Bays are often tied together using diagonal bracing in order to maintain rigidity of the building's structure.

Trusses used to support roofs are selected on the basis of the span, the slope, and the roof material. Some of the more common types of trusses used are shown in Fig. 3–3. In particular, the scissors truss, Fig. 3–3*a*, can be used for short spans that require overhead clearance. The Howe and Pratt trusses, Figs. 3–3*b* and 3–3*c*, are used for roofs of moderate span, about 60 ft (18 m) to 100 ft (30 m). If larger spans are required to support the roof, the fan truss or Fink truss may be used, Figs. 3–3*d* and 3–3*e*. These trusses may be built with a cambered bottom cord such as that shown in Fig. 3–3*f*. If a flat roof or nearly flat roof is to be selected, the Warren truss, Fig. 3–3*g*, is often used. Also, the Howe and Pratt trusses may be modified for flat roofs. Sawtooth trusses, Fig. 3–3*h*, are often used where column spacing is not objectionable and uniform lighting is important. A textile mill would be an example. The bowstring truss, Fig. 3–3*i*, is sometimes selected for garages and small airplane hangars; and the arched truss, Fig. 3–3*j*, although relatively expensive, can be used for high rises and long spans such as field houses, gymnasiums, and so on.

Although more decorative than structural, these simple Pratt trusses are used for the entrance of a building.

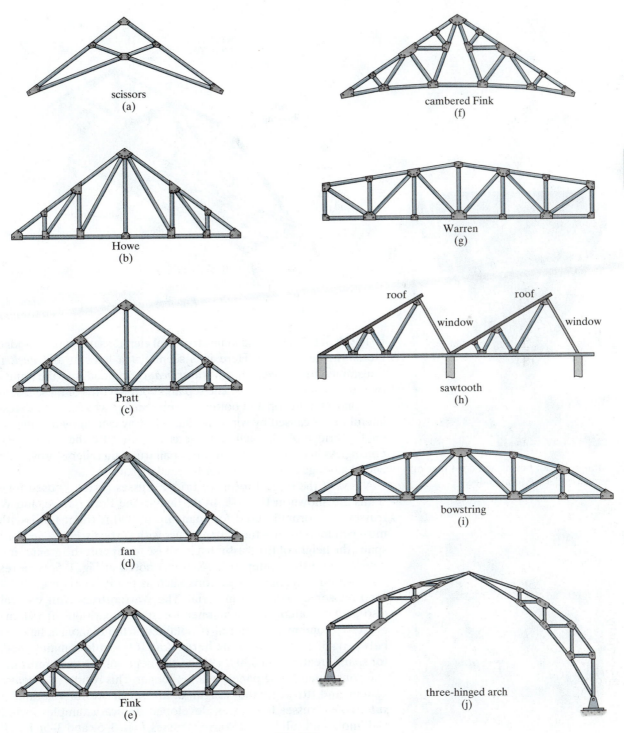

scissors
(a)

Howe
(b)

Pratt
(c)

fan
(d)

Fink
(e)

cambered Fink
(f)

Warren
(g)

roof   roof

window   window

sawtooth
(h)

bowstring
(i)

three-hinged arch
(j)

Fig. 3–3

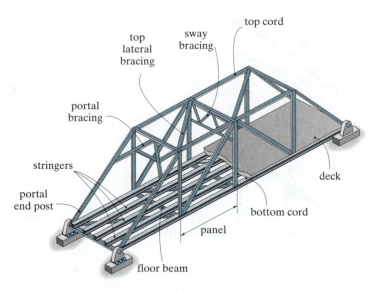

Fig. 3–4

Stringers, floor beams, and cross bracing shown under a typical bridge.

Parker trusses are used to support this bridge.

Bridge Trusses.   The main structural elements of a typical bridge truss are shown in Fig. 3–4. Here it is seen that a load on the **deck** is first transmitted to **stringers**, then to **floor beams**, and finally to the joints of the two supporting side trusses. The top and bottom cords of these side trusses are connected by top and bottom *lateral bracing,* which serves to resist the lateral forces caused by wind and the sidesway caused by moving vehicles on the bridge. Additional stability is provided by the *portal* and *sway bracing.* As in the case of many long-span trusses, a roller is provided at one end of a bridge truss to allow for thermal expansion.

A few of the typical forms of bridge trusses currently used for single spans are shown in Fig. 3–5. In particular, the Pratt, Howe, and Warren trusses are normally used for spans up to 200 ft (61 m) in length. The most common form is the Warren truss with verticals, Fig. 3–5c. For larger spans, the height of the truss must increase to support the greater moment developed in the center of the span as noted in Fig. 1–5. As a result, a truss with a polygonal upper cord, such as the Parker truss, Fig. 3–5d, is used for some savings in material. The Warren truss with verticals can also be fabricated in this manner for spans up to 300 ft (91 m). The greatest economy of material is obtained if the diagonals have a slope between 45° and 60° with the horizontal. If this rule is maintained, then for spans greater than 300 ft (91 m), the depth of the truss must increase and consequently the panels will get longer. This results in a heavy deck system and, to keep the weight of the deck within tolerable limits, **subdivided** trusses have been developed. Typical examples include the Baltimore and subdivided Warren trusses, Figs. 3–5e and 3–5f. Finally, the K-truss shown in Fig. 3–5g can also be used in place of a subdivided truss, since it accomplishes the same purpose.

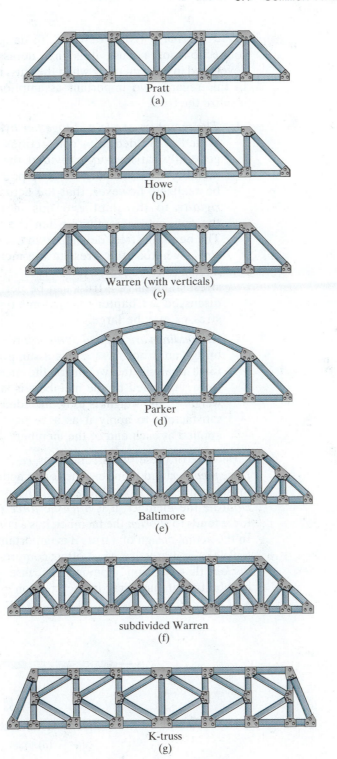

Pratt
(a)

Howe
(b)

Warren (with verticals)
(c)

Parker
(d)

Baltimore
(e)

subdivided Warren
(f)

K-truss
(g)

Fig. 3–5

Assumptions for Design.   To design both the members and the connections of a truss, it is first necessary to determine the *force* developed in each member when the truss is subjected to a given loading. In this regard, two important assumptions will be made in order to idealize the truss.

1. *The members are joined together by smooth pins*. In cases where bolted or welded joint connections are used, this assumption is generally satisfactory provided the center lines of the joining members are concurrent at a point, as in Fig. 3–1. It should be realized, however, that the actual connections do give some *rigidity* to the joint and this in turn introduces bending of the connected members when the truss is subjected to a load. The bending stress developed in the members is called *secondary stress,* whereas the stress in the members of the idealized truss, having pin-connected joints, is called *primary stress*. A secondary stress analysis of a truss can be performed using a computer, as discussed in Chapter 16. For some types of truss geometries these stresses may be large.

2. *All loadings are applied at the joints*. In most situations, such as for bridge and roof trusses, this assumption is true. Frequently in the force analysis, the weight of the members is neglected, since the force supported by the members is large in comparison with their weight. If the weight is to be included in the analysis, it is generally satisfactory to apply it as a vertical force, half of its magnitude applied at each end of the member.

Because of these two assumptions, *each truss member acts as an axial force member,* and therefore the forces acting at the ends of the member must be directed along the axis of the member. If the force tends to *elongate* the member, it is a *tensile force (T),* Fig. 3–6a; whereas if the force tends to *shorten* the member, it is a *compressive force (C),* Fig. 3–6b. In the actual design of a truss it is important to state whether the force is tensile or compressive. Most often, compression members must be made *thicker* than tension members, because of the buckling or sudden instability that may occur in compression members.

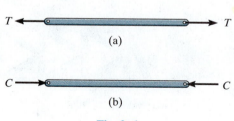

(a)

(b)

Fig. 3–6

3.2 Classification of Coplanar Trusses

Before beginning the force analysis of a truss, it is important to classify the truss as simple, compound, or complex, and then to be able to specify its determinacy and stability.

Simple Truss.   To prevent collapse, the framework of a truss must be rigid. Obviously, the four-bar frame *ABCD* in Fig. 3–7 will collapse unless a diagonal, such as *AC*, is added for support. The simplest framework that is rigid or stable is a *triangle*. Consequently, a ***simple truss*** is constructed by starting with a basic triangular element, such as *ABC* in Fig. 3–8, and connecting two members (*AD* and *BD*) to form an additional element. Thus it is seen that as each additional element of two members is placed on the truss, the number of joints is increased by one.

Fig. 3–7

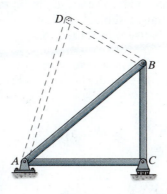

Fig. 3–8

An example of a simple truss is shown in Fig. 3–9, where the basic "stable" triangular element is *ABC*, from which the remainder of the joints, *D*, *E*, and *F*, are established in alphabetical sequence. For this method of construction, however, it is important to realize that simple trusses *do not* have to consist entirely of triangles. An example is shown in Fig. 3–10, where starting with triangle *ABC*, bars *CD* and *AD* are added to form joint *D*. Finally, bars *BE* and *DE* are added to form joint *E*.

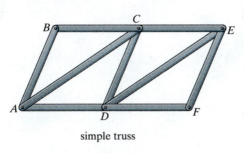

simple truss

Fig. 3–9

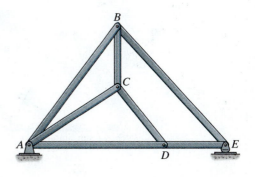

simple truss

Fig. 3–10

Compound Truss.   A *compound truss* is formed by connecting two or more simple trusses together. Quite often this type of truss is used to support loads acting over a *large span,* since it is cheaper to construct a somewhat lighter compound truss than to use a heavier single simple truss.

There are three ways in which simple trusses are joined together to form a compound truss. The trusses may be connected by a common joint and bar. An example is given in Fig. 3–11*a,* where the shaded truss *ABC* is connected to the shaded truss *CDE* in this manner. The trusses may be joined by three bars, as in the case of the shaded truss *ABC* is connected to the larger truss *DEF,* Fig. 3–11*b.* And finally, the trusses may be joined where bars of a large simple truss, called the *main truss,* have been replaced by simple trusses, called *secondary trusses.* An example is shown in Fig. 3–11*c,* where dashed members of the main truss *ABCDE* have been replaced by the secondary shaded trusses. If this truss carried roof loads, the use of the secondary trusses might be more economical, since the dashed members may be subjected to excessive bending, whereas the secondary trusses can better transfer the load.

Complex Truss.   A *complex truss* is one that cannot be classified as being either simple or compound. The truss in Fig. 3–12 is an example.

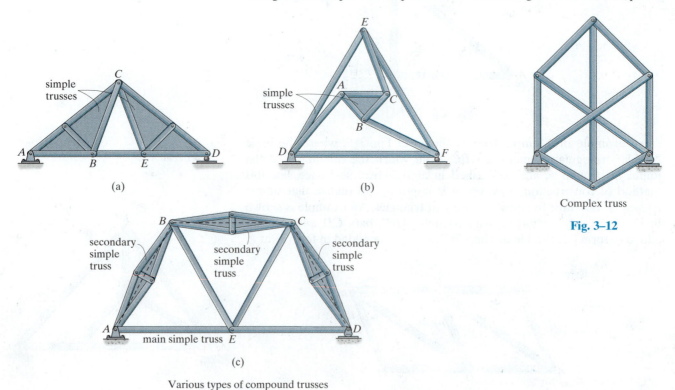

simple
trusses

C

A   *D*

B   *E*

(a)

simple
trusses

E

A
C

B

D   *F*

(b)

Complex truss

Fig. 3–12

secondary
simple
truss

secondary
simple
truss

secondary
simple
truss

B   *C*

A   *D*

main simple truss *E*

(c)

Various types of compound trusses

Fig. 3–11

Determinacy.   For any problem in truss analysis, it should be realized that the total number of *unknowns* includes the forces in *b* number of bars of the truss and the total number of external support reactions *r*. Since the truss members are all straight axial force members lying in the same plane, the force system acting at each joint is *coplanar and concurrent*. Consequently, rotational or moment equilibrium is automatically satisfied at the joint (or pin), and it is only necessary to satisfy $\Sigma F_x = 0$ and $\Sigma F_y = 0$ to ensure translational or force equilibrium. Therefore, only two equations of equilibrium can be written for each joint, and if there are *j* number of joints, the total number of equations available for solution is *2j*. By simply comparing the total number of unknowns $(b + r)$ with the total number of available equilibrium equations, it is therefore possible to specify the determinacy for either a simple, compound, or complex truss. We have

$$\begin{array}{ll} b + r = 2j & \text{statically determinate} \\ b + r > 2j & \text{statically indeterminate} \end{array} \qquad (3\text{–}1)$$

In particular, the *degree of indeterminacy* is specified by the difference in the numbers $(b + r) - 2j$.

Stability.   If $b + r < 2j$, a truss will be *unstable*, that is, a loading can be applied to the truss that causes it to collapse, since there will be an insufficient number of bars or reactions to constrain all the joints. Also, a truss can be unstable if it is statically determinate or statically indeterminate. In this case the stability will have to be determined either by inspection or by a force analysis.

External Stability.   As stated in Sec. 2.5, a *structure (or truss) is externally unstable if all of its reactions are concurrent or parallel*. For example, if a horizontal force is applied to the top cord of each of the two trusses in Fig. 3–13, each truss will be externally unstable, since the support reactions have lines of action that are either concurrent or parallel.

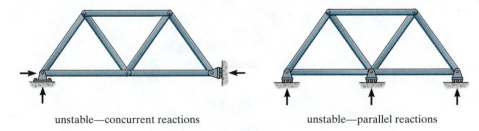

unstable—concurrent reactions   unstable—parallel reactions

Fig. 3–13

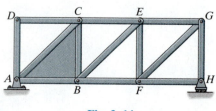

Fig. 3–14

3

Internal Stability. The internal stability of a truss can often be checked by careful inspection of the arrangement of its members. If it can be determined that each joint is held fixed so that it cannot move in a "rigid body" sense with respect to the other joints, then the truss will be stable. Notice that *a simple truss will always be internally stable,* since by the nature of its construction it requires starting from a basic triangular element and adding successive "rigid elements," each containing two additional members and a joint. The truss in Fig. 3–14 exemplifies this construction, where, starting with the shaded triangle element *ABC*, the successive joints *D, E, F, G, H* have been added.

If a truss is constructed so that it does not hold its joints in a fixed position, it will be unstable or have a "critical form." An obvious example of this is shown in Fig. 3–15, where it can be seen that no restraint or fixity is provided between joints *C* and *F* or *B* and *E*, and so the truss will collapse under a vertical load.

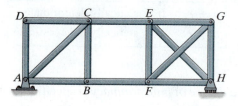

Fig. 3–15

To determine the internal stability of a *compound truss*, it is necessary to identify the way in which the simple trusses are connected together. For example, the compound truss in Fig. 3–16 is unstable since the inner simple truss *ABC* is connected to the outer simple truss *DEF* using three bars, *AD, BE,* and *CF*, which are *concurrent* at point *O*. Thus an external load can be applied to joint *A, B,* or *C* and cause the truss *ABC* to rotate slightly.

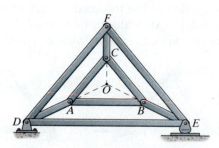

Fig. 3–16

If a truss is identified as *complex,* it may not be possible to tell by inspection if it is stable. For example, it can be shown by the analysis discussed in Sec. 3.7 that the complex truss in Fig. 3–17 is unstable or has a "critical form" only if the dimension $d = d'$. If $d \neq d'$ it is stable.

The instability of any form of truss, be it simple, compound, or complex, may also be noticed by using a computer to solve the $2j$ simultaneous equations written for all the joints of the truss. If inconsistent results are obtained, the truss will be unstable or have a critical form.

If a computer analysis is not performed, the methods discussed previously can be used to check the stability of the truss. To summarize, if the truss has b bars, r external reactions, and j joints, then if

$$
\begin{aligned}
b + r &< 2j \quad &&\text{unstable} \\
b + r &\geq 2j \quad &&\text{unstable if truss support reactions} \\
& &&\text{are concurrent or parallel or if} \\
& &&\text{some of the components of the} \\
& &&\text{truss form a collapsible mechanism}
\end{aligned}
\tag{3–2}
$$

Bear in mind, however, that *if a truss is unstable, it does not matter whether it is statically determinate or indeterminate.* Obviously, the use of an unstable truss is to be avoided in practice.

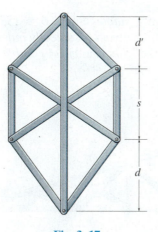

Fig. 3–17

EXAMPLE 3.1

Classify each of the trusses in Fig. 3–18 as stable, unstable, statically determinate, or statically indeterminate. The trusses are subjected to arbitrary external loadings that are assumed to be known and can act anywhere on the trusses.

SOLUTION

Fig. 3–18a. *Externally stable,* since the reactions are not concurrent or parallel. Since $b = 19$, $r = 3$, $j = 11$, then $b + r = 2j$ or $22 = 22$. Therefore, the truss is *statically determinate*. By inspection the truss is *internally stable.*

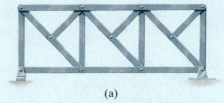

(a)

Fig. 3–18

Fig. 3–18b. *Externally stable.* Since $b = 15$, $r = 4$, $j = 9$, then $b + r > 2j$ or $19 > 18$. The truss is *statically indeterminate* to the first degree. By inspection the truss is *internally stable.*

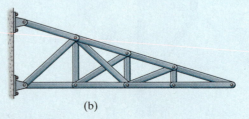

(b)

Fig. 3–18c. *Externally stable.* Since $b = 9$, $r = 3$, $j = 6$, then $b + r = 2j$ or $12 = 12$. The truss is *statically determinate*. By inspection the truss is *internally stable*.

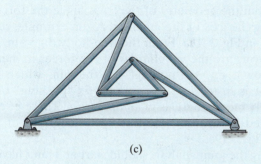

(c)

Fig. 3–18d. *Externally stable.* Since $b = 12$, $r = 3$, $j = 8$, then $b + r < 2j$ or $15 < 16$. The truss is *internally unstable*.

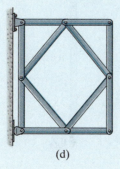

(d)

3.3   The Method of Joints

If a truss is in equilibrium, then each of its joints must also be in equilibrium. Hence, the **method of joints** consists of satisfying the equilibrium conditions $\Sigma F_x = 0$ and $\Sigma F_y = 0$ for the forces exerted *on the pin* at each joint of the truss.

When using the method of joints, it is necessary to draw each joint's free-body diagram before applying the equilibrium equations. Recall that the *line of action* of each member force acting on the joint is *specified* from the geometry of the truss, since the force in a member passes along the axis of the member. As an example, consider joint B of the truss in Fig. 3–19a. From the free-body diagram, Fig. 3–19b, the only unknowns are the *magnitudes* of the forces in members BA and BC. As shown, $\mathbf{F}_{BA}$ is "pulling" on the pin, which indicates that member BA is in *tension*, whereas $\mathbf{F}_{BC}$ is "pushing" on the pin, and consequently member BC is in *compression*. These effects are clearly demonstrated by drawing the free-body diagrams of the connected members, Fig. 3–19c.

In all cases, the joint analysis should start at a joint having at least one known force and at most two unknown forces, as in Fig. 3–19b. In this way, application of $\Sigma F_x = 0$ and $\Sigma F_y = 0$ yields two algebraic equations that can be solved for the two unknowns. When applying these equations, the correct sense of an unknown member force can be determined using one of two possible methods.

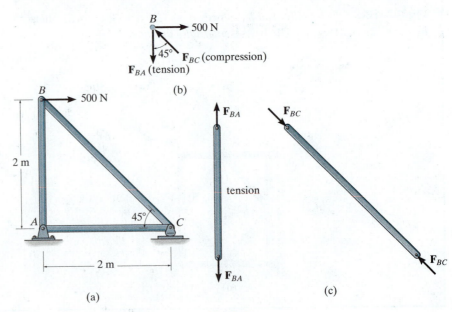

Fig. 3–19

1. *Always assume the unknown member forces acting on the joint's free-body diagram to be in tension, i.e., "pulling" on the pin. If this is done, then numerical solution of the equilibrium equations will yield positive scalars for members in tension and negative scalars for members in compression.* Once an unknown member force is found, use its *correct* magnitude and sense (T or C) on subsequent joint free-body diagrams.

2. *The correct sense of direction of an unknown member force can, in many cases, be determined "by inspection."* For example, F_{BC} in Fig. 3–19b must push on the pin (compression) since its horizontal component, $F_{BC} \sin 45°$, must balance the 500-N force ($\Sigma F_x = 0$). Likewise, F_{BA} is a tensile force since it balances the vertical component, $F_{BC} \cos 45°$ ($\Sigma F_y = 0$). In more complicated cases, the sense of an unknown member force can be *assumed*; then, after applying the equilibrium equations, the assumed sense can be verified from the numerical results. A *positive* answer indicates that the sense is *correct*, whereas a *negative* answer indicates that the sense shown on the free-body diagram must be *reversed*. This is the method we will use in the example problems which follow.

Procedure for Analysis

The following procedure provides a means for analyzing a truss using the method of joints.

- Draw the free-body diagram of a joint having at least one known force and at most two unknown forces. (If this joint is at one of the supports, it may be necessary to calculate the external reactions at the supports by drawing a free-body diagram of the entire truss.)

- Use one of the two methods previously described for establishing the sense of an unknown force.

- The x and y axes should be oriented such that the forces on the free-body diagram can be easily resolved into their x and y components. Apply the two force equilibrium equations $\Sigma F_x = 0$ and $\Sigma F_y = 0$, solve for the two unknown member forces, and verify their correct directional sense.

- Continue to analyze each of the other joints, where again it is necessary to choose a joint having at most two unknowns and at least one known force.

- Once the force in a member is found from the analysis of a joint at one of its ends, the result can be used to analyze the forces acting on the joint at its other end. Remember, a member in *compression* "pushes" on the joint and a member in *tension* "pulls" on the joint.

EXAMPLE 3.2

Determine the force in each member of the roof truss shown in the photo. The dimensions and loadings are shown in Fig. 3–20a. State whether the members are in tension or compression.

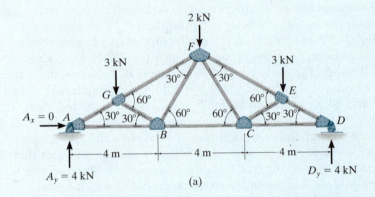

(a)

SOLUTION

Only the forces in half the members have to be determined, since the truss is symmetric with respect to *both* loading and geometry.

Joint A, Fig. 3–20b.   We can start the analysis at joint A. Why? The free-body diagram is shown in Fig. 3–20b.

$$+\uparrow\Sigma F_y = 0; \quad 4 - F_{AG}\sin 30° = 0 \qquad F_{AG} = 8 \text{ kN (C)} \qquad \textit{Ans.}$$

$$\xrightarrow{+}\Sigma F_x = 0; \quad F_{AB} - 8\cos 30° = 0 \qquad F_{AB} = 6.928 \text{ kN (T)} \qquad \textit{Ans.}$$

Joint G, Fig. 3–20c.   In this case note how the orientation of the x, y axes avoids simultaneous solution of equations.

$$+\nwarrow\Sigma F_y = 0; \quad F_{GB}\sin 60° - 3\cos 30° = 0$$

$$F_{GB} = 3.00 \text{ kN (C)} \qquad \textit{Ans.}$$

$$+\nearrow\Sigma F_x = 0; \quad 8 - 3\sin 30° - 3.00\cos 60° - F_{GF} = 0$$

$$F_{GF} = 5.00 \text{ kN (C)} \qquad \textit{Ans.}$$

Joint B, Fig. 3–20d.

$$+\uparrow\Sigma F_y = 0; \quad F_{BF}\sin 60° - 3.00\sin 30° = 0$$

$$F_{BF} = 1.73 \text{ kN (T)} \qquad \textit{Ans.}$$

$$\xrightarrow{+}\Sigma F_x = 0; \quad F_{BC} + 1.73\cos 60° + 3.00\cos 30° - 6.928 = 0$$

$$F_{BC} = 3.46 \text{ kN (T)} \qquad \textit{Ans.}$$

(b)

(c)

(d)

Fig. 3–20

EXAMPLE 3.3

Determine the force in each member of the scissors truss shown in Fig. 3–21a. State whether the members are in tension or compression. The reactions at the supports are given.

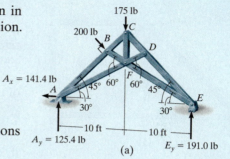

SOLUTION

The truss will be analyzed in the following sequence:

Joint E, Fig. 3–21b.   Note that simultaneous solution of equations is avoided by the x, y axes orientation.

$+\nearrow\Sigma F_y = 0;$ $191.0 \cos 30° - F_{ED} \sin 15° = 0$

$$F_{ED} = 639.1 \text{ lb (C)} \qquad Ans.$$

$+\searrow\Sigma F_x = 0;$ $639.1 \cos 15° - F_{EF} - 191.0 \sin 30° = 0$

$$F_{EF} = 521.8 \text{ lb (T)} \qquad Ans.$$

Joint D, Fig. 3–21c.

$+\swarrow\Sigma F_x = 0;$ $-F_{DF} \sin 75° = 0$ $F_{DF} = 0$ $Ans.$

$+\nwarrow\Sigma F_y = 0;$ $-F_{DC} + 639.1 = 0$ $F_{DC} = 639.1 \text{ lb (C)}$ $Ans.$

Joint C, Fig. 3–21d.

$\xrightarrow{+}\Sigma F_x = 0;$ $F_{CB} \sin 45° - 639.1 \sin 45° = 0$

$$F_{CB} = 639.1 \text{ lb (C)} \qquad Ans.$$

$+\uparrow\Sigma F_y = 0;$ $-F_{CF} - 175 + 2(639.1) \cos 45° = 0$

$$F_{CF} = 728.8 \text{ lb (T)} \qquad Ans.$$

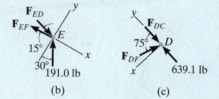

Joint B, Fig. 3–21e.

$+\nwarrow\Sigma F_y = 0;$ $F_{BF} \sin 75° - 200 = 0$ $F_{BF} = 207.1 \text{ lb (C)}$ $Ans.$

$+\swarrow\Sigma F_x = 0;$ $639.1 + 207.1 \cos 75° - F_{BA} = 0$

$$F_{BA} = 692.7 \text{ lb (C)} \qquad Ans.$$

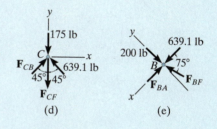

Joint A, Fig. 3–21f.

$\xrightarrow{+}\Sigma F_x = 0;$ $F_{AF} \cos 30° - 692.7 \cos 45° - 141.4 = 0$

$$F_{AF} = 728.9 \text{ lb (T)} \qquad Ans.$$

$+\uparrow\Sigma F_y = 0;$ $125.4 - 692.7 \sin 45° + 728.9 \sin 30° = 0$ check

Notice that since the reactions have been calculated, a further check of the calculations can be made by analyzing the last joint F. Try it and find out.

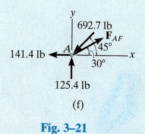

Fig. 3–21

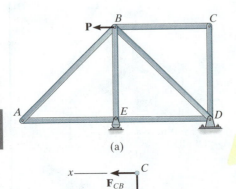

(a)

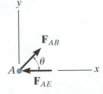

$$\xrightarrow{+}\; \Sigma F_x = 0;\; F_{CB} = 0$$

$$+\!\downarrow\! \Sigma F_y = 0;\; F_{CD} = 0$$

(b)

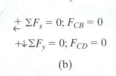

$$+\uparrow \Sigma F_y = 0;\; F_{AB} \sin \theta = 0$$
$$F_{AB} = 0 \;(\text{since } \sin \theta \neq 0)$$

$$\xrightarrow{+}\; \Sigma F_x = 0;\; -F_{AE} + 0 = 0$$
$$F_{AE} = 0$$

(c)

Fig. 3–22

3.4 Zero-Force Members

Truss analysis using the method of joints is greatly simplified if one is able to first determine those members that support *no loading*. These *zero-force members* may be necessary for the stability of the truss during construction and to provide support if the applied loading is changed. The zero-force members of a truss can generally be determined by inspection of the joints, and they occur in two cases.

Case 1.   Consider the truss in Fig. 3–22a. The two members at joint C are connected together at a right angle *and* there is no external load on the joint. The free-body diagram of joint C, Fig. 3–22b, indicates that the force in each member must be zero in order to maintain equilibrium. Furthermore, as in the case of joint A, Fig. 3–22c, this must be true regardless of the angle, say θ, between the members.

Case 2.   Zero-force members also occur at joints having a geometry as joint D in Fig. 3–23a. Here *no external load acts on the joint,* so that a force summation in the y direction, Fig. 3–23b, which is perpendicular to the two collinear members, requires that $F_{DF} = 0$. Using this result, FC is also a zero-force member, as indicated by the force analysis of joint F, Fig. 3–23c.

In summary, then, if only two non-collinear members form a truss joint and no external load or support reaction is applied to the joint, the members must be zero-force members, Case 1. Also, if three members form a truss joint for which two of the members are collinear, the third member is a zero-force member, provided no external force or support reaction is applied to the joint, Case 2. Particular attention should be directed to these conditions of joint geometry and loading, since the analysis of a truss can be considerably simplified by *first* spotting the zero-force members.

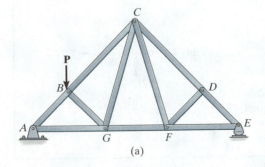

(a)

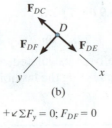

(b)

$$+\!\swarrow\! \Sigma F_y = 0;\; F_{DF} = 0$$

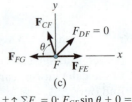

(c)

$$+\uparrow \Sigma F_y = 0;\; F_{CF} \sin \theta + 0 = 0$$
$$F_{CF} = 0 \;(\text{since } \sin \theta \neq 0)$$

Fig. 3–23

EXAMPLE   3.4

Using the method of joints, indicate all the members of the truss shown in Fig. 3–24a that have zero force.

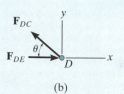

(b)

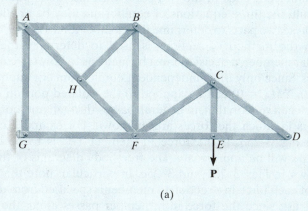

(a)

Fig. 3–24

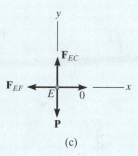

(c)

SOLUTION

Looking for joints similar to those discussed in Figs. 3–22 and 3–23, we have

Joint D, Fig. 3–24b.

$+\uparrow\Sigma F_y = 0;$ $F_{DC}\sin\theta = 0$ $F_{DC} = 0$ *Ans.*

$\xrightarrow{+}\Sigma F_x = 0;$ $F_{DE} + 0 = 0$ $F_{DE} = 0$ *Ans.*

Joint E, Fig. 3–24c.

$\xleftarrow{+}\Sigma F_x = 0;$ $F_{EF} = 0$ *Ans.*

(Note that $F_{EC} = P$ and an analysis of joint C would yield a force in member CF.)

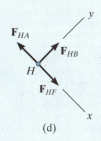

(d)

Joint H, Fig. 3–24d.

$+\nearrow\Sigma F_y = 0;$ $F_{HB} = 0$ *Ans.*

Joint G, Fig. 3–24e. The rocker support at G can only exert an x component of force on the joint, i.e., $\mathbf{G}_x$. Hence,

$+\uparrow\Sigma F_y = 0;$ $F_{GA} = 0$ *Ans.*

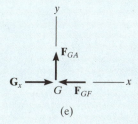

(e)

3.5   The Method of Sections

If the forces in only a few members of a truss are to be found, the method of sections generally provides the most direct means of obtaining these forces. The **method of sections** consists of passing an *imaginary section* through the truss, thus cutting it into two parts. Provided the entire truss is in equilibrium, each of the two parts must also be in equilibrium; and as a result, the three equations of equilibrium may be applied to either one of these two parts to determine the member forces at the "cut section."

When the method of sections is used to determine the force in a particular member, a decision must be made as to how to "cut" or section the truss. Since only *three* independent equilibrium equations ($\Sigma F_x = 0$, $\Sigma F_y = 0$, $\Sigma M_O = 0$) can be applied to the isolated portion of the truss, try to select a section that, in general, passes through not more than *three* members in which the forces are unknown. For example, consider the truss in Fig. 3–25a. If the force in member *GC* is to be determined, section *aa* will be appropriate. The free-body diagrams of the two parts are shown in Figs. 3–25b and 3–25c. In particular, note that the line of action of each force in a sectioned member is specified from the *geometry* of the truss, since the force in a member passes along the axis of the member. Also, the member forces acting on one part of the truss are equal but opposite to those acting on the other part—Newton's third law. As shown, members assumed to be in *tension* (*BC* and *GC*) are subjected to a "pull," whereas the member in *compression* (*GF*) is subjected to a "push."

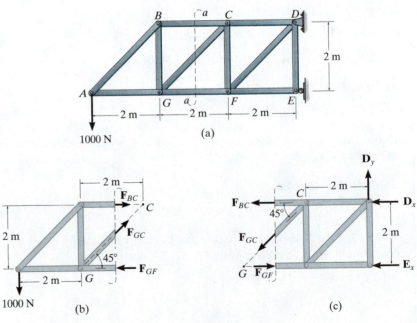

(a)

(b)

(c)

Fig. 3–25

The three unknown member forces $\mathbf{F}_{BC}$, $\mathbf{F}_{GC}$, and $\mathbf{F}_{GF}$ can be obtained by applying the three equilibrium equations to the free-body diagram in Fig. 3–25b. If, however, the free-body diagram in Fig. 3–25c is considered, the three support reactions $\mathbf{D}_x$, $\mathbf{D}_y$, and $\mathbf{E}_x$ will have to be determined *first*. Why? (This, of course, is done in the usual manner by considering a free-body diagram of the *entire truss*.) When applying the equilibrium equations, consider ways of writing the equations so as to yield a *direct solution* for each of the unknowns, rather than having to solve simultaneous equations. For example, summing moments about C in Fig. 3–25b would yield a direct solution for $\mathbf{F}_{GF}$ since $\mathbf{F}_{BC}$ and $\mathbf{F}_{GC}$ create zero moment about C. Likewise, $\mathbf{F}_{BC}$ can be obtained directly by summing moments about G. Finally, $\mathbf{F}_{GC}$ can be found directly from a force summation in the vertical direction, since $\mathbf{F}_{GF}$ and $\mathbf{F}_{BC}$ have no vertical components.

As in the method of joints, there are two ways in which one can determine the correct sense of an unknown member force.

1. *Always assume that the unknown member forces at the cut section are in tension, i.e., "pulling" on the member.* By doing this, the numerical solution of the equilibrium equations will yield *positive scalars for members in tension and negative scalars for members in compression*.

2. *The correct sense of an unknown member force can in many cases be determined "by inspection."* For example, $\mathbf{F}_{BC}$ is a tensile force as represented in Fig. 3–25b, since moment equilibrium about G requires that $\mathbf{F}_{BC}$ create a moment opposite to that of the 1000-N force. Also, $\mathbf{F}_{GC}$ is tensile since its vertical component must balance the 1000-N force. In more complicated cases, the sense of an unknown member force may be *assumed*. If the solution yields a *negative* scalar, it indicates that the force's sense is *opposite* to that shown on the free-body diagram. This is the method we will use in the example problems which follow.

A truss bridge being constructed over Lake Shasta in northern California.

© Hank deLespinasse Studios, Inc.

Procedure for Analysis

The following procedure provides a means for applying the method of sections to determine the forces in the members of a truss.

Free-Body Diagram

- Make a decision as to how to "cut" or section the truss through the members where forces are to be determined.

- Before isolating the appropriate section, it may first be necessary to determine the truss's *external* reactions, so that the three equilibrium equations are used *only* to solve for member forces at the cut section.

- Draw the free-body diagram of that part of the sectioned truss which has the least number of forces on it.

- Use one of the two methods described above for establishing the sense of an unknown force.

Equations of Equilibrium

- Moments should be summed about a point that lies at the intersection of the lines of action of two unknown forces; in this way, the third unknown force is determined directly from the equation.

- If two of the unknown forces are *parallel*, forces may be summed *perpendicular* to the direction of these unknowns to determine *directly* the third unknown force.

An example of a Warren truss (with verticals). See page 87.

EXAMPLE   **3.5**

Determine the force in members GJ and CO of the roof truss shown in the photo. The dimensions and loadings are shown in Fig. 3–26a. State whether the members are in tension or compression. The reactions at the supports have been calculated.

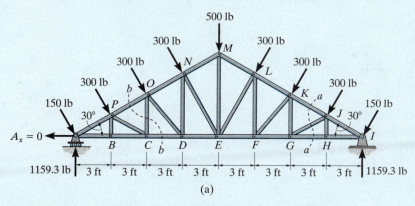

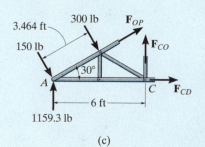

Fig. 3–26

(a)

SOLUTION

Member *CF*.

Free-Body Diagram.   The force in member GJ can be obtained by considering the section aa in Fig. 3–26a. The free-body diagram of the right part of this section is shown in Fig. 3–26b.

Equations of Equilibrium.   A direct solution for $\mathbf{F}_{GJ}$ can be obtained by applying $\Sigma M_I = 0$. Why? For simplicity, slide $\mathbf{F}_{GJ}$ to point G (principle of transmissibility), Fig. 3–26b. Thus,

$$\zeta + \Sigma M_I = 0; \qquad -F_{GJ} \sin 30°(6) + 300(3.464) = 0$$

$$F_{GJ} = 346 \text{ lb (C)} \qquad\qquad Ans.$$

(b)

Member *GC*.

Free-Body Diagram.   The force in CO can be obtained by using section bb in Fig. 3–26a. The free-body diagram of the left portion of the section is shown in Fig. 3–26c.

Equations of Equilibrium.   Moments will be summed about point A in order to eliminate the unknowns $\mathbf{F}_{OP}$ and $\mathbf{F}_{CD}$.

$$\zeta + \Sigma M_A = 0; \qquad -300(3.464) + F_{CO}(6) = 0$$

$$F_{CO} = 173 \text{ lb (T)} \qquad\qquad Ans.$$

(c)

EXAMPLE 3.6

Determine the force in members *GF* and *GD* of the truss shown in Fig. 3–27*a*. State whether the members are in tension or compression. The reactions at the supports have been calculated.

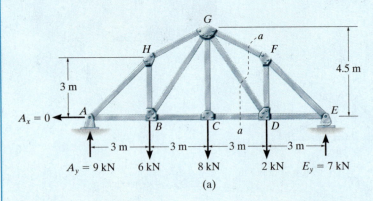

(a)

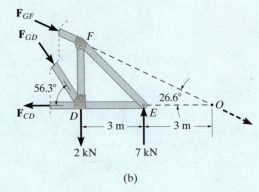

(b)

Fig. 3–27

SOLUTION

Free-Body Diagram. Section *aa* in Fig. 3–27*a* will be considered. Why? The free-body diagram to the right of this section is shown in Fig. 3–27*b*. The distance *EO* can be determined by proportional triangles or realizing that member *GF* drops vertically $4.5 - 3 = 1.5$ m in 3 m, Fig. 3–27*a*. Hence to drop 4.5 m from *G* the distance from *C* to *O* must be 9 m. Also, the angles that $\mathbf{F}_{GD}$ and $\mathbf{F}_{GF}$ make with the horizontal are $\tan^{-1}(4.5/3) = 56.3°$ and $\tan^{-1}(4.5/9) = 26.6°$, respectively.

Equations of Equilibrium. The force in *GF* can be determined directly by applying $\Sigma M_D = 0$. Why? For the calculation use the principle of transmissibility and slide $\mathbf{F}_{GF}$ to point *O*. Thus,

$$\zeta + \Sigma M_D = 0; \qquad -F_{GF}\sin 26.6°(6) + 7(3) = 0$$
$$F_{GF} = 7.83 \text{ kN (C)} \qquad\qquad \textit{Ans.}$$

The force in *GD* is determined directly by applying $\Sigma M_O = 0$. For simplicity use the principle of transmissibility and slide $\mathbf{F}_{GD}$ to *D*. Hence,

$$\zeta + \Sigma M_O = 0; \qquad -7(3) + 2(6) + F_{GD}\sin 56.3°(6) = 0$$
$$F_{GD} = 1.80 \text{ kN (C)} \qquad\qquad \textit{Ans.}$$

EXAMPLE 3.7

Determine the force in members BC and MC of the K-truss shown in Fig. 3–28a. State whether the members are in tension or compression. The reactions at the supports have been calculated.

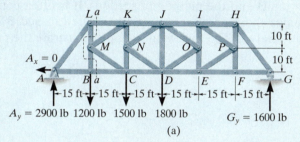

(a)

SOLUTION

Free-Body Diagram. Although section aa shown in Fig. 3–28a cuts through four members, it is possible to solve for the force in member BC using this section. The free-body diagram of the left portion of the truss is shown in Fig. 3–28b.

Equations of Equilibrium. Summing moments about point L eliminates *three* of the unknowns, so that

$$\zeta + \Sigma M_L = 0; \qquad -2900(15) + F_{BC}(20) = 0$$
$$F_{BC} = 2175 \text{ lb (T)} \qquad\qquad\qquad Ans.$$

Free-Body Diagrams. The force in MC can be obtained indirectly by first obtaining the force in MB from vertical force equilibrium of joint B, Fig. 3–28c, i.e., $F_{MB} = 1200$ lb (T). Then from the free-body diagram in Fig. 3–28b.

$$+\uparrow \Sigma F_y = 0; \qquad 2900 - 1200 + 1200 - F_{ML} = 0$$
$$F_{ML} = 2900 \text{ lb (T)}$$

Using these results, the free-body diagram of joint M is shown in Fig. 3–28d.

Equations of Equilibrium.

$$\xrightarrow{+} \Sigma F_x = 0; \qquad \left(\frac{3}{\sqrt{13}}\right)F_{MC} - \left(\frac{3}{\sqrt{13}}\right)F_{MK} = 0$$

$$+\uparrow \Sigma F_y = 0; \qquad 2900 - 1200 - \left(\frac{2}{\sqrt{13}}\right)F_{MC} - \left(\frac{2}{\sqrt{13}}\right)F_{MK} = 0$$

$$F_{MK} = 1532 \text{ lb (C)} \qquad F_{MC} = 1532 \text{ lb (T)} \qquad Ans.$$

Sometimes, as in this example, application of both the method of sections and the method of joints leads to the most direct solution to the problem.

It is also possible to solve for the force in MC by using the result for F_{BC}. In this case, pass a vertical section through LK, MK, MC, and BC, Fig. 3–28a. Isolate the left section and apply $\Sigma M_K = 0$.

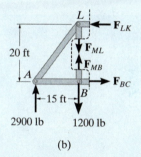

(b)

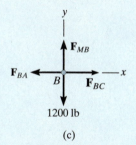

(c)

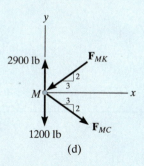

(d)

Fig. 3–28

3.6   Compound Trusses

In Sec. 3.2 it was stated that compound trusses are formed by connecting two or more simple trusses together either by bars or by joints. Occasionally this type of truss is best analyzed by applying *both* the method of joints and the method of sections. It is often convenient to first recognize the type of construction as listed in Sec. 3.2 and then perform the analysis. The following examples illustrate the procedure.

EXAMPLE | **3.8**

Indicate how to analyze the compound truss shown in Fig. 3–29a. The reactions at the supports have been calculated.

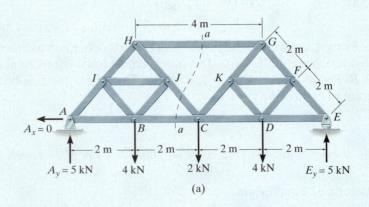

(a)

SOLUTION

The truss is a compound truss since the simple trusses *ACH* and *CEG* are connected by the pin at *C* and the bar *HG*.

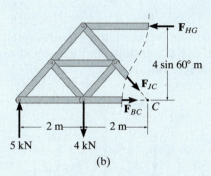

(b)

Section *aa* in Fig. 3–29*a* cuts through bar *HG* and two other members having unknown forces. A free-body diagram for the left part is shown in Fig. 3–29*b*. The force in *HG* is determined as follows:

$$\zeta + \Sigma M_C = 0; \qquad -5(4) + 4(2) + F_{HG}(4 \sin 60°) = 0$$

$$F_{HG} = 3.46 \text{ kN (C)}$$

We can now proceed to determine the force in each member of the simple trusses using the method of joints. For example, the free-body diagram of *ACH* is shown in Fig. 3–29*c*. The joints of this truss can be analyzed in the following sequence:

Joint A: Determine the force in *AB* and *AI*.
Joint H: Determine the force in *HI* and *HJ*.
Joint I: Determine the force in *IJ* and *IB*.
Joint B: Determine the force in *BC* and *BJ*.
Joint J: Determine the force in *JC*.

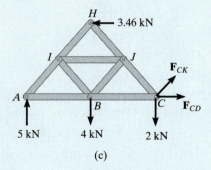

(c)

Fig. 3–29

EXAMPLE   3.9

Compound roof trusses are used in a garden center, as shown in the photo. They have the dimensions and loading shown in Fig. 3–30a. Indicate how to analyze this truss.

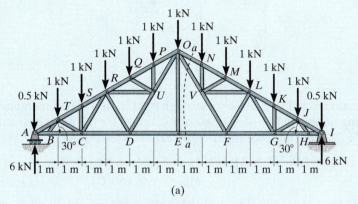

(a)

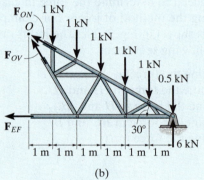

(b)

Fig. 3–30

SOLUTION

We can obtain the force in *EF* by using section *aa* in Fig. 3–30a. The free-body diagram of the right segment is shown in Fig. 3–30b

$$\zeta + \Sigma M_O = 0; \qquad -1(1) - 1(2) - 1(3) - 1(4) - 1(5) - 0.5(6)$$
$$+ 6(6) - F_{EF}(6 \tan 30°) = 0$$
$$F_{EF} = 5.20 \text{ kN (T)} \qquad \qquad Ans.$$

By inspection notice that *BT*, *EO*, and *HJ* are zero-force members since $+\uparrow \Sigma F_y = 0$ at joints *B*, *E*, and *H*, respectively. Also, by applying $+\nwarrow \Sigma F_y = 0$ (perpendicular to *AO*) at joints *P*, *Q*, *S*, and *T*, we can directly determine the force in members *PU*, *QU*, *SC*, and *TC*, respectively.

EXAMPLE 3.10

Indicate how to analyze the compound truss shown in Fig. 3–31a. The reactions at the supports have been calculated.

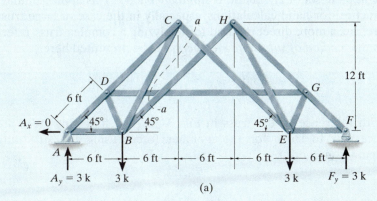

(a)

Fig. 3–31

SOLUTION

The truss may be classified as a compound truss since the simple trusses $ABCD$ and $FEHG$ are connected by three nonparallel or nonconcurrent bars, namely, CE, BH, and DG.

Using section aa in Fig. 3–31a we can determine the force in each connecting bar. The free-body diagram of the left part of this section is shown in Fig. 3–31b. Hence,

$$\zeta + \Sigma M_B = 0; \quad -3(6) - F_{DG}(6 \sin 45°) + F_{CE} \cos 45°(12) \\ + F_{CE} \sin 45°(6) = 0 \tag{1}$$

$$+\uparrow \Sigma F_y = 0; \quad 3 - 3 - F_{BH} \sin 45° + F_{CE} \sin 45° = 0 \tag{2}$$

$$\xrightarrow{+} \Sigma F_x = 0; \quad -F_{BH} \cos 45° + F_{DG} - F_{CE} \cos 45° = 0 \tag{3}$$

From Eq. (2), $F_{BH} = F_{CE}$; then solving Eqs. (1) and (3) simultaneously yields

$$F_{BH} = F_{CE} = 2.68 \text{ k (C)} \qquad F_{DG} = 3.78 \text{ k (T)}$$

Analysis of each connected simple truss can now be performed using the method of joints. For example, from Fig. 3–31c, this can be done in the following sequence.

Joint A: Determine the force in AB and AD.
Joint D: Determine the force in DC and DB.
Joint C: Determine the force in CB.

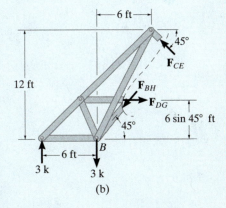

(b)

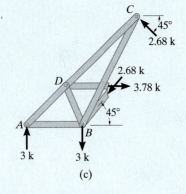

(c)

3.7   Complex Trusses

The member forces in a complex truss can be determined using the method of joints; however, the solution will require writing the two equilibrium equations for each of the j joints of the truss and then solving the complete set of $2j$ equations *simultaneously*.* This approach may be impractical for hand calculations, especially in the case of large trusses. Therefore, a more direct method for analyzing a complex truss, referred to as the *method of substitute members*, will be presented here.

Procedure for Analysis

With reference to the truss in Fig. 3–32a, the following steps are necessary to solve for the member forces using the substitute member method.

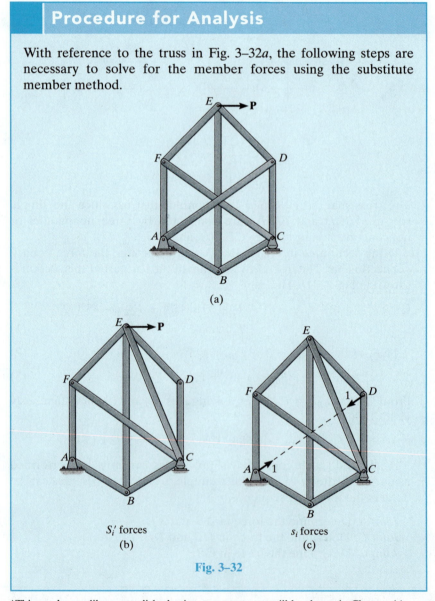

(a)

S_i' forces
(b)

s_i forces
(c)

Fig. 3–32

*This can be readily accomplished using a computer as will be shown in Chapter 14.

Reduction to Stable Simple Truss

Determine the reactions at the supports and begin by imagining how to analyze the truss by the method of joints, i.e., progressing from joint to joint and solving for each member force. If a joint is reached where there are *three unknowns*, remove one of the members at the joint and replace it by an *imaginary* member elsewhere in the truss. By doing this, reconstruct the truss to be a stable simple truss.

For example, in Fig. 3–32a it is observed that each joint will have three *unknown* member forces acting on it. Hence we will remove member AD and replace it with the imaginary member EC, Fig. 3–32b. This truss can now be analyzed by the method of joints for the two types of loading that follow.

External Loading on Simple Truss

Load the simple truss with the actual loading **P**, then determine the force S'_i in each member i. In Fig. 3–32b, provided the reactions have been determined, one could start at joint A to determine the forces in AB and AF, then joint F to determine the forces in FE and FC, then joint D to determine the forces in DE and DC (both of which are zero), then joint E to determine EB and EC, and finally joint B to determine the force in BC.

Remove External Loading from Simple Truss

Consider the simple truss without the external load **P**. Place equal but opposite collinear *unit loads* on the truss at the two joints from which the member was removed. If these forces develop a force s_i in the ith truss member, then by proportion an unknown force x in the removed member would exert a force xs_i in the ith member.

From Fig. 3–32c the equal but opposite unit loads will create *no reactions* at A and C when the equations of equilibrium are applied to the entire truss. The s_i forces can be determined by analyzing the joints in the same sequence as before, namely, joint A, then joints F, D, E, and finally B.

Superposition

If the effects of the above two loadings are combined, the force in the ith member of the truss will be

$$S_i = S'_i + xs_i \tag{1}$$

In particular, for the substituted member EC in Fig. 3–32b the force $S_{EC} = S'_{EC} + xs_{EC}$. Since member EC does not actually exist on the original truss, we will choose x to have a magnitude such that it yields *zero force* in EC. Hence,

$$S'_{EC} + xs_{EC} = 0 \tag{2}$$

or $x = -S'_{EC}/s_{EC}$. Once the value of x has been determined, the force in the other members i of the complex truss can be determined from Eq. (1).

EXAMPLE | 3.11

Determine the force in each member of the complex truss shown in Fig. 3–33a. Assume joints *B*, *F*, and *D* are on the same horizontal line. State whether the members are in tension or compression.

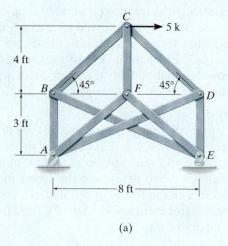

(a)

Fig. 3–33

SOLUTION

Reduction to Stable Simple Truss. By inspection, each joint has three unknown member forces. A joint analysis can be performed by hand if, for example, member *CF* is removed and member *DB* substituted, Fig. 3–33b. The resulting truss is stable and will not collapse.

External Loading on Simple Truss. As shown in Fig. 3–33b, the support reactions on the truss have been determined. Using the method of joints, we can first analyze joint *C* to find the forces in members *CB* and *CD*; then joint *F*, where it is seen that *FA* and *FE* are zero-force members; then joint *E* to determine the forces in members *EB* and *ED*; then joint *D* to determine the forces in *DA* and *DB*; then finally joint *B* to determine the force in *BA*. Considering tension as positive and compression as negative, these S_i' forces are recorded in column 2 of Table 1.

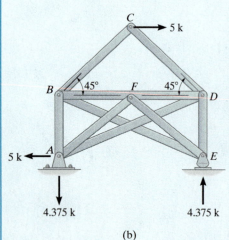

(b)

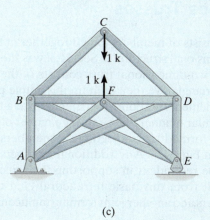

(c)

Remove External Loading from Simple Truss. The unit load acting on the truss is shown in Fig. 3–33c. These equal but opposite forces create no external reactions on the truss. The joint analysis follows the same sequence as discussed previously, namely, joints C, F, E, D, and B. The results of the s_i force analysis are recorded in column 3 of Table 1.

Superposition. We require

$$S_{DB} = S'_{DB} + xs_{DB} = 0$$

Substituting the data for S'_{DB} and s_{DB}, where S'_{DB} is negative since the force is compressive, we have

$$-2.50 + x(1.167) = 0 \qquad x = 2.143$$

The values of xs_i are recorded in column 4 of Table 1, and the actual member forces $S_i = S'_i + xs_i$ are listed in column 5.

TABLE 1				
Member	S_i	s_i	xs_i	S_i
CB	3.54	−0.707	−1.52	2.02 (T)
CD	−3.54	−0.707	−1.52	5.05 (C)
FA	0	0.833	1.79	1.79 (T)
FE	0	0.833	1.79	1.79 (T)
EB	0	−0.712	−1.53	1.53 (C)
ED	−4.38	−0.250	−0.536	4.91 (C)
DA	5.34	−0.712	−1.53	3.81 (T)
DB	−2.50	1.167	2.50	0
BA	2.50	−0.250	−0.536	1.96 (T)
CB				2.14 (T)

P

Fig. 3–34

3.8   Space Trusses

A **space truss** consists of members joined together at their ends to form a stable three-dimensional structure. In Sec. 3.2 it was shown that the simplest form of a stable two-dimensional truss consists of the members arranged in the form of a triangle. We then built up the simple plane truss from this basic triangular element by adding two members at a time to form further elements. In a similar manner, the simplest element of a stable space truss is a **tetrahedron**, formed by connecting six members together with four joints as shown in Fig. 3–34. Any additional members added to this basic element would be redundant in supporting the force **P**. A simple space truss can be built from this basic tetrahedral element by adding three additional members and another joint forming multiconnected tetrahedrons.

Determinacy and Stability.   Realizing that in three dimensions there are three equations of equilibrium available for each joint ($\Sigma F_x = 0$, $\Sigma F_y = 0$, $\Sigma F_z = 0$), then for a space truss with j number of joints, $3j$ equations are available. If the truss has b number of bars and r number of reactions, then like the case of a planar truss (Eqs. 3–1 and 3–2) we can write

$$
\begin{array}{lll}
b + r < 3j & \text{unstable truss} & \\
b + r = 3j & \text{statically determinate—check stability} & \text{(3–3)} \\
b + r > 3j & \text{statically indeterminate—check stability} &
\end{array}
$$

The *external stability* of the space truss requires that the support reactions keep the truss in force and moment equilibrium along and about any and all axes. This can sometimes be checked by inspection, although if the truss is unstable a solution of the equilibrium equations will give inconsistent results. *Internal stability* can sometimes be checked by careful inspection of the member arrangement. Provided each joint is held fixed by its supports or connecting members, so that it cannot move with respect to the other joints, the truss can be classified as internally stable. Also, if we do a force analysis of the truss and obtain inconsistent results, then the truss configuration will be unstable or have a "critical form."

Assumptions for Design.   The members of a space truss may be treated as axial-force members provided the external loading is applied at the joints and the joints consist of ball-and-socket connections. This assumption is justified provided the joined members at a connection intersect at a common point and the weight of the members can be neglected. In cases where the weight of a member is to be included in the analysis, it is generally satisfactory to apply it as a vertical force, half of its magnitude applied to each end of the member.

For the force analysis the supports of a space truss are generally modeled as a short link, plane roller joint, slotted roller joint, or a ball-and-socket joint. Each of these supports and their reactive force components are shown in Table 3.1.

The roof of this pavilion is supported using a system of space trusses.

TABLE 3.1 Supports and Their Reactive Force Components

(1)

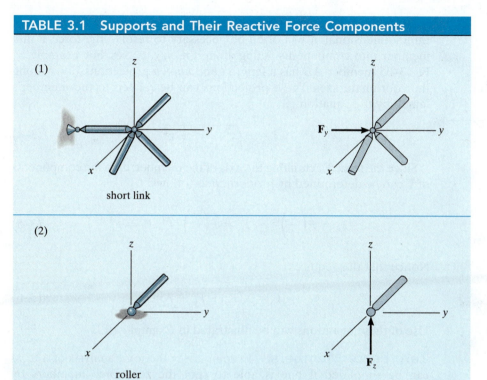

short link

(2)

roller

(3)

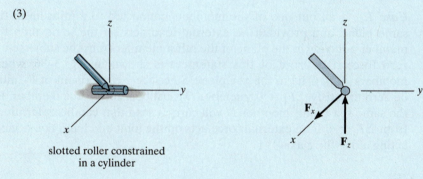

slotted roller constrained
in a cylinder

(4)

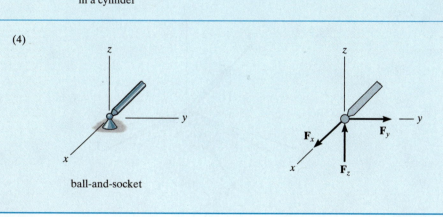

ball-and-socket

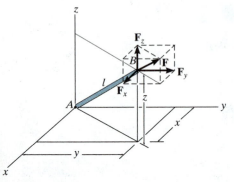

Fig. 3–35

x, y, z, Force Components.

Since the analysis of a space truss is three-dimensional, it will often be necessary to resolve the force **F** in a member into components acting along the x, y, z axes. For example, in Fig. 3–35 member AB has a length l and *known* projections x, y, z along the coordinate axes. These projections can be related to the member's length by the equation

$$l = \sqrt{x^2 + y^2 + z^2} \tag{3–4}$$

Since the force **F** acts along the axis of the member, then the components of **F** can be determined by *proportion* as follows:

$$F_x = F\left(\frac{x}{l}\right) \qquad F_y = F\left(\frac{y}{l}\right) \qquad F_z = F\left(\frac{z}{l}\right) \tag{3–5}$$

Notice that this requires

$$F = \sqrt{F_x^2 + F_y^2 + F_z^2} \tag{3–6}$$

Use of these equations will be illustrated in Example 3.12.

Zero-Force Members.

In some cases the joint analysis of a truss can be simplified if one is able to spot the zero-force members by recognizing two common cases of joint geometry.

Case 1. If all but one of the members connected to a joint lie in the same plane, and provided no external load acts on the joint, then the member not lying in the plane of the other members must be subjected to zero force. The proof of this statement is shown in Fig. 3–36, where members A, B, C lie in the x–y plane. Since the z component of $\mathbf{F}_D$ must be zero to satisfy $\Sigma F_z = 0$, member D must be a zero-force member. By the same reasoning, member D will carry a load that can be determined from $\Sigma F_z = 0$ if an external force acts on the joint and has a component acting along the z axis.

Because of their cost effectiveness, towers such as these are often used to support multiple electric transmission lines.

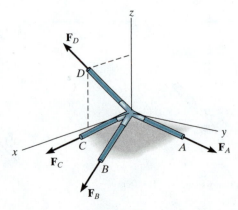

Fig. 3–36

Case 2. If it has been determined that all but two of several members connected at a joint support zero force, then the two remaining members must also support zero force, provided they do not lie along the same line. This situation is illustrated in Fig. 3–37, where it is known that A and C are zero-force members. Since $\mathbf{F}_D$ is collinear with the y axis, then application of $\Sigma F_x = 0$ or $\Sigma F_z = 0$ requires the x or z component of $\mathbf{F}_B$ to be zero. Consequently, $F_B = 0$. This being the case, $F_D = 0$ since $\Sigma F_y = 0$.

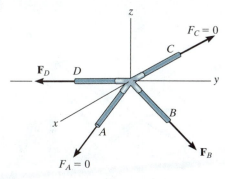

Fig. 3–37

Particular attention should be directed to the foregoing two cases of joint geometry and loading, since the analysis of a space truss can be considerably simplified by first spotting the zero-force members.

Procedure for Analysis

Either the method of sections or the method of joints can be used to determine the forces developed in the members of a space truss.

Method of Sections

If only a *few* member forces are to be determined, the method of sections may be used. When an imaginary section is passed through a truss and the truss is separated into two parts, the force system acting on either one of the parts must satisfy the six scalar equilibrium equations: $\Sigma F_x = 0$, $\Sigma F_y = 0$, $\Sigma F_z = 0$, $\Sigma M_x = 0$, $\Sigma M_y = 0$, $\Sigma M_z = 0$. By proper choice of the section and axes for summing forces and moments, many of the unknown member forces in a space truss can be computed *directly*, using a single equilibrium equation. In this regard, recall that the *moment* of a force about an axis is *zero* provided *the force is parallel to the axis or its line of action passes through a point on the axis.*

Method of Joints

Generally, if the forces in *all* the members of the truss must be determined, the method of joints is most suitable for the analysis. When using the method of joints, it is necessary to solve the three scalar equilibrium equations $\Sigma F_x = 0$, $\Sigma F_y = 0$, $\Sigma F_z = 0$ at each joint. Since it is relatively easy to draw the free-body diagrams and apply the equations of equilibrium, the method of joints is very consistent in its application.

EXAMPLE 3.12

Determine the force in each member of the space truss shown in Fig. 3–38a. The truss is supported by a ball-and-socket joint at A, a slotted roller joint at B, and a cable at C.

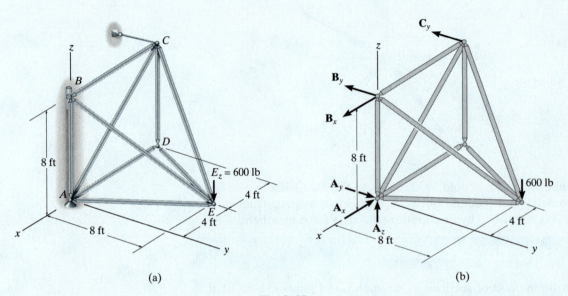

Fig. 3–38

SOLUTION

The truss is statically determinate since $b + r = 3j$ or $9 + 6 = 3(5)$, Fig. 3–38b.

Support Reactions.   We can obtain the support reactions from the free-body diagram of the entire truss, Fig. 3–38b, as follows:

$$\Sigma M_y = 0; \qquad -600(4) + B_x(8) = 0 \qquad B_x = 300 \text{ lb}$$
$$\Sigma M_z = 0; \qquad\qquad\qquad\quad C_y = 0$$
$$\Sigma M_x = 0; \qquad B_y(8) - 600(8) = 0 \qquad B_y = 600 \text{ lb}$$
$$\Sigma F_x = 0; \qquad\quad 300 - A_x = 0 \qquad A_x = 300 \text{ lb}$$
$$\Sigma F_y = 0; \qquad\quad A_y - 600 = 0 \qquad A_y = 600 \text{ lb}$$
$$\Sigma F_z = 0; \qquad\quad A_z - 600 = 0 \qquad A_z = 600 \text{ lb}$$

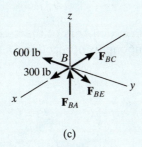

(c)

Joint B. We can begin the method of joints at B since there are three unknown member forces at this joint, Fig. 3–38c. The components of F_{BE} can be determined by proportion to the length of member BE, as indicated by Eqs. 3–5. We have

$\Sigma F_y = 0;$ $-600 + F_{BE}\left(\frac{8}{12}\right) = 0$ $F_{BE} = 900\,\text{lb (T)}$ *Ans.*

$\Sigma F_x = 0;$ $300 - F_{BC} - 900\left(\frac{4}{12}\right) = 0$ $F_{BC} = 0$ *Ans.*

$\Sigma F_z = 0;$ $F_{BA} - 900\left(\frac{8}{12}\right) = 0$ $F_{BA} = 600\,\text{lb (C)}$ *Ans.*

Joint A. Using the result for $F_{BA} = 600\,\text{lb (C)}$, the free-body diagram of joint A is shown in Fig. 3–38d. We have

$\Sigma F_z = 0;$ $600 - 600 + F_{AC}\sin 45° = 0$

$F_{AC} = 0$ *Ans.*

$\Sigma F_y = 0;$ $-F_{AE}\left(\frac{2}{\sqrt{5}}\right) + 600 = 0$

$F_{AE} = 670.8\,\text{lb (C)}$ *Ans.*

$\Sigma F_x = 0;$ $-300 + F_{AD} + 670.8\left(\frac{1}{\sqrt{5}}\right) = 0$

$F_{AD} = 0$ *Ans.*

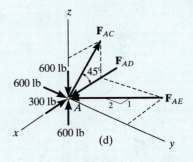

(d)

Joint D. By inspection the members at joint D, Fig. 3–38a, support zero force, since the arrangement of the members is similar to either of the two cases discussed in reference to Figs. 3–36 and 3–37. Also, from Fig. 3–38e,

$\Sigma F_x = 0;$ $F_{DE} = 0$ *Ans.*

$\Sigma F_z = 0;$ $F_{DC} = 0$ *Ans.*

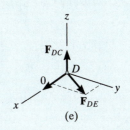

(e)

Joint C. By observation of the free-body diagram, Fig. 3–38f,

$F_{CE} = 0$ *Ans.*

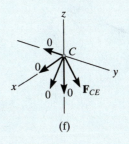

(f)

3

EXAMPLE 3.13

Determine the zero-force members of the truss shown in Fig. 3–39a. The supports exert components of reaction on the truss as shown.

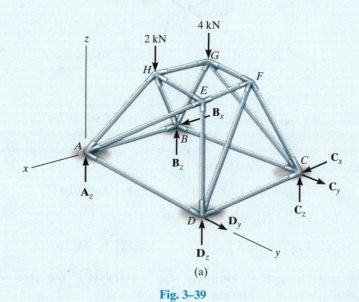

(a)

Fig. 3–39

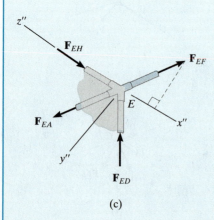

(b)

(c)

SOLUTION

The free-body diagram, Fig. 3–39a, indicates there are eight unknown reactions for which only six equations of equilibrium are available for solution. Although this is the case, the reactions can be determined, since $b + r = 3j$ or $16 + 8 = 3(8)$.

To spot the zero-force members, we must compare the conditions of joint geometry and loading to those of Figs. 3–36 and 3–37. Consider joint F, Fig. 3–39b. Since members FC, FD, FE lie in the $x' - y'$ plane and FG is not in this plane, FG is a zero-force member. ($\Sigma F_{z'} = 0$ must be satisfied.) In the same manner, from joint E, Fig. 3–39c, EF *is a zero-force member*, since it does not lie in the $y'' - z''$ plane. ($\Sigma F_{x''} = 0$ must be satisfied.) Returning to joint F, Fig. 3–39b, it can be seen that $F_{FD} = F_{FC} = 0$, since $F_{FE} = F_{FG} = 0$, and there are no external forces acting on the joint. Use this procedure to show that AB is a zero force member.

The numerical force analysis of the joints can now proceed by analyzing joint G ($F_{GF} = 0$) to determine the forces in GH, GB, GC. Then analyze joint H to determine the forces in HE, HB, HA; joint E to determine the forces in EA, ED; joint A to determine the forces in AB, AD, and A_z; joint B to determine the force in BC and B_x, B_z; joint D to determine the force in DC and D_y, D_z; and finally, joint C to determine C_x, C_y, C_z.

FUNDAMENTAL PROBLEMS

F3–1. Determine the force in each member of the truss and state whether it is in tension or compression.

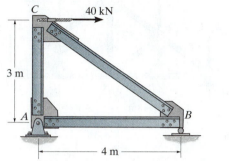

F3–1

F3–2. Determine the force in each member of the truss and state whether it is in tension or compression.

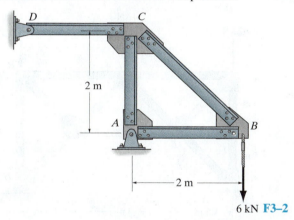

6 kN F3–2

F3–3. Determine the force in each member of the truss and state whether it is in tension or compression.

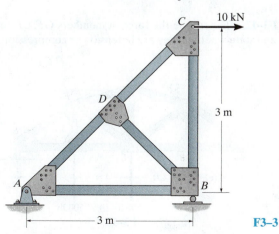

F3–3

F3–4. Determine the force in each member of the truss and state whether it is in tension or compression.

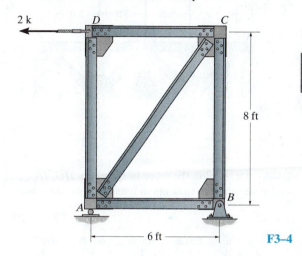

F3–4

F3–5. Determine the force in each member of the truss and state whether it is in tension or compression.

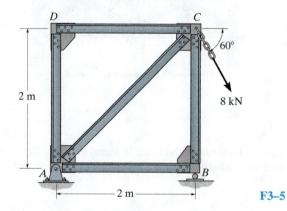

F3–5

F3–6. Determine the force in each member of the truss and state whether it is in tension or compression.

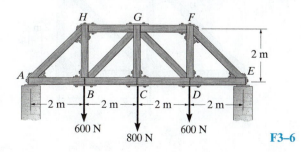

F3–6

F3–7.   Determine the force in members *HG*, *BG*, and *BC* and state whether they are in tension or compression.

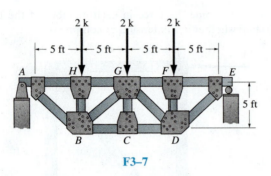

F3–7

F3–10.   Determine the force in members *GF*, *CF*, and *CD* and state whether they are in tension or compression.

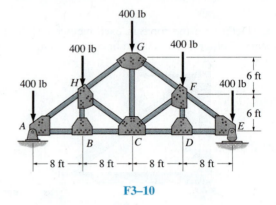

F3–10

F3–8.   Determine the force in members *HG*, *HC*, and *BC* and state whether they are in tension or compression.

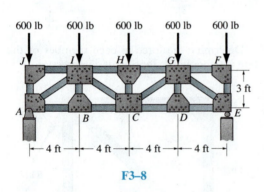

F3–8

F3–9.   Determine the force in members *ED*, *BD*, and *BC* and state whether they are in tension or compression.

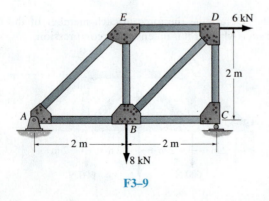

F3–9

F3–11.   Determine the force in members *FE*, *FC*, and *BC* and state whether they are in tension or compression.

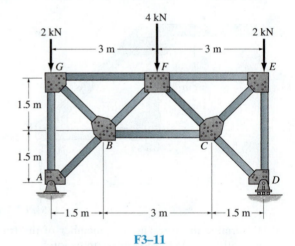

F3–11

F3–12.   Determine the force in members *GF*, *CF*, and *CD* and state whether they are in tension or compression.

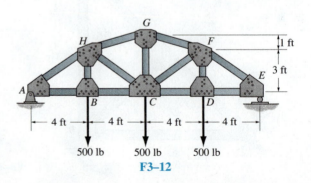

F3–12

PROBLEMS

Sec. 3.1–3.2

3–1. Classify each of the following trusses as statically determinate, statically indeterminate, or unstable. If indeterminate, state its degree.

3–2. Classify each of the following trusses as statically determinate, indeterminate, or unstable. If indeterminate, state its degree.

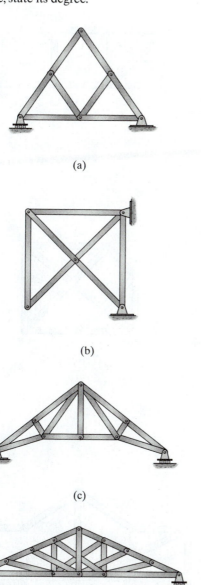

(a)

(b)

(c)

(d)

Prob. 3–1

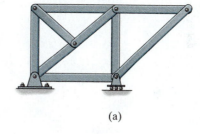

(a)

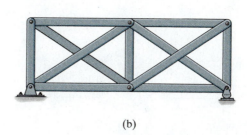

(b)

Prob. 3–2

3–3. Classify each of the following trusses as stable, unstable, statically determinate, or statically indeterminate. If indeterminate state its degree.

*3–4. Classify each of the following trusses as statically determinate, statically indeterminate, or unstable.

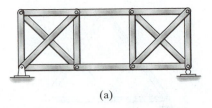

(a)

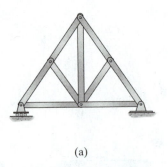

(a)

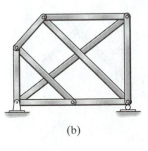

(b)

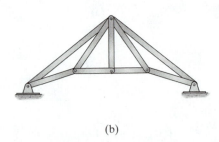

(b)

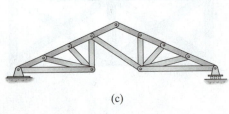

(c)

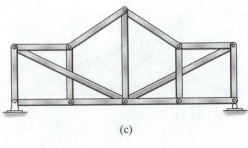

(c)

Prob. 3–3

Prob. 3–4

Sec. 3.3–3.4

3–5. Determine the force in each member of the truss. State if the members are in tension or compression. Assume all members are pin connected.

3–7. Determine the force in each member of the truss. State whether the members are in tension or compression. Set $P = 8$ kN.

***3–8.** If the maximum force that any member can support is 8 kN in tension and 6 kN in compression, determine the maximum force P that can be supported at joint D.

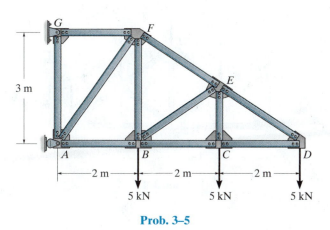

Prob. 3–5

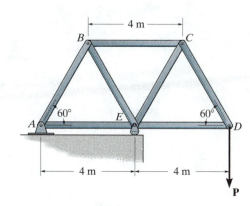

Prob. 3–7/8

3–6. Determine the force in each member of the truss. State if the members are in tension or compression.

3–9. Determine the force in each member of the truss. State if the members are in tension or compression.

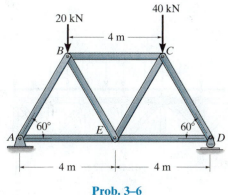

Prob. 3–6

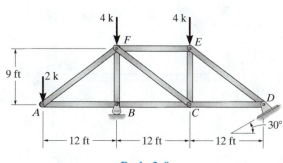

Prob. 3–9

3–10.   Determine the force in each member of the truss. State if the members are in tension or compression.

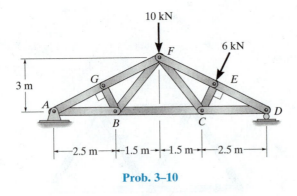

10 kN

6 kN

Prob. 3–10

*3–12.   Determine the force in each member of the truss. State if the members are in tension or compression. Assume all members are pin connected.

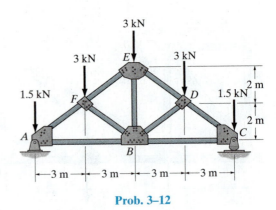

3 kN

3 kN   3 kN

1.5 kN   1.5 kN

Prob. 3–12

3–11.   Specify the type of compound truss and determine the force in each member. State if the members are in tension or compression. Assume the members are pin connected.

3–13.   The truss shown is used to support the floor deck. The uniform load on the deck is 2.5 k/ft. This load is transferred from the deck to the floor beams, which rest on the top joints of the truss. Determine the force in each member of the truss, and state if the members are in tension or compression. Assume all members are pin connected.

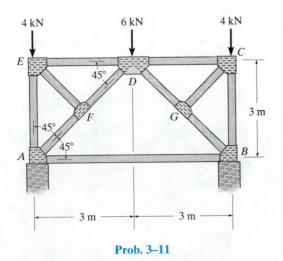

4 kN   6 kN   4 kN

Prob. 3–11

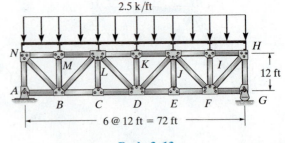

2.5 k/ft

Prob. 3–13

3–14. Determine the force in each member of the truss. Indicate if the members are in tension or compression. Assume all members are pin connected.

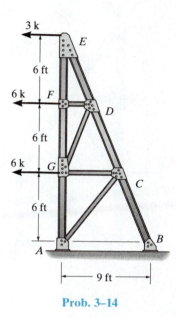

Prob. 3–14

3–15. Determine the force in each member of the truss. State if the members are in tension or compression. Assume all members are pin connected.

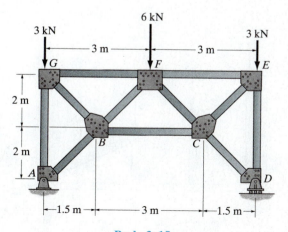

Prob. 3–15

*3–16. The members of the truss have a mass of 5 kg/m. Lifting is done using a cable connected to joints E and G. Determine the largest member force and specify if it is in tension or compression. Assume half the weight of each member can be applied as a force acting at each joint.

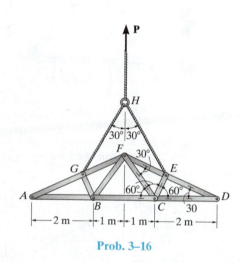

Prob. 3–16

3–17. Determine the force in each member of the truss. State if the members are in tension or compression.

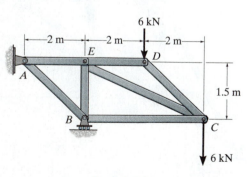

Prob. 3–17

Sec. 3.5–3.6

3–18. Determine the force in members ED, BD and BC of the truss and indicate if the members are in tension or compression.

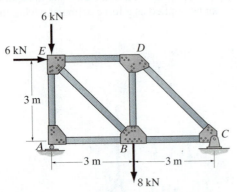

Prob. 3–18

3–19. Determine the force in members JK, JN, and CD. State if the members are in tension or compression. Identify all the zero-force members.

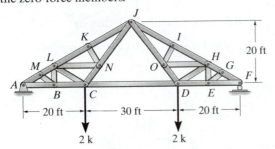

Prob. 3–19

*3–20.** Determine the force in members FC, BC, and FE. State if the members are in tension or compression. Assume all members are pin connected.

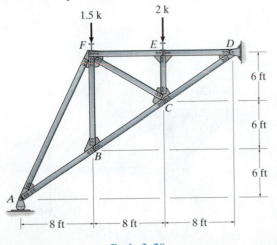

Prob. 3–20

3–21. Determine the force in members FG, GD, CD and GA of the truss. State if the members are in tension or compression.

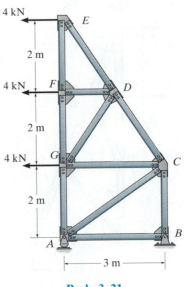

Prob. 3–21

3–22. Determine the force in members HG, HC, HB and AB of the truss. State if the members are in tension or compression. Assume all members are pin connected.

3–23. Determine the force in members GF, GC, HC and BC of the truss. State if the members are in tension or compression. Assume all members are pin connected.

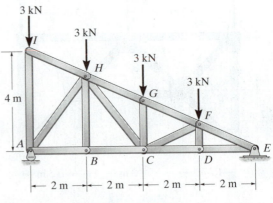

Probs. 3–22/23

*3–24. Specify the type of compound truss and determine the force in members *JH*, *BJ*, and *BI*. State if the members are in tension or compression. The internal angle between any two members is 60°. The truss is pin supported at *A* and roller supported at *F*. Assume all members are pin connected.

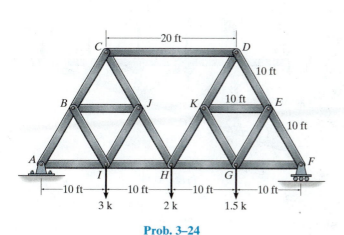

Prob. 3–24

3–26. Determine the force in members *HI*, *ID*, and *DC* of the truss and state if the members are in tension or compression.

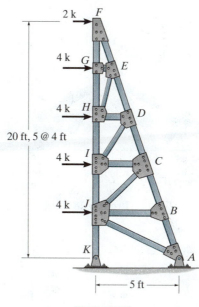

Prob. 3–26

3–25. Determine the forces in members *JI*, *JD*, and *DE* of the truss. State if the members are in tension or compression.

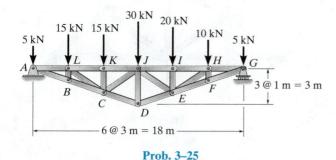

Prob. 3–25

3–27. Determine the force in members *GF*, *FB*, and *BC* of the *Fink* truss and state if the members are in tension or compression.

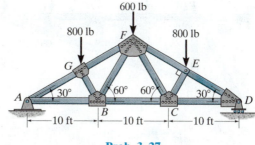

Prob. 3–27

*3–28. Specify the type of compound truss and determine the forces in members *JH, IH,* and *CD*. State if the members are in tension or compression. Assume all members are pin connected.

3–30. Determine the force in members *JI, IC,* and *CD* of the truss. State if the members are in tension or compression. Assume all members are pin connected.

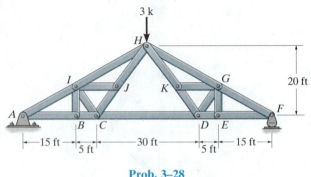

Prob. 3–28

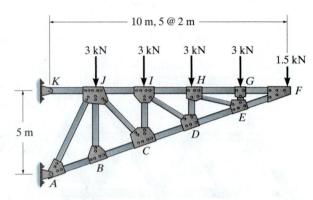

Prob. 3–30

3–29. Determine the force in members *IH, ID,* and *CD* of the truss. State if the members are in tension or compression. Assume all members are pin connected.

3–31. Determine the forces in members *GH, HC,* and *BC* of the truss. State if the members are in tension or compression. Assume all members are pin connected.

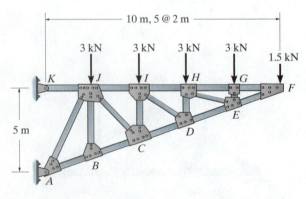

Prob. 3–29

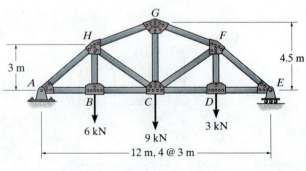

Prob. 3–31

Sec. 3.7

*3–32. Determine the forces in all the members of the complex truss. State if the members are in tension or compression. *Hint:* Substitute member *AD* with one placed between *E* and *C*.

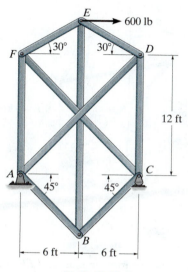

Prob. 3–32

3–33. Determine the forces in all the members of the lattice (complex) truss. State if the members are in tension or compression. *Hint:* Substitute member *JE* by one placed between *K* and *F*.

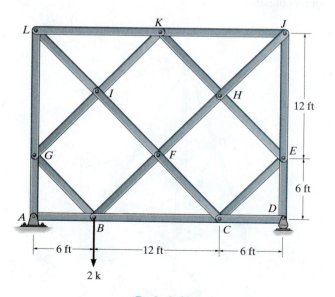

Prob. 3–33

3–34. Determine the force in each member and state if the members are in tension or compression.

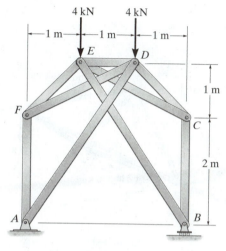

Prob. 3–34

3–35. Determine the forces in all the members of the complex truss. State if the members are in tension or compression. *Hint:* Substitute member *AB* with one placed between *C* and *E*.

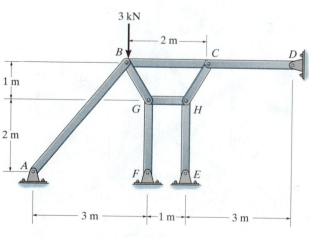

Prob. 3–35

Sec. 3.8

*3–36. Determine the force in the members of the space truss, and state whether they are in tension or compression.

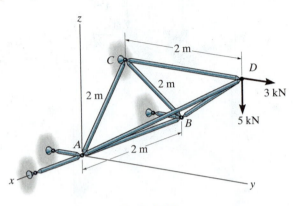

Prob. 3–36

3–37. Determine the force in each member of the space truss and state if the members are in tension or compression. The truss is supported by a ball-and-socket joint at *A* and short links at *B* and *C*.

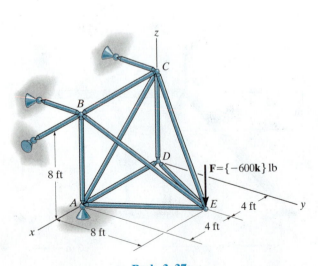

Prob. 3–37

3–38. Determine the force in each member of the space truss and state if the members are in tension or compression. The truss is supported by ball-and-socket joints at *C, D, E,* and *G*. *Note:* Although this truss is indeterminate to the first degree, a solution is possible due to symmetry of geometry and loading.

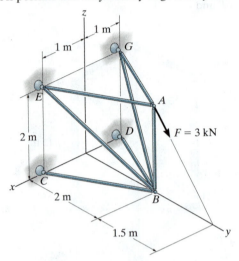

Prob. 3–38

3–39. Determine the force in members *FE* and *ED* of the space truss and state if the members are in tension or compression. The truss is supported by a ball-and-socket joint at *C* and short links at *A* and *B*.

*3–40. Determine the force in members *GD, GE, GF* and *FD* of the space truss and state if the members are in tension or compression.

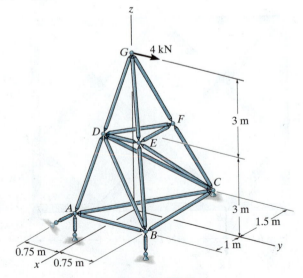

Probs. 3–39/40

3–41. Determine the reactions and the force in each member of the space truss. Indicate if the members are in tension or compression.

3–42. Determine the force in members $AB, BD,$ and FE of the space truss and state if the members are in tension or compression.

3–43. Determine the force in members AF, AE and FD of the space truss and state if the members are in tension or compression.

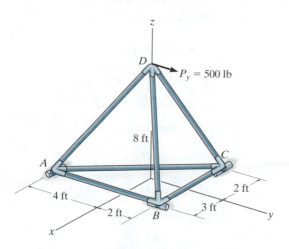

Prob. 3–41

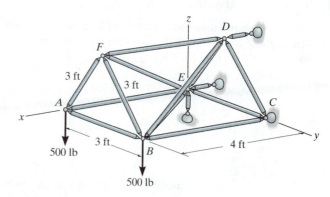

Probs. 3–42/43

PROJECT PROBLEM

3–1P. The Pratt roof trusses are uniformly spaced every 15 ft. The deck, roofing material, and the purlins have an average weight of 5.6 lb/ft². The building is located in New York where the anticipated snow load is 20 lb/ft² and the anticipated ice load is 8 lb/ft². These loadings occur over the horizontal projected area of the roof. Determine the force in each member due to dead load, snow, and ice loads. Neglect the weight of the truss members and assume A is pinned and E is a roller.

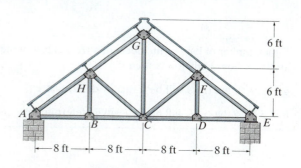

Project Prob. 3–1P

CHAPTER REVIEW

Trusses are composed of slender members joined together at their end points to form a series of triangles.

For analysis we assume the members are pin connected, and the loads are applied at the joints. Thus, the members will either be in tension or compression.

Trusses can be classified in three ways:

Simple trusses are formed by starting with an initial triangular element and connecting to it two other members and a joint to form a second triangle, etc.

simple truss

compound truss

Compound trusses are formed by connecting together two or more simple trusses using a common joint and/or additional member.

compound truss

Complex trusses are those that cannot be classified as either simple or compound.

complex truss

If the number of bars or members of a truss is *b*, and there are *r* reactions and *j* joints, then if

$b + r = 2j$ the truss will be statically determinate

$b + r > 2j$ the truss will be statically indeterminate

The truss will be externally unstable if the reactions are concurrent or parallel.

unstable concurrent reactions

Internal stability can be checked by counting the number of bars b, reactions r, and joints j.

If $b + r < 2j$ the truss is unstable.

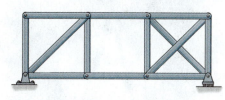

unstable parallel reactions

If $b + r \geq 2j$ it may still be unstable, so it becomes necessary to inspect the truss and look for bar arrangements that form a parallel mechanism, without forming a triangular element.

unstable internally

Planar trusses can be analyzed by the *method of joints*. This is done by selecting each joint in sequence, having at most one known force and at least two unknowns. The free-body diagram of each joint is constructed and two force equations of equilibrium, $\Sigma F_x = 0$, $\Sigma F_y = 0$, are written and solved for the unknown member forces.

The method of sections requires passing a section through the truss and then drawing a free-body diagram of one of its sectioned parts. The member forces cut by the section are then found from the three equations of equilibrium. Normally a single unknown can be found if one sums moments about a point that eliminates the two other forces.

Compound and complex trusses can also be analyzed by the method of joints and the method of sections. The "method of substitute members" can be used to obtain a direct solution for the force in a particular member of a complex truss.

Chapter 4

© Gari Wyn Williams/Alamy

The simply supported beams and girders of this building frame were designed to resist the internal shear and moment acting throughout their lengths.

Internal Loadings Developed in Structural Members

Before a structural member can be proportioned, it is necessary to determine the force and moment that act within it. In this chapter we will develop the methods for finding these loadings at specified points along a member's axis and for showing the variation graphically using the shear and moment diagrams. Applications are given for both beams and frames.

4.1 Internal Loadings at a Specified Point

As discussed in Sec. 2.4, the internal load at a specified point in a member can be determined by using the **method of sections**. In general, this loading for a coplanar structure will consist of a normal force N, shear force V, and bending moment M.* It should be realized, however, that these loadings actually represent the *resultants* of the *stress distribution* acting over the member's cross-sectional area at the cut section. Once the resultant internal loadings are known, the magnitude of the stress can be determined provided an assumed distribution of stress over the cross-sectional area is specified.

---

*Three-dimensional frameworks can also be subjected to a *torsional moment*, which tends to twist the member about its axis.

Sign Convention.   Before presenting a method for finding the internal normal force, shear force, and bending moment, we will need to establish a sign convention to define their "positive" and "negative" values.* Although the choice is arbitrary, the sign convention to be adopted here has been widely accepted in structural engineering practice, and is illustrated in Fig. 4–1*a*. On the *left-hand face* of the cut member the normal force **N** acts to the right, the internal shear force **V** acts downward, and the moment **M** acts counterclockwise. In accordance with Newton's third law, an equal but opposite normal force, shear force, and bending moment must act on the right-hand face of the member at the section. Perhaps an easy way to remember this sign convention is to isolate a small segment of the member and note that *positive normal force tends to elongate the segment*, Fig. 4–1*b*; *positive shear tends to rotate the segment clockwise*, Fig. 4–1*c*; and *positive bending moment tends to bend the segment concave upward*, so as to "hold water," Fig. 4–1*d*.

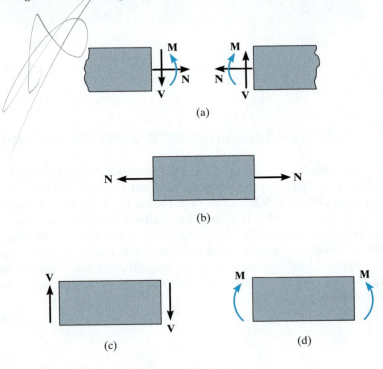

Fig. 4–1

---

*This will be convenient later, in Secs. 4.2 and 4.3, where we will express V and M as functions of x and then *plot* these functions. Having a sign convention is similar to assigning coordinate directions x positive to the right and y positive upward when plotting a function $y = f(x)$.

Procedure for Analysis

The following procedure provides a means for applying the method of sections to determine the internal normal force, shear force, and bending moment at a specific location in a structural member.

Support Reactions

- Before the member is "cut" or sectioned, it may be necessary to determine the member's support reactions so that the equilibrium equations are used only to solve for the internal loadings when the member is sectioned.

- If the member is part of a pin-connected structure, the pin reactions can be determined using the methods of Sec. 2.6.

Free-Body Diagram

- Keep all distributed loadings, couple moments, and forces acting on the member in their *exact location,* then pass an imaginary section through the member, perpendicular to its axis at the point where the internal loading is to be determined.

- After the section is made, draw a free-body diagram of the segment that has the least number of loads on it. At the section indicate the unknown resultants **N**, **V**, and **M** acting in their *positive* directions (Fig. 4–1*a*).

Equations of Equilibrium

- Moments should be summed at the section about axes that pass through the *centroid* of the member's cross-sectional area, in order to eliminate the unknowns **N** and **V** and thereby obtain a direct solution for **M**.

- If the solution of the equilibrium equations yields a quantity having a negative magnitude, the assumed directional sense of the quantity is opposite to that shown on the free-body diagram.

These hammerhead piers are tapered due to the greater shear and moment they must resist at their center.

EXAMPLE 4.1

The building roof shown in the photo has a weight of 1.8 kN/m^2 and is supported on 8-m long simply supported beams that are spaced 1 m apart. Each beam, shown in Fig. 4–2b transmits its loading to two girders, located at the front and back of the building. Determine the internal shear and moment in the front girder at point C, Fig. 4–2a. Neglect the weight of the members.

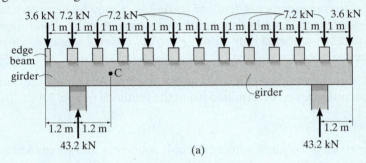

(a)

SOLUTION

Support Reactions. The roof loading is transmitted to each beam as a one-way slab ($L_2/L_1 = 8 \text{ m}/1 \text{ m} = 8 > 2$). The tributary loading on each interior beam is therefore $(1.8 \text{ kN/m}^2)(1 \text{ m}) = 1.8 \text{ kN/m}$. (The two edge beams support 0.9 kN/m.) From Fig. 4–2b, the reaction of each interior beam on the girder is $(1.8 \text{ kN/m})(8 \text{ m})/2 = 7.2 \text{ kN}$.

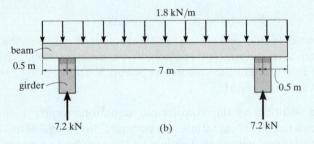

(b)

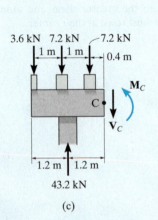

(c)

Fig. 4–2

Free-Body Diagram. The free-body diagram of the girder is shown in Fig. 4–2a. Notice that each column reaction is

$$[(2(3.6 \text{ kN}) + 11(7.2 \text{ kN})]/2 = 43.2 \text{ kN}$$

The free-body diagram of the left girder segment is shown in Fig. 4–2c. Here the internal loadings are assumed to act in their positive directions.

Equations of Equilibrium.

$$+\uparrow \Sigma F_y = 0; \qquad 43.2 - 3.6 - 2(7.2) - V_C = 0 \qquad V_C = 25.2 \text{ kN} \qquad \textit{Ans.}$$

$$\zeta + \Sigma M_C = 0; \quad M_C + 7.2(0.4) + 7.2(1.4) + 3.6(2.4) - 43.2(1.2) = 0 \qquad M_C = 30.2 \text{ kN} \cdot \text{m} \qquad \textit{Ans.}$$

EXAMPLE 4.2

Determine the internal shear and moment acting at a section passing through point C in the beam shown in Fig. 4–3a.

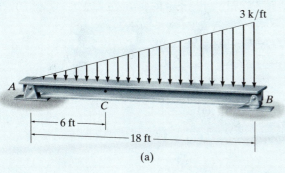

(a)

(b)

Fig. 4–3

SOLUTION

Support Reactions. Replacing the distributed load by its resultant force and computing the reactions yields the results shown in Fig. 4–3b.

Free-Body Diagram. Segment AC will be considered since it yields the simplest solution, Fig. 4–3c. The distributed load intensity at C is computed by proportion, that is,

$$w_C = (6 \text{ ft}/18 \text{ ft})(3 \text{ k}/\text{ft}) = 1 \text{ k}/\text{ft}$$

Equations of Equilibrium.

$$+\uparrow \Sigma F_y = 0; \qquad 9 - 3 - V_C = 0 \qquad\qquad V_C = 6 \text{ k} \qquad \textit{Ans.}$$

$$\zeta + \Sigma M_C = 0; \qquad -9(6) + 3(2) + M_C = 0 \qquad M_C = 48 \text{ k} \cdot \text{ft} \qquad \textit{Ans.}$$

This problem illustrates the importance of *keeping* the distributed loading on the beam until *after* the beam is sectioned. If the beam in Fig. 4–3b were sectioned at C, the effect of the distributed load on segment AC would not be recognized, and the result $V_C = 9 \text{ k}$ and $M_C = 54 \text{ k} \cdot \text{ft}$ would be wrong.

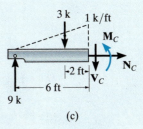

(c)

EXAMPLE 4.3

The 9-k force in Fig. 4–4a is supported by the floor panel DE, which in turn is simply supported at its ends by floor beams. These beams transmit their loads to the simply supported girder AB. Determine the internal shear and moment acting at point C in the girder.

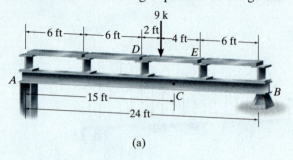

(a)

Fig. 4–4

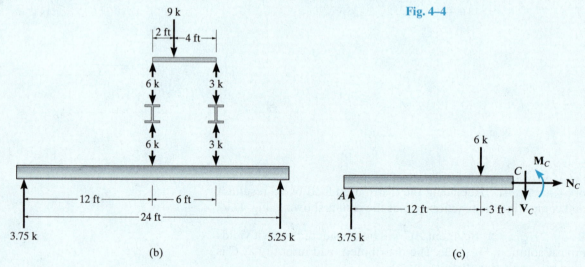

(b)   (c)

SOLUTION

Support Reactions. Equilibrium of the floor panel, floor beams, and girder is shown in Fig. 4–4b. It is advisable to check these results.

Free-Body Diagram. The free-body diagram of segment AC of the girder will be used since it leads to the simplest solution, Fig. 4–4c. Note that there are *no loads* on the floor beams supported by AC.

Equations of Equilibrium.

$$+\uparrow \Sigma F_y = 0; \qquad 3.75 - 6 - V_C = 0 \qquad V_C = -2.25 \text{ k} \qquad \textit{Ans.}$$

$$\zeta + \Sigma M_C = 0; \quad -3.75(15) + 6(3) + M_C = 0 \qquad M_C = 38.25 \text{ k} \cdot \text{ft} \qquad \textit{Ans.}$$

4.2   Shear and Moment Functions

The design of a beam requires a detailed knowledge of the *variations* of the internal shear force V and moment M acting at each point along the axis of the beam. The internal normal force is generally not considered for two reasons: (1) in most cases the loads applied to a beam act perpendicular to the beam's axis and hence produce only an internal shear force and bending moment, and (2) for design purposes the beam's resistance to shear, and particularly to bending, is more important than its ability to resist normal force. An important exception to this occurs, however, when beams are subjected to compressive axial forces, since the buckling or instability that may occur has to be investigated.

The variations of V and M as a function of the position x of an arbitrary point along the beam's axis can be obtained by using the method of sections discussed in Sec. 4.1. Here, however, it is necessary to locate the imaginary section or cut at an arbitrary distance x from one end of the beam rather than at a specific point.

In general, the internal shear and moment functions will be discontinuous, or their slope will be discontinuous, at points where the type or magnitude of the distributed load changes or where concentrated forces or couple moments are applied. Because of this, shear and moment functions must be determined for each region of the beam located *between* any two discontinuities of loading. For example, coordinates x_1, x_2, and x_3 will have to be used to describe the variation of V and M throughout the length of the beam in Fig. 4–5a. These coordinates will be valid only within regions from A to B for x_1, from B to C for x_2, and from C to D for x_3. Although each of these coordinates has the same origin, as noted here, this does not have to be the case. Indeed, it may be easier to develop the shear and moment functions using coordinates x_1, x_2, x_3 having origins at A, B, and D as shown in Fig. 4–5b. Here x_1 and x_2 are positive to the right and x_3 is positive to the left.

Additional reinforcement, provided by vertical plates called *stiffeners*, is used over the pin and rocker supports of these bridge girders. Here the reactions will cause large internal shear in the girders and the stiffeners will prevent localized buckling of the girder flanges or web. Also, note the tipping of the rocker support caused by the thermal expansion of the bridge deck.

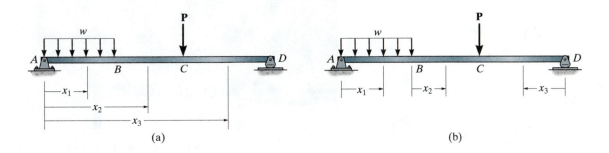

(a)

(b)

Fig. 4–5

4

Procedure for Analysis

The following procedure provides a method for determining the variation of shear and moment in a beam as a function of position x.

Support Reactions

- Determine the support reactions on the beam and resolve all the external forces into components acting perpendicular and parallel to the beam's axis.

Shear and Moment Functions

- Specify separate coordinates x and associated origins, extending into regions of the beam between concentrated forces and/or couple moments, or where there is a discontinuity of distributed loading.

- Section the beam perpendicular to its axis at each distance x, and from the free-body diagram of one of the segments determine the unknowns V and M at the cut section as functions of x. On the free-body diagram, V and M should be shown acting in their *positive directions*, in accordance with the sign convention given in Fig. 4–1.

- V is obtained from $\Sigma F_y = 0$ and M is obtained by summing moments about the point S located at the cut section, $\Sigma M_s = 0$.

- The results can be checked by noting that $dM/dx = V$ and $dV/dx = w$, where w is positive when it acts upward, away from the beam. These relationships are developed in Sec. 4.3.

The joists, beams, and girders used to support this floor can be designed once the internal shear and moment are known throughout their lengths.

EXAMPLE 4.4

Determine the shear and moment in the beam shown in Fig. 4–6a as a function of x.

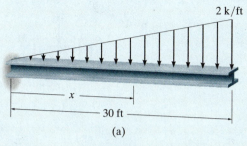

2 k/ft

x

30 ft

(a)

Fig. 4–6

SOLUTION

Support Reactions. For the purpose of computing the support reactions, the distributed load is replaced by its resultant force of 30 k, Fig. 4–6b. It is important to remember, however, that this resultant is not the actual load on the beam.

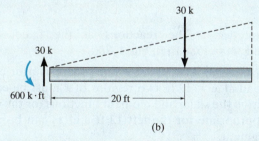

30 k

30 k

600 k · ft   20 ft

(b)

Shear and Moment Functions. A free-body diagram of the beam segment of length x is shown in Fig. 4–6c. Note that the intensity of the triangular load at the section is found by proportion; that is, $w/x = 2/30$ or $w = x/15$. With the load intensity known, the resultant of the distributed loading is found in the usual manner as shown in the figure. Thus,

$$+\uparrow \Sigma F_y = 0; \quad 30 - \frac{1}{2}\left(\frac{x}{15}\right)x - V = 0$$

$$V = 30 - 0.0333x^2 \qquad Ans.$$

$$\zeta + \Sigma M_S = 0; \quad 600 - 30x + \left[\frac{1}{2}\left(\frac{x}{15}\right)x\right]\frac{x}{3} + M = 0$$

$$M = -600 + 30x - 0.0111x^3 \qquad Ans.$$

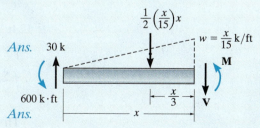

$\frac{1}{2}\left(\frac{x}{15}\right)x$

$w = \frac{x}{15}$ k/ft

30 k

M

600 k · ft

$\frac{x}{3}$

V

x

(c)

Note that $dM/dx = V$ and $dV/dx = -x/15 = w$, which serves as a check of the results.

EXAMPLE 4.5

Determine the shear and moment in the beam shown in Fig. 4–7a as a function of x.

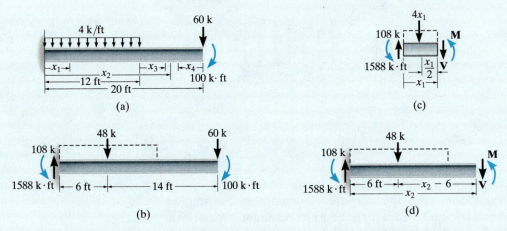

Fig. 4–7

SOLUTION

Support Reactions. The reactions at the fixed support are $V = 108$ k and $M = 1588$ k · ft, Fig. 4–7b.

Shear and Moment Functions. Since there is a discontinuity of distributed load at $x = 12$ ft, two regions of x must be considered in order to describe the shear and moment functions for the entire beam. Here x_1 is appropriate for the left 12 ft and x_2 can be used for the remaining segment.

$0 \leq x_1 \leq 12$ ft. Notice that V and M are shown in the positive directions, Fig. 4–7c.

$+\uparrow \Sigma F_y = 0;$ $108 - 4x_1 - V = 0,$ $V = 108 - 4x_1$ *Ans.*

$\zeta + \Sigma M_S = 0;$ $1588 - 108x_1 + 4x_1 \left(\dfrac{x_1}{2}\right) + M = 0$

$$M = -1588 + 108x_1 - 2x_1^2$$ *Ans.*

12 ft $\leq x_2 \leq 20$ ft, Fig. 4–7d.

$+\uparrow \Sigma F_y = 0;$ $108 - 48 - V = 0,$ $V = 60$ *Ans.*

$\zeta + \Sigma M_S = 0;$ $1588 - 108x_2 + 48(x_2 - 6) + M = 0$

$$M = 60x_2 - 1300$$ *Ans.*

These results can be partially checked by noting that when $x_2 = 20$ ft, then $V = 60$ k and $M = -100$ k · ft. Also, note that $dM/dx = V$ and $dV/dx = w$.

EXAMPLE 4.6

Determine the shear and moment in the beam shown in Fig. 4–8a as a function of x.

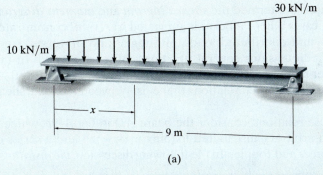

(a)

Fig. 4–8

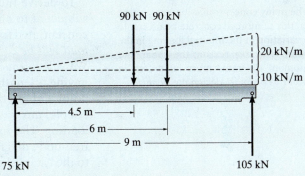

(b)

SOLUTION

Support Reactions. To determine the support reactions, the distributed load is divided into a triangular and rectangular loading, and these loadings are then replaced by their resultant forces. These reactions have been computed and are shown on the beam's free-body diagram, Fig. 4–8b.

Shear and Moment Functions. A free-body diagram of the cut section is shown in Fig. 4–8c. As above, the trapezoidal loading is replaced by rectangular and triangular distributions. Note that the intensity of the triangular load at the cut is found by proportion. The resultant force of each distributed loading and its location are indicated. Applying the equilibrium equations, we have

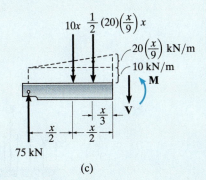

(c)

$$+\uparrow \Sigma F_y = 0; \qquad 75 - 10x - \left[\frac{1}{2}(20)\left(\frac{x}{9}\right)x\right] - V = 0$$

$$V = 75 - 10x - 1.11x^2 \qquad \textit{Ans.}$$

$$\zeta + \Sigma M_S = 0; \qquad -75x + (10x)\left(\frac{x}{2}\right) + \left[\frac{1}{2}(20)\left(\frac{x}{9}\right)x\right]\frac{x}{3} + M = 0$$

$$M = 75x - 5x^2 - 0.370x^3 \qquad \textit{Ans.}$$

4.3 Shear and Moment Diagrams for a Beam

If the variations of V and M as functions of x obtained in Sec. 4.2 are plotted, the graphs are termed the **shear diagram** and **moment diagram**, respectively. In cases where a beam is subjected to *several* concentrated forces, couples, and distributed loads, plotting V and M versus x can become quite tedious since several functions must be plotted. In this section a simpler method for constructing these diagrams is discussed—a method based on differential relations that exist between the load, shear, and moment.

To derive these relations, consider the beam AD in Fig. 4–9a, which is subjected to an arbitrary distributed loading $w = w(x)$ and a series of concentrated forces and couples. In the following discussion, *the distributed load will be considered positive when the loading acts upward* as shown. We will consider the free-body diagram for a small segment of the beam having a length Δx, Fig. 4–9b. Since this segment has been chosen at a point x along the beam that is *not* subjected to a concentrated force or couple, any results obtained will not apply at points of concentrated loading. The internal shear force and bending moment shown on the free-body diagram are assumed to act in the *positive direction* according to the established sign convention, Fig. 4–1. Note that both the shear force and moment acting on the right face must be increased by a small, finite amount in order to keep the segment in equilibrium. The distributed loading has been replaced by a concentrated force $w(\Delta x)$ that acts at a fractional distance $\epsilon(\Delta x)$ from the right end, where $0 < \epsilon < 1$. (For example, if w is uniform or constant, then $w(\Delta x)$ will act at $\frac{1}{2}\Delta x$, so $\epsilon = \frac{1}{2}$.) Applying the equations of equilibrium, we have

$$+\uparrow\Sigma F_y = 0; \quad V + w(\Delta x) - (V + \Delta V) = 0$$
$$\Delta V = w(\Delta x)$$
$$\zeta+\Sigma M_O = 0; \quad -V\Delta x - M - w(\Delta x)\epsilon(\Delta x) + (M + \Delta M) = 0$$
$$\Delta M = V(\Delta x) + w\epsilon(\Delta x)^2$$

The many concentrated loadings acting on this reinforced concrete beam create a variation of the internal loading in the beam. For this reason, the shear and moment diagrams must be drawn in order to properly design the beam.

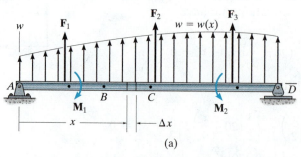

(a)

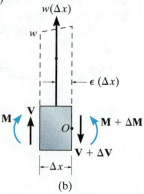

(b)

Fig. 4–9

Dividing by Δx and taking the limit as $\Delta x \rightarrow 0$, these equations become

$$\frac{dV}{dx} = w$$

$$\left.\begin{array}{c}\text{Slope of} \\ \text{Shear Diagram}\end{array}\right\} = \left\{\begin{array}{l}\text{Intensity of} \\ \text{Distributed Load}\end{array}\right. \qquad (4\text{–}1)$$

$$\frac{dM}{dx} = V$$

$$\left.\begin{array}{c}\text{Slope of} \\ \text{Moment Diagram}\end{array}\right\} = \left\{\text{Shear}\right. \qquad (4\text{–}2)$$

As noted, Eq. 4–1 states that *the slope of the shear diagram at a point* *(dV/dx) is equal to the intensity of the distributed load w at the point.* Likewise, Eq. 4–2 states that *the slope of the moment diagram (dM/dx) is* *equal to the intensity of the shear at the point.*

Equations 4–1 and 4–2 can be "integrated" from one point to another between concentrated forces or couples (such as from B to C in Fig. 4–9a), in which case

$$\Delta V = \int w\,dx$$

$$\left.\begin{array}{c}\text{Change in} \\ \text{Shear}\end{array}\right\} = \left\{\begin{array}{l}\text{Area under} \\ \text{Distributed Loading} \\ \text{Diagram}\end{array}\right. \qquad (4\text{–}3)$$

and

$$\Delta M = \int V\,dx$$

$$\left.\begin{array}{c}\text{Change in} \\ \text{Moment}\end{array}\right\} = \left\{\begin{array}{l}\text{Area under} \\ \text{Shear Diagram}\end{array}\right. \qquad (4\text{–}4)$$

As noted, Eq. 4–3 states that *the change in the shear between any two* *points on a beam equals the area under the distributed loading diagram* *between the points.* Likewise, Eq. 4–4 states that *the change in the moment* *between the two points equals the area under the shear diagram between the* *points.* If the areas under the load and shear diagrams are easy to compute, Eqs. 4–3 and 4–4 provide a method for determining the numerical values of the shear and moment at various points along a beam.

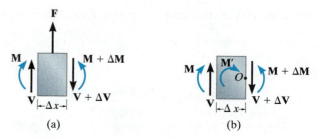

Fig. 4–10

From the derivation it should be noted that Eqs. 4–1 and 4–3 cannot be used at points where a concentrated force acts, since these equations do not account for the sudden change in shear at these points. Similarly, because of a discontinuity of moment Eqs. 4–2 and 4–4 cannot be used at points where a couple moment is applied. In order to account for these two cases, we must consider the free-body diagrams of differential elements of the beam in Fig. 4–9a which are located at concentrated force and couple moments. Examples of these elements are shown in Figs. 4–10a and 4–10b, respectively. From Fig. 4–10a it is seen that force equilibrium requires the change in shear to be

$$+\uparrow \Sigma F_y = 0; \qquad\qquad \Delta V = F \qquad\qquad (4\text{–}5)$$

Thus, when **F** acts *upward* on the beam, ΔV is positive, so that the shear diagram shows a "jump" *upward.* Likewise, if **F** acts *downward,* the jump (ΔV) is *downward.* From Fig. 4–10b, letting $\Delta x \rightarrow 0$, moment equilibrium requires the change in moment to be

$$\zeta + \Sigma M_O = 0; \qquad\qquad \Delta M = M' \qquad\qquad (4\text{–}6)$$

In this case, if an external couple moment **M'** is applied *clockwise,* ΔM is positive, so that the moment diagram jumps *upward,* and when **M** acts *counterclockwise,* the jump (ΔM) must be *downward.*

Table 4.1 illustrates application of Eqs. 4–1, 4–2, 4–5, and 4–6 to some common loading cases assuming V and M retain positive values. The slope at various points on each curve is indicated. None of these results should be memorized; rather, each should be studied carefully so that one becomes fully aware of how the shear and moment diagrams can be constructed on the basis of knowing the variation of the slope from the load and shear diagrams, respectively. It would be well worth the time and effort to self-test your understanding of these concepts by covering over the shear and moment diagram columns in the table and then trying to reconstruct these diagrams on the basis of knowing the loading.

TABLE 4.1   Relationship between Loading, Shear, and Moment

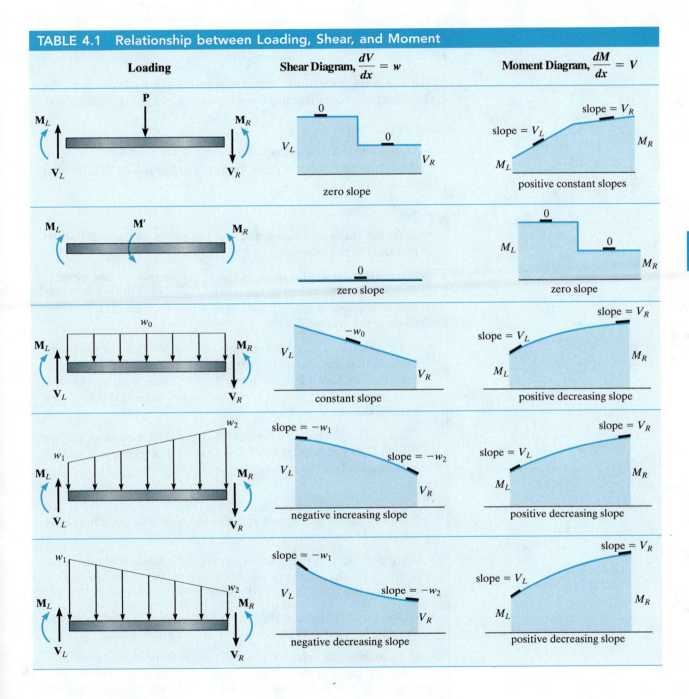

Loading	Shear Diagram, $\dfrac{dV}{dx} = w$	Moment Diagram, $\dfrac{dM}{dx} = V$

Procedure for Analysis

The following procedure provides a method for constructing the shear and moment diagrams for a beam using Eqs. 4–1 through 4–6.

Support Reactions

- Determine the support reactions and resolve the forces acting on the beam into components which are perpendicular and parallel to the beam's axis.

Shear Diagram

- Establish the V and x axes and plot the values of the shear at the two *ends* of the beam.

- Since $dV/dx = w$, the *slope* of the *shear diagram* at any point is equal to the intensity of the *distributed loading* at the point. (Note that w is positive when it acts upward.) Begin by sketching the slope at the end points.

- If a numerical value of the shear is to be determined at the point, one can find this value either by using the method of sections as discussed in Sec. 4.1 or by using Eq. 4–3, which states that the *change in the shear force* is equal to the *area under the distributed loading diagram*.

- Since $w(x)$ is *integrated* to obtain $V(x)$, if $w(x)$ is a curve of degree n, then $V(x)$ will be a curve of degree $n + 1$. For example, if $w(x)$ is uniform, $V(x)$ will be linear.

Moment Diagram

- Establish the M and x axes and plot the values of the moment at the ends of the beam.

- Since $dM/dx = V$, the *slope* of the *moment diagram* at any point is equal to the intensity of the *shear* at the point. Begin by sketching the slope at the end points.

- At the point where the shear is zero, $dM/dx = 0$, and therefore this may be a point of maximum or minimum moment.

- If the numerical value of the moment is to be determined at a point, one can find this value either by using the method of sections as discussed in Sec. 4.1 or by using Eq. 4–4, which states that the *change in the moment* is equal to the *area under the shear diagram*.

- Since $V(x)$ is *integrated* to obtain $M(x)$, if $V(x)$ is a curve of degree n, then $M(x)$ will be a curve of degree $n + 1$. For example, if $V(x)$ is linear, $M(x)$ will be parabolic.

EXAMPLE 4.7

The two horizontal members of the power line support frame are subjected to the cable loadings shown in Fig. 4–11a. Draw the shear and moment diagrams for each member.

SOLUTION

Support Reactions.   Each pole exerts a force of 6 kN on each member as shown on the free-body diagram.

Shear Diagram.   The end points $x = 0$, $V = -4$ kN and $x = 6$ m, $V = 4$ kN are plotted first, Fig. 4–11b. As indicated, the shear between each concentrated force is *constant* since $w = dV/dx = 0$. The shear just to the right of point B (or C and D) can be determined by the method of sections. Fig. 4–11d. The shear diagram can also be established by "following the load" on the free-body diagram. Beginning at A the 4 kN load acts downward so $V_A = -4$ kN. No load acts between A and B so the slope is zero and the shear is constant. At B the 6 kN force acts upward, so the shear jumps up 6 kN, from -4 kN to $+2$ kN. Again, the slope remains constant until it reaches the 4 kN load, where the downward force of 4 kN drops the shear from 2 to -2, etc.

Moment Diagram.   The moment at the end points $x = 0$, $M = 0$ and $x = 6$ m, $M = 0$ is plotted first, Fig. 4–11c. The slope of the moment diagram within each 1.5-m-long region is constant because V is constant. Specific values of the moment, such as at C, can be determined by the method of sections, Fig. 4–11d, or by finding the change in moment by the area under the shear diagram. For example, since $M_A = 0$ at A, then at C, $M_C = M_A + \Delta M_{AC} = 0 + (-4)(1.5) + 2(1.5) = -3$ kN·m.

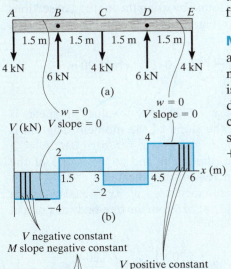

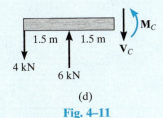

Fig. 4–11

EXAMPLE 4.8

Draw the shear and moment diagrams for the beam in Fig. 4–12a.

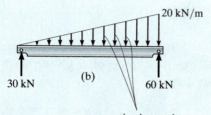

(b)

30 kN   60 kN

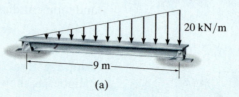

20 kN/m

←——— 9 m ———→

(a)

Fig. 4–12

w negative increasing
V slope negative increasing

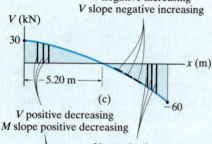

V (kN)

30

←—5.20 m—→

x (m)

(c)

−60

V positive decreasing
M slope positive decreasing

V negative increasing
M slope negative increasing

M (kN·m)   104

x (m)

(d)

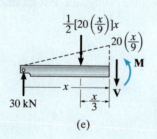

$\frac{1}{2}[20(\frac{x}{9})]x$

$20(\frac{x}{9})$

M

x

$\frac{x}{3}$ **V**

30 kN

(e)

SOLUTION

Support Reactions. The reactions have been calculated and are shown on the free-body diagram of the beam, Fig. 4–12b.

Shear Diagram. The end points $x = 0$, $V = +30$ kN and $x = 9$ m, $V = -60$ kN are first plotted. Note that the shear diagram *starts* with zero slope since $w = 0$ at $x = 0$, and ends with a slope of $w = -20$ kN/m.

The point of zero shear can be found by using the method of sections from a beam segment of length x, Fig. 4–12e. We require $V = 0$, so that

$$+\uparrow \Sigma F_y = 0; \qquad 30 - \frac{1}{2}\left[20\left(\frac{x}{9}\right)\right]x = 0 \qquad x = 5.20 \text{ m}$$

Moment Diagram. For $0 < x < 5.20$ m the value of shear is positive but decreasing and so the slope of the moment diagram is also positive and decreasing $(dM/dx = V)$. At $x = 5.20$ m, $dM/dx = 0$. Likewise for 5.20 m $< x < 9$ m, the shear and so the slope of the moment diagram are negative increasing as indicated.

The maximum value of moment is at $x = 5.20$ m since $dM/dx = V = 0$ at this point, Fig. 4–12d. From the free-body diagram in Fig. 4–12e we have

$$\zeta + \Sigma M_S = 0; \quad -30(5.20) + \frac{1}{2}\left[20\left(\frac{5.20}{9}\right)\right](5.20)\left(\frac{5.20}{3}\right) + M = 0$$

$$M = 104 \text{ kN} \cdot \text{m}$$

EXAMPLE   4.9

Draw the shear and moment diagrams for each of the beams shown in Fig. 4–13.

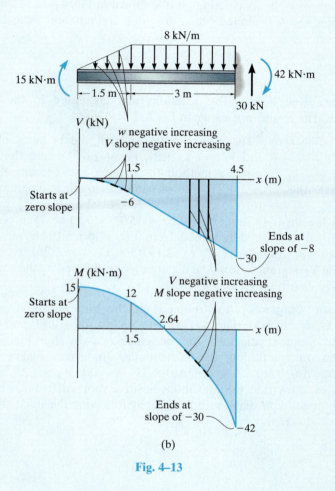

Fig. 4–13

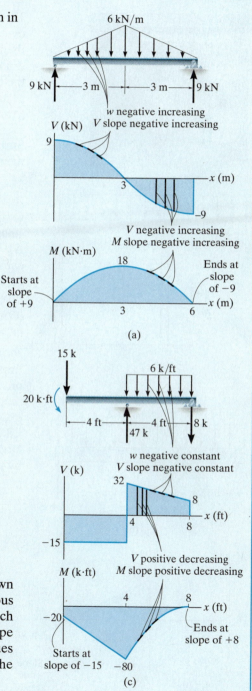

SOLUTION

In each case the support reactions have been calculated and are shown in the top figures. Following the techniques outlined in the previous examples, the shear and moment diagrams are shown under each beam. Carefully notice how they were established, based on the slope and moment, where $dV/dx = w$ and $dM/dx = V$. Calculated values are found using the method of sections or finding the areas under the load or shear diagrams.

EXAMPLE 4.10

The beam shown in the photo is used to support a portion of the overhang for the entranceway of the building. The idealized model for the beam with the load acting on it is shown in Fig. 4–14a. Assume B is a roller and C is pinned. Draw the shear and moment diagrams for the beam.

SOLUTION

Support Reactions. The reactions are calculated in the usual manner. The results are shown in Fig. 4–14b.

Shear Diagram. The shear at the ends of the beam is plotted first, i.e., $V_A = 0$ and $V_C = -2.19$ kN, Fig. 4–14c. To find the shear to the left of B use the method of sections for segment AB, or calculate the area under the distributed loading diagram, i.e., $\Delta V = V_B - 0 = -10(0.75)$, $V_{B^-} = -7.50$ kN. The support reaction causes the shear to jump up $-7.50 + 15.31 = 7.81$ kN. The point of zero shear can be determined from the slope -10 kN/m, or by proportional triangles, $7.81/x = 2.19/(1 - x)$, $x = 0.781$ m. Notice how the V diagram follows the negative slope, defined by the constant negative distributed loading.

Moment Diagram. The moment at the end points is plotted first, $M_A = M_C = 0$, Fig. 4–14d. The values of -2.81 and 0.239 on the moment diagram can be calculated by the method of sections, or by finding the areas under the shear diagram. For example, $\Delta M = M_B - 0 = \frac{1}{2}(-7.50)(0.75) = -2.81$, $M_B = -2.81$ kN·m. Likewise, show that the maximum positive moment is 0.239 kN·m. Notice how the M diagram is formed, by following the slope, defined by the V diagram.

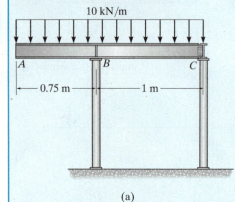

(a)

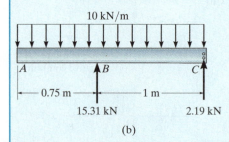

(b)

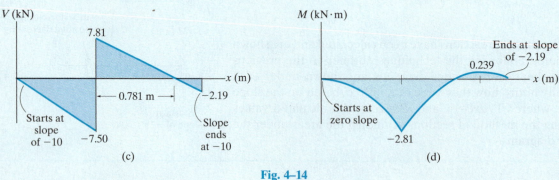

(c)   (d)

Fig. 4–14

EXAMPLE 4.11

Draw the shear and moment diagrams for the compound beam shown in Fig. 4–15a. Assume the supports at A and C are rollers and B and E are pin connections.

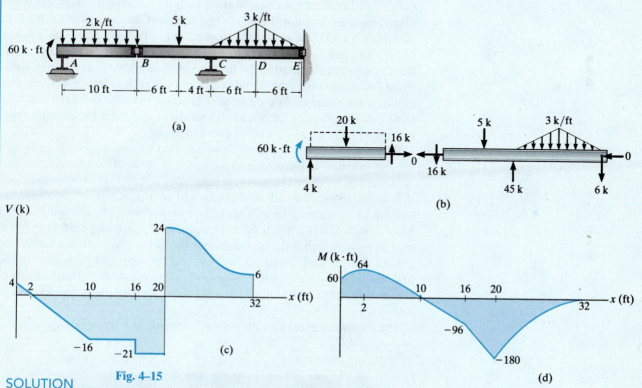

Fig. 4–15

SOLUTION

Support Reactions. Once the beam segments are disconnected from the pin at B, the support reactions can be calculated as shown in Fig. 4–15b.

Shear Diagram. As usual, we start by plotting the end shear at A and E, Fig. 4–15c. The shape of the V diagram is formed by following its slope, defined by the loading. Try to establish the values of shear using the appropriate areas under the load diagram (w curve) to find the change in shear. The zero value for shear at $x = 2$ ft can either be found by proportional triangles, or by using statics, as was done in Fig. 4–12e of Example 4.8.

Moment Diagram. The end moments $M_A = 60 \text{ k} \cdot \text{ft}$ and $M_E = 0$ are plotted first, Fig. 4–15d. Study the diagram and note how the various curves are established using $dM/dx = V$. Verify the numerical values for the peaks using statics or by calculating the appropriate areas under the shear diagram to find the change in moment.

4.4   Shear and Moment Diagrams for a Frame

Recall that a *frame* is composed of several connected members that are either fixed or pin connected at their ends. The design of these structures often requires drawing the shear and moment diagrams for each of the members. To analyze any problem, we can use the procedure for analysis outlined in Sec. 4.3. This requires first determining the reactions at the frame supports. Then, using the method of sections, we find the axial force, shear force, and moment acting at the ends of each member. Provided all loadings are resolved into components acting parallel and perpendicular to the member's axis, the shear and moment diagrams for each member can then be drawn as described previously.

When drawing the moment diagram, one of two sign conventions is used in practice. In particular, if the frame is made of *reinforced concrete,* designers often draw the moment diagram positive on the tension side of the frame. In other words, if the moment produces tension on the outer surface of the frame, the moment diagram is drawn positive on this side. Since concrete has a low tensile strength, it will then be possible to tell at a glance on which side of the frame the reinforcement steel must be placed. In this text, however, we will use the opposite sign convention and *always draw the moment diagram positive on the compression side of the member*. This convention follows that used for beams discussed in Sec. 4.1.

The following examples illustrate this procedure numerically.

The simply supported girder of this concrete building frame was designed by first drawing its shear and moment diagrams.

EXAMPLE | 4.12

Draw the moment diagram for the tapered frame shown in Fig. 4–16a. Assume the support at A is a roller and B is a pin.

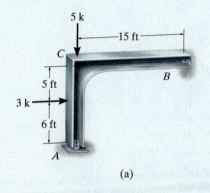

(a)

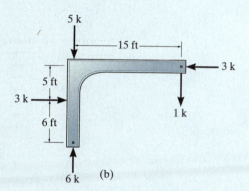

(b)

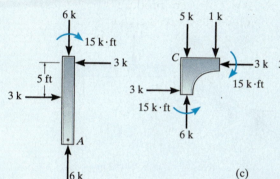

(c)

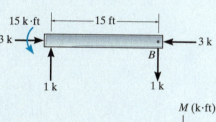

Fig. 4–16

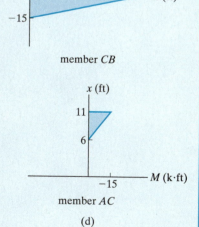

member CB

member AC

(d)

SOLUTION

Support Reactions.   The support reactions are shown on the free-body diagram of the entire frame, Fig. 4–16b. Using these results, the frame is then sectioned into two members, and the internal reactions at the joint ends of the members are determined, Fig. 4–16c. Note that the external 5-k load is shown only on the free-body diagram of the joint at C.

Moment Diagram.   In accordance with our positive sign convention, and using the techniques discussed in Sec. 4.3, the moment diagrams for the frame members are shown in Fig. 4–16d.

EXAMPLE   4.13

Draw the shear and moment diagrams for the frame shown in Fig. 4–17a. Assume A is a pin, C is a roller, and B is a fixed joint.

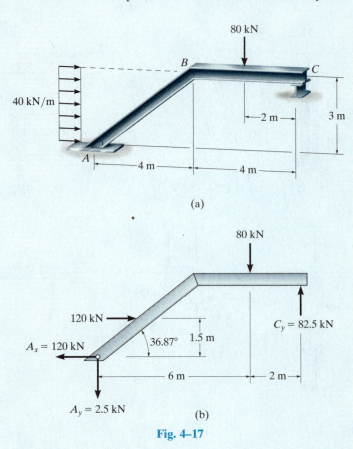

(a)

(b)

Fig. 4–17

SOLUTION

Support Reactions.   The free-body diagram of the entire frame is shown in Fig. 4–17b. Here the distributed load, which represents wind loading, has been replaced by its resultant, and the reactions have been computed. The frame is then sectioned at joint B and the internal loadings at B are determined, Fig. 4–17c. As a check, equilibrium is satisfied at joint B, which is also shown in the figure.

Shear and Moment Diagrams.   The components of the distributed load, $(72 \text{ kN})/(5 \text{ m}) = 14.4 \text{ kN/m}$ and $(96 \text{ kN})/(5 \text{ m}) = 19.2 \text{ kN/m}$, are shown on member AB, Fig. 4–17d. The associated shear and moment diagrams are drawn for each member as shown in Figs. 4–17d and 4–17e.

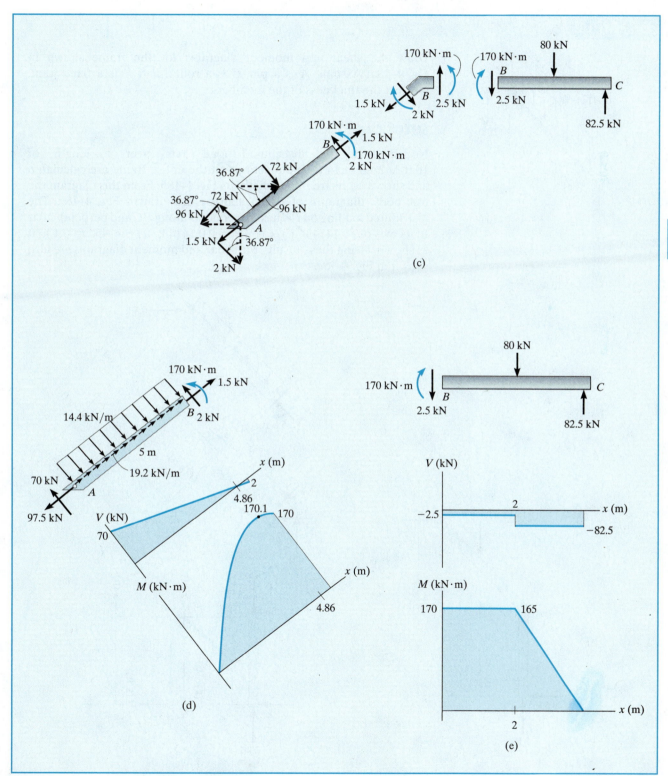

(c)

(d)

(e)

EXAMPLE 4.14

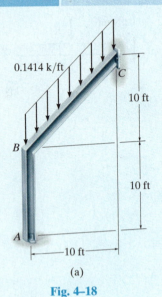

0.1414 k/ft

C

10 ft

B

10 ft

A

10 ft

(a)

Fig. 4–18

Draw the shear and moment diagrams for the frame shown in Fig. 4–18a. Assume A is a pin, C is a roller, and B is a fixed joint. Neglect the thickness of the members.

SOLUTION

Notice that the distributed load acts over a length of $10 \text{ ft } \sqrt{2} = 14.14 \text{ ft}$. The reactions on the entire frame are calculated and shown on its free-body diagram, Fig. 4–18b. From this diagram the free-body diagrams of each member are drawn, Fig. 4–18c. The distributed loading on BC has components along BC and perpendicular to its axis of $(0.1414 \text{ k/ft}) \cos 45° = (0.1414 \text{ k/ft}) \sin 45° = 0.1 \text{ k/ft}$ as shown. Using these results, the shear and moment diagrams are also shown in Fig. 4–18c.

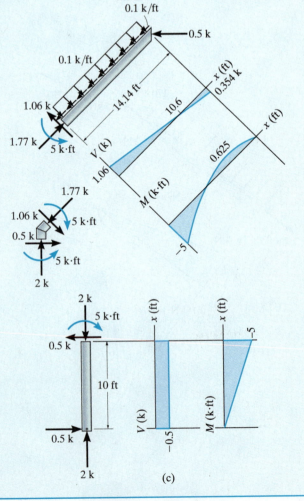

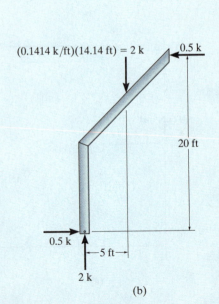

(0.1414 k/ft)(14.14 ft) = 2 k

0.5 k

20 ft

0.5 k

5 ft

2 k

(b)

(c)

4.5   Moment Diagrams Constructed by the Method of Superposition

Since beams are used primarily to resist bending stress, it is important that the moment diagram accompany the solution for their design. In Sec. 4.3 the moment diagram was constructed by *first* drawing the shear diagram. If we use the principle of superposition, however, each of the loads on the beam can be treated separately and the moment diagram can then be constructed in a series of parts rather than a single and sometimes complicated shape. It will be shown later in the text that this can be particularly advantageous when applying geometric deflection methods to determine both the deflection of a beam and the reactions on statically indeterminate beams.

Most loadings on beams in structural analysis will be a combination of the loadings shown in Fig. 4–19. Construction of the associated moment diagrams has been discussed in Example 4.8.

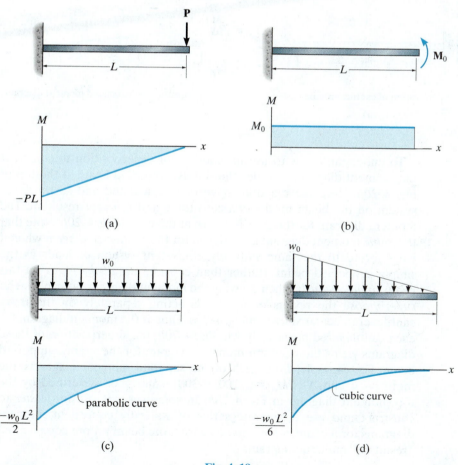

Fig. 4–19

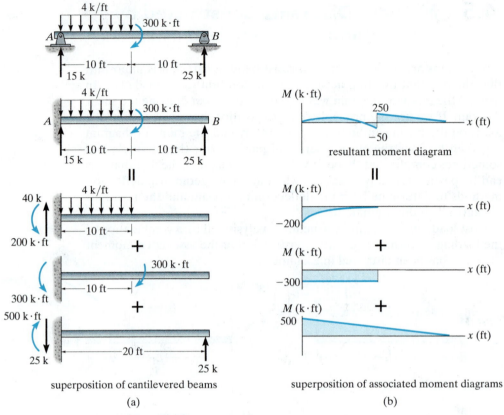

superposition of cantilevered beams

(a)

superposition of associated moment diagrams

(b)

Fig. 4–20

To understand how to use the method of superposition to construct the moment diagram consider the simply supported beam at the top of Fig. 4–20a. Here the reactions have been calculated and so the force system on the beam produces a zero force and moment resultant. The moment diagram for this case is shown at the top of Fig. 4–20b. Note that this *same* moment diagram is produced for the *cantilevered beam* when it is subjected to the same statically equivalent system of loads as the simply supported beam. Rather than considering *all the loads* on this beam *simultaneously* when drawing the moment diagram, we can instead *superimpose* the results of the loads acting separately on the three cantilevered beams shown in Fig. 4–20a. Thus, if the moment diagram for each cantilevered beam is drawn, Fig. 4–20b, the superposition of these diagrams yields the resultant moment diagram for the simply supported beam. For example, from each of the separate moment diagrams, the moment at end A is $M_A = -200 - 300 + 500 = 0$, as verified by the top moment diagram in Fig. 4–20b. In some cases it is often *easier* to construct and use a separate series of statically equivalent moment diagrams for a beam, *rather than* construct the beam's more complicated "resultant" moment diagram.

In a similar manner, we can also simplify construction of the "resultant" moment diagram for a beam by using a superposition of "simply supported" beams. For example, the loading on the beam shown at the top of Fig. 4–21a is equivalent to the beam loadings shown below it. Consequently, the separate moment diagrams for each of these three beams can be used *rather than* drawing the resultant moment diagram shown in Fig. 4–21b.

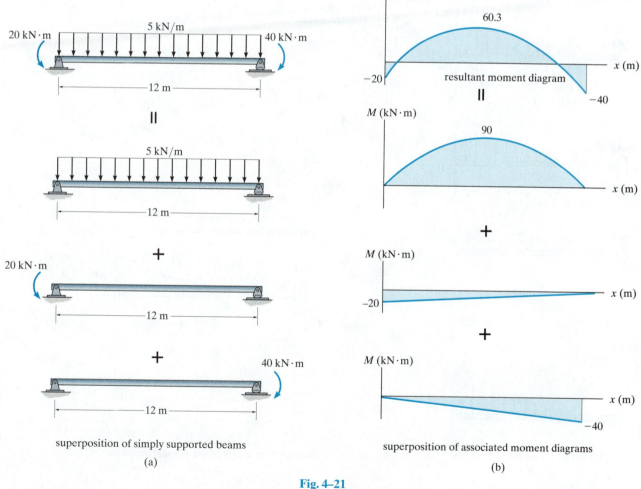

superposition of simply supported beams

(a)

superposition of associated moment diagrams

(b)

Fig. 4–21

EXAMPLE   4.15

Draw the moment diagrams for the beam shown at the top of Fig. 4–22*a* using the method of superposition. Consider the beam to be cantilevered from the support at *B*.

SOLUTION

If the beam were supported as a cantilever from *B*, it would be subjected to the statically equivalent loadings shown in Fig. 4–22*a*. The superimposed three cantilevered beams are shown below it together with their associated moment diagrams in Fig. 4–22*b*. (As an aid to their construction, refer to Fig. 4–19.) Although *not needed here,* the sum of these diagrams will yield the resultant moment diagram for the beam. For practice, try drawing this diagram and check the results.

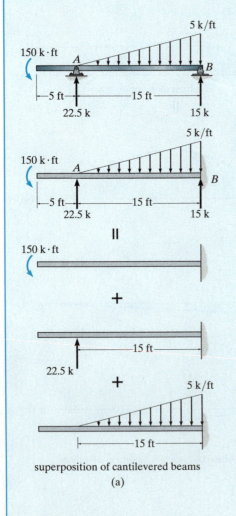

superposition of cantilevered beams
(a)

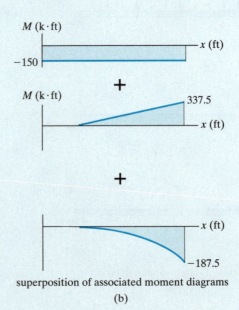

superposition of associated moment diagrams
(b)

Fig. 4–22

EXAMPLE 4.16

Draw the moment diagrams for the beam shown at the top of Fig. 4–23a using the method of superposition. Consider the beam to be cantilevered from the pin at A.

SOLUTION

The superimposed cantilevered beams are shown in Fig. 4–23a together with their associated moment diagrams, Fig. 4–23b. Notice that the reaction at the pin (22.5 k) is not considered since it produces no moment diagram. As an exercise verify that the resultant moment diagram is given at the top of Fig. 4–23b.

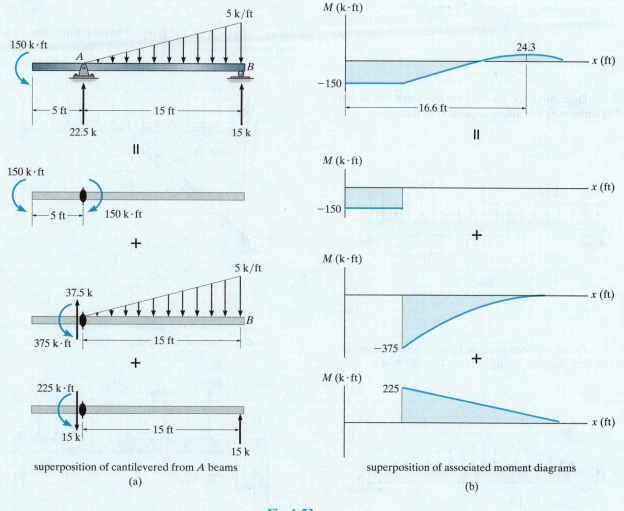

superposition of cantilevered from A beams

(a)

superposition of associated moment diagrams

(b)

Fig. 4–23

FUNDAMENTAL PROBLEMS

F4–1.   Determine the internal normal force, shear force, and bending moment acting at point C in the beam.

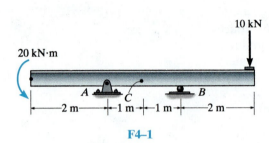

F4–1

F4–2.   Determine the internal normal force, shear force, and bending moment acting at point C in the beam.

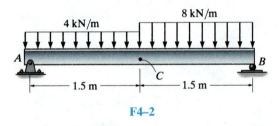

F4–2

F4–3.   Determine the internal normal force, shear force, and bending moment acting at point C in the beam.

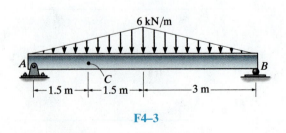

F4–3

F4–4.   Determine the internal normal force, shear force, and bending moment acting at point C in the beam.

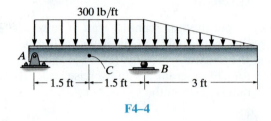

F4–4

F4–5.   Determine the internal normal force, shear force, and bending moment acting at point C in the beam.

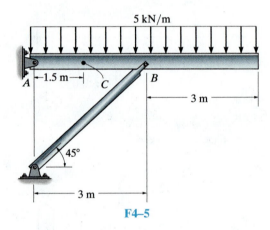

F4–5

F4–6.   Determine the internal normal force, shear force, and bending moment acting at point C in the beam.

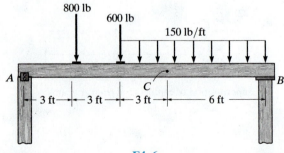

F4–6

F4–7. Determine the internal shear and moment in the beam as a function of x.

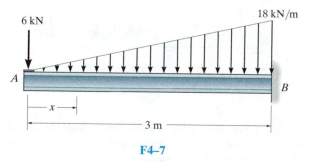

F4–7

F4–8. Determine the internal shear and moment in the beam as a function of x.

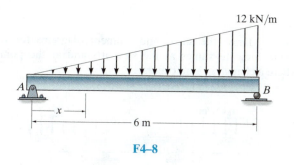

F4–8

F4–9. Determine the internal shear and moment in the beam as a function of x throughout the beam.

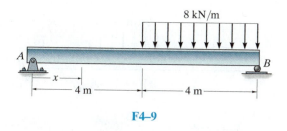

F4–9

F4–10. Determine the internal shear and moment in the beam as a function of x throughout the beam.

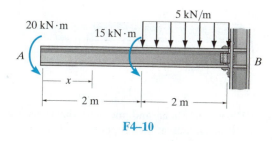

F4–10

F4–11. Determine the internal shear and moment in the beam as a function of x throughout the beam.

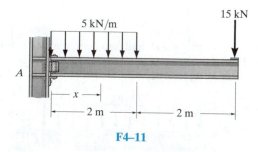

F4–11

F4–12. Determine the internal shear and moment in the beam as a function of x throughout the beam.

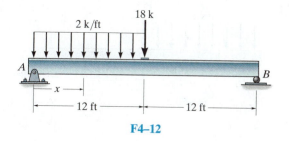

F4–12

4

F4–13. Draw the shear and moment diagrams for the beam. Indicate values at the supports and at the points where a change in load occurs.

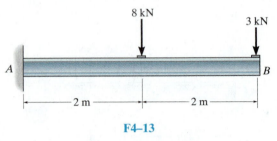

8 kN

3 kN

A   B

2 m   2 m

F4–13

F4–14. Draw the shear and moment diagrams for the beam. Indicate values at the supports and at the points where a change in load occurs.

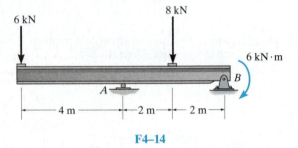

6 kN

8 kN

6 kN·m

A   B

4 m   2 m   2 m

F4–14

F4–15. Draw the shear and moment diagrams for the beam. Indicate values at the supports and at the points where a change in load occurs.

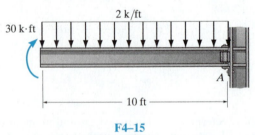

2 k/ft

30 k·ft

A

10 ft

F4–15

F4–16. Draw the shear and moment diagrams for the beam. Indicate values at the supports and at the points where a change in load occurs.

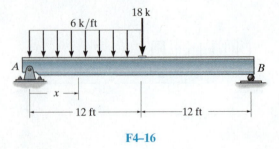

18 k

6 k/ft

A   B

x

12 ft   12 ft

F4–16

F4–17. Draw the shear and moment diagrams for the beam. Indicate values at the supports and at the points where a change in load occurs.

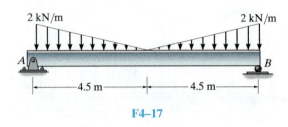

2 kN/m   2 kN/m

A   B

4.5 m   4.5 m

F4–17

F4–18. Draw the shear and moment diagrams for the beam. Indicate values at the supports and at the points where a change in load occurs.

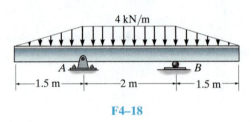

4 kN/m

A   B

1.5 m   2 m   1.5 m

F4–18

F4–19. Draw the shear and moment diagrams for the beam. Indicate values at the supports and at the points where a change in load occurs.

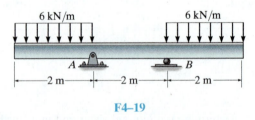

6 kN/m   6 kN/m

A   B

2 m   2 m   2 m

F4–19

F4–20. Draw the shear and moment diagrams for the beam. Indicate values at the supports and at the points where a change in load occurs.

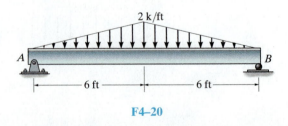

2 k/ft

A   B

6 ft   6 ft

F4–20

F4–21. Draw the moment diagrams for the frame. Assume the frame is pin connected at *B*.

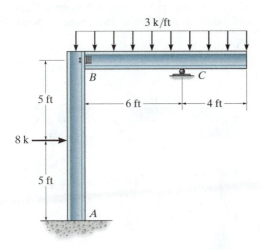

3 k/ft

B

C

5 ft

6 ft

4 ft

8 k

5 ft

A

F4–21

F4–23. Draw the moment diagrams for the frame. Assume the frame is pinned at *C* and the members are fixed connected at *B*.

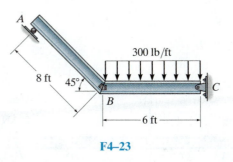

A

300 lb/ft

8 ft   45°

B

C

6 ft

F4–23

F4–22. Draw the moment diagrams for the frame. Assume the frame is pin connected at *A*, *B*, and *C* and fixed connected at *E* and *D*.

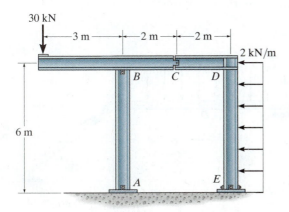

30 kN

3 m   2 m   2 m

2 kN/m

B   *C*   *D*

6 m

A   *E*

F4–22

F4–24. Draw the moment diagrams for the frame. Assume the frame is pin connected at *A*, *B*, and *C*.

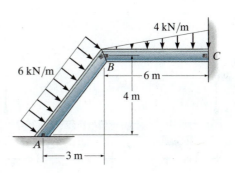

4 kN/m

C

6 kN/m   *B*

6 m

4 m

A

3 m

F4–24

PROBLEMS

4

Sec. 4.1

4–1. Determine the internal normal force, shear force, and bending moment in the beam at points C and D. Assume the support at A is a pin and B is a roller.

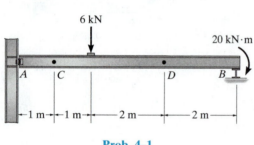

Prob. 4–1

4–2. Determine the internal normal force, shear force, and bending moment at point C.

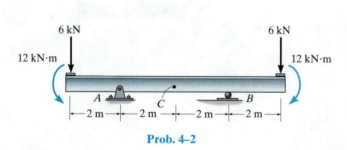

Prob. 4–2

4–3. The boom DF of the jib crane and the column DE have a uniform weight of 50 lb/ft. If the hoist and load weigh 300 lb, determine the internal normal force, shear force, and bending moment in the crane at points A, B, and C.

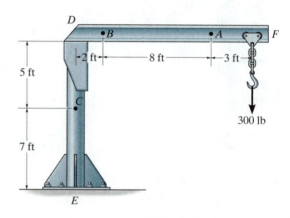

Prob. 4–3

*4–4. Determine the internal normal force, shear force, and bending moment at point D. Take $w = 150 \text{ N/m}$.

4–5. The beam AB will fail if the maximum internal moment at D reaches $800 \text{ N} \cdot \text{m}$ or the normal force in member BC becomes 1500 N. Determine the largest load w it can support.

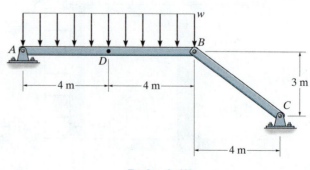

Probs. 4–4/5

4–6. Determine the internal normal force, shear force, and bending moment in the beam at points C and D. Assume the support at A is a roller and B is a pin.

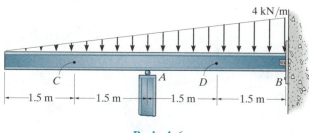

Prob. 4–6

4–7. Determine the internal normal force, shear force, and bending moment acting at point C, located just to the right of the 12-kN force.

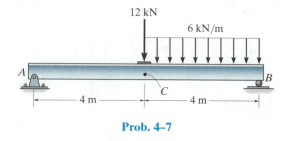

Prob. 4–7

*4–8. Determine the internal normal force, shear force, and bending moment in the beam at points C and D. Assume the support at A is a roller and B is a pin.

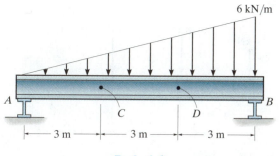

Prob. 4–8

4–9. Determine the internal normal force, shear force, and bending moment at point C.

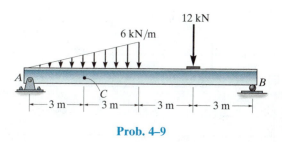

Prob. 4–9

4–10. Determine the internal normal force, shear force, and bending moment in the beam at points B and C. The support at A is a roller and D is pinned.

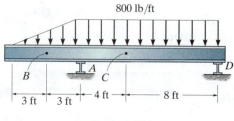

Prob. 4–10

4–11. Determine the internal normal force, shear force, and bending moment in the beam at point C, located just to the left of the 800-lb force.

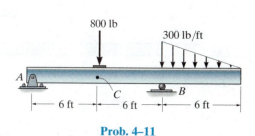

Prob. 4–11

Sec. 4.2

*4–12. Determine the shear and moment throughout the beam as a function of x.

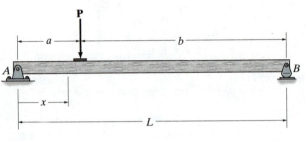

Prob. 4–12

4–13. Draw the shear and moment diagrams for the beam. Also, express the shear and moment in the beam as a function of x within the region $2\text{ m} < x < 10\text{ m}$.

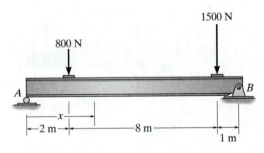

Prob. 4–13

4–14. Determine the shear and moment throughout the beam as a function of x.

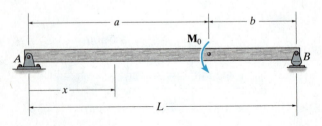

Prob. 4–14

4–15. Determine the shear and moment in the beam as a function of x, where $2\text{ m} < x < 4\text{ m}$.

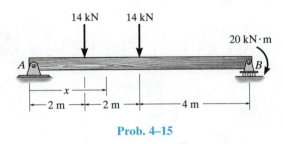

Prob. 4–15

*4–16. Determine the shear and moment throughout the beam as a function of x.

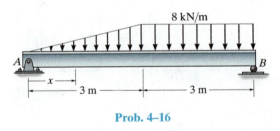

Prob. 4–16

4–17. Determine the shear and moment throughout the beam as a function of x.

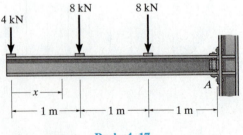

Prob. 4–17

4–18. Determine the shear and moment throughout the beam as functions of x.

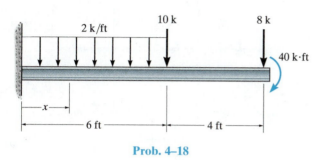

Prob. 4–18

4–19. Determine the shear and moment throughout the beam as functions of x.

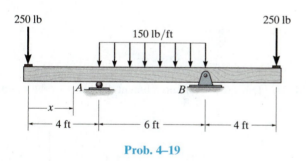

Prob. 4–19

*4–20. Determine the shear and moment in the beam as functions of x.

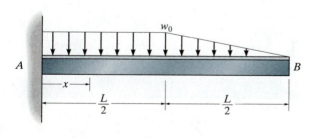

Prob. 4–20

4–21. Determine the shear and moment in the beam as a function of x.

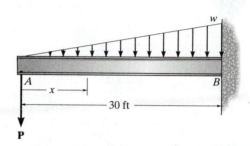

Prob. 4–21

4–22. Determine the shear and moment throughout the beam as functions of x.

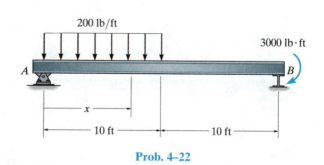

Prob. 4–22

Sec. 4.3

4–23. Draw the shear and moment diagrams for the beam.

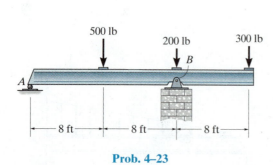

Prob. 4–23

*4–24. Draw the shear and moment diagrams for the beam.

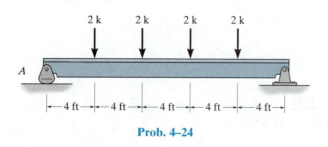

Prob. 4–24

4–25. Draw the shear and moment diagrams for the beam.

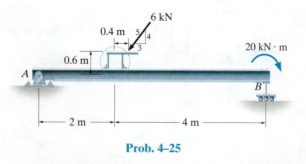

Prob. 4–25

4–26. Draw the shear and moment diagrams for the beam. Assume the support at A is a roller.

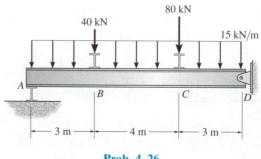

Prob. 4–26

4–27. Draw the shear and moment diagrams for the beam.

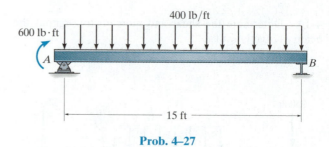

Prob. 4–27

*4–28. Draw the shear and moment diagrams for the beam.

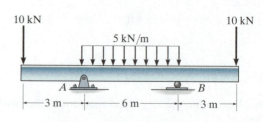

Prob. 4–28

4–29. Draw the shear and moment diagrams for the beam.

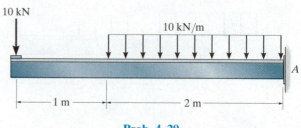

Prob. 4–29

4–30. Draw the shear and moment diagrams for the beam.

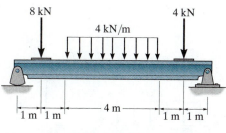

8 kN   4 kN
4 kN/m
1 m 1 m   4 m   1 m 1 m

Prob. 4–30

4–31. Draw the shear and moment diagrams for the compound beam.

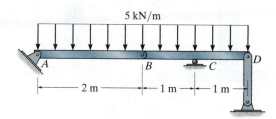

5 kN/m
A   B   C   D
2 m   1 m   1 m

Prob. 4–31

*4–32. Draw the shear and moment diagrams for the beam.

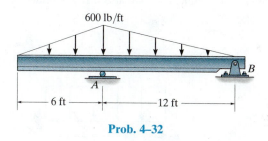

600 lb/ft
A   B
6 ft   12 ft

Prob. 4–32

4–33. Draw the shear and moment diagrams for the beam.

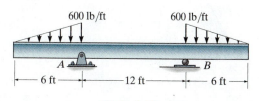

600 lb/ft   600 lb/ft
A   B
6 ft   12 ft   6 ft

Prob. 4–33

4–34. Draw the shear and moment diagrams for the beam.

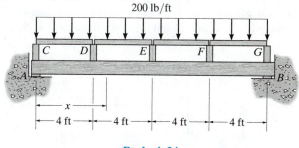

200 lb/ft
C   D   E   F   G
A   B
4 ft   4 ft   4 ft   4 ft
x

Prob. 4–34

4–35. Draw the shear and moment diagrams for the compound beam.

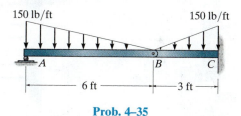

150 lb/ft   150 lb/ft
A   B   C
6 ft   3 ft

Prob. 4–35

*4–36. Draw the shear and moment diagrams for the beam.

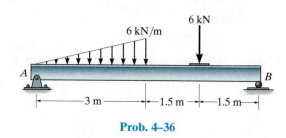

6 kN
6 kN/m
A   B
3 m   1.5 m   1.5 m

Prob. 4–36

4–37. Draw the shear and moment diagrams for the beam. Assume the support at B is a pin.

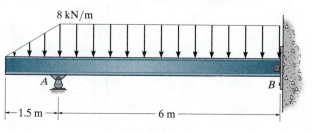

8 kN/m
A   B
1.5 m   6 m

Prob. 4–37

Sec. 4.4

4–38. Draw the shear and moment diagrams for each of the three members of the frame. Assume the frame is pin connected at $A, C,$ and D and there is a fixed joint at B.

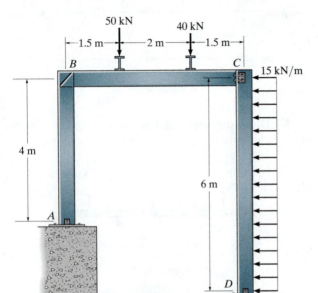

Prob. 4–38

4–39. Draw the shear and moment diagrams for each member of the frame.

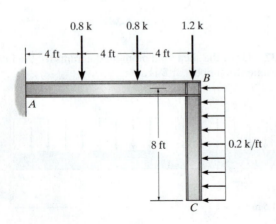

Prob. 4–39

*4–40. Draw the shear and moment diagrams for each member of the frame. Assume A is a rocker, and D is pinned.

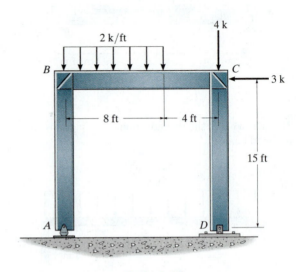

Prob. 4–40

4–41. Draw the shear and moment diagrams for each member of the frame. The joint at B is fixed connected.

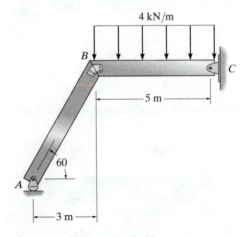

Prob. 4–41

4–42. Draw the shear and moment diagrams for each member of the frame. Assume A is fixed, the joint at B is a pin, and support C is a roller.

*4–44.** Draw the shear and moment diagrams for each member of the frame. Assume the joints at A, B, and C are pin connected.

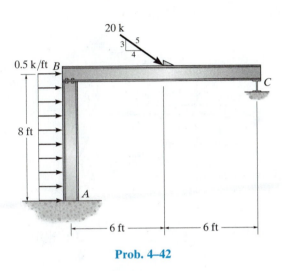

Prob. 4–42

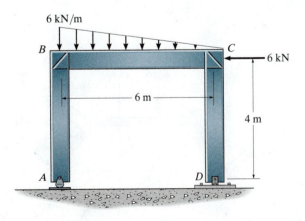

Prob. 4–44

4–43. Draw the shear and moment diagrams for each member of the frame. Assume the frame is roller supported at A and pin supported at C.

4–45. Draw the shear and moment diagrams for each member of the frame.

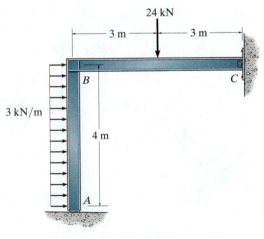

Prob. 4–43

Prob. 4–45

4

4–46. Draw the shear and moment diagrams for each member of the frame. Assume joints B and C are fixed connected.

*4–48. Leg BC on the framework can be designed to extend either outward as shown, or inward with the support C positioned below the center 2-k load. Draw the moment diagrams for the frame in each case, to make a comparison of the two designs.

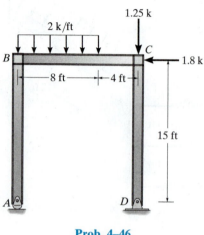

Prob. 4–46

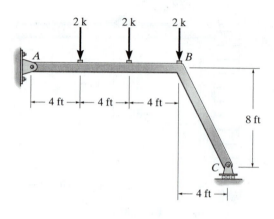

Prob. 4–48

4–49. Draw the shear and moment diagrams for each member of the frame. The members are pin connected at A, B, and C.

4–47. Draw the shear and moment diagrams for each member of the frame.

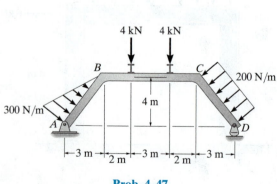

Prob. 4–47

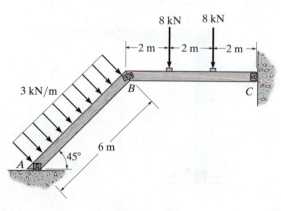

Prob. 4–49

4

Sec. 4.5

4–50. Draw the moment diagrams for the beam using the method of superposition. The beam is cantilevered from A.

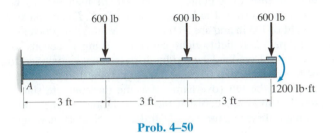

Prob. 4–50

4–51. Draw the moment diagrams for the beam using the method of superposition.

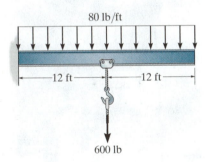

Prob. 4–51

*4–52. Draw the moment diagrams for the beam using the method of superposition. Consider the beam to be simply supported. Assume A is a pin and B is a roller.

4–53. Solve Prob. 4–52 by considering the beam to be cantilevered from the support at A.

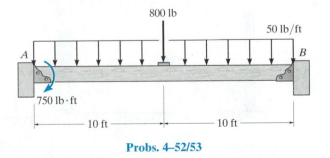

Probs. 4–52/53

4–54. Draw the moment diagrams for the beam using the method of superposition. Consider the beam to be cantilevered from the pin support at A.

4–55. Draw the moment diagrams for the beam using the method of superposition. Consider the beam to be cantilevered from the rocker at B.

*4–56. Draw the moment diagrams for beam using the method of superposition. Consider the beam to be cantilevered from end C.

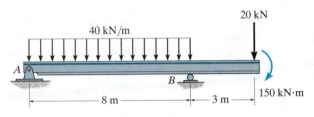

Probs. 4–54/55/56

4–57. Draw the moment diagrams for the beam using the method of superposition. Consider the beam to be cantilevered from end C.

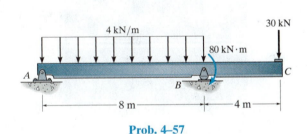

Prob. 4–57

4–58. Draw the moment diagrams for the beam using the method of superposition. Consider the beam to be cantilevered from the support at A.

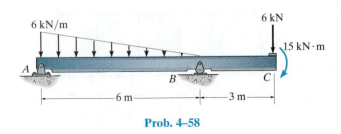

Prob. 4–58

PROJECT PROBLEMS

4–1P. The balcony located on the third floor of a motel is shown in the photo. It is constructed using a 4-in.-thick concrete (plain stone) slab which rests on the four simply supported floor beams, two cantilevered side girders *AB* and *HG*, and the front and rear girders. The idealized framing plan with average dimensions is shown in the adjacent figure. According to local code, the balcony live load is 45 psf. Draw the shear and moment diagrams for the front girder *BG* and a side girder *AB*. Assume the front girder is a channel that has a weight of 25 lb/ft and the side girders are wide flange sections that have a weight of 45 lb/ft. Neglect the weight of the floor beams and front railing. For this solution treat each of the five slabs as two-way slabs.

4–2P. The canopy shown in the photo provides shelter for the entrance of a building. Consider all members to be simply supported. The bar joists at *C, D, E, F* each have a weight of 135 lb and are 20 ft long. The roof is 4 in. thick and is to be plain lightweight concrete having a density of 102 lb/ft^3. Live load caused by drifting snow is assumed to be trapezoidal, with 60 psf at the right (against the wall) and 20 psf at the left (overhang). Assume the concrete slab is simply supported between the joists. Draw the shear and moment diagrams for the side girder *AB*. Neglect its weight.

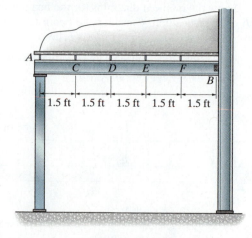

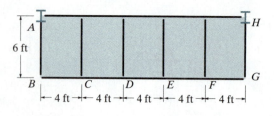

Prob. 4–1P

Prob. 4–2P

CHAPTER REVIEW

Structural members subjected to planar loads support an internal normal force **N**, shear force **V**, and bending moment **M**. To find these values at a specific point in a member, the method of sections must be used. This requires drawing a free-body diagram of a segment of the member, and then applying the three equations of equilibrium. Always show the three internal loadings on the section in their positive directions.

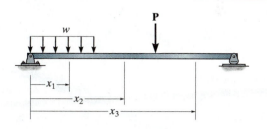

positive sign convention

The internal shear and moment can be expressed as a function of x along the member by establishing the origin at a fixed point (normally at the left end of the member, and then using the method of sections, where the section is made a distance x from the origin). For members subjected to several loads, different x coordinates must extend between the loads.

Shear and moment diagrams for structural members can be drawn by plotting the shear and moment functions. They also can be plotted using the two graphical relationships.

$$\frac{dV}{dx} = w$$

$$\left.\begin{array}{c}\text{Slope of}\\\text{Shear Diagram}\end{array}\right\} = \left\{\begin{array}{c}\text{Intensity of}\\\text{Distributed Load}\end{array}\right.$$

$$\frac{dM}{dx} = V$$

$$\left.\begin{array}{c}\text{Slope of}\\\text{Moment Diagram}\end{array}\right\} = \left\{\text{Shear}\right.$$

Note that a point of zero shear locates the point of maximum moment since $V = dM/dx = 0$.

A force acting downward on the beam will cause the shear diagram to jump downwards, and a counterclockwise couple moment will cause the moment diagram to jump downwards.

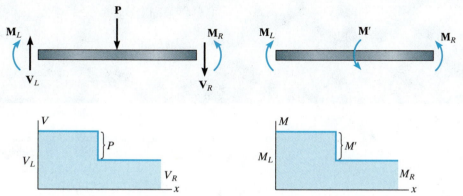

Using the method of superposition, the moment diagrams for a member can be represented by a series of simpler shapes. The shapes represent the moment diagram for each of the separate loadings. The resultant moment diagram is then the algebraic addition of the separate diagrams.

Chapter 5

© Oleksiy Maksymenko Photography/Alamy

This is an example of a parabolic through-arch bridge, because it the deck of the bridge runs through it.

Cables and Arches

Cables and arches often form the main load-carrying element in many types of structures, and in this chapter we will discuss some of the important aspects related to their structural analysis. The chapter begins with a general discussion of cables, followed by an analysis of cables subjected to a concentrated load and to a uniform distributed load. Since most arches are statically indeterminate, only the special case of a three-hinged arch will be considered. The analysis of this structure will provide some insight regarding the fundamental behavior of all arched structures.

5.1 Cables

Cables are often used in engineering structures for support and to transmit loads from one member to another. When used to support suspension roofs, bridges, and trolley wheels, cables form the main load-carrying element in the structure. In the force analysis of such systems, the weight of the cable itself may be neglected; however, when cables are used as guys for radio antennas, electrical transmission lines, and derricks, the cable weight may become important and must be included in the structural analysis. Two cases will be considered in the sections that follow: a cable subjected to concentrated loads and a cable subjected to a distributed load. Provided these loadings are coplanar with the cable, the requirements for equilibrium are formulated in an identical manner.

When deriving the necessary relations between the force in the cable and its slope, we will make the assumption that the cable is *perfectly flexible* and *inextensible*. Due to its flexibility, the cable offers no resistance to shear or bending and, therefore, the force acting in the cable is always tangent to the cable at points along its length. Being inextensible, the cable has a constant length both before and after the load is applied. As a result, once the load is applied, the geometry of the cable remains fixed, and the cable or a segment of it can be treated as a rigid body.

5.2   Cable Subjected to Concentrated Loads

When a cable of negligible weight supports several concentrated loads, the cable takes the form of several straight-line segments, each of which is subjected to a constant tensile force. Consider, for example, the cable shown in Fig. 5–1. Here θ specifies the angle of the cable's *cord AB*, and L is the cable's span. If the distances L_1, L_2, and L_3 and the loads $\mathbf{P}_1$ and $\mathbf{P}_2$ are known, then the problem is to determine the *nine unknowns consisting of the tension in each of the three* segments, the *four* components of reaction at A and B, and the sags y_C and y_D at the *two* points C and D. For the solution we can write *two* equations of force equilibrium at each of points A, B, C, and D. This results in a total of *eight equations*. To complete the solution, it will be necessary to know something about the geometry of the cable in order to obtain the necessary ninth equation. For example, if the cable's total *length* $\mathscr{L}$ is specified, then the Pythagorean theorem can be used to relate $\mathscr{L}$ to each of the three segmental lengths, written in terms of θ, y_C, y_D, L_1, L_2, and L_3. Unfortunately, this type of problem cannot be solved easily by hand. Another possibility, however, is to specify one of the sags, either y_C or y_D, instead of the cable length. By doing this, the equilibrium equations are then sufficient for obtaining the unknown forces and the remaining sag. Once the sag at each point of loading is obtained, $\mathscr{L}$ can then be determined by trigonometry.

When performing an equilibrium analysis for a problem of this type, the forces in the cable can also be obtained by writing the equations of equilibrium for the entire cable or any portion thereof. The following example numerically illustrates these concepts.

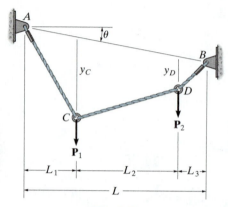

Fig. 5–1

The deck of this harp cable-stayed bridge is supported by a series of cables attached at various points along the deck and pylons.

EXAMPLE 5.1

Determine the tension in each segment of the cable shown in Fig. 5–2a. Also, what is the dimension h?

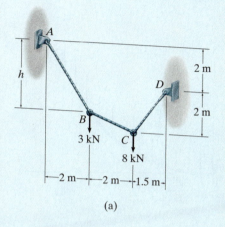

(a)

SOLUTION

By inspection, there are four unknown external reactions (A_x, A_y, D_x, and D_y) and three unknown cable tensions, one in each cable segment. These seven unknowns along with the sag h can be determined from the eight available equilibrium equations ($\Sigma F_x = 0$, $\Sigma F_y = 0$) applied to points A through D.

A more direct approach to the solution is to recognize that the slope of cable CD is specified, and so a free-body diagram of the entire cable is shown in Fig. 5–2b. We can obtain the tension in segment CD as follows:

$$\zeta + \Sigma M_A = 0;$$

$$T_{CD}(3/5)(2 \text{ m}) + T_{CD}(4/5)(5.5 \text{ m}) - 3 \text{ kN}(2 \text{ m}) - 8 \text{ kN}(4 \text{ m}) = 0$$

$$T_{CD} = 6.79 \text{ kN} \qquad \textit{Ans.}$$

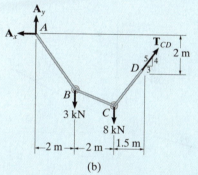

(b)

Now we can analyze the equilibrium of points C and B in sequence. Point C (Fig. 5–2c);

$$\xrightarrow{+} \Sigma F_x = 0; \quad 6.79 \text{ kN} (3/5) - T_{BC} \cos \theta_{BC} = 0$$

$$+\uparrow \Sigma F_y = 0; \quad 6.79 \text{ kN} (4/5) - 8 \text{ kN} + T_{BC} \sin \theta_{BC} = 0$$

$$\theta_{BC} = 32.3° \qquad T_{BC} = 4.82 \text{ kN} \qquad \textit{Ans.}$$

Point B (Fig. 5–2d);

$$\xrightarrow{+} \Sigma F_x = 0; \quad -T_{BA} \cos \theta_{BA} + 4.82 \text{ kN} \cos 32.3° = 0$$

$$+\uparrow \Sigma F_y = 0; \quad T_{BA} \sin \theta_{BA} - 4.82 \text{ kN} \sin 32.3° - 3 \text{ kN} = 0$$

$$\theta_{BA} = 53.8° \qquad T_{BA} = 6.90 \text{ kN} \qquad \textit{Ans.}$$

Hence, from Fig. 5–2a,

$$h = (2 \text{ m}) \tan 53.8° = 2.74 \text{ m} \qquad \textit{Ans.}$$

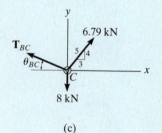

(c)

(d)

Fig. 5–2

5

5.3   Cable Subjected to a Uniform Distributed Load

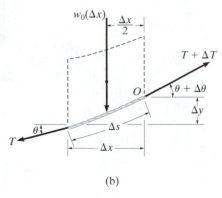

(a)

(b)

Fig. 5–3

Cables provide a very effective means of supporting the dead weight of girders or bridge decks having very long spans. A suspension bridge is a typical example, in which the deck is suspended from the cable using a series of close and equally spaced hangers.

In order to analyze this problem, we will first determine the shape of a cable subjected to a uniform *horizontally* distributed vertical load w_0, Fig. 5–3a. Here the x, y axes have their origin located at the lowest point on the cable, such that the slope is zero at this point. The free-body diagram of a small segment of the cable having a length Δs is shown in Fig. 5–3b. Since the tensile force in the cable changes continuously in both magnitude and direction along the cable's length, this change is denoted on the free-body diagram by ΔT. The distributed load is represented by its resultant force $w_0 \Delta x$, which acts at $\Delta x/2$ from point O. Applying the equations of equilibrium yields

$$\xrightarrow{+} \Sigma F_x = 0; \quad -T \cos \theta + (T + \Delta T) \cos(\theta + \Delta \theta) = 0$$

$$+\uparrow \Sigma F_y = 0; \quad -T \sin \theta - w_0(\Delta x) + (T + \Delta T) \sin(\theta + \Delta \theta) = 0$$

$$\zeta + \Sigma M_O = 0; \quad w_0(\Delta x)(\Delta x/2) - T \cos \theta \, \Delta y + T \sin \theta \, \Delta x = 0$$

Dividing each of these equations by Δx and taking the limit as $\Delta x \to 0$, and hence $\Delta y \to 0$, $\Delta \theta \to 0$, and $\Delta T \to 0$, we obtain

$$\frac{d(T \cos \theta)}{dx} = 0 \qquad (5-1)$$

$$\frac{d(T \sin \theta)}{dx} = w_0 \qquad (5-2)$$

$$\frac{dy}{dx} = \tan \theta \qquad (5-3)$$

Integrating Eq. 5–1, where $T = F_H$ at $x = 0$, we have:

$$T \cos \theta = F_H \qquad (5-4)$$

which indicates the horizontal component of force at *any point* along the cable remains *constant*.

Integrating Eq. 5–2, realizing that $T \sin \theta = 0$ at $x = 0$, gives

$$T \sin \theta = w_0 x \qquad (5-5)$$

Dividing Eq. 5–5 by Eq. 5–4 eliminates T. Then using Eq. 5–3, we can obtain the slope at any point,

$$\tan \theta = \frac{dy}{dx} = \frac{w_0 x}{F_H} \qquad (5-6)$$

Performing a second integration with $y = 0$ at $x = 0$ yields

$$y = \frac{w_0}{2F_H}x^2 \tag{5–7}$$

This is the equation of a *parabola*. The constant F_H may be obtained by using the boundary condition $y = h$ at $x = L$. Thus,

$$F_H = \frac{w_0 L^2}{2h} \tag{5–8}$$

Finally, substituting into Eq. 5–7 yields

$$y = \frac{h}{L^2}x^2 \tag{5–9}$$

From Eq. 5–4, the maximum tension in the cable occurs when θ is maximum; i.e., at $x = L$. Hence, from Eqs. 5–4 and 5–5,

$$T_{\max} = \sqrt{F_H^2 + (w_0 L)^2} \tag{5–10}$$

Or, using Eq. 5–8, we can express $T_{\max}$ in terms of w_0, i.e.,

$$T_{\max} = w_0 L \sqrt{1 + (L/2h)^2} \tag{5–11}$$

Realize that we have neglected the weight of the cable, which is *uniform* along the *length* of the cable, and not along its horizontal projection. Actually, a cable subjected to its own weight and free of any other loads will take the form of a ***catenary curve***. However, if the sag-to-span ratio is small, which is the case for most structural applications, this curve closely approximates a parabolic shape, as determined here.

From the results of this analysis, it follows that a cable will *maintain a parabolic shape,* provided the dead load of the deck for a suspension bridge or a suspended girder will be *uniformly distributed* over the horizontal projected length of the cable. Hence, if the girder in Fig. 5–4a is supported by a series of hangers, which are close and uniformly spaced, the load in each hanger must be the *same* so as to ensure that the cable has a parabolic shape.

Using this assumption, we can perform the structural analysis of the girder or any other framework which is freely suspended from the cable. In particular, if the girder is simply supported as well as supported by the cable, the analysis will be statically indeterminate to the first degree, Fig. 5–4b. However, if the girder has an internal pin at some intermediate point along its length, Fig. 5–4c, then this provides a condition of zero moment, and so a determinate structural analysis of the girder can be performed.

The Verrazano-Narrows Bridge at the entrance to New York Harbor has a main span of 4260 ft (1.30 km).

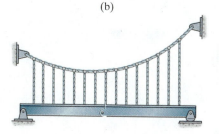

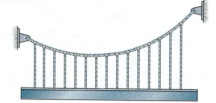

(a)

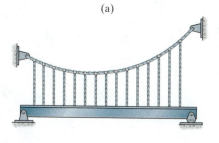

(b)

(c)

Fig. 5–4

EXAMPLE | 5.2

The cable in Fig. 5–5a supports a girder which weighs 850 lb/ft. Determine the tension in the cable at points A, B, and C.

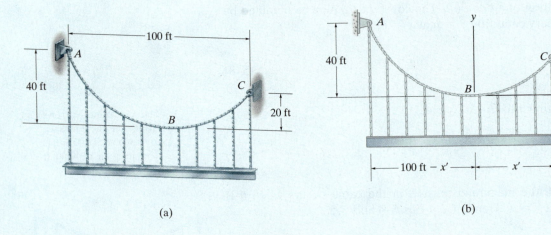

(a)   (b)

Fig. 5–5

SOLUTION

The origin of the coordinate axes is established at point B, the lowest point on the cable, where the slope is zero, Fig. 5–5b. From Eq. 5–7, the parabolic equation for the cable is:

$$y = \frac{w_0}{2F_H}x^2 = \frac{850 \text{ lb/ft}}{2F_H}x^2 = \frac{425}{F_H}x^2 \qquad (1)$$

Assuming point C is located x' from B, we have

$$20 = \frac{425}{F_H}x'^2$$

$$F_H = 21.25x'^2 \qquad (2)$$

Also, for point A,

$$40 = \frac{425}{F_H}[-(100 - x')]^2$$

$$40 = \frac{425}{21.25x'^2}[-(100 - x')]^2$$

$$x'^2 + 200x' - 10\,000 = 0$$

$$x' = 41.42 \text{ ft}$$

Thus, from Eqs. 2 and 1 (or Eq. 5–6) we have

$$F_H = 21.25(41.42)^2 = 36\ 459.2 \text{ lb}$$

$$\frac{dy}{dx} = \frac{850}{36\ 459.2}x = 0.02331x \qquad\qquad (3)$$

At point A,

$$x = -(100 - 41.42) = -58.58 \text{ ft}$$

$$\tan \theta_A = \left.\frac{dy}{dx}\right|_{x=-58.58} = 0.02331(-58.58) = -1.366$$

$$\theta_A = -53.79°$$

Using Eq. 5–4,

$$T_A = \frac{F_H}{\cos \theta_A} = \frac{36\ 459.2}{\cos(-53.79°)} = 61.7 \text{ k} \qquad\qquad Ans.$$

At point B, $x = 0$,

$$\tan \theta_B = \left.\frac{dy}{dx}\right|_{x=0} = 0, \qquad \theta_B = 0°$$

$$T_B = \frac{F_H}{\cos \theta_B} = \frac{36\ 459.2}{\cos 0°} = 36.5 \text{ k} \qquad\qquad Ans.$$

At point C,

$$x = 41.42 \text{ ft}$$

$$\tan \theta_C = \left.\frac{dy}{dx}\right|_{x=41.42} = 0.02331(41.42) = 0.9657$$

$$\theta_C = 44.0°$$

$$T_C = \frac{F_H}{\cos \theta_C} = \frac{36\ 459.2}{\cos 44.0°} = 50.7 \text{ k} \qquad\qquad Ans.$$

EXAMPLE 5.3

The suspension bridge in Fig. 5–6a is constructed using the two stiffening trusses that are pin connected at their ends C and supported by a pin at A and a rocker at B. Determine the maximum tension in the cable IH. The cable has a parabolic shape and the bridge is subjected to the single load of 50 kN.

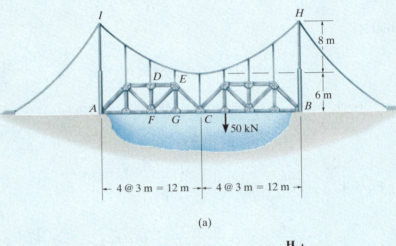

(a)

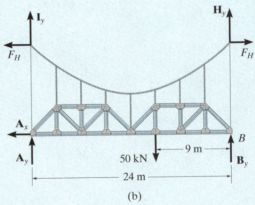

(b)

Fig. 5–6

SOLUTION

The free-body diagram of the cable-truss system is shown in Fig. 5–6b. According to Eq. 5–4 ($T \cos \theta = F_H$), the horizontal component of cable tension at I and H must be constant, F_H. Taking moments about B, we have

$$\zeta + \Sigma M_B = 0; \quad -I_y(24 \text{ m}) - A_y(24 \text{ m}) + 50 \text{ kN}(9 \text{ m}) = 0$$

$$I_y + A_y = 18.75$$

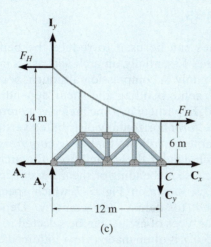

(c)

If only half the suspended structure is considered, Fig. 5–6c, then summing moments about the pin at C, we have

$$\zeta + \Sigma M_C = 0; \quad F_H(14 \text{ m}) - F_H(6 \text{ m}) - I_y(12 \text{ m}) - A_y(12 \text{ m}) = 0$$

$$I_y + A_y = 0.667F_H$$

From these two equations,

$$18.75 = 0.667F_H$$

$$F_H = 28.125 \text{ kN}$$

To obtain the maximum tension in the cable, we will use Eq. 5–11, but first it is necessary to determine the value of an assumed uniform distributed loading w_0 from Eq. 5–8:

$$w_0 = \frac{2F_H h}{L^2} = \frac{2(28.125 \text{ kN})(8 \text{ m})}{(12 \text{ m})^2} = 3.125 \text{ kN/m}$$

Thus, using Eq. 5–11, we have

$$T_{max} = w_0 L \sqrt{1 + (L/2h)^2}$$

$$= 3.125(12 \text{ m}) \sqrt{1 + (12 \text{ m}/2(8 \text{ m}))^2}$$

$$= 46.9 \text{ kN} \qquad\qquad\qquad Ans.$$

5.4   Arches

Like cables, arches can be used to reduce the bending moments in long-span structures. Essentially, an arch acts as an inverted cable, so it receives its load mainly in compression although, because of its rigidity, it must also resist some bending and shear depending upon how it is loaded and shaped. In particular, if the arch has a ***parabolic shape*** and it is subjected to a *uniform* horizontally distributed vertical load, then from the analysis of cables it follows that *only compressive forces* will be resisted by the arch. Under these conditions the arch shape is called a ***funicular arch*** because no bending or shear forces occur within the arch.

    A typical arch is shown in Fig. 5–7, which specifies some of the nomenclature used to define its geometry. Depending upon the application, several types of arches can be selected to support a loading. A ***fixed arch***, Fig. 5–8a, is often made from reinforced concrete. Although it may require less material to construct than other types of arches, it must have solid foundation abutments since it is indeterminate to the third degree and, consequently, additional stresses can be introduced into the arch due to relative settlement of its supports. A ***two-hinged arch***, Fig. 5–8b, is commonly made from metal or timber. It is indeterminate to the first degree, and although it is not as rigid as a fixed arch, it is somewhat insensitive to settlement. We could make this structure statically determinate by replacing one of the hinges with a roller. Doing so, however, would remove the capacity of the structure to resist bending along its span, and as a result it would serve as a curved beam, and *not* as an arch. A ***three-hinged arch***, Fig. 5–8c, which is also made from metal or timber, is statically determinate. Unlike statically indeterminate arches, it is not affected by settlement or temperature changes. Finally, if two-and three-hinged arches are to be constructed without the need for larger foundation abutments and if clearance is not a problem, then the supports can be connected with a tie rod, Fig. 5–8d. A ***tied arch*** allows the structure to behave as a rigid unit, since the tie rod carries the horizontal component of thrust at the supports. It is also unaffected by relative settlement of the supports.

extrados (or back) — crown
springline
intrados (or soffit)
centerline
rise
haunch
abutment

Fig. 5–7

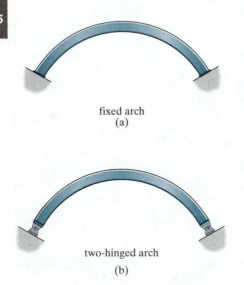

fixed arch
(a)

two-hinged arch
(b)

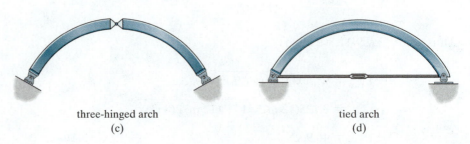

three-hinged arch
(c)

tied arch
(d)

Fig. 5–8

5.5   Three-Hinged Arch

To provide some insight as to how arches transmit loads, we will now consider the analysis of a three-hinged arch such as the one shown in Fig. 5–9a. In this case, the third hinge is located at the crown and the supports are located at different elevations. In order to determine the reactions at the supports, the arch is disassembled and the free-body diagram of each member is shown in Fig. 5–9b. Here there are six unknowns for which six equations of equilibrium are available. One method of solving this problem is to apply the moment equilibrium equations about points A and B. Simultaneous solution will yield the reactions C_x and C_y. The support reactions are then determined from the force equations of equilibrium. Once obtained, the internal normal force, shear, and moment loadings at any point along the arch can be found using the method of sections. Here, of course, the section should be taken perpendicular to the axis of the arch at the point considered. For example, the free-body diagram for segment AD is shown in Fig. 5–9c.

Three-hinged arches can also take the form of two pin-connected trusses, each of which would replace the arch ribs AC and CB in Fig. 5–9a. The analysis of this form follows the same procedure outlined above. The following examples numerically illustrate these concepts.

This three-hinge truss arch is used to support a pedestrian walkway. This bridge is referred to as a "through-arch bridge."

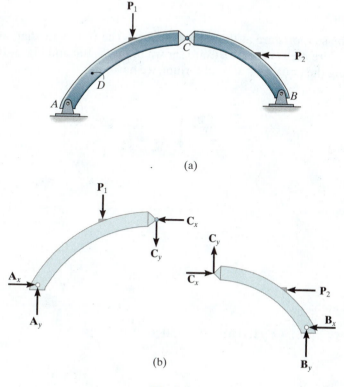

(a)

(b)

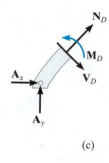

(c)

Fig. 5–9

EXAMPLE 5.4

The three-hinged open-spandrel deck arch bridge like the one shown in the photo has a parabolic shape. If this arch were to support a uniform load and have the dimensions shown in Fig. 5–10a, show that the arch is subjected *only to axial compression* at any intermediate point such as point D. Assume the load is uniformly transmitted to the arch ribs.

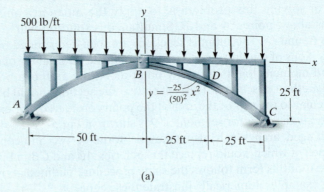

(a)

Fig. 5–10

SOLUTION

Here the supports are at the same elevation. The free-body diagrams of the entire arch and part BC are shown in Fig. 5–10b and Fig. 5–10c. Applying the equations of equilibrium, we have:

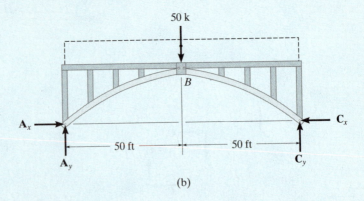

(b)

Entire arch:

$$\zeta + \Sigma M_A = 0; \quad C_y(100 \text{ ft}) - 50 \text{ k}(50 \text{ ft}) = 0$$

$$C_y = 25 \text{ k}$$

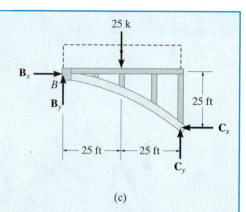

(c)

Arch segment BC:

$$\zeta + \Sigma M_B = 0; \quad -25 \text{ k}(25 \text{ ft}) + 25 \text{ k}(50 \text{ ft}) - C_x(25 \text{ ft}) = 0$$

$$C_x = 25 \text{ k}$$

$$\xrightarrow{+} \Sigma F_x = 0; \quad B_x = 25 \text{ k}$$

$$+\uparrow \Sigma F_y = 0; \quad B_y - 25 \text{ k} + 25 \text{ k} = 0$$

$$B_y = 0$$

A section of the arch taken through point D, $x = 25$ ft, $y = -25(25)^2/(50)^2 = -6.25$ ft, is shown in Fig. 5–10d. The slope of the segment at D is

$$\tan \theta = \frac{dy}{dx} = \frac{-50}{(50)^2} x \bigg|_{x=25 \text{ ft}} = -0.5$$

$$\theta = -26.6°$$

Applying the equations of equilibrium, Fig. 5–10d we have

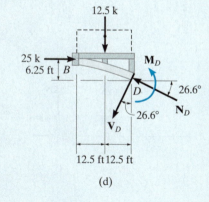

(d)

$$\xrightarrow{+} \Sigma F_x = 0; \quad 25 \text{ k} - N_D \cos 26.6° - V_D \sin 26.6° = 0$$

$$+\uparrow \Sigma F_y = 0; \quad -12.5 \text{ k} + N_D \sin 26.6° - V_D \cos 26.6° = 0$$

$$\zeta + \Sigma M_D = 0; \quad M_D + 12.5 \text{ k}(12.5 \text{ ft}) - 25 \text{ k}(6.25 \text{ ft}) = 0$$

$$N_D = 28.0 \text{ k} \qquad Ans.$$

$$V_D = 0 \qquad Ans.$$

$$M_D = 0 \qquad Ans.$$

Note: If the arch had a different shape or if the load were nonuniform, then the internal shear and moment would be nonzero. Also, if a simply supported beam were used to support the distributed loading, it would have to resist a maximum bending moment of $M = 625$ k · ft. By comparison, it is more efficient to structurally resist the load in direct compression (although one must consider the possibility of buckling) than to resist the load by a bending moment.

EXAMPLE 5.5

The three-hinged tied arch is subjected to the loading shown in Fig. 5–11a. Determine the force in members *CH* and *CB*. The dashed member *GF* of the truss is intended to carry no force.

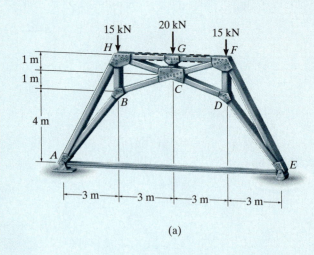

(a)

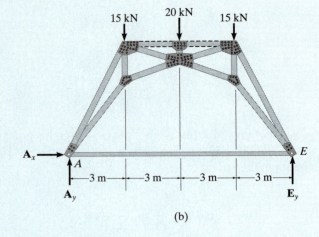

(b)

Fig. 5–11

SOLUTION

The support reactions can be obtained from a free-body diagram of the entire arch, Fig. 5–11b:

$$\zeta + \Sigma M_A = 0; \quad E_y(12\,\text{m}) - 15\,\text{kN}(3\,\text{m}) - 20\,\text{kN}(6\,\text{m}) - 15\,\text{kN}(9\,\text{m}) = 0$$

$$E_y = 25\,\text{kN}$$

$$\xrightarrow{+} \Sigma F_x = 0; \quad A_x = 0$$

$$+\uparrow \Sigma F_y = 0; \quad A_y - 15\,\text{kN} - 20\,\text{kN} - 15\,\text{kN} + 25\,\text{kN} = 0$$

$$A_y = 25\,\text{kN}$$

The force components acting at joint *C* can be determined by considering the free-body diagram of the left part of the arch, Fig. 5–11c. First, we determine the force F_{AE}:

$$\zeta + \Sigma M_C = 0; \quad F_{AE}(5\,\text{m}) - 25\,\text{kN}(6\,\text{m}) + 15\,\text{kN}(3\,\text{m}) = 0$$

$$F_{AE} = 21.0\,\text{kN}$$

(c)

Then,

$$\overset{+}{\rightarrow}\Sigma F_x = 0; \quad -C_x + 21.0 \text{ kN} = 0, \quad C_x = 21.0 \text{ kN}$$

$$+\uparrow\Sigma F_y = 0; \quad 25 \text{ kN} - 15 \text{ kN} - 20 \text{ kN} + C_y = 0, \quad C_y = 10 \text{ kN}$$

To obtain the forces in CH and CB, we can use the method of joints as follows:

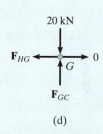

(d)

Joint G; Fig. 5–11d,

$$+\uparrow\Sigma F_y = 0; \quad F_{GC} - 20 \text{ kN} = 0$$

$$F_{GC} = 20 \text{ kN (C)}$$

Joint C; Fig. 5–11e,

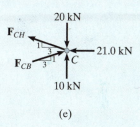

(e)

$$\overset{+}{\rightarrow}\Sigma F_x = 0; \quad F_{CB}\left(\frac{3}{\sqrt{10}}\right) - 21.0 \text{ kN} - F_{CH}\left(\frac{3}{\sqrt{10}}\right) = 0$$

$$+\uparrow\Sigma F_y = 0; \quad F_{CB}\left(\frac{1}{\sqrt{10}}\right) + F_{CH}\left(\frac{1}{\sqrt{10}}\right) - 20 \text{ kN} + 10 \text{ kN} = 0$$

Thus,

$$F_{CB} = 26.9 \text{ kN (C)} \qquad \textit{Ans.}$$

$$F_{CH} = 4.74 \text{ kN (T)} \qquad \textit{Ans.}$$

Note: Tied arches are sometimes used for bridges, as in the case of this through-arch bridge. Here the deck is supported by suspender bars that transmit their load to the arch. The deck is in tension so that it supports the actual thrust or horizontal force at the ends of the arch.

EXAMPLE 5.6

The three-hinged trussed arch shown in Fig. 5–12a supports the symmetric loading. Determine the required height h_1 of the joints B and D, so that the arch takes a funicular shape. Member HG is intended to carry no force.

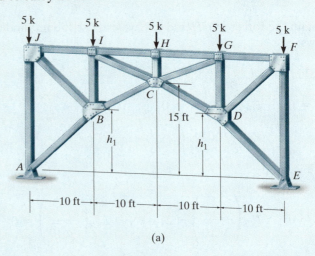

(a)

SOLUTION

For a symmetric loading, the funicular shape for the arch must be *parabolic* as indicated by the dashed line (Fig. 5–12b). Here we must find the equation which fits this shape. With the x, y axes having an origin at C, the equation is of the form $y = -cx^2$. To obtain the constant c, we require

$$-(15 \text{ ft}) = -c(20 \text{ ft})^2$$

$$c = 0.0375/\text{ft}$$

Therefore,

$$y_D = -(0.0375/\text{ft})(10 \text{ ft})^2 = -3.75 \text{ ft}$$

So that from Fig. 5–12a,

$$h_1 = 15 \text{ ft} - 3.75 \text{ ft} = 11.25 \text{ ft} \qquad \textit{Ans.}$$

Using this value, if the method of joints is now applied to the truss, the results will show that the top cord and diagonal members will all be zero-force members, and the symmetric loading will be supported *only by the bottom cord* members AB, BC, CD, and DE of the truss.

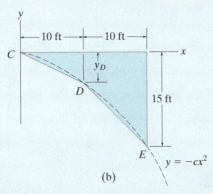

(b)

Fig. 5–12

PROBLEMS

Sec. 5.1–5.2

5–1. Determine the tension in each segment of the cable and the distance y_D.

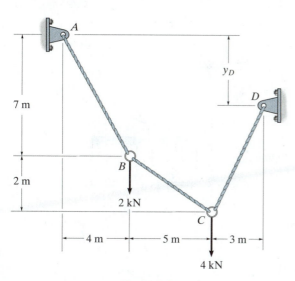

Prob. 5–1

5–2. The cable supports the loading shown. Determine the magnitude of the vertical force **P** so that $y_C = 6$ ft.

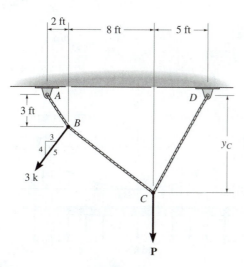

Prob. 5–2

5–3. Determine the forces P_1 and P_2 needed to hold the cable in the position shown, i.e., so segment BC remains horizontal.

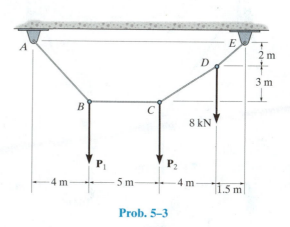

Prob. 5–3

*5–4. The cable supports the loading shown. Determine the distance x and the tension in cable BC. Set $P = 100$ lb.

5–5. The cable supports the loading shown. Determine the magnitude of the horizontal force **P** so that $x = 6$ ft.

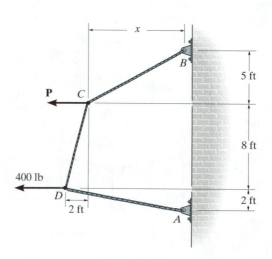

Probs. 5–4/5

5–6. Determine the tension in each segment of the cable and the cable's total length.

Sec. 5.3

*5–8. The cable supports the uniform load of $w_0 = 600$ lb/ft. Determine the tension in the cable at each support A and B.

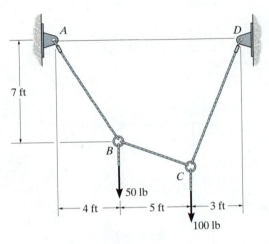

Prob. 5–6

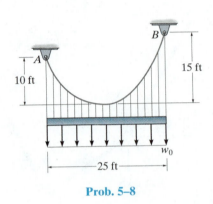

Prob. 5–8

5–9. Determine the maximum and minimum tension in the cable.

5–7. The cable supports the three loads shown. Determine the magnitude of $\mathbf{P}_1$ if $P_2 = 3$ kN and $y_B = 0.8$ m. Also find the sag y_D.

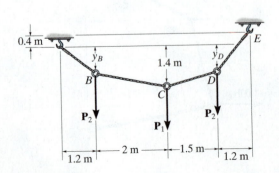

Prob. 5–7

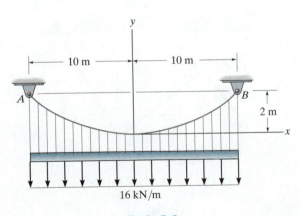

Prob. 5–9

5–10. The cable is subjected to a uniform loading of $w = 250\,lb/ft$. Determine the maximum and minimum tension in the cable.

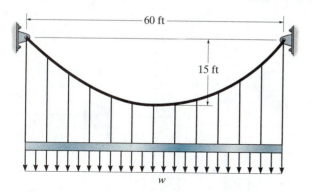

Prob. 5–10

5–11. The cable is subjected to a uniform loading of $w = 250\,lb/ft$. Determine the maximum and minimum tension in the cable.

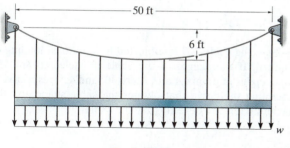

Prob. 5–11

*5–12. The cable will break when the maximum tension reaches $T_{max} = 12\,kN$. Determine the uniform distributed load w required to develop this maximum tension.

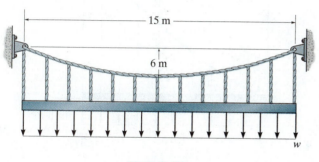

Prob. 5–12

5–13. The trusses are pin connected and suspended from the parabolic cable. Determine the maximum force in the cable when the structure is subjected to the loading shown.

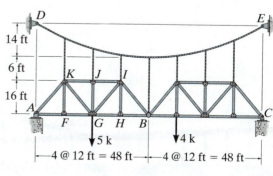

Prob. 5–13

5–14. Determine the maximum and minimum tension in the parabolic cable and the force in each of the hangers. The girder is subjected to the uniform load and is pin connected at B.

5–15. Draw the shear and moment diagrams for the pin connected girders AB and BC. The cable has a parabolic shape.

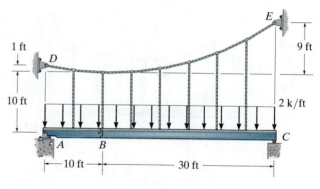

Probs. 5–14/15

*5–16. The cable will break when the maximum tension reaches $T_{max} = 5000$ kN. Determine the maximum uniform distributed load w required to develop this maximum tension.

5–17. The cable is subjected to a uniform loading of $w = 60$ kN/m. Determine the maximum and minimum tension in cable.

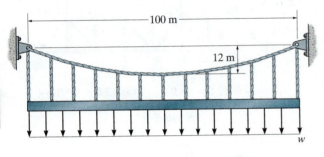

Probs. 5–16/17

5–18. The beams AB and BC are supported by the cable that has a parabolic shape. Determine the tension in the cable at points D, F, and E.

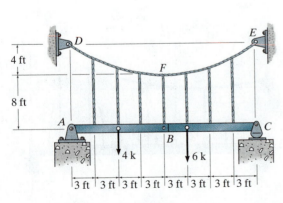

Prob. 5–18

5–19. The beams AB and BC are supported by the cable that has a parabolic shape. Determine the tension in the cable at points D, F, and E, and the force in each of the equally spaced hangers.

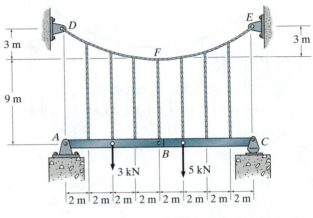

Prob. 5–19

*5–20. The cable AB is subjected to a uniform loading of 300 lb/ft. If the weight of the cable is neglected and the slope angles at points A and B are 30° and 45°, respectively, determine the curve that defines the cable shape and the maximum tension developed in the cable.

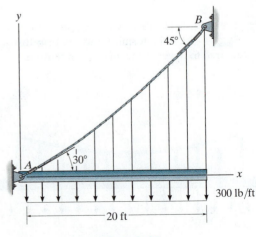

Prob. 5–20

Sec. 5.4–5.5

5–21. Determine the horizontal and vertical components of reaction at A, B, and C of the three-hinged arch. Assume A, B, and C are pin connected.

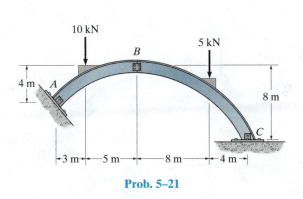

Prob. 5–21

5–22. Determine the magnitudes of the resultant forces at the pins A, B, and C of the three-hinged arched roof truss.

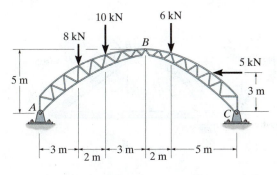

Prob. 5–22

5–23. The three-hinged spandrel arch is subjected to the loading shown. Determine the internal moment in the arch at point D.

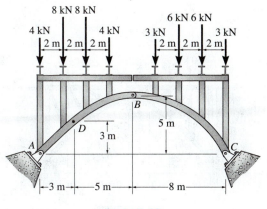

Prob. 5–23

***5–24.** The tied three-hinged truss arch is subjected to the loading shown. Determine the components of reaction at A and C, and the tension in the tie rod.

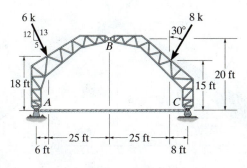

Prob. 5–24

5

5–25. The bridge is constructed as a *three-hinged trussed arch*. Determine the horizontal and vertical components of reaction at the hinges (pins) at A, B, and C. The dashed member DE is intended to carry *no* force.

5–26. Determine the design heights h_1, h_2, and h_3 of the bottom cord of the truss so the three-hinged trussed arch responds as a funicular arch.

*5–28.** Determine the horizontal and vertical components of reaction at A, B, and C of the three-hinged arch. Assume A, B, and C are pin connected.

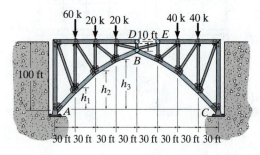

Probs. 5–25/26

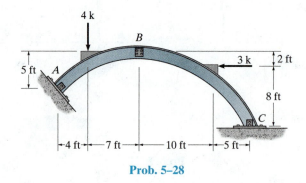

Prob. 5–28

5–27. The laminated-wood three-hinged arch is subjected to the loading shown. Determine the horizontal and vertical components of reactions at the pins A, B, and C, and draw the moment diagram for member AB.

5–29. The arch structure is subjected to the loading shown. Determine the horizontal and vertical components of reaction at A and C, and the force in member AC.

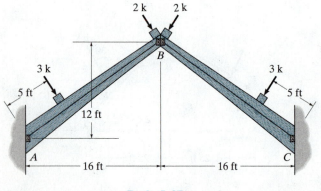

Prob. 5–27

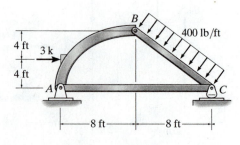

Prob. 5–29

CHAPTER REVIEW

Cables support their loads in tension if we consider them perfectly flexible.

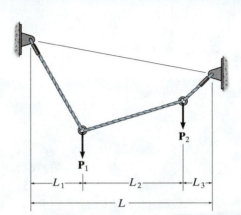

If the cable is subjected to concentrated loads then the force acting in each cable segment is determined by applying the equations of equilibrium to the free-body diagram of groups of segments of the cable or to the joints where the forces are applied.

If the cable supports a uniform load over a projected horizontal distance, then the shape of the cable takes the form of a parabola.

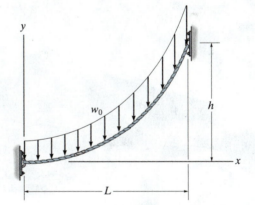

Arches are designed primarily to carry a compressive force. A parabolic shape is required to support a uniform loading distributed over its horizontal projection.

Three-hinged arches are statically determinate and can be analyzed by separating the two members and applying the equations of equilibrium to each member.

three-hinged arch

Chapter 6

© Asb63/Fotolia

The train produces a moving load that must be considered using influence lines when designing the girders of this bridge.

Influence Lines for Statically Determinate Structures

Influence lines have important application for the design of structures that resist large live loads. In this chapter we will discuss how to draw the influence line for a statically determinate structure. The theory is applied to structures subjected to a distributed load or a series of concentrated forces, and specific applications to floor girders and bridge trusses are given. The determination of the absolute maximum live shear and moment in a member is discussed at the end of the chapter.

6.1   Influence Lines

In the previous chapters we developed techniques for analyzing the forces in structural members due to *dead* or *fixed loads*. It was shown that the *shear* and *moment diagrams* represent the most descriptive methods for displaying the variation of these loads in a member. If a structure is subjected to a *live* or *moving load,* however, the variation of the shear and bending moment in the member is best described using the **influence line**. An influence line represents the variation of either the reaction, shear, or moment, at a *specific point* in a member as a concentrated force moves over the member. Once this line is constructed, one can tell at a glance where the moving load should be placed on the structure so that it creates the greatest influence at the specified point. Furthermore, the magnitude of the associated reaction, shear, or moment at the point can then be calculated from the ordinates of the influence-line diagram. For these reasons, influence lines play an important part in the design of bridges, industrial crane rails, conveyors, and other structures where loads move across their span.

Although the procedure for constructing an influence line is rather basic, one should clearly be aware of the *difference* between constructing an influence line and constructing a shear or moment diagram. Influence lines represent the effect of a *moving load* only at a *specified point* on a member, whereas shear and moment diagrams represent the effect of *fixed loads* at *all points* along the axis of the member.

Procedure for Analysis

Either of the following two procedures can be used to construct the influence line at a specific point P in a member for any function (reaction, shear, or moment). For both of these procedures we will choose the moving force to have a *dimensionless magnitude of unity.**

Tabulate Values

- Place a unit load at various locations, x, along the member, and at *each* location use statics to determine the value of the function (reaction, shear, or moment) at the specified point.

- If the influence line for a vertical force *reaction* at a point on a beam is to be constructed, consider the reaction to be *positive* at the point when it acts *upward* on the beam.

- If a shear or moment influence line is to be drawn for a point, take the shear or moment at the point as positive according to the same sign convention used for drawing shear and moment diagrams. (See Fig. 4–1.)

- All statically determinate beams will have influence lines that consist of straight line segments. After some practice one should be able to minimize computations and locate the unit load *only* at points representing the *end points* of each line segment.

- To avoid errors, it is recommended that one first construct a table, listing "unit load at x" versus the corresponding value of the function calculated at the specific point; that is, "reaction R," "shear V," or "moment M." Once the load has been placed at various points along the span of the member, the tabulated values can be plotted and the influence-line segments constructed.

Influence-Line Equations

- The influence line can also be constructed by placing the unit load at a *variable* position x on the member and then computing the value of R, V, or M at the point as a function of x. In this manner, the equations of the various line segments composing the influence line can be determined and plotted.

*The reason for this choice will be explained in Sec. 6.2.

EXAMPLE | 6.1

Construct the influence line for the vertical reaction at A of the beam in Fig. 6–1a.

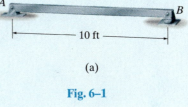

(a)

Fig. 6–1

SOLUTION

Tabulate Values.   A unit load is placed on the beam at each selected point x and the value of A_y is calculated by summing moments about B. For example, when $x = 2.5$ ft and $x = 5$ ft, see Figs. 6–1b and 6–1c, respectively. The results for A_y are entered in the table, Fig. 6–1d. A plot of these values yields the influence line for the reaction at A, Fig. 6–1e. This line or diagram gives the reaction at A as the unit load moves from one position to the next along the beam.

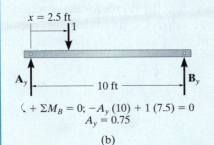

$\zeta + \Sigma M_B = 0;\ -A_y(10) + 1(7.5) = 0$
$A_y = 0.75$

(b)

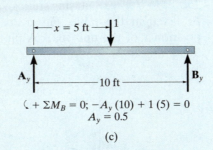

$\zeta + \Sigma M_B = 0;\ -A_y(10) + 1(5) = 0$
$A_y = 0.5$

(c)

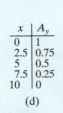

x	A_y
0	1
2.5	0.75
5	0.5
7.5	0.25
10	0

(d)

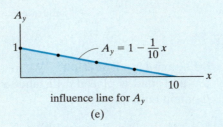

$A_y = 1 - \dfrac{1}{10}x$

influence line for A_y

(e)

Influence-Line Equation.   Rather than placing the unit load at *specific points* on the beam and plotting the values of A_y for each point, we can place the unit load a variable distance x from A, Fig. 6–1f. The reaction A_y as a function of x can be determined from

$$\zeta + \Sigma M_B = 0;\quad -A_y(10) + (10 - x)(1) = 0$$

$$A_y = 1 - \tfrac{1}{10}x$$

This is the equation of the influence line, plotted in Fig. 6–1e.

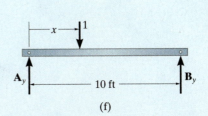

(f)

EXAMPLE 6.2

Construct the influence line for the vertical reaction at B of the beam in Fig. 6–2a.

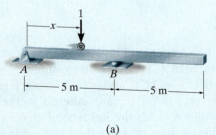

(a)

Fig. 6–2

SOLUTION

Tabulate Values.   Using statics, verify that the values for the reaction B_y listed in the table, Fig. 6–2b, are correctly computed for each position x of the unit load. A plot of the values yields the influence line in Fig. 6–2c.

x	B_y
0	0
2.5	0.5
5	1
7.5	1.5
10	2

(b)

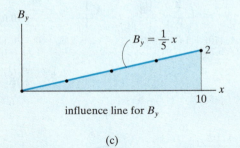

influence line for B_y

(c)

Influence-Line Equation.   We must determine the reaction B_y when the unit load is placed a distance x from A, Fig. 6–2d. Applying the moment equation about A,

$$\zeta + \Sigma M_A = 0; \quad B_y(5) - 1(x) = 0$$

$$B_y = \tfrac{1}{5} x$$

This is plotted in Fig. 6–2c.

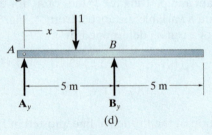

(d)

EXAMPLE 6.3

Construct the influence line for the shear at point C of the beam in Fig. 6–3a.

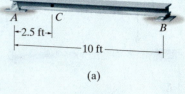

(a)

Fig. 6–3

SOLUTION

Tabulate Values. At each selected position x of the unit load, the method of sections is used to calculate the value of V_C. Note in particular that the unit load must be placed *just to the left* ($x = 2.5^-$) and *just to the right* ($x = 2.5^+$) of point C since the shear is discontinuous at C, Figs. 6–3b and 6–3c. A plot of the values in Fig. 6–3d yields the influence line for the shear at C, Fig. 6–3e. Here the diagram gives the shear at C as the unit load moves from one position to the next along the beam.

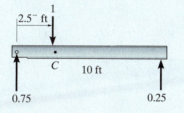

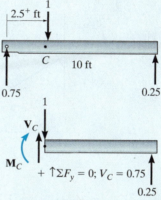

x	V_C
0	0
2.5^-	-0.25
2.5^+	0.75
5	0.5
7.5	0.25
10	0

(d)

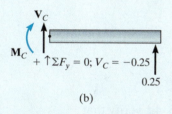

(b)

(c)

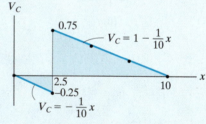

influence line for V_C

(e)

Influence-Line Equations. Here two equations have to be determined, since there are two segments for the influence line due to the discontinuity of shear at C. First we determine A_y, Fig. 6–3f. Using the result we then obtain the internal shear at C for each region, Fig. 6–3g. The equations are plotted in Fig. 6–3e.

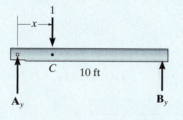

$\downarrow + \Sigma M_B = 0; \quad -A_y(10) + 1(10 - x) = 0$

$$A_y = 1 - \frac{1}{10}x$$

(f)

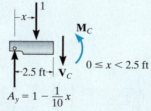

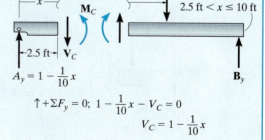

$\uparrow + \Sigma F_y = 0; \quad 1 - \frac{1}{10}x - 1 - V_C = 0$

$$V_C = -\frac{1}{10}x$$

$A_y = 1 - \frac{1}{10}x$

$\uparrow + \Sigma F_y = 0; \quad 1 - \frac{1}{10}x - V_C = 0$

$$V_C = 1 - \frac{1}{10}x$$

(g)

EXAMPLE 6.4

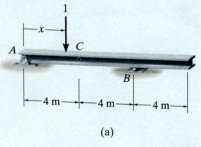

(a)

Fig. 6–4

Construct the influence line for the shear at point C of the beam in Fig. 6–4a.

SOLUTION

Tabulate Values.   Using statics and the method of sections, verify that the values of the shear V_C at point C in Fig. 6–4b correspond to each position x of the unit load on the beam. A plot of the values in Fig. 6–4b yields the influence line in Fig. 6–4c.

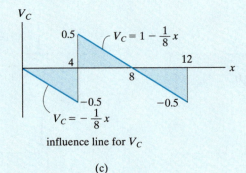

influence line for V_C

(c)

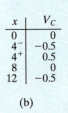

x	V_C
0	0
4^-	-0.5
4^+	0.5
8	0
12	-0.5

(b)

Influence-Line Equations.   From Fig. 6–4d, verify that

$$V_C = -\tfrac{1}{8}x \qquad 0 \le x < 4 \text{ m}$$

$$V_C = 1 - \tfrac{1}{8}x \qquad 4 \text{ m} < x \le 12 \text{ m}$$

These equations are plotted in Fig. 6–4c.

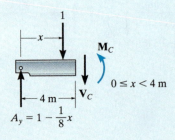

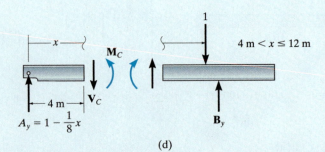

(d)

EXAMPLE 6.5

Construct the influence line for the moment at point C of the beam in Fig. 6–5a.

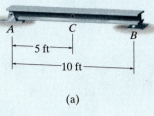

(a)

Fig. 6–5

SOLUTION

Tabulate Values. At each selected position of the unit load, the value of M_C is calculated using the method of sections. For example, see Fig. 6–5b for $x = 2.5$ ft. A plot of the values in Fig. 6–5c yields the influence line for the moment at C, Fig. 6–5d. This diagram gives the moment at C as the unit load moves from one position to the next along the beam.

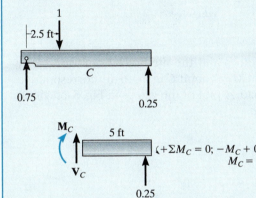

(b)

x	M_C
0	0
2.5	1.25
5	2.5
7.5	1.25
10	0

(c)

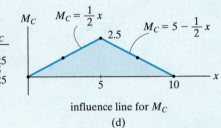

influence line for M_C

(d)

Influence-Line Equations. The two line segments for the influence line can be determined using $\Sigma M_C = 0$ along with the method of sections shown in Fig. 6–5e. These equations when plotted yield the influence line shown in Fig. 6–5d.

$\zeta + \Sigma M_C = 0;$ $M_C + 1(5 - x) - \left(1 - \tfrac{1}{10}x\right)5 = 0$ $\zeta + \Sigma M_C = 0;$ $M_C - \left(1 - \tfrac{1}{10}x\right)5 = 0$

$M_C = \tfrac{1}{2}x$ $0 \le x < 5$ ft $M_C = 5 - \tfrac{1}{2}x$ 5 ft $< x \le 10$ ft

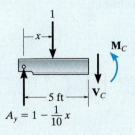

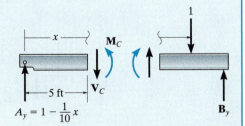

(e)

EXAMPLE 6.6

Construct the influence line for the moment at point C of the beam in Fig. 6–6a.

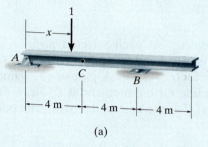

(a)

Fig. 6–6

SOLUTION

Tabulate Values. Using statics and the method of sections, verify that the values of the moment M_C at point C in Fig. 6–6b correspond to each position x of the unit load. A plot of the values in Fig. 6–6b yields the influence line in Fig. 6–6c.

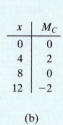

x	M_C
0	0
4	2
8	0
12	−2

(b)

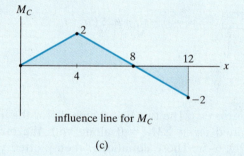

influence line for M_C

(c)

Influence-Line Equations.   From Fig. 6–6d verify that

$$M_C = \tfrac{1}{2}x \qquad\qquad 0 \le x < 4 \text{ m}$$

$$M_C = 4 - \tfrac{1}{2}x \qquad 4 \text{ m} < x \le 12 \text{ m}$$

These equations are plotted in Fig. 6–6c.

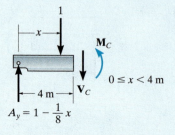

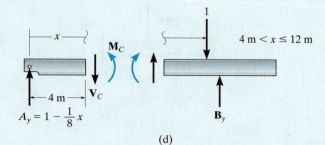

(d)

6.2 Influence Lines for Beams

Since beams (or girders) often form the main load-carrying elements of a floor system or bridge deck, it is important to be able to construct the influence lines for the reactions, shear, or moment at any specified point in a beam.

Loadings. Once the influence line for a function (reaction, shear, or moment) has been constructed, it will then be possible to position the live loads on the beam which will produce the maximum value of the function. Two types of loadings will now be considered.

Concentrated Force. Since the numerical values of a function for an influence line are determined using a dimensionless unit load, then for any concentrated force **F** acting on the beam at any position x, *the value of the function can be found by multiplying the ordinate of the influence line at the position x by the magnitude of* **F**. For example, consider the influence line for the reaction at A for the beam AB, Fig. 6–7. If the *unit load* is at $x = \frac{1}{2}L$, the reaction at A is $A_y = \frac{1}{2}$ as indicated from the influence line. Hence, if the force F lb is at this same point, the reaction is $A_y = \left(\frac{1}{2}\right)(F)$ lb. Of course, this same value can also be determined by statics. Obviously, the *maximum influence* caused by **F** occurs when it is placed on the beam at the same location as the *peak* of the influence line—in this case at $x = 0$, where the reaction would be $A_y = (1)(F)$ lb.

Uniform Load. Consider a portion of a beam subjected to a uniform load w_0, Fig. 6–8. As shown, each dx segment of this load creates a concentrated force of $dF = w_0\,dx$ on the beam. If dF is located at x, where the beam's influence-line ordinate for some function (reaction, shear, moment) is y, then the value of the function is $(dF)(y) = (w_0\,dx)y$. The effect of all the concentrated forces $d\mathbf{F}$ is determined by integrating over the entire length of the beam, that is, $\int w_0 y\,dx = w_0 \int y\,dx$. Also, since $\int y\,dx$ is equivalent to the *area* under the influence line, then, in general, *the value of a function caused by a uniform distributed load is simply the area under the influence line for the function multiplied by the intensity of the uniform load.* For example, in the case of a uniformly loaded beam shown in Fig. 6–9, the reaction A_y can be determined from the influence line as $A_y = (\text{area})(w_0) = \left[\frac{1}{2}(1)(L)\right]w_0 = \frac{1}{2}w_0L$. This value can of course also be determined from statics.

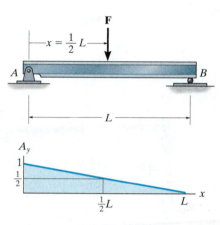

influence line for A_y

Fig. 6–7

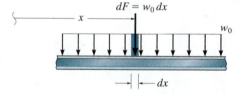

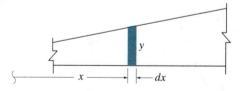

influence line for function

Fig. 6–8

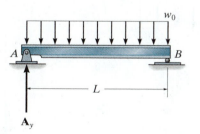 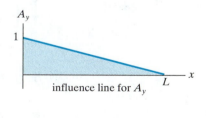

influence line for A_y

Fig. 6–9

EXAMPLE 6.7

Determine the maximum *positive* shear that can be developed at point
C in the beam shown in Fig. 6–10*a* due to a concentrated moving load
of 4000 lb and a uniform moving load of 2000 lb/ft.

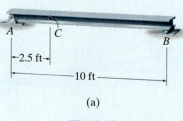

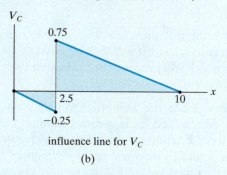

influence line for V_C

(a)

(b)

Fig. 6–10

SOLUTION

The influence line for the shear at *C* has been established in
Example 6.3 and is shown in Fig. 6–10*b*.

Concentrated Force.   The maximum positive shear at *C* will occur
when the 4000-lb force is located at $x = 2.5^+$ ft, since this is the positive
peak of the influence line. The ordinate of this peak is +0.75; so that

$$V_C = 0.75(4000 \text{ lb}) = 3000 \text{ lb}$$

Uniform Load.   The uniform moving load creates the maximum
positive influence for V_C when the load acts on the beam between
$x = 2.5^+$ ft and $x = 10$ ft, since within this region the influence line
has a positive area. The magnitude of $\mathbf{V}_C$ due to this loading is

$$V_C = \left[\tfrac{1}{2}(10 \text{ ft} - 2.5 \text{ ft})(0.75) \right] 2000 \text{ lb/ft} = 5625 \text{ lb}$$

Total Maximum Shear at C.

$$(V_C)_{\text{max}} = 3000 \text{ lb} + 5625 \text{ lb} = 8625 \text{ lb} \qquad \textit{Ans.}$$

Notice that once the *positions* of the loads have been established using
the influence line, Fig. 6–10*c*, this value of $(V_C)_{\text{max}}$ can *also* be determined
using statics and the method of sections. Show that this is the case.

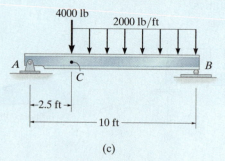

(c)

EXAMPLE 6.8

The frame structure shown in Fig. 6–11a is used to support a hoist for transferring loads for storage at points underneath it. It is anticipated that the load on the dolly is 3 kN and the beam CB has a mass of 24 kg/m. Assume the dolly has negligible size and can travel the entire length of the beam. Also, assume A is a pin and B is a roller. Determine the maximum vertical support reactions at A and B and the maximum moment in the beam at D.

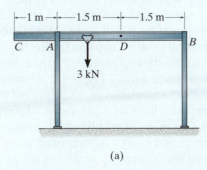

(a)

SOLUTION

Maximum Reaction at A. We first draw the influence line for A_y, Fig. 6–11b. Specifically, when a unit load is at A the reaction at A is 1 as shown. The ordinate at C, is 1.33. Here the maximum value for A_y occurs when the dolly is at C. Since the dead load (beam weight) must be placed over the entire length of the beam, we have,

$$(A_y)_{max} = 3000(1.33) + 24(9.81)\left[\tfrac{1}{2}(4)(1.33)\right]$$

$$= 4.63 \text{ kN} \qquad\qquad Ans.$$

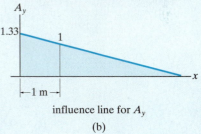

influence line for A_y

(b)

Maximum Reaction at B. The influence line (or beam) takes the shape shown in Fig. 6–11c. The values at C and B are determined by statics. Here the dolly must be at B. Thus,

$$(B_y)_{max} = 3000(1) + 24(9.81)\left[\tfrac{1}{2}(3)(1)\right] + 24(9.81)\left[\tfrac{1}{2}(1)(-0.333)\right]$$

$$= 3.31 \text{ kN} \qquad\qquad Ans.$$

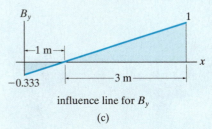

influence line for B_y

(c)

Maximum Moment at D. The influence line has the shape shown in Fig. 6–11d. The values at C and D are determined from statics. Here,

$$(M_D)_{max} = 3000(0.75) + 24(9.81)\left[\tfrac{1}{2}(1)(-0.5)\right] + 24(9.81)\left[\tfrac{1}{2}(3)(0.75)\right]$$

$$= 2.46 \text{ kN} \cdot \text{m} \qquad\qquad Ans.$$

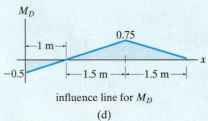

influence line for M_D

(d)

Fig. 6–11

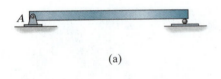

(a)

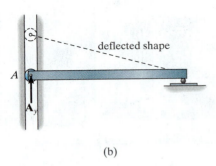

deflected shape

A

$\mathbf{A}_y$

(b)

6

6.3   Qualitative Influence Lines

In 1886, Heinrich Müller-Breslau developed a technique for rapidly constructing the shape of an influence line. Referred to as the **Müller-Breslau principle**, *it states that the influence line for a function (reaction, shear, or moment) is to the same scale as the deflected shape of the beam when the beam is acted upon by the function.* In order to draw the deflected shape properly, the capacity of the beam to resist the applied function must be *removed* so the beam can deflect when the function is applied. For example, consider the beam in Fig. 6–12a. If the shape of the influence line for the *vertical reaction* at A is to be determined, the pin is first replaced by a *roller guide* as shown in Fig. 6–12b. A roller guide is necessary since the beam must still resist a horizontal force at A but *no vertical force.* When the positive (upward) force $\mathbf{A}_y$ is then applied at A, the beam deflects to the dashed position,* which represents the general shape of the influence line for A_y, Fig. 6–12c. (Numerical values for this specific case have been calculated in Example 6.1.) If the shape of the influence line for the *shear* at C is to be determined, Fig. 6–13a, the connection at C may be symbolized by a *roller guide* as shown in Fig. 6–13b. This device will resist a moment and axial force but no *shear.*[†] Applying a positive shear force $\mathbf{V}_C$ to the beam at C and allowing the beam to deflect to the dashed position, we find the influence-line shape as shown in Fig. 6–13c. Finally, if the shape of the influence line for the *moment* at C, Fig. 6–14a, is to be determined, an internal *hinge* or *pin* is placed at C, since this connection resists axial and shear forces but *cannot resist a moment,* Fig. 6–14b. Applying positive moments $\mathbf{M}_C$ to the beam, the beam then deflects to the dashed position, which is the shape of the influence line, Fig. 6–14c.

The proof of the Müller-Breslau principle can be established using the principle of virtual work. Recall that *work* is the product of either *a linear*

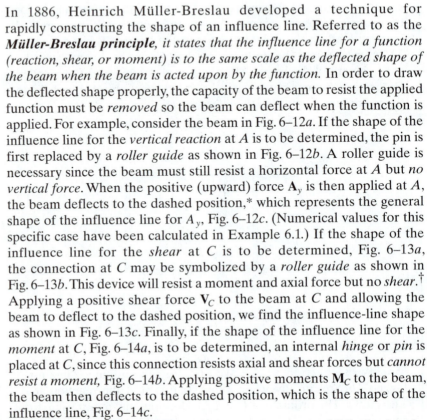

A_y

influence line for A_y

x

(c)

Fig. 6–12

Design of this bridge girder is based on influence lines that must be constructed for this train loading.

*Throughout the discussion all deflected positions are drawn to an exaggerated scale to illustrate the concept.

[†]Here the rollers *symbolize* supports that carry loads both in tension or compression. See Table 2.1, support (2).

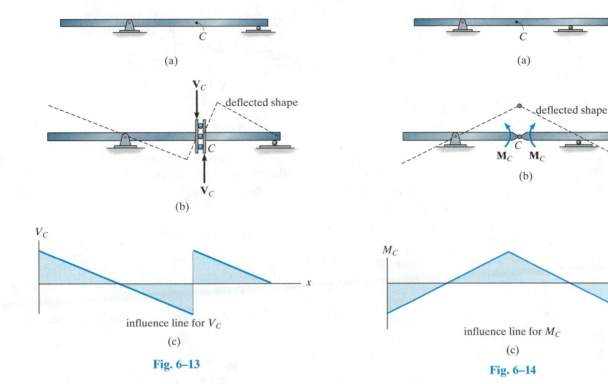

Fig. 6–13

influence line for V_C

(c)

influence line for M_C

(c)

Fig. 6–14

displacement and force in the direction of the displacement or *a rotational displacement and moment in the direction of the displacement*. If a rigid body (beam) is in equilibrium, the sum of all the forces and moments on it must be equal to zero. Consequently, if the body is given an *imaginary* or *virtual displacement*, the work done by *all* these forces and couple moments must also be equal to zero. Consider, for example, the simply supported beam shown in Fig. 6–15a, which is subjected to a unit load placed at an arbitrary point along its length. If the beam is given a virtual (or imaginary) displacement δy at the support A, Fig. 6–15b, then only the support reaction $\mathbf{A}_y$ and the unit load do virtual work. Specifically, A_y does positive work $A_y \, \delta y$ and the unit load does negative work, $-1\delta y'$. (The support at B does not move and therefore the force at B does no work.) Since the beam is in equilibrium and therefore does not actually move, the virtual work sums to zero, i.e.,

$$A_y \, \delta y - 1 \, \delta y' = 0$$

If δy is set equal to 1, then

$$A_y = \delta y'$$

In other words, the value of A_y represents the ordinate of the influence line at the position of the unit load. Since this value is equivalent to the displacement $\delta y'$ at the position of the unit load, it shows that the *shape* of the influence line for the reaction at A has been established. This proves the Müller-Breslau principle for reactions.

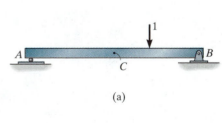

(a)

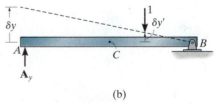

(b)

Fig. 6–15

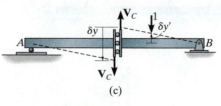

(c)

Fig. 6–15

In the same manner, if the beam is sectioned at C, and the beam undergoes a virtual displacement δy at this point, Fig. 6–15c, then only the internal shear at C and the unit load do work. Thus, the virtual work equation is

$$V_C\, \delta y - 1\, \delta y' = 0$$

Again, if $\delta y = 1$, then

$$V_C = \delta y'$$

and the *shape* of the influence line for the shear at C has been established.

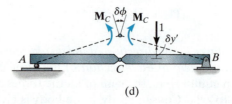

(d)

Lastly, assume a hinge or pin is introduced into the beam at point C, Fig. 6–15d. If a virtual rotation $\delta\phi$ is introduced at the pin, virtual work will be done only by the internal moment and the unit load. So

$$M_C\, \delta\phi - 1\, \delta y' = 0$$

Setting $\delta\phi = 1$, it is seen that

$$M_C = \delta y'$$

which indicates that the deflected beam has the same *shape* as the influence line for the internal moment at point C (see Fig. 6–14).

Obviously, the Müller-Breslau principle provides a quick method for establishing the *shape* of the influence line. Once this is known, the ordinates at the peaks can be determined by using the basic method discussed in Sec. 6.1. Also, by simply knowing the general shape of the influence line, it is possible to *locate* the live load on the beam and then determine the maximum value of the function by *using statics*. Example 6.12 illustrates this technique.

For each beam in Fig. 6–16a through 6–16c, sketch the influence line for the vertical reaction at A.

SOLUTION

The support is replaced by a roller guide at A since it will resist $\mathbf{A}_x$, but not $\mathbf{A}_y$. The force $\mathbf{A}_y$ is then applied.

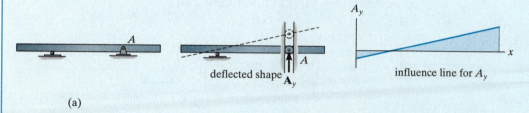

(a)

Fig. 6–16

Again, a roller guide is placed at A and the force $\mathbf{A}_y$ is applied.

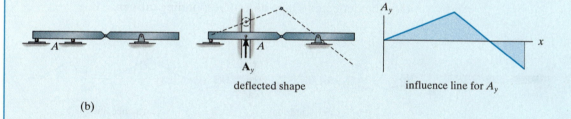

(b)

A *double-roller guide* must be used at A in this case, since this type of support will resist both a moment $\mathbf{M}_A$ at the fixed support and axial load $\mathbf{A}_x$, but will not resist $\mathbf{A}_y$.

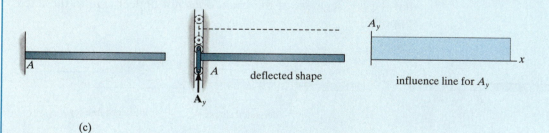

(c)

EXAMPLE | 6.10

For each beam in Figs. 6–17a through 6–17c, sketch the influence line for the shear at B.

SOLUTION

The roller guide is introduced at B and the positive shear V_B is applied. Notice that the right segment of the beam will *not deflect* since the roller at A actually constrains the beam from moving vertically, either up or down. [See support (2) in Table 2.1.]

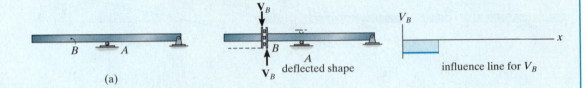

(a)

Fig. 6–17

Placing the roller guide at B and applying the positive shear at B yields the deflected shape and corresponding influence line.

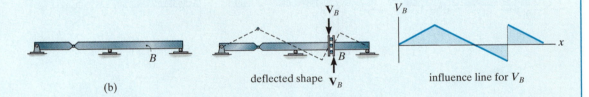

(b)

Again, the roller guide is placed at B, the positive shear is applied, and the deflected shape and corresponding influence line are shown. Note that the left segment of the beam does not deflect, due to the fixed support.

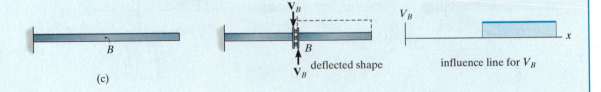

(c)

EXAMPLE | 6.11

For each beam in Figs. 6–18a through Fig. 6–18c, sketch the influence line for the moment at B.

SOLUTION

A hinge is introduced at B and positive moments $\mathbf{M}_B$ are applied to the beam. The deflected shape and corresponding influence line are shown.

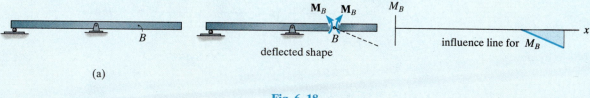

(a)

Fig. 6–18

Placing a hinge at B and applying positive moments M_B to the beam yields the deflected shape and influence line.

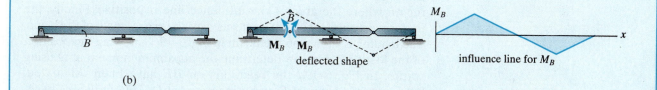

(b)

With the hinge and positive moment at B the deflected shape and influence line are shown. The left segment of the beam is constrained from moving due to the fixed wall at A.

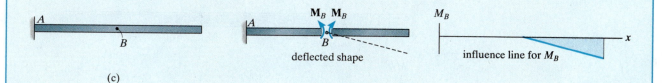

(c)

EXAMPLE 6.12

Determine the maximum positive moment that can be developed at point D in the beam shown in Fig. 6–19a due to a concentrated moving load of 4000 lb, a uniform moving load of 300 lb/ft, and a beam weight of 200 lb/ft.

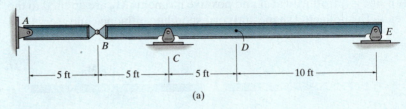

(a)

Fig. 6–19

SOLUTION

A hinge is placed at D and positive moments $\mathbf{M}_D$ are applied to the beam. The deflected shape and corresponding influence line are shown in Fig. 6–19b. Immediately one recognizes that the concentrated moving load of 4000 lb creates a maximum *positive* moment at D when it is placed at D, i.e., the peak of the influence line. Also, the uniform moving load of 300 lb/ft must extend from C to E in order to cover the region where the area of the influence line is positive. Finally, the uniform *dead weight* of 200 lb/ft acts over the *entire length* of the beam. The loading is shown on the beam in Fig. 6–19c. Knowing the position of the loads, we can now determine the maximum moment at D using statics. In Fig. 6–19d the reactions on BE have been calculated. Sectioning the beam at D and using segment DE, Fig. 6–19e, we have

$$\zeta + \Sigma M_D = 0; \qquad -M_D - 5000(5) + 4750(10) = 0$$

$$M_D = 22\,500 \text{ lb} \cdot \text{ft} = 22.5 \text{ k} \cdot \text{ft} \qquad \textit{Ans.}$$

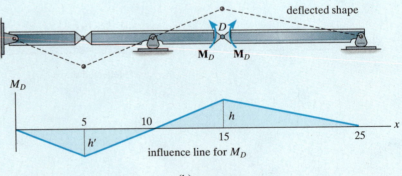

(b)

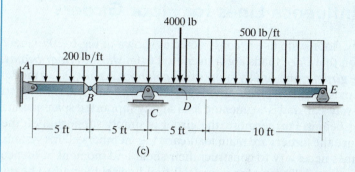

(c)

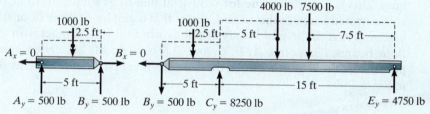

(d)

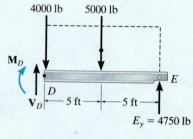

(e)

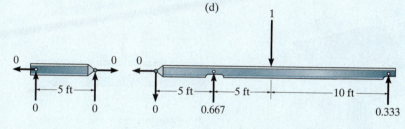

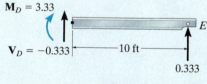

(f)

This problem can also be worked by using *numerical values* for the influence line as in Sec. 6.1. Actually, by inspection of Fig. 6–19b, only the peak value h at D must be determined. This requires placing a unit load on the beam at D in Fig. 6–19a and then solving for the internal moment in the beam at D. The calculations are shown in Fig. 6–19f. Thus $M_D = h = 3.33$. By proportional triangles, $h'/(10-5) = 3.33/(15-10)$ or $h' = 3.33$. Hence, with the loading on the beam as in Fig. 6–19c, using the areas and peak values of the influence line, Fig. 6–19b, we have

$$M_D = 500\left[\frac{1}{2}(25-10)(3.33)\right] + 4000(3.33) - 200\left[\frac{1}{2}(10)(3.33)\right]$$

$$= 22\,500 \text{ lb} \cdot \text{ft} = 22.5 \text{ k} \cdot \text{ft} \qquad \qquad \textit{Ans.}$$

6.4   Influence Lines for Floor Girders

Occasionally, floor systems are constructed as shown in Fig. 6–20a, where it can be seen that floor loads are transmitted from *slabs* to *floor beams*, then to *side girders*, and finally supporting *columns*. An idealized model of this system is shown in plane view, Fig. 6–20b. Here the slab is assumed to be a one-way slab and is segmented into simply supported spans resting on the floor beams. Furthermore, the girder is simply supported on the columns. Since the girders are main load-carrying members in this system, it is sometimes necessary to construct their shear and moment influence lines. This is especially true for industrial buildings subjected to heavy concentrated loads. In this regard, notice that a unit load on the floor slab is transferred to the girder only at points where it is in contact with the floor beams, i.e., points *A*, *B*, *C*, and *D*. These points are called *panel points*, and the region between these points is called a *panel*, such as *BC* in Fig. 6–20b.

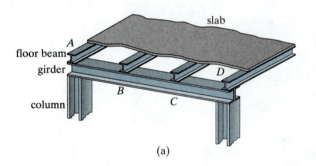

(a)

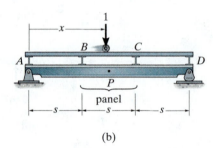

(b)

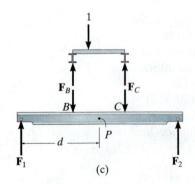

(c)

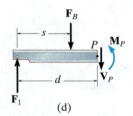

(d)

Fig. 6–20

The influence line for a specified point on the girder can be determined using the same statics procedure as in Sec. 6.1; i.e., place the unit load at various points x on the floor slab and always compute the function (shear or moment) at the specified point P in the girder, Fig. 6–20b. Plotting these values versus x yields the influence line for the function at P. In particular, the value for the internal moment in a girder panel will depend upon where point P is chosen for the influence line, since the magnitude of $\mathbf{M}_P$ depends upon the point's location from the end of the girder. For example, if the unit load acts on the floor slab as shown in Fig. 6–20c, one first finds the reactions $\mathbf{F}_B$ and $\mathbf{F}_C$ on the slab, then calculates the support reactions $\mathbf{F}_1$ and $\mathbf{F}_2$ on the girder. The internal moment at P is then determined by the method of sections, Fig. 6–20d. This gives $M_P = F_1 d - F_B(d - s)$. Using a similar analysis, the internal shear $\mathbf{V}_P$ can be determined. In this case, however, $\mathbf{V}_P$ will be *constant* throughout the panel $BC(V_P = F_1 - F_B)$ and so it does not depend upon the exact location d of P within the panel. For this reason, influence lines for shear in floor girders are specified for *panels* in the girder and not specific points along the girder. The shear is then referred to as *panel shear*. It should also be noted that since the girder is affected only by the loadings transmitted by the floor beams, the unit load is generally placed at each floor-beam location to establish the necessary data used to draw the influence line.

The following numerical examples should clarify the force analysis.

The design of the floor system of this warehouse building must account for critical locations of storage materials on the floor. Influence lines must be used for this purpose. (*Photo courtesy of Portland Cement Association.*)

EXAMPLE   6.13

Draw the influence line for the shear in panel CD of the floor girder in Fig. 6–21a.

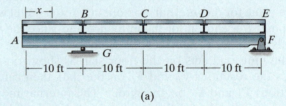

(a)

Fig. 6–21

SOLUTION

Tabulate Values.   The unit load is placed at each floor beam location and the shear in panel CD is calculated. A table of the results is shown in Fig. 6–21b. The details for the calculations when $x = 0$ and $x = 20$ ft are given in Figs. 6–21c and 6–21d, respectively. Notice how in each case the reactions of the floor beams on the girder are calculated first, followed by a determination of the girder support reaction at F ($\mathbf{G}_y$ is not needed), and finally, a segment of the girder is considered and the internal panel shear V_{CD} is calculated. As an exercise, verify the values for V_{CD} when $x = 10$ ft, 30 ft, and 40 ft.

x	V_{CD}
0	0.333
10	0
20	−0.333
30	0.333
40	0

(b)

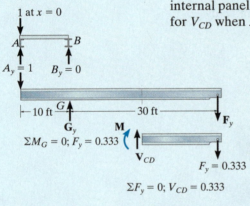

(c)

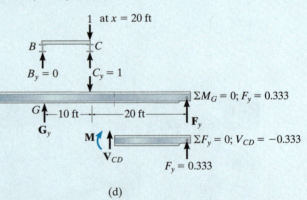

(d)

Influence Line.   When the tabular values are plotted and the points connected with straight line segments, the resulting influence line for V_{CD} is as shown in Fig. 6–21e.

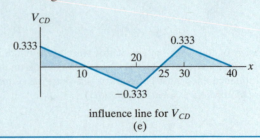

influence line for V_{CD}

(e)

EXAMPLE 6.14

Draw the influence line for the moment at point F for the floor girder in Fig. 6–22a.

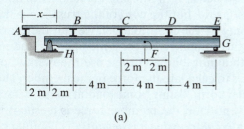

x	M_F
0	0
2	0.429
4	0.857
8	2.571
10	2.429
12	2.286
16	0

(a)   (b)

Fig. 6–22

SOLUTION

Tabulate Values. The unit load is placed at $x = 0$ and each panel point thereafter. The corresponding values for M_F are calculated and shown in the table, Fig. 6–22b. Details of the calculations for $x = 2$ m are shown in Fig. 6–22c. As in the previous example, it is first necessary to determine the reactions of the floor beams on the girder, followed by a determination of the girder support reaction $\mathbf{G}_y$ ($\mathbf{H}_y$ is not needed), and finally segment GF of the girder is considered and the internal moment $\mathbf{M}_F$ is calculated. As an exercise, determine the other values of M_F listed in Fig. 6–22b.

Influence Line. A plot of the tabular values yields the influence line for M_F, Fig. 6–22d.

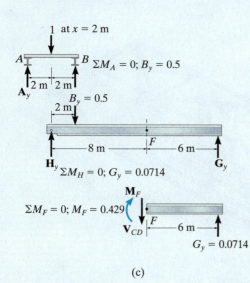

(c)

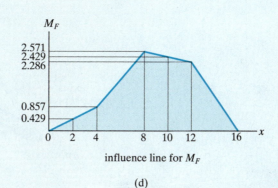

influence line for M_F

(d)

6.5   Influence Lines for Trusses

The members of this truss bridge were designed using influence lines in accordance with the AASHTO specifications.

Trusses are often used as primary load-carrying elements for bridges. Hence, for design it is important to be able to construct the influence lines for each of its members. As shown in Fig. 6–23, the loading on the bridge deck is transmitted to stringers, which in turn transmit the loading to floor beams and then to the *joints* along the bottom cord of the truss. Since the truss members are affected only by the joint loading, we can therefore obtain the ordinate values of the influence line for a member by loading each joint along the deck with a unit load and then use the method of joints or the method of sections to calculate the force in the member. The data can be arranged in tabular form, listing "unit load at joint" versus "force in member." As a convention, if the member force is *tensile* it is considered a *positive* value; if it is *compressive* it is *negative*. The influence line for the member is constructed by plotting the data and drawing straight lines between the points.

The following examples illustrate the method of construction.

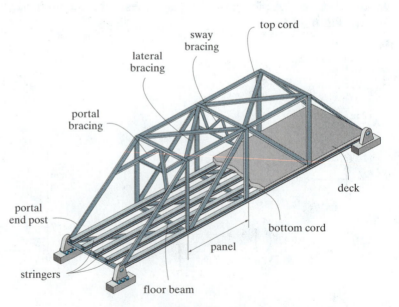

Fig. 6–23

6

EXAMPLE 6.15

Draw the influence line for the force in member *GB* of the bridge truss shown in Fig. 6–24*a*.

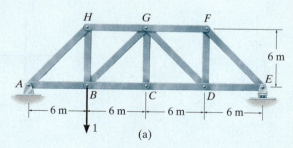

(a)

Fig. 6–24

SOLUTION

Tabulate Values.   Here each successive joint at the bottom cord is loaded with a unit load and the force in member *GB* is calculated using the method of sections, Fig. 6–24*b*. For example, placing the unit load at $x = 6$ m (joint *B*), the support reaction at *E* is calculated first, Fig. 6–24*a*, then passing a section through *HG*, *GB*, *BC* and isolating the right segment, the force in *GB* is determined, Fig. 6–24*c*. In the same manner, determine the other values listed in the table.

Influence Line.   Plotting the tabular data and connecting the points yields the influence line for member *GB*, Fig. 6–24*d*. Since the influence line extends over the entire span of the truss, member *GB* is referred to as a *primary member*. This means *GB* is subjected to a force regardless of where the bridge deck (roadway) is loaded, except, of course, at $x = 8$ m. The point of zero force, $x = 8$ m, is determined by similar triangles between $x = 6$ m and $x = 12$ m, that is, $(0.354 + 0.707)/(12 - 6) = 0.354/x'$, $x' = 2$ m, so $x = 6 + 2 = 8$ m.

x	F_{GB}
0	0
6	0.354
12	−0.707
18	−0.354
24	0

(b)

F_{HG} 45°

F_{GB}

F_{BC}

$\Sigma F_y = 0$; $0.25 - F_{GB} \sin 45° = 0$
$F_{GB} = 0.354$

0.25

(c)

F_{GB}

0.354

8   12   18   24 x
6

−0.354

−0.707

influence line for F_{GB}

(d)

EXAMPLE 6.16

Draw the influence line for the force in member CG of the bridge truss shown in Fig. 6–25a.

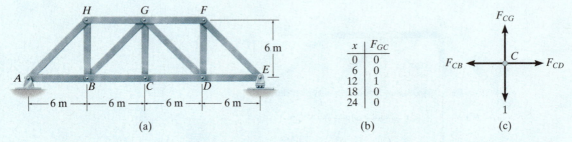

x	F_{GC}
0	0
6	0
12	1
18	0
24	0

(a)   (b)   (c)

Fig. 6–25

SOLUTION

Tabulate Values. A table of unit-load position at the joints of the bottom cord versus the force in member CG is shown in Fig. 6–25b. These values are easily obtained by isolating joint C, Fig. 6–25c. Here it is seen that CG is a zero-force member unless the unit load is applied at joint C, in which case $F_{CG} = 1$ (T).

Influence Line. Plotting the tabular data and connecting the points yields the influence line for member CG as shown in Fig. 6–25d. In particular, notice that when the unit load is at $x = 9$ m, the force in member CG is $F_{CG} = 0.5$. This situation requires the unit load to be placed on the bridge deck *between* the joints. The transference of this load from the deck to the truss is shown in Fig. 6–25e. From this one can see that indeed $F_{CG} = 0.5$ by analyzing the equilibrium of joint C, Fig. 6–25f. Since the influence line for CG does *not* extend over the entire span of the truss, Fig. 6–25d, member CG is referred to as a *secondary member*.

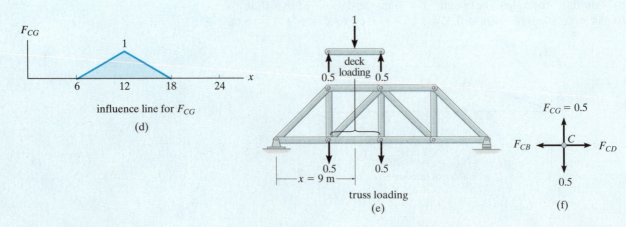

influence line for F_{CG}

(d)

truss loading

(e)

$F_{CG} = 0.5$

(f)

EXAMPLE 6.17

In order to determine the maximum force in each member of the Warren truss, shown in the photo, we must first draw the influence lines for each of its members. If we consider a similar truss as shown in Fig. 6–26a, determine the largest force that can be developed in member BC due to a moving force of 25 k and a moving distributed load of 0.6 k/ft. The loading is applied at the top cord.

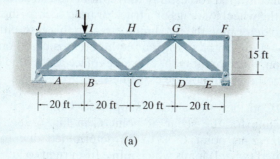

(a)

x	F_{BC}
0	0
20	1
40	0.667
60	0.333
80	0

(b)

Fig. 6–26

SOLUTION

Tabulate Values. A table of unit-load position x at the joints along the top cord versus the force in member BC is shown in Fig. 6–26b. The method of sections can be used for the calculations. For example, when the unit load is at joint I ($x = 20$ ft), Fig. 6–26a, the reaction E_y is determined first ($E_y = 0.25$). Then the truss is sectioned through BC, IC, and HI, and the right segment is isolated, Fig. 6–26c. One obtains F_{BC} by summing moments about point I, to eliminate F_{HI} and F_{IC}. In a similar manner determine the other values in Fig. 6–26b.

Influence Line. A plot of the tabular values yields the influence line, Fig. 6–26d. By inspection, BC is a primary member. Why?

Concentrated Live Force. The largest force in member BC occurs when the moving force of 25 k is placed at $x = 20$ ft. Thus,

$$F_{BC} = (1.00)(25) = 25.0 \text{ k}$$

Distributed Live Load. The uniform live load must be placed over the entire deck of the truss to create the largest tensile force in BC.* Thus,

$$F_{BC} = \left[\tfrac{1}{2}(80)(1.00)\right]0.6 = 24.0 \text{ k}$$

Total Maximum Force.

$$(F_{BC})_{max} = 25.0 \text{ k} + 24.0 \text{ k} = 49.0 \text{ k} \qquad Ans.$$

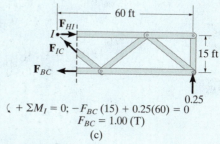

$$\zeta + \Sigma M_I = 0; \ -F_{BC}(15) + 0.25(60) = 0$$
$$F_{BC} = 1.00 \text{ (T)}$$

(c)

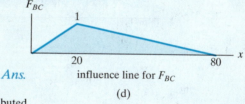

influence line for F_{BC}

(d)

*The largest *tensile* force in member GB of Example 6.15 is created when the distributed load acts on the deck of the truss from $x = 0$ to $x = 8$ m, Fig. 6–24d.

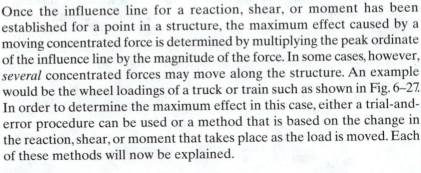

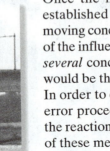

As the train passes over this girder bridge the engine and its cars will exert vertical reactions on the girder. These along with the dead load of the bridge must be considered for design.

6.6   Maximum Influence at a Point due to a Series of Concentrated Loads

Once the influence line for a reaction, shear, or moment has been established for a point in a structure, the maximum effect caused by a moving concentrated force is determined by multiplying the peak ordinate of the influence line by the magnitude of the force. In some cases, however, *several* concentrated forces may move along the structure. An example would be the wheel loadings of a truck or train such as shown in Fig. 6–27. In order to determine the maximum effect in this case, either a trial-and-error procedure can be used or a method that is based on the change in the reaction, shear, or moment that takes place as the load is moved. Each of these methods will now be explained.

Shear.   Consider the simply supported beam with the associated influence line for the shear at point *C* in Fig. 6–28a. The maximum *positive shear* at point *C* is to be determined due to the series of concentrated (wheel) loads which move from right to left over the beam. The critical loading will occur when one of the loads is placed *just to the right* of point *C*, which is coincident with the positive peak of the influence line. By trial and error each of three possible cases can therefore be investigated, Fig. 6–28b. We have

Case 1: $(V_C)_1 = 1(0.75) + 4(0.625) + 4(0.5) = 5.25$ k

Case 2: $(V_C)_2 = 1(-0.125) + 4(0.75) + 4(0.625) = 5.375$ k

Case 3: $(V_C)_3 = 1(0) + 4(-0.125) + 4(0.75) = 2.5$ k

Case 2, with the 1-k force located 5^+ ft from the left support, yields the largest value for V_C and therefore represents the critical loading. Actually, investigation of Case 3 is unnecessary, since by inspection such an arrangement of loads would yield a value of $(V_C)_3$ that would be less than $(V_C)_2$.

Fig. 6–27

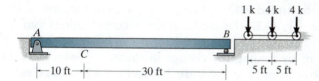

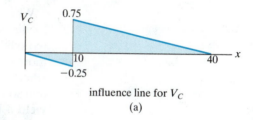

influence line for V_C

(a)

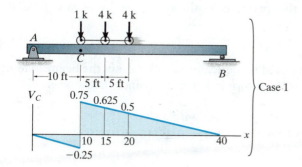

Case 1

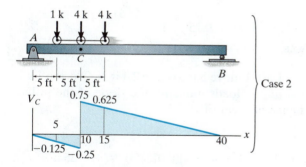

Case 2

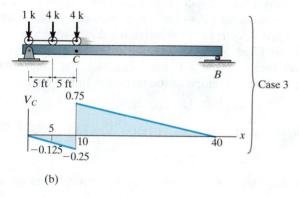

Case 3

(b)

Fig. 6–28

When many concentrated loads act on the span, as in the case of the loadings shown in Fig. 6–27, the trial-and-error computations can be tedious. Instead, the critical position of the loads can be determined in a more direct manner by finding the change in shear, ΔV, which occurs when the loads move from Case 1 to Case 2, then from Case 2 to Case 3, and so on. As long as each computed ΔV is *positive*, the new position will yield a larger shear in the beam at C than the previous position. Each movement is investigated until a negative change in shear is calculated. When this occurs, the previous position of the loads will give the critical value.

To illustrate this method numerically consider again the beam, loading, and influence line for V_C, shown in Fig. 6–28a. Since the slope is

$$s = 0.75/(40 - 10) = 0.25/10 = 0.025$$

and the jump at C has a magnitude of

$$0.75 + 0.25 = 1,$$

then when the loads of Case 1 move 5 ft to Case 2, Fig. 6–28b, the 1-k load jumps *down* (-1) and *all* the loads move *up* the slope of the influence line. This causes a change of shear of

$$\Delta V_{1-2} = 1(-1) + [1 + 4 + 4](0.025)(5) = +0.125 \text{ k}$$

Since this result is positive, Case 2 will yield a larger value for V_C than Case 1. [Compare the answers for $(V_C)_1$ and $(V_C)_2$ previously calculated, where indeed $(V_C)_2 = (V_C)_1 + 0.125$.] Investigating ΔV_{2-3}, which occurs when Case 2 moves to Case 3, Fig. 6–28b, we must account for the downward (negative) jump of the 4-k load and the 5-ft horizontal movement of all the loads *up* the slope of the influence line. We have

$$\Delta V_{2-3} = 4(-1) + (1 + 4 + 4)(0.025)(5) = -2.875 \text{ k}$$

Since ΔV_{2-3} is negative, Case 2 is the position of the critical loading, as determined previously.

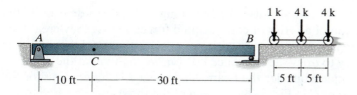

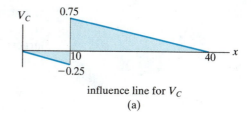

influence line for V_C

(a)

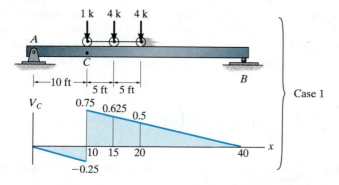

Case 1

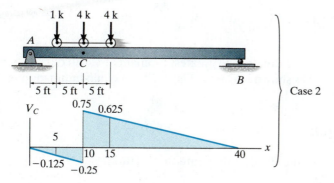

Case 2

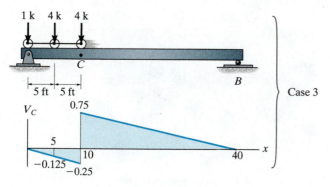

Case 3

Fig. 6–28 (repeated)

Moment. We can also use the foregoing methods to determine the critical position of a series of concentrated forces so that they create the largest internal moment at a specific point in a structure. Of course, it is first necessary to draw the influence line for the moment at the point.

As an example, consider the beam, loading, and influence line for the moment at point C in Fig. 6–29a. If each of the three concentrated forces is placed on the beam, coincident with the peak of the influence line, we will obtain the greatest influence from each force. The three cases of loading are shown in Fig. 6–29b. When the loads of Case 1 are moved 4 ft to the left to Case 2, it is observed that the 2-k load *decreases* ΔM_{1-2}, since the *slope* $(7.5/10)$ is *downward*, Fig. 6–29a. Likewise, the 4-k and 3-k forces cause an *increase* of ΔM_{1-2}, since the *slope* $[7.5/(40 - 10)]$ is *upward*. We have

$$\Delta M_{1-2} = -2\left(\frac{7.5}{10}\right)(4) + (4 + 3)\left(\frac{7.5}{40 - 10}\right)(4) = 1.0 \text{ k} \cdot \text{ft}$$

Since ΔM_{1-2} is positive, we must further investigate moving the loads 6 ft from Case 2 to Case 3.

$$\Delta M_{2-3} = -(2 + 4)\left(\frac{7.5}{10}\right)(6) + 3\left(\frac{7.5}{40 - 10}\right)(6) = -22.5 \text{ k} \cdot \text{ft}$$

Here the change is negative, so the greatest moment at C will occur when the beam is loaded as shown in Case 2, Fig. 6–29c. The maximum moment at C is therefore

$$(M_C)_{\text{max}} = 2(4.5) + 4(7.5) + 3(6.0) = 57.0 \text{ k} \cdot \text{ft}$$

The following examples further illustrate this method.

The girders of this bridge must resist the maximum moment caused by the weight of this jet plane as it passes over it.

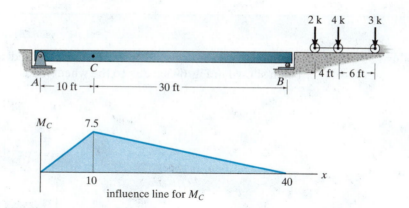

influence line for M_C

(a)

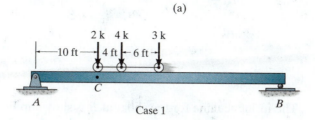

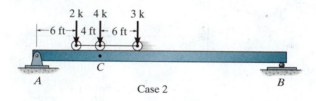

Case 1

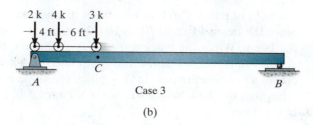

Case 2

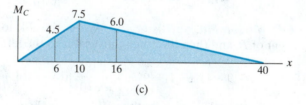

Case 3

(b)

(c)

Fig. 6–29

EXAMPLE 6.18

Determine the maximum positive shear created at point B in the beam shown in Fig. 6–30a due to the wheel loads of the moving truck.

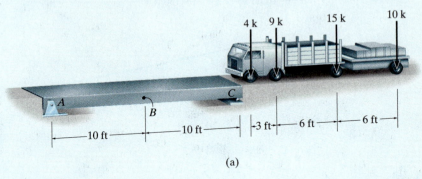

(a)

Fig. 6–30

SOLUTION

The influence line for the shear at B is shown in Fig. 6–30b.

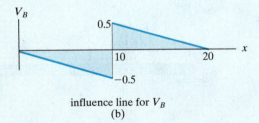

influence line for V_B
(b)

3-ft Movement of 4-k Load.   Imagine that the 4-k load acts just to the right of point B so that we obtain its maximum positive influence. Since the beam segment BC is 10 ft long, the 10-k load is not as yet on the beam. When the truck moves 3 ft to the left, the 4-k load jumps *downward* on the influence line 1 unit and the 4-k, 9-k, and 15-k loads create a positive increase in ΔV_B, since the slope is upward to the left. Although the 10-k load also moves forward 3 ft, it is still not on the beam. Thus,

$$\Delta V_B = 4(-1) + (4 + 9 + 15)\left(\frac{0.5}{10}\right)3 = +0.2 \text{ k}$$

6-ft Movement of 9-k Load.   When the 9-k load acts just to the right of B, and then the truck moves 6 ft to the left, we have

$$\Delta V_B = 9(-1) + (4 + 9 + 15)\left(\frac{0.5}{10}\right)(6) + 10\left(\frac{0.5}{10}\right)(4) = +1.4 \text{ k}$$

Note in the calculation that the 10-k load only moves 4 ft on the beam.

6-ft Movement of 15-k Load.   If the 15-k load is positioned just to the right of B and then the truck moves 6 ft to the left, the 4-k load moves only 1 ft until it is off the beam, and likewise the 9-k load moves only 4 ft until it is off the beam. Hence,

$$\Delta V_B = 15(-1) + 4\left(\frac{0.5}{10}\right)(1) + 9\left(\frac{0.5}{10}\right)(4) + (15 + 10)\left(\frac{0.5}{10}\right)(6)$$

$$= -5.5 \text{ k}$$

Since ΔV_B is now negative, the correct position of the loads occurs when the 15-k load is just to the right of point B, Fig. 6–30c. Consequently,

$$(V_B)_{max} = 4(-0.05) + 9(-0.2) + 15(0.5) + 10(0.2)$$
$$= 7.5 \text{ k} \qquad\qquad Ans.$$

In practice one also has to consider motion of the truck from left to right and then choose the maximum value between these two situations.

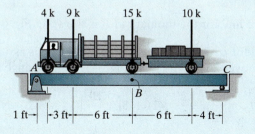

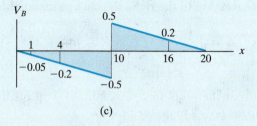

(c)

EXAMPLE 6.19

Determine the maximum positive moment created at point B in the beam shown in Fig. 6–31a due to the wheel loads of the crane.

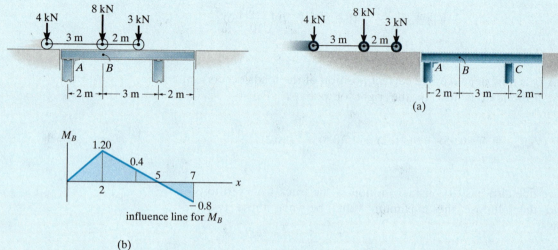

(a)

(b)

influence line for M_B

Fig. 6–31

SOLUTION

The influence line for the moment at B is shown in Fig. 6–31b.

2-m Movement of 3-kN Load.   If the 3-kN load is assumed to act at B and then moves 2 m to the right, Fig. 6–31b, the change in moment is

$$\Delta M_B = -3\left(\frac{1.20}{3}\right)(2) + 8\left(\frac{1.20}{2}\right)(2) = 7.20 \text{ kN} \cdot \text{m}$$

Why is the 4-kN load not included in the calculations?

3-m Movement of 8-kN Load.   If the 8-kN load is assumed to act at B and then moves 3 m to the right, the change in moment is

$$\Delta M_B = -3\left(\frac{1.20}{3}\right)(3) - 8\left(\frac{1.20}{3}\right)(3) + 4\left(\frac{1.20}{2}\right)(2)$$

$$= -8.40 \text{ kN} \cdot \text{m}$$

Notice here that the 4-kN load was initially 1 m off the beam, and therefore moves only 2 m on the beam.

Since there is a sign change in ΔM_B, the correct position of the loads for maximum positive moment at B occurs when the 8-kN force is at B, Fig. 6–31b. Therefore,

$$(M_B)_{\text{max}} = 8(1.20) + 3(0.4) = 10.8 \text{ kN} \cdot \text{m} \qquad \textit{Ans.}$$

EXAMPLE   6.20

Determine the maximum compressive force developed in member *BG* of the side truss in Fig. 6–32a due to the right side wheel loads of the car and trailer. Assume the loads are applied directly to the truss and move only to the right.

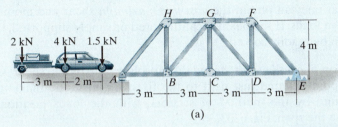

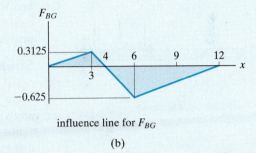

(a)

Fig. 6–32

influence line for F_{BG}

(b)

SOLUTION

The influence line for the force in member *BG* is shown in Fig. 6–32b. Here a trial-and-error approach for the solution will be used. Since we want the greatest negative (compressive) force in *BG*, we begin as follows:

1.5-kN Load at Point C.   In this case

$$F_{BG} = 1.5 \text{ kN}(-0.625) + 4(0) + 2 \text{ kN}\left(\frac{0.3125}{3 \text{ m}}\right)(1 \text{ m})$$

$$= -0.729 \text{ kN}$$

4-kN Load at Point C.   By inspection this would seem a more reasonable case than the previous one.

$$F_{BG} = 4 \text{ kN}(-0.625) + 1.5 \text{ kN}\left(\frac{-0.625}{6 \text{ m}}\right)(4 \text{ m}) + 2 \text{ kN}(0.3125)$$

$$= -2.50 \text{ kN}$$

2-kN Load at Point C.   In this case all loads will create a compressive force in *BC*.

$$F_{BG} = 2 \text{ kN}(-0.625) + 4 \text{ kN}\left(\frac{-0.625}{6 \text{ m}}\right)(3 \text{ m}) + 1.5 \text{ kN}\left(\frac{-0.625}{6 \text{ m}}\right)(1 \text{ m})$$

$$= -2.66 \text{ kN} \qquad\qquad\qquad\qquad Ans.$$

Since this final case results in the largest answer, the critical loading occurs when the 2-kN load is at *C*.

6.7   Absolute Maximum Shear and Moment

In Sec. 6.6 we developed the methods for computing the maximum shear and moment at a *specified point* in a beam due to a series of concentrated moving loads. A more general problem involves the determination of both the *location of the point* in the beam *and the position of the loading* on the beam so that one can obtain the *absolute maximum* shear and moment caused by the loads. If the beam is cantilevered or simply supported, this problem can be readily solved.

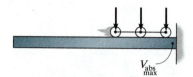

Fig. 6–33

Shear.   For a *cantilevered beam* the absolute maximum shear will occur at a point located just next to the fixed support. The maximum shear is found by the method of sections, with the loads positioned anywhere on the span, Fig. 6–33.

For *simply supported beams* the absolute maximum shear will occur just next to one of the supports. For example, if the loads are equivalent, they are positioned so that the first one in sequence is placed close to the support, as in Fig. 6–34.

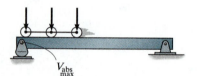

Fig. 6–34

Moment.   The absolute maximum moment for a *cantilevered beam* occurs at the same point where absolute maximum shear occurs, although in this case the concentrated loads should be positioned at the *far end* of the beam, as in Fig. 6–35.

For a *simply supported beam* the critical position of the loads and the associated absolute maximum moment cannot, in general, be determined by inspection. We can, however, determine the position analytically. For purposes of discussion, consider a beam subjected to the forces F_1, F_2, F_3 shown in Fig. 6–36a. Since the moment diagram for a series of concentrated forces consists of straight line segments having peaks at each force, the absolute maximum moment will occur under one of the forces. Assume this maximum moment occurs under F_2. The position of the loads F_1, F_2, F_3 on the beam will be specified by the distance x, measured from F_2 to the beam's centerline as shown. To determine a specific value of x, we first obtain the resultant force of the system, F_R, and its distance

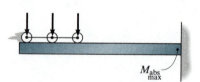

Fig. 6–35

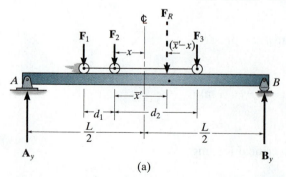

(a)

(b)

Fig. 6–36

$\overline{x}'$ measured from $\mathbf{F}_2$. Once this is done, moments are summed about B, which yields the beam's left reaction, $\mathbf{A}_y$, that is,

$$\Sigma M_B = 0; \qquad A_y = \frac{1}{L}(F_R)\left[\frac{L}{2} - (\overline{x}' - x)\right]$$

If the beam is sectioned just to the left of $\mathbf{F}_2$, the resulting free-body diagram is shown in Fig. 6–36b. The moment $\mathbf{M}_2$ under $\mathbf{F}_2$ is therefore

$$\Sigma M = 0; \qquad M_2 = A_y\left(\frac{L}{2} - x\right) - F_1 d_1$$

$$= \frac{1}{L}(F_R)\left[\frac{L}{2} - (\overline{x}' - x)\right]\left(\frac{L}{2} - x\right) - F_1 d_1$$

$$= \frac{F_R L}{4} - \frac{F_R \overline{x}'}{2} - \frac{F_R x^2}{L} + \frac{F_R x \overline{x}'}{L} - F_1 d_1$$

The absolute maximum moment in this girder bridge is the result of the moving concentrated loads caused by the wheels of these train cars. The cars must be in the critical position, and the location of the point in the girder where the absolute maximum moment occurs must be identified.

For maximum M_2 we require

$$\frac{dM_2}{dx} = \frac{-2F_R x}{L} + \frac{F_R \overline{x}'}{L} = 0$$

or

$$x = \frac{\overline{x}'}{2}$$

Hence, we may conclude that the *absolute maximum moment in a simply supported beam occurs under one of the concentrated forces, such that this force is positioned on the beam so that it and the resultant force of the system are equidistant from the beam's centerline.* Since there are a series of loads on the span (for example, $\mathbf{F}_1$, $\mathbf{F}_2$, $\mathbf{F}_3$ in Fig. 6–36a), this principle will have to be applied to each load in the series and the corresponding maximum moment computed. By comparison, the largest moment is the absolute maximum. As a general rule, though, the absolute maximum moment often occurs under the largest force lying nearest the resultant force of the system.

Envelope of Maximum Influence-Line Values.

Rules or formulations for determining the absolute maximum shear or moment are difficult to establish for beams supported in ways other than the cantilever or simple support discussed here. An elementary way to proceed to solve this problem, however, requires constructing influence lines for the shear or moment at selected points along the entire length of the beam and then computing the maximum shear or moment in the beam for each point using the methods of Sec. 6.6. These values when plotted yield an "envelope of maximums," from which both the absolute maximum value of shear or moment and its location can be found. Obviously, a computer solution for this problem is desirable for complicated situations, since the work can be rather tedious if carried out by hand calculations.

EXAMPLE 6.21

Determine the absolute maximum moment in the simply supported bridge deck shown in Fig. 6–37a.

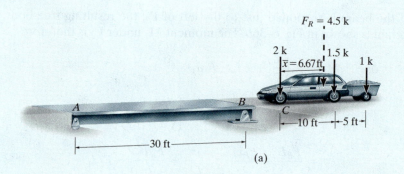

(a)

SOLUTION

The magnitude and position of the resultant force of the system are determined first, Fig. 6–37a. We have

$$+\downarrow F_R = \Sigma F; \qquad F_R = 2 + 1.5 + 1 = 4.5 \text{ k}$$

$$\zeta + M_{R_C} = \Sigma M_C; \qquad 4.5\bar{x} = 1.5(10) + 1(15)$$

$$\bar{x} = 6.67 \text{ ft}$$

Let us first assume the absolute maximum moment occurs under the 1.5-k load. The load and the resultant force are positioned equidistant from the beam's centerline, Fig. 6–37b. Calculating A_y first, Fig. 6–37b, we have

$$\zeta + \Sigma M_B = 0; \qquad -A_y(30) + 4.5(16.67) = 0 \qquad A_y = 2.50 \text{ k}$$

Now using the left section of the beam, Fig. 6–37c, yields

$$\zeta + \Sigma M_S = 0; \qquad -2.50(16.67) + 2(10) + M_S = 0$$

$$M_S = 21.7 \text{ k} \cdot \text{ft}$$

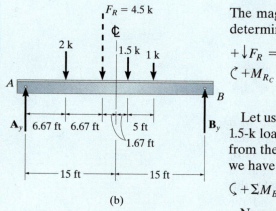

(b)

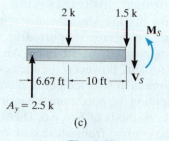

(c)

Fig. 6–37

There is a possibility that the absolute maximum moment may occur under the 2-k load, since 2 k > 1.5 k and $\mathbf{F}_R$ is between both 2 k and 1.5 k. To investigate this case, the 2-k load and $\mathbf{F}_R$ are positioned equidistant from the beam's centerline, Fig. 6–37d. Show that $A_y = 1.75$ k as indicated in Fig. 6–37e and that

$$M_S = 20.4 \text{ k} \cdot \text{ft}$$

By comparison, the absolute maximum moment is

$$M_S = 21.7 \text{ k} \cdot \text{ft} \qquad\qquad Ans.$$

which occurs under the 1.5-k load, when the loads are positioned on the beam as shown in Fig. 6–37b.

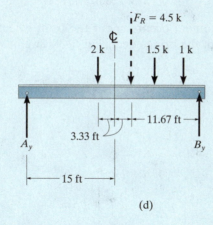

(d)

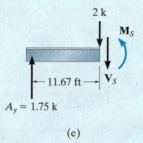

(e)

EXAMPLE 6.22

The truck has a mass of 2 Mg and a center of gravity at G as shown in Fig. 6–38a. Determine the absolute maximum moment developed in the simply supported bridge deck due to the truck's weight. The bridge has a length of 10 m.

SOLUTION

As noted in Fig. 6–38a, the weight of the truck, $2(10^3)\,kg(9.81\ m/s^2) = 19.62\,kN$, and the wheel reactions have been calculated by statics. Since the largest reaction occurs at the front wheel, we will select this wheel along with the resultant force and position them *equidistant* from the centerline of the bridge, Fig. 6–38b. Using the resultant force rather than the wheel loads, the vertical reaction at B is then

$$\zeta + \Sigma M_A = 0; \qquad B_y(10) - 19.62(4.5) = 0$$
$$B_y = 8.829\ kN$$

The maximum moment occurs under the front wheel loading. Using the right section of the bridge deck, Fig. 6–38c, we have

$$\zeta + \Sigma M_S = 0; \qquad 8.829(4.5) - M_s = 0$$
$$M_s = 39.7\ kN \cdot m \qquad\qquad Ans.$$

19.62 kN

G

2 m   1 m

6.54 kN   13.08 kN

(a)

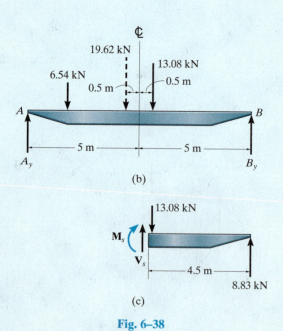

(b)

(c)

Fig. 6–38

FUNDAMENTAL PROBLEMS

F6–1. Use the Müller-Breslau principle to sketch the influence lines for the vertical reaction at A, the shear at C, and the moment at C.

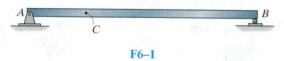

F6–1

F6–2. Use the Müller-Breslau principle to sketch the influence lines for the vertical reaction at A, the shear at D, and the moment at B.

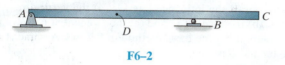

F6–2

F6–3. Use the Müller-Breslau principle to sketch the influence lines for the vertical reaction at A, the shear at D, and the moment at D.

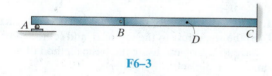

F6–3

F6–4. Use the Müller-Breslau principle to sketch the influence lines for the vertical reaction at A, the shear at B, and the moment at B.

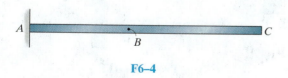

F6–4

F6–5. Use the Müller-Breslau principle to sketch the influence lines for the vertical reaction at A, the shear at C, and the moment at C.

F6–5

F6–6. Use the Müller-Breslau principle to sketch the influence lines for the vertical reaction at A, the shear just to the left of the roller support at E, and the moment at A.

F6–6

F6–7. The beam supports a distributed live load of 1.5 kN/m and single concentrated load of 8 kN. The dead load is 2 kN/m. Determine (a) the maximum positive moment at C, (b) the maximum positive shear at C.

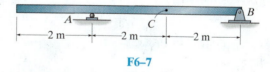

F6–7

F6–8. The beam supports a distributed live load of 2 kN/m and single concentrated load of 6 kN. The dead load is 4 kN/m. Determine (a) the maximum vertical positive reaction at C, (b) the maximum negative moment at A.

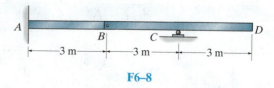

F6–8

6

PROBLEMS

Sec. 6.1–6.3

6–1. Draw the influence lines for (a) the moment at C, (b) the vertical reaction at B, and (c) the shear at C. Assume B is a fixed support. Solve this problem using the basic method of Sec. 6.1.

6–2. Solve Prob. 6–1 using the Müller-Breslau principle.

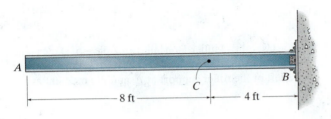

Probs. 6–1/2

6–3. Draw the influence lines for (a) the vertical reaction at A, (b) the moment at C, and (c) the shear just to the left of the support at B. Solve this problem using the basic method of Sec. 6.1.

*6–4. Solve Prob. 6–3 using the Müller-Breslau principle.

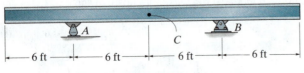

Probs. 6–3/4

6–5. Draw the influence lines for (a) the vertical reaction at B, (b) the shear just to the right of the rocker at A, and (c) the moment at C. Solve this problem using the basic method of Sec. 6.1.

6–6. Solve Prob. 6–5 using Müller-Breslau's principle.

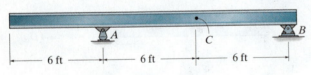

Probs. 6–5/6

6–7. Draw the influence line for (a) the moment at B, (b) the shear at C, and (c) the vertical reaction at B. Solve this problem using the basic method of Sec. 6.1. *Hint:* The support at A resists only a horizontal force and a bending moment.

*6–8. Solve Prob. 6–7 using the Müller-Breslau principle.

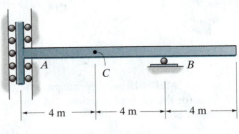

Probs. 6–7/8

6–9. Draw the influence line for (a) the vertical reaction at A, (b) the shear at C, and (c) the moment at C. Solve this problem using the basic method of Sec. 6.1. Assume A is a roller and B is a pin.

6–10. Solve Prob. 6–9 using Müller-Breslau's principle.

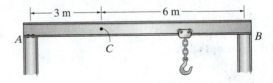

Probs. 6–9/10

6–11. Draw the influence lines for (a) the vertical reaction at B, (b) the shear just to the left of B, and (c) the moment at B. Solve this problem using the basic method of Sec. 6.1. Assume A is a pin and B is a roller.

*6–12. Solve Prob. 6–11 using Müller-Breslau's principle.

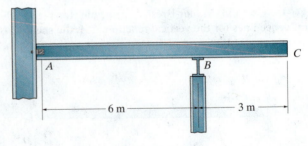

Probs. 6–11/12

6–13. Draw the influence line for (a) the moment at C, (b) the shear just to the right of the support at B, and (c) the vertical reaction at B. Solve this problem using the basic method of Sec. 6.1. Assume A is a pin and B is a roller.

6–14. Solve Prob. 6–13 using the Müller-Breslau principle.

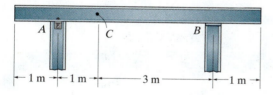

Probs. 6–13/14

6–15. The beam is subjected to a uniform dead load of 1.2 kN/m and a single live load of 40 kN. Determine (a) the maximum moment created by these loads at C, and (b) the maximum positive shear at C. Assume A is a pin, and B is a roller.

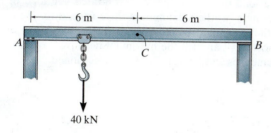

Prob. 6–15

*6–16. Draw the influence line for (a) the force in the cable BC, (b) the vertical reaction at A, and (c) the moment at D.

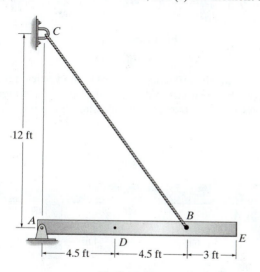

Prob. 6–16

6–17. A uniform live load of 300 lb/ft and a single live concentrated force of 1500 lb are to be placed on the beam. The beam has a weight of 150 lb/ft. Determine (a) the maximum vertical reaction at support B, and (b) the maximum negative moment at point B. Assume the support at A is a pin and B is a roller.

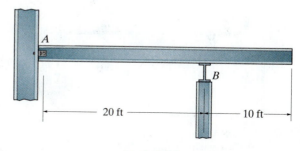

Prob. 6–17

6–18. The beam supports a uniform dead load of 0.4 k/ft, a live load of 1.5 k/ft, and a single live concentrated force of 8 k. Determine (a) the maximum positive moment at C, and (b) the maximum positive vertical reaction at B. Assume A is a roller and B is a pin.

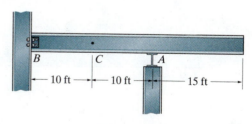

Prob. 6–18

6–19. The beam is used to support a dead load of 0.6 k/ft, a live load of 2 k/ft and a concentrated live load of 8 k. Determine (a) the maximum positive (upward) reaction at A, (b) the maximum positive moment at C, and (c) the maximum positive shear just to the right of the support at A. Assume the support at A is a pin and B is a roller.

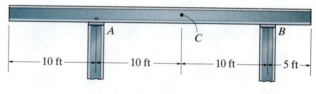

Prob. 6–19

*6–20. The compound beam is subjected to a uniform dead load of 1.5 kN/m and a single live load of 10 kN. Determine (a) the maximum negative moment created by these loads at A, and (b) the maximum positive shear at B. Assume A is a fixed support, B is a pin, and C is a roller.

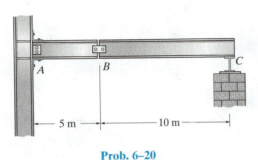

Prob. 6–20

6–21. Where should a single 500-lb live load be placed on the beam so it causes the largest moment at D? What is this moment? Assume the support at A is fixed, B is pinned, and C is a roller.

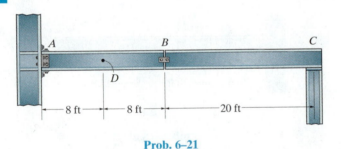

Prob. 6–21

6–22. The beam is subjected to a uniform live load of 1.2 kN/m, a dead load of 0.5 kN/m, and a single live load of 40 kN. Determine (a) the maximum positive moment created by these loads at E, and (b) the maximum positive shear at E. Assume A and C are roller, and B is a short link.

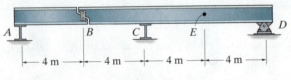

Prob. 6–22

6–23. The beam supports a uniform dead load of 500 N/m and a single live concentrated force of 3000 N. Determine (a) the maximum negative moment at E, and (b) the maximum positive shear at E. Assume the support at A is a pin, B and D are rollers, and C is a pin.

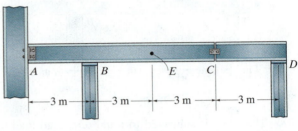

Prob. 6–23

*6–24. The beam is used to support a dead load of 400 lb/ft, a live load of 2 k/ft, and a concentrated live load of 8 k. Determine (a) the maximum positive vertical reaction at A, (b) the maximum positive shear just to the right of the support at A, and (c) the maximum negative moment at C. Assume A is a roller, C is fixed, and B is pinned.

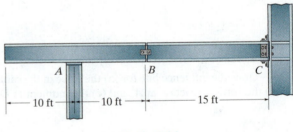

Prob. 6–24

6–25. Draw the influence line for (a) the vertical reaction at A, (b) the shear just to the right of B, and (c) the moment at A. Assume A is fixed, C is a roller, and B is a pin.

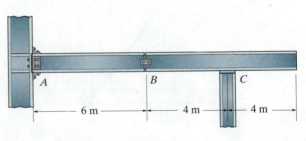

Prob. 6–25

Sec. 6.4

6–26. A uniform live load of 1.8 kN/m and a single concentrated live force of 4 kN are placed on the floor beams. Determine (a) the maximum positive shear in panel BC of the girder and (b) the maximum moment in the girder at G.

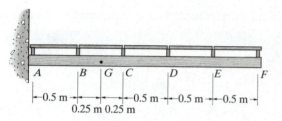

Prob. 6–26

6–27. Draw the influence line for the moment at E in the girder. Determine the maximum positive moment in the girder at E if a single concentrated live force of 5 kN and a uniform live load of 1.5 kN/m can be placed on the floor beams. Assume A is a pin and D is a roller.

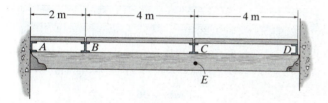

Prob. 6–27

*6–28. A uniform live load of 2 k/ft and a single concentrated live force of 6 k are placed on the floor beams. If the beams also support a uniform dead load of 350 lb/ft, determine (a) the maximum positive shear in panel CD of the girder and (b) the maximum negative moment in the girder at D. Assume the support at C is a roller and E is a pin.

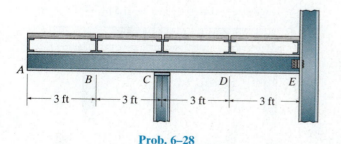

Prob. 6–28

6–29. Draw the influence line for (a) the shear in panel BC of the girder, and (b) the moment at D.

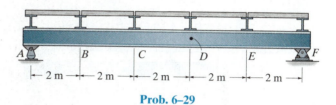

Prob. 6–29

6–30. A uniform live load of 250 lb/ft and a single concentrated live force of 1.5 k are to be placed on the floor beams. Determine (a) the maximum positive shear in panel AB, and (b) the maximum moment at D. Assume only vertical reaction occur at the supports.

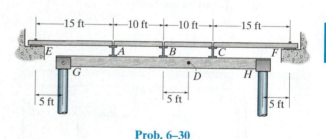

Prob. 6–30

6–31. A uniform live load of 16 kN/m and a single concentrated live force of 34 kN are placed on the top beams. If the beams also support a uniform dead load of 3 kN/m, determine (a) the maximum positive shear in panel BC of the girder and (b) the maximum positive moment in the girder at C. Assume B is a roller and D is a pin.

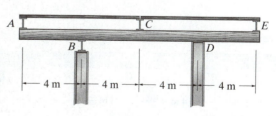

Prob. 6–31

*6–32. A uniform live load of 0.4 k/ft and a single concentrated live force of 6 k are placed on the floor beams. Determine (a) the maximum positive shear in panel *EF* of the girder, and (b) the maximum positive moment at *H*.

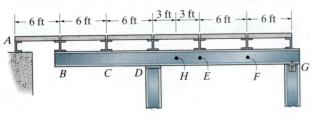

Prob. 6–32

6–33. A uniform live load of 300 lb/ft and a single concentrated live force of 2 k are to be placed on the floor beams. Determine (a) the maximum negative shear in panel *AB*, and (b) the maximum negative moment at *B*. Assume the supports at *A* and *E* are pins and the pipe columns only exert vertical reactions on the beams.

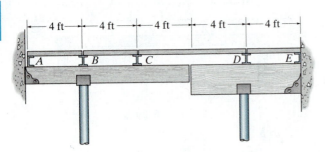

Prob. 6–33

6–34. A uniform live load of 8 kN/m and a single concentrated live force of 12 kN are placed on the floor beams. If the beams also support a uniform dead load of 400 N/m, determine (a) the maximum positive shear in panel *CD* of the girder and (b) the maximum positive moment in the girder at *C*.

Prob. 6–34

6–35. Draw the influence line for the shear in panel *CD* of the girder. Determine the maximum negative live shear in panel *CD* due to a uniform live load of 500 lb/ft acting on the top beams.

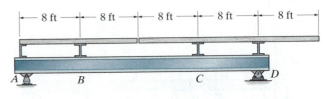

Prob. 6–35

*6–36. A uniform live load of 6.5 kN/m and a single concentrated live force of 15 kN are placed on the floor beams. If the beams also support a uniform dead load of 600 N/m, determine (a) the maximum positive shear in panel *CD* of the girder and (b) the maximum positive moment in the girder at *D*.

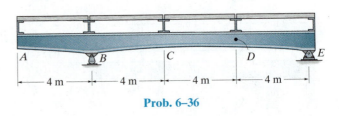

Prob. 6–36

6–37. A uniform live load of 1.75 kN/m and a single concentrated live force of 8 kN are placed on the floor beams. If the beams also support a uniform dead load of 250 N/m, determine (a) the maximum negative shear in panel *BC* of the girder and (b) the maximum positive moment at *B*.

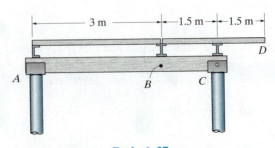

Prob. 6–37

Sec. 6.5

6–38. Draw the influence line for the force in (a) member *KJ* and (b) member *CJ*.

6–39. Draw the influence line for the force in (a) member *JI*, (b) member *IE*, and (c) member *EF*.

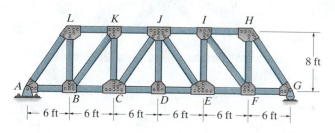

Probs. 6–38/39

*6–40. Draw the influence line for the force in member *KJ*.

6–41. Draw the influence line for the force in member *JE*.

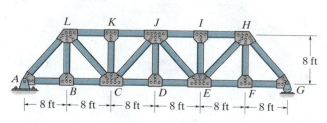

Probs. 6–40/41

6–42. Draw the influence line for the force in member *CD*.

6–43. Draw the influence line for the force in member *JK*.

*6–44. Draw the influence line for the force in member *DK*.

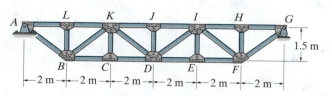

Probs. 6–42/43/44

6–45. Draw the influence line for the force in (a) member *EH* and (b) member *JE*.

6–46. Draw the influence line for the force in member *JI*.

6–47. Draw the influence line for the force in member *AL*.

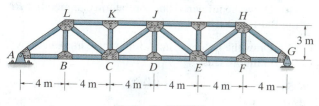

Probs. 6–45/46/47

*6–48. Draw the influence line for the force in member *BC* of the Warren truss. Indicate numerical values for the peaks. All members have the same length.

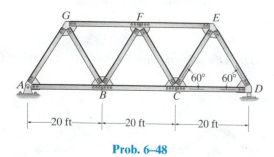

Prob. 6–48

6–49. Draw the influence line for the force in member *CD* of the Baltimore truss.

6–50. Draw the influence line for the force in member *PG* of the Baltimore truss.

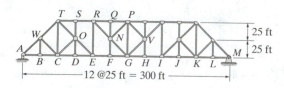

Probs. 6–49/50

6–51. Draw the influence line for the force in member RQ of the Baltimore truss.

*6–52. Draw the influence line for the force in member TC of the Baltimore truss.

6–53. Draw the influence line for the force in member NP of the Baltimore truss.

6–54. Draw the influence line for the force in member RN of the Baltimore truss.

6–55. Draw the influence line for the force in member NG of the Baltimore truss.

*6–56. Draw the influence line for the force in member CO of the Baltimore truss.

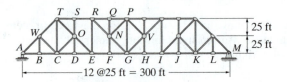

Probs. 6–51/52/53/54/55/56

6–57. Draw the influence line for the force in member CD.

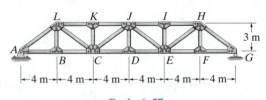

Prob. 6–57

6–58. Draw the influence line for the force in member KJ.

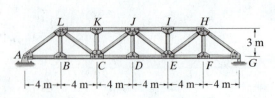

Prob. 6–58

6–59. Draw the influence line for the force in member GD, then determine the maximum force (tension or compression) that can be developed in this member due to a uniform live load of 3 kN/m that acts on the bridge deck along the bottom cord of the truss.

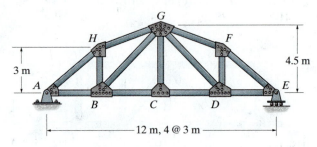

Prob. 6–59

*6–60. Draw the influence line for the force in member CD, and then determine the maximum force (tension or compression) that can be developed in this member due to a uniform live load of 800 lb/ft which acts along the bottom cord of the truss.

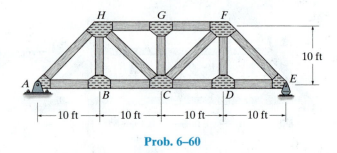

Prob. 6–60

6–61. Draw the influence line for the force in member CF, and then determine the maximum force (tension or compression) that can be developed in this member due to a uniform live load of 800 lb/ft which is transmitted to the truss along the bottom cord.

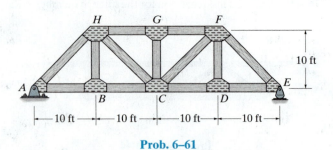

Prob. 6–61

Sec. 6.6

6–62. Determine the maximum moment at point C on the single girder caused by the moving dolly that has a mass of 2 Mg and a mass center at G. Assume A is a roller.

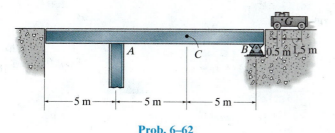

Prob. 6–62

6–63. Determine the maximum moment in the suspended rail at point B if the rail supports the load of 2.5 k on the trolley.

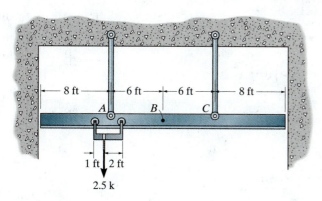

Prob. 6–63

*6–64. Determine the maximum live moment at C caused by the moving loads.

6–65. Determine the maximum live shear at C caused by the moving loads.

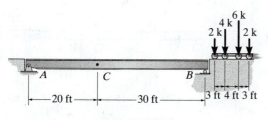

Probs. 6–64/65

6–66. Determine the maximum positive moment at the splice C on the side girder caused by the moving load which travels along the center of the bridge.

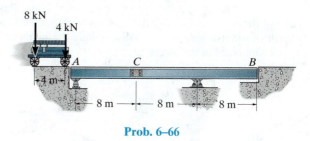

Prob. 6–66

6–67. Determine the maximum moment at C caused by the moving load.

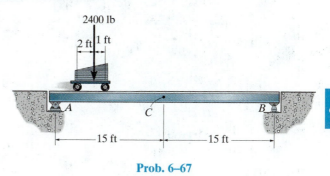

Prob. 6–67

*6–68. The truck and trailer exerts the wheel reactions shown on the deck of the girder bridge. Determine the largest moment it exerts in the splice at C. Assume the truck travels in *either direction* along the *center* of the deck, and therefore transfers *half* of the load shown to each of the two side girders. Assume the splice is a fixed connection and, like the girder, can support both shear and moment.

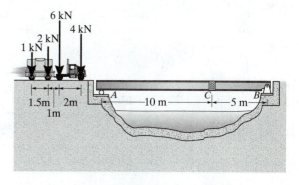

Prob. 6–68

6–69. Draw the influence line for the force in member *GF* of the bridge truss. Determine the maximum live force (tension or compression) that can be developed in the member due to a 5-k truck having the wheel loads shown. Assume the truck can travel in *either direction* along the *center* of the deck, so that *half* the load shown is transferred to each of the two side trusses. Also assume the members are pin connected at the gusset plates.

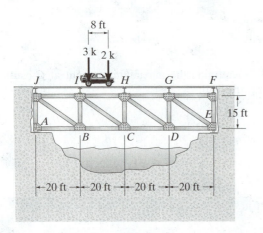

Prob. 6–69

6–70. The 9-k truck exerts the wheel reactions shown on the deck of a girder bridge. Determine (a) the largest live shear it creates in the splice at *C*, and (b) the largest moment it exerts in the splice. Assume the truck travels in *either direction* along the *center* of the deck, and therefore transfers *half* of the load shown in each of the two side girders. Assume the splice is a fixed connection and, like the girder, can support both shear and moment.

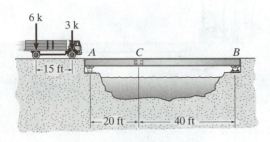

Prob. 6–70

Sec. 6.7

6–71. The truck has a mass of 4 Mg and mass center at G_1, and the trailer has a mass of 1 Mg and mass center at G_2. Determine the absolute maximum live moment developed in the bridge.

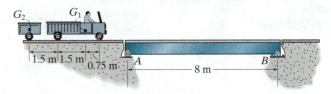

Prob. 6–71

*6–72. The truck has a mass of 4-Mg and mass center at *G*. Determine the absolute maximum live moment developed in the bridge.

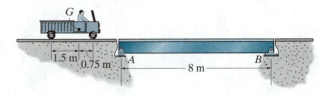

Prob. 6–72

6–73. Determine the absolute maximum live moment in the girder bridge due to the loading shown. The load is applied directly to the girder.

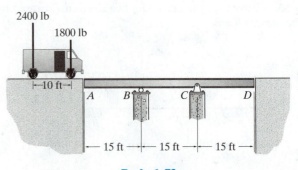

Prob. 6–73

6–74. Determine the absolute maximum live shear and absolute maximum live moment in the jib beam AB due to the crane loading. The end constraints require $0.1 \text{ m} \le x \le 3.9 \text{ m}$.

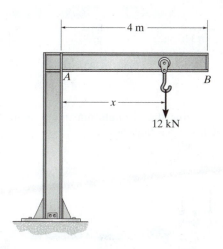

Prob. 6–74

6–75. The maximum wheel loadings for the wheels of a crane that is used in an industrial building are given. The crane travels along the runway girders that are simply supported on columns. Determine (a) the absolute maximum shear in an intermediate girder AB, and (b) the absolute maximum moment in the girder.

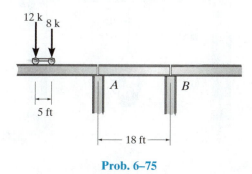

Prob. 6–75

*6–76. The cart has a mass of 2 Mg and center of mass at G. Determine the maximum live moment created in the side girder at C as it crosses the bridge. Assume the cart travels along the *center* of the deck, so that *half* the load shown is transferred to each of the two side girders.

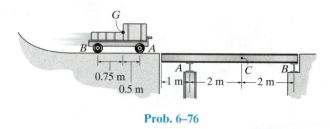

Prob. 6–76

6–77. Determine the absolute maximum shear in the beam due to the loading shown.

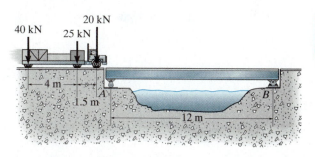

Prob. 6–77

6–78. Determine the absolute maximum moment in the beam due to the loading shown.

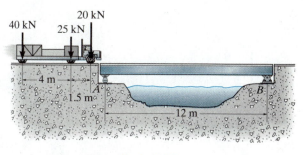

Prob. 6–78

6–79. Determine the absolute maximum shear in the bridge girder due to the loading shown.

***6–80.** Determine the absolute maximum moment in the bridge girder due to the loading shown.

6–82. Determine the absolute maximum shear in the beam due to the loading shown.

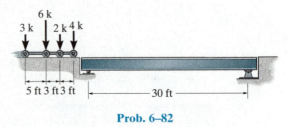

Prob. 6–82

6–83. Determine the absolute maximum moment in the bridge due to the loading shown.

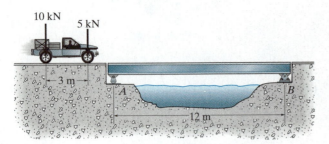

Probs. 6–79/80

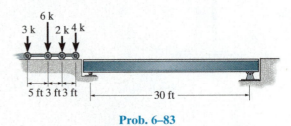

Prob. 6–83

***6–84.** The trolley rolls at *C* and *D* along the bottom and top flange of beam *AB*. Determine the absolute maximum moment developed in the beam if the load supported by the trolley is 2 k. Assume the support at *A* is a pin and at *B* a roller.

6–81. Determine the absolute maximum live moment in the bridge.

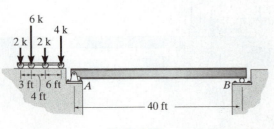

Prob. 6–81

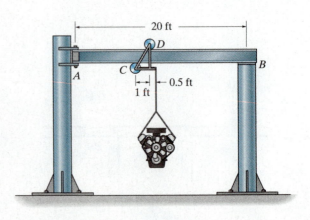

Prob. 6–84

PROJECT PROBLEMS

6–1P. The chain hoist on the wall crane can be placed anywhere along the boom (0.1 m < x < 3.4 m) and has a rated capacity of 28 kN. Use an impact factor of 0.3 and determine the absolute maximum bending moment in the boom and the maximum force developed in the tie rod BC. The boom is pinned to the wall column at its left end A. Neglect the size of the trolley at D.

6–2P. A simply supported pedestrian bridge is to be constructed in a city park and two designs have been proposed as shown in case a and case b. The truss members are to be made from timber. The deck consists of 1.5-m-long planks that have a mass of 20 kg/m². A local code states the live load on the deck is required to be 5 kPa with an impact factor of 0.2. Consider the deck to be simply supported on stringers. Floor beams then transmit the load to the bottom joints of the truss. (See Fig. 6–23.) In each case find the member subjected to the largest tension and largest compression load and suggest why you would choose one design over the other. Neglect the weights of the truss members.

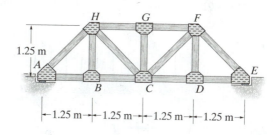

case *a*

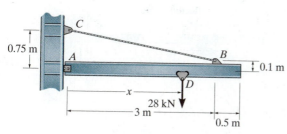

Prob. 6–1P

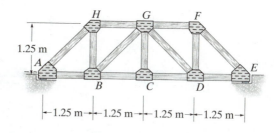

case *b*

Prob. 6–2P

CHAPTER REVIEW

An influence line indicates the value of a reaction, shear, or moment at a specific point on a member, as a unit load moves over the member.

Once the influence line for a reaction, shear, or moment (function) is constructed, then it will be possible to locate the live load on the member to produce the maximum positive or negative value of the function.

A concentrated live force is applied at the positive (negative) peaks of the influence line. The value of the function is then equal to the product of the influence line ordinate and the magnitude of the force.

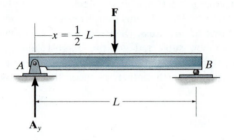

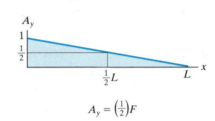

$$A_y = \left(\tfrac{1}{2}\right)F$$

A uniform distributed load extends over a positive (negative) region of the influence line. The value of the function is then equal to the product of the area under the influence line for the region and the magnitude of the uniform load.

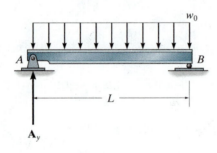

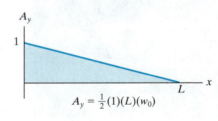

$$A_y = \tfrac{1}{2}(1)(L)(w_0)$$

The general shape of the influence line can be established using the Müller-Breslau principle, which states that the influence line for a reaction, shear, or moment is to the same scale as the deflected shape of the member when it is acted upon by the reaction, shear, or moment.

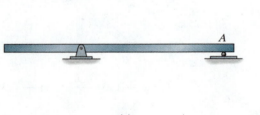

(a)

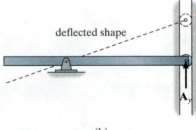

(b)

Influence lines for floor girders and trusses can be established by placing the unit load at each panel point or joint, and calculating the value of the required reaction, shear, or moment.

When a series of concentrated loads pass over the member, then the various positions of the load on the member have to be considered to determine the largest shear or moment in the member. In general, place the loadings so that each contributes its maximum influence, as determined by multiplying each load by the ordinate of the influence line. This process of finding the actual position can be done using a trial-and-error procedure, or by finding the change in either the shear or moment when the loads are moved from one position to another. Each moment is investigated until a negative change in the shear or moment occurs. Once this happens the previous position will define the critical loading.

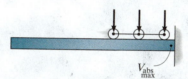

Absolute maximum *shear* in a cantilever or simply supported beam will occur at a support, when one of the loads is placed next to the support.

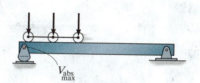

Absolute maximum *moment* in a cantilevered beam occurs when the series of concentrated loads are placed at the farthest point away from the fixed support.

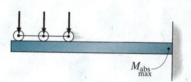

To determine the absolute maximum moment in a simply supported beam, the resultant of the force system is first determined. Then it, along with one of the concentrated forces in the system is positioned so that these two forces are equidistant from the centerline of the beam. The maximum moment then occurs under the selected force. Each force in the system is selected in this manner, and by comparison the largest for all these cases is the absolute maximum moment.(f)

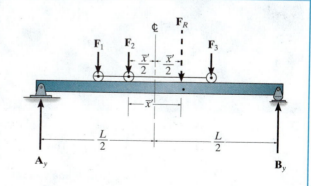

Chapter 7

© Chris Aschenbrener/Alamy

The portal frame of this bridge must resist the lateral loads caused by traffic and wind. Because this frame is statically indeterminate, an approximate analysis can be made of the loadings to design the members, before a more exact structural analysis is considered.

Approximate Analysis of Statically Indeterminate Structures

In this chapter we will present some of the approximate methods used to analyze statically indeterminate trusses and frames. These methods were developed on the basis of structural behavior, and their accuracy in most cases compares favorably with more exact methods of analysis. Although not all types of structural forms will be discussed here, it is hoped that enough insight is gained from the study of these methods so that one can judge what would be the best approximations to make when performing an approximate force analysis of a statically indeterminate structure.

7.1 Use of Approximate Methods

When a *model* is used to represent any structure, the analysis of it must satisfy *both* the conditions of equilibrium and compatibility of displacement at the joints. As will be shown in later chapters of this text, the compatibility conditions for a *statically indeterminate* structure can be related to the loads provided we know the material's modulus of elasticity and the size and shape of the members. For an initial design, however, we will *not* know a member's size, and so a statically indeterminate analysis cannot be considered. For analysis a simpler model of the structure must be developed, one that is statically determinate. Once this model is specified, the analysis of it is called an ***approximate analysis***. By performing an approximate analysis, a preliminary design of the members of a structure can be made, and when this is complete, the more exact indeterminate analysis can then be performed and the design refined. An approximate analysis also provides insight as to a structure's behavior under load and is beneficial when checking a more exact analysis or when time, money, or capability are not available for performing the more exact analysis.

Realize that, in a general sense, all methods of structural analysis are approximate, simply because the actual conditions of loading, geometry, material behavior, and joint resistance at the supports are never known in an *exact sense*. In this text, however, the statically indeterminate analysis of a structure will be referred to as an *exact analysis,* and the simpler statically determinate analysis will be referred to as the *approximate analysis*.

7.2   Trusses

A common type of truss often used for lateral bracing of a building or for the top and bottom cords of a bridge is shown in Fig. 7–1a. (Also see Fig. 3–4.) When used for this purpose, this truss is not considered a primary element for the support of the structure, and as a result it is often analyzed by approximate methods. In the case shown, it will be noticed that if a diagonal is removed from each of the three panels, it will render the truss statically determinate. Hence, the truss is statically indeterminate to the third degree (using Eq. 3–1, $b + r > 2j$, or $16 + 3 > 8(2)$) and therefore we must make three assumptions regarding the bar forces in order to reduce the truss to one that is statically determinate. These assumptions can be made with regard to the cross-diagonals, realizing that when one diagonal in a panel is in tension the corresponding cross-diagonal will be in compression. This is evident from Fig. 7–1b, where the "panel shear" **V** is carried by the *vertical component* of tensile force in member *a* and the *vertical component* of compressive force in member *b*. Two methods of analysis are generally acceptable.

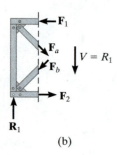

(b)

Fig. 7–1

Method 1:   If the diagonals are intentionally designed to be *long and slender,* it is reasonable to assume that they *cannot support a compressive force;* otherwise, they may easily buckle. Hence the panel shear is resisted entirely by the *tension diagonal,* whereas the *compressive diagonal is assumed to be a zero-force member*.

Method 2:   If the diagonal members are intended to be constructed from large rolled sections such as angles or channels, they may be equally capable of supporting a tensile and compressive force. Here we will assume that the tension and compression diagonals each carry *half* the panel shear.

Both of these methods of approximate analysis are illustrated numerically in the following examples.

An approximate method can be used to determine the forces in the cross bracing in each panel of this bascule railroad bridge. Here the cross members are thin and so we can assume they carry no compressive force.

EXAMPLE 7.1

Determine (approximately) the forces in the members of the truss shown in Fig. 7–2a. The diagonals are to be designed to support both tensile and compressive forces, and therefore each is assumed to carry half the panel shear. The support reactions have been computed.

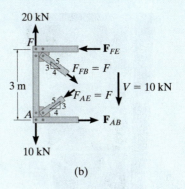

(b)

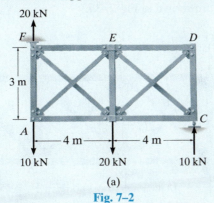

(a)

Fig. 7–2

SOLUTION

By inspection the truss is statically indeterminate to the second degree. The two assumptions require the tensile and compressive diagonals to carry equal forces, that is, $F_{FB} = F_{AE} = F$. For a vertical section through the left panel, Fig. 7–2b, we have

$+\uparrow \Sigma F_y = 0;$ $20 - 10 - 2(\frac{3}{5})F = 0$ $F = 8.33$ kN   *Ans.*

so that

$$F_{FB} = 8.33 \text{ kN (T)} \qquad Ans.$$

$$F_{AE} = 8.33 \text{ kN (C)} \qquad Ans.$$

$\zeta + \Sigma M_A = 0;$ $-8.33(\frac{4}{5})(3) + F_{FE}(3) = 0$ $F_{FE} = 6.67$ kN (C)   *Ans.*

$\zeta + \Sigma M_F = 0;$ $-8.33(\frac{4}{5})(3) + F_{AB}(3) = 0$ $F_{AB} = 6.67$ kN (T)   *Ans.*

From joint A, Fig. 7–2c,

$+\uparrow \Sigma F_y = 0;$ $F_{AF} - 8.33(\frac{3}{5}) - 10 = 0$ $F_{AF} = 15$ kN (T)   *Ans.*

A vertical section through the right panel is shown in Fig. 7–2d. Show that

$$F_{DB} = 8.33 \text{ kN (T)}, \quad F_{ED} = 6.67 \text{ kN (C)} \qquad Ans.$$

$$F_{EC} = 8.33 \text{ kN (C)}, \quad F_{BC} = 6.67 \text{ kN (T)} \qquad Ans.$$

Furthermore, using the free-body diagrams of joints D and E, Figs. 7–2e and 7–2f, show that

$$F_{DC} = 5 \text{ kN (C)} \qquad Ans.$$

$$F_{EB} = 10 \text{ kN (T)} \qquad Ans.$$

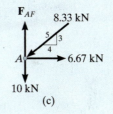

(c)

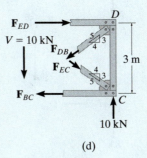

(d)

(e)

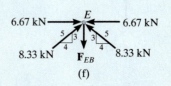

(f)

7

EXAMPLE 7.2

Cross bracing is used to provide lateral support for this bridge deck due to the wind and unbalanced traffic loads. Determine (approximately) the forces in the members of this truss. Assume the diagonals are slender and therefore will not support a compressive force. The loads and support reactions are shown in Fig. 7–3a.

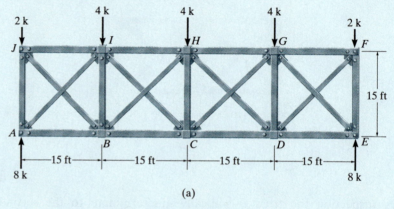

(a)

Fig. 7–3

SOLUTION

By inspection the truss is statically indeterminate to the fourth degree. Thus the four assumptions to be used require that each compression diagonal sustain zero force. Hence, from a vertical section through the left panel, Fig. 7–3b, we have

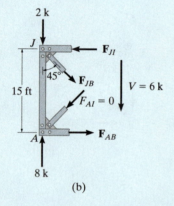

(b)

$$F_{AI} = 0 \qquad \text{Ans.}$$

$$+\uparrow \Sigma F_y = 0; \qquad 8 - 2 - F_{JB} \cos 45° = 0$$

$$F_{JB} = 8.49 \text{ k (T)} \qquad \text{Ans.}$$

$$\zeta + \Sigma M_A = 0; \qquad -8.49 \sin 45°(15) + F_{JI}(15) = 0$$

$$F_{JI} = 6 \text{ k (C)} \qquad \text{Ans.}$$

$$\zeta + \Sigma M_J = 0; \qquad -F_{AB}(15) = 0$$

$$F_{AB} = 0 \qquad \text{Ans.}$$

From joint A, Fig. 7–3c,

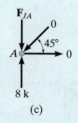

(c)

$$F_{JA} = 8 \text{ k (C)} \qquad \text{Ans.}$$

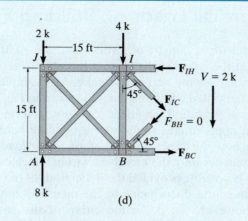

(d)

A vertical section of the truss through members $IH, IC, BH,$ and BC is shown in Fig. 7–3d. The panel shear is $V = \Sigma F_y = 8 - 2 - 4 = 2$ k. We require

$$F_{BH} = 0 \qquad \text{Ans.}$$

$+\uparrow \Sigma F_y = 0;$ $\qquad 8 - 2 - 4 - F_{IC} \cos 45° = 0$

$$F_{IC} = 2.83 \text{ k (T)} \qquad \text{Ans.}$$

$\zeta + \Sigma M_B = 0;$ $\qquad -8(15) + 2(15) - 2.83 \sin 45°(15) + F_{IH}(15) = 0$

$$F_{IH} = 8 \text{ k (C)} \qquad \text{Ans.}$$

$\zeta + \Sigma M_I = 0;$ $\qquad -8(15) + 2(15) + F_{BC}(15) = 0$

$$F_{BC} = 6 \text{ k (T)} \qquad \text{Ans.}$$

From joint B, Fig. 7–3e,

$+\uparrow \Sigma F_y = 0;$ $\qquad 8.49 \sin 45° - F_{BI} = 0$

$$F_{BI} = 6 \text{ k (C)} \qquad \text{Ans.}$$

The forces in the other members can be determined by symmetry, except F_{CH}; however, from joint C, Fig. 7–3f, we have

$+\uparrow \Sigma F_y = 0;$ $\qquad 2(2.83 \sin 45°) - F_{CH} = 0$

$$F_{CH} = 4 \text{ k (C)} \qquad \text{Ans.}$$

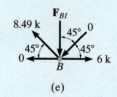

(e)

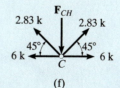

(f)

7.3   Vertical Loads on Building Frames

Building frames often consist of girders that are *rigidly connected* to columns so that the entire structure is better able to resist the effects of lateral forces due to wind and earthquake. An example of such a rigid framework, often called a building bent, is shown in Fig. 7–4.

In practice, a structural engineer can use several techniques for performing an approximate analysis of a building bent. Each is based upon knowing how the structure *will deform under load*. One technique would be to consider only the members within a localized region of the structure. This is possible provided the deflections of the members within the region cause little disturbance to the members outside the region. Most often, however, the deflection curve of the entire structure is considered. From this, the approximate location of points of inflection, that is, the points where the member changes its curvature, can be specified. These points can be considered as *pins* since there is zero moment within the member at the points of inflection. We will use this idea in this section to analyze the forces on building frames due to vertical loads, and in Secs. 7.5 and 7.6 an approximate analysis for frames subjected to lateral loads will be presented. Since the frame can be subjected to both of these loadings simultaneously, then, provided the material remains elastic, the resultant loading is determined by superposition.

Assumptions for Approximate Analysis.

Consider a typical girder located within a building bent and subjected to a uniform vertical load, as shown in Fig. 7–5a. The column supports at A and B will each exert three reactions on the girder, and therefore the girder will be statically indeterminate to the third degree (6 reactions – 3 equations of equilibrium). To make the girder statically determinate, an approximate analysis will therefore require three assumptions. If the columns are extremely stiff, no rotation at A and B will occur, and the deflection curve for the girder will look like that shown in Fig. 7–5b. Using one of

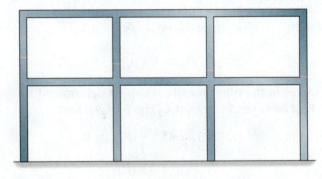

typical building frame

Fig. 7–4

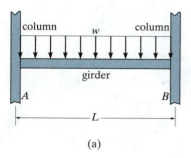

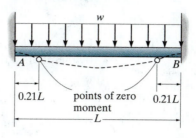

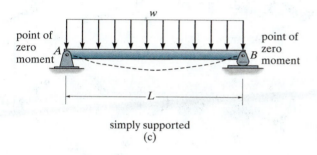

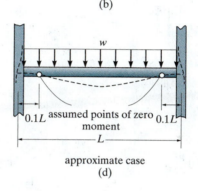

Fig. 7–5

the methods presented in Chapters 9 through 11, an exact analysis reveals that for this case inflection points, or points of zero moment, occur at $0.21L$ from each support. If, however, the column connections at A and B are very flexible, then like a simply supported beam, zero moment will occur at the supports, Fig. 7–5c. In reality, however, the columns will provide some flexibility at the supports, and therefore we will assume that zero moment occurs at the *average point* between the two extremes, i.e., at $(0.21L + 0)/2 \approx 0.1L$ from each support, Fig. 7–5d. Furthermore, an exact analysis of frames supporting vertical loads indicates that the axial forces in the girder are negligible.

In summary then, each girder of length L may be modeled by a simply supported span of length $0.8L$ resting on two cantilevered ends, each having a length of $0.1L$, Fig. 7–5e. The following three assumptions are incorporated in this model:

1. There is zero moment in the girder, $0.1L$ from the left support.

2. There is zero moment in the girder, $0.1L$ from the right support.

3. The girder does not support an axial force.

By using statics, the internal loadings in the girders can now be obtained and a preliminary design of their cross sections can be made. Be aware, however, that this method has this *limited application*. For example, it cannot be extended to give the force and moment reactions at the supports on the frame. The intent here is simply to approximate the loadings within the girders. The following example illustrates this numerically.

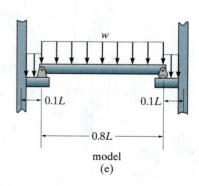

model
(e)

EXAMPLE 7.3

Determine (approximately) the moment at the joints E and C caused by members EF and CD of the building bent in Fig. 7–6a.

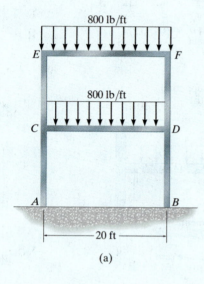

(a)

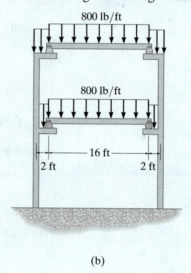

(b)

Fig. 7–6

SOLUTION

For an approximate analysis the frame is modeled as shown in Fig. 7–6b. Note that the cantilevered spans supporting the center portion of the girder have a length of $0.1L = 0.1(20) = 2$ ft. Equilibrium requires the end reactions for the center portion of the girder to be 6400 lb, Fig. 7–6c. The cantilevered spans are then subjected to a reaction moment of

$$M = 1600(1) + 6400(2) = 14\,400 \text{ lb·ft} = 14.4 \text{ k·ft} \qquad Ans.$$

This approximate moment, with opposite direction, acts on the joints at E and C, Fig. 7–6a. Using the results, the approximate moment diagram for one of the girders is shown in Fig. 7–6d.

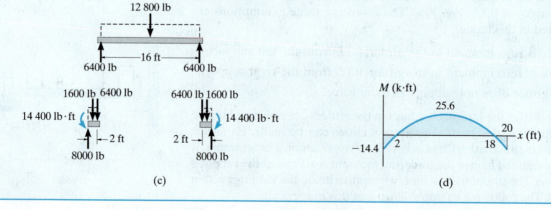

(c)

(d)

7.4 Portal Frames and Trusses

Frames. *Portal frames* are frequently used over the entrance of a bridge* and as a main stiffening element in building design in order to transfer horizontal or lateral forces applied at the top of the frame to the foundation. On bridges, these frames resist the forces caused by wind, earthquake, and unbalanced traffic loading on the bridge deck. Portals can be pin supported, fixed supported, or supported by partial fixity. The approximate analysis of each case will now be discussed for a simple three-member portal.

Pin Supported. A typical pin-supported portal frame is shown in Fig. 7–7a. Since four unknowns exist at the supports but only three equilibrium equations are available for solution, this structure is statically indeterminate to the first degree. Consequently, only one assumption must be made to reduce the frame to one that is statically determinate.

The elastic deflection of the portal is shown in Fig. 7–7b. This diagram indicates that a point of inflection, that is, where the moment changes from positive bending to negative bending, is located *approximately* at the girder's midpoint. Since the moment in the girder is zero at this point, we can *assume* a hinge exists there and then proceed to determine the reactions at the supports using statics. If this is done, it is found that the horizontal reactions (shear) at the base of each column are *equal* and the other reactions are those indicated in Fig. 7–7c. Furthermore, the moment diagrams for this frame are indicated in Fig. 7–7d.

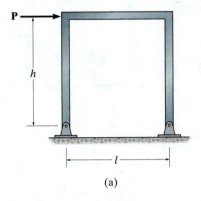

(a)

Fig. 7–7

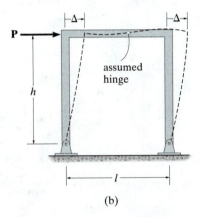

(b)

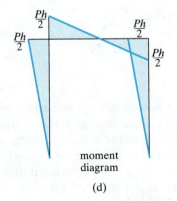

moment diagram

(d)

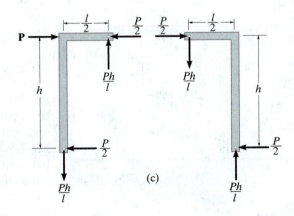

(c)

*See Fig. 3–4.

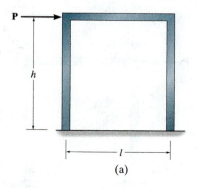

(a)

Fixed Supported. Portals with two fixed supports, Fig. 7–8*a*, are statically indeterminate to the third degree since there are a total of six unknowns at the supports. If the vertical members have equal lengths and cross-sectional areas, the frame will deflect as shown in Fig. 7–8*b*. For this case we will *assume* points of inflection occur at the midpoints of all three members, and therefore hinges are placed at these points. The reactions and moment diagrams for each member can therefore be determined by dismembering the frame at the hinges and applying the equations of equilibrium to each of the four parts. The results are shown in Fig. 7–8*c*. Note that, as in the case of the pin-connected portal, the horizontal reactions (shear) at the base of each column are *equal*. The moment diagram for this frame is indicated in Fig. 7–8*d*.

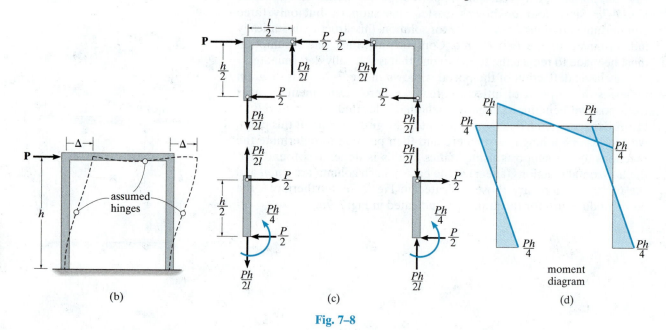

(b)

(c)

(d)

Fig. 7–8

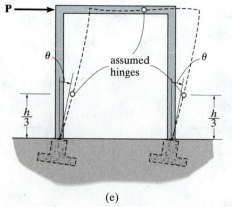

(e)

Fig. 7–9

Partial Fixity. Since it is both difficult and costly to construct a perfectly fixed support or foundation for a portal frame, it is conservative and somewhat realistic to assume a slight rotation occurs at the supports, Fig. 7–9. As a result, the points of inflection on the columns lie somewhere between the case of having a pin-supported portal, Fig. 7–7*b*, where the "inflection points" are at the supports (base of columns), and a fixed-supported portal, Fig. 7–8*b*, where the inflection points are at the center of the columns. Many engineers arbitrarily define the location at $h/3$, Fig. 7–9, and therefore place hinges at these points, and also at the center of the girder.

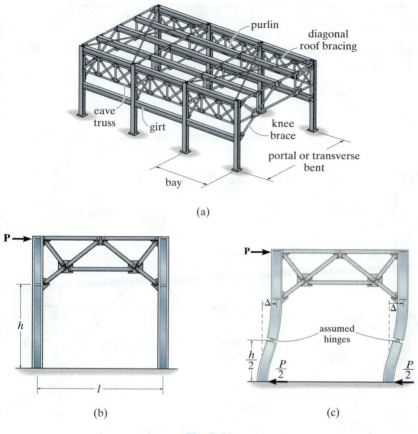

(a)

(b)   (c)

Fig. 7–10

Trusses. When a portal is used to span large distances, a truss may be used in place of the horizontal girder. Such a structure is used on large bridges and as transverse bents for large auditoriums and industrial buildings. An example of a typical industrial building along with the terminology of its members is shown in Fig. 7–10a. If we consider the approximate analysis of one of its portal frames, Fig. 7–10b, then the suspended roof truss is assumed to be pin connected at its points of attachment to the columns. Furthermore, the truss keeps the columns straight within the region of attachment when the portal is subjected to the sidesway Δ, Fig. 7–10c. Consequently, we can analyze trussed portals using the same assumptions as those used for simple portal frames. For pin-supported columns, assume the horizontal reactions (shear) are equal, as in Fig. 7–7c. For fixed-supported columns loaded at their top, assume the horizontal reactions are equal and an inflection point (or hinge) occurs on each column, measured midway between the base of the column and the *lowest point* of truss member connection to the column. See Fig. 7–8c and Fig. 7–10c.

The following example illustrates how to determine the forces in the members of a trussed portal using the approximate method of analysis described above.

EXAMPLE   7.4

Determine by approximate methods the forces acting in the members of the Warren portal shown in Fig. 7–11a.

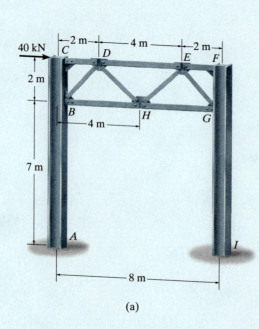

(a)

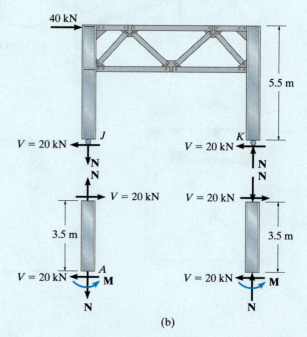

(b)

Fig. 7–11

SOLUTION

The truss portion B, C, F, G acts as a rigid unit. Since the supports are fixed, a point of inflection is assumed to exist $7\,\text{m}/2 = 3.5\,\text{m}$ above A and I, and equal horizontal reactions or shear act at the base of the columns, i.e., $\Sigma F_x = 0$; $V = 40\,\text{kN}/2 = 20\,\text{kN}$. With these assumptions, we can separate the structure at the hinges J and K, Fig. 7–11b, and determine the reactions on the columns as follows:

Lower Half of Column

$\zeta + \Sigma M_A = 0;$ $M - 3.5(20) = 0$ $M = 70\,\text{kN}\cdot\text{m}$

Upper Portion of Column

$\zeta + \Sigma M_J = 0;$ $-40(5.5) + N(8) = 0$ $N = 27.5\,\text{kN}$

Using the method of sections, Fig. 7–11c, we can now proceed to obtain the forces in members CD, BD, and BH.

$$+\uparrow\Sigma F_y = 0; \quad -27.5 + F_{BD}\sin 45° = 0 \qquad F_{BD} = 38.9 \text{ kN (T) } Ans.$$

$$\zeta+\Sigma M_B = 0; \quad -20(3.5) - 40(2) + F_{CD}(2) = 0 \quad F_{CD} = 75 \text{ kN (C) } Ans.$$

$$\zeta+\Sigma M_D = 0; \quad F_{BH}(2) - 20(5.5) + 27.5(2) = 0 \quad F_{BH} = 27.5 \text{ kN (T) } Ans.$$

In a similar manner, show that one obtains the results on the free-body diagram of column FGI in Fig. 7–11d. Using these results, we can now find the force in each of the other truss members of the portal using the method of joints.

Joint D, Fig. 7–11e

$$+\uparrow\Sigma F_y = 0; \quad F_{DH}\sin 45° - 38.9\sin 45° = 0 \quad F_{DH} = 38.9 \text{ kN (C) } Ans.$$

$$\underset{\rightarrow}{+}\Sigma F_x = 0; \quad 75 - 2(38.9\cos 45°) - F_{DE} = 0 \quad F_{DE} = 20 \text{ kN (C) } Ans.$$

Joint H, Fig. 7–11f

$$+\uparrow\Sigma F_y = 0; \quad F_{HE}\sin 45° - 38.9\sin 45° = 0 \quad F_{HE} = 38.9 \text{ kN (T) } Ans.$$

These results are summarized in Fig. 7–11g.

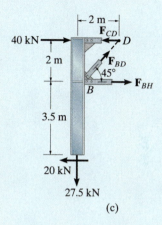

(c)

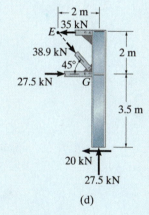

(d)

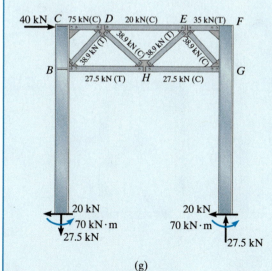

(g)

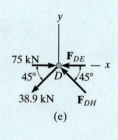

(e)

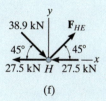

(f)

7.5   Lateral Loads on Building Frames: Portal Method

In Sec. 7.4 we discussed the action of lateral loads on portal frames and found that for a frame fixed supported at its base, points of inflection occur at approximately the center of each girder and column and the columns carry equal shear loads, Fig. 7–8. A building bent deflects in the same way as a portal frame, Fig. 7–12a, and therefore it would be appropriate to assume inflection points occur at the center of the columns and girders. If we consider each bent of the frame to be composed of a series of portals, Fig. 7–12b, then as a further assumption, the *interior columns* would represent the effect of *two portal columns* and would therefore carry twice the shear V as the two exterior columns.

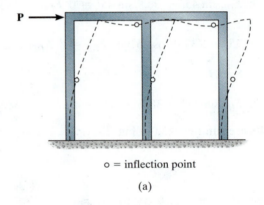

o = inflection point

(a)

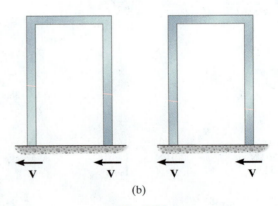

(b)

Fig. 7–12

In summary, the portal method for analyzing fixed-supported building frames requires the following assumptions:

1. A hinge is placed at the center of each girder, since this is assumed to be a point of zero moment.

2. A hinge is placed at the center of each column, since this is assumed to be a point of zero moment.

3. At a given floor level the shear at the interior column hinges is twice that at the exterior column hinges, since the frame is considered to be a superposition of portals.

These assumptions provide an adequate reduction of the frame to one that is statically determinate yet stable under loading.

By comparison with the more exact statically indeterminate analysis, *the **portal method** is most suitable for buildings having low elevation and uniform framing.* The reason for this has to do with the structure's action under load. In this regard, *consider the frame as acting like a cantilevered beam* that is fixed to the ground. Recall from mechanics of materials that *shear resistance* becomes more important in the design of *short* beams, whereas *bending* is more important if the beam is *long*. (See Sec. 7.6.) The portal method is based on the assumption related to shear as stated in item 3 above.

The following examples illustrate how to apply the portal method to analyze a building bent.

The portal method of analysis can be used to (approximately) perform a lateral-load analysis of this single-story frame.

EXAMPLE 7.5

Determine (approximately) the reactions at the base of the columns of the frame shown in Fig. 7–13a. Use the portal method of analysis.

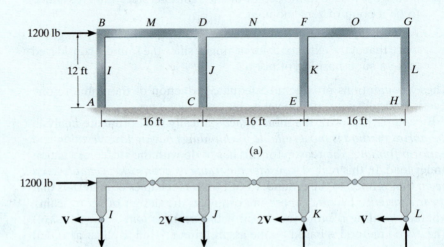

(a)

(b)

Fig. 7–13

SOLUTION

Applying the first two assumptions of the portal method, we place hinges at the centers of the girders and columns of the frame, Fig. 7–13a. A section through the column hinges at I, J, K, L yields the free-body diagram shown in Fig. 7–13b. Here the third assumption regarding the column shears applies. We require

$$\xrightarrow{+} \Sigma F_x = 0; \qquad 1200 - 6V = 0 \qquad V = 200 \text{ lb}$$

Using this result, we can now proceed to dismember the frame at the hinges and determine their reactions. *As a general rule, always start this analysis at the corner or joint where the horizontal load is applied.* Hence, the free-body diagram of segment *IBM* is shown in Fig. 7–13c. The three reaction components at the hinges I_y, M_x, and M_y are determined by applying $\Sigma M_M = 0, \Sigma F_x = 0, \Sigma F_y = 0$, respectively. The adjacent segment *MJN* is analyzed next, Fig. 7–13d, followed by segment *NKO*, Fig. 7–13e, and finally segment *OGL*, Fig. 7–13f. Using these results, the free-body diagrams of the columns with their support reactions are shown in Fig. 7–13g.

If the horizontal segments of the girders in Figs. 7–13c, d, e and f are considered, show that the moment diagram for the girder looks like that shown in Fig. 7–13h.

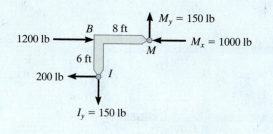

(c)

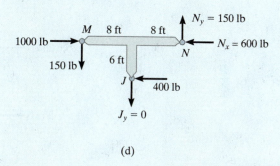

(d)

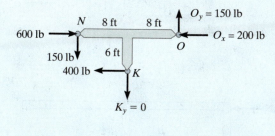

(e)

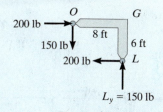

(f)

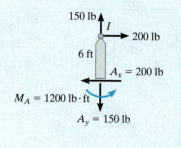
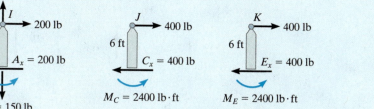
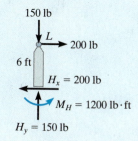

(g)

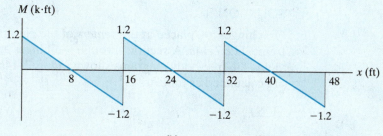

(h)

EXAMPLE 7.6

Determine (approximately) the reactions at the base of the columns of the frame shown in Fig. 7–14a. Use the portal method of analysis.

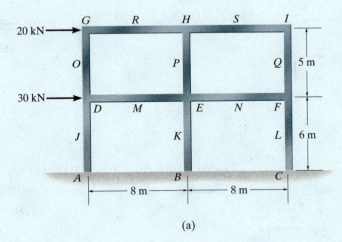

(a)

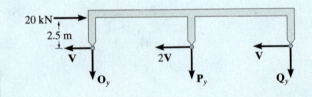

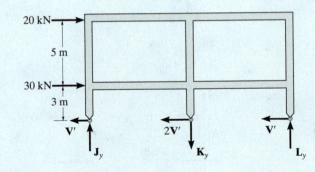

(b)

Fig. 7–14

SOLUTION

First hinges are placed at the *centers* of the girders and columns of the frame, Fig. 7–14a. A section through the hinges at O, P, Q and J, K, L yields the free-body diagrams shown in Fig. 7–14b. The column shears are calculated as follows:

$$\xrightarrow{+} \Sigma F_x = 0; \qquad 20 - 4V = 0 \qquad V = 5 \text{ kN}$$

$$\xrightarrow{+} \Sigma F_x = 0; \qquad 20 + 30 - 4V' = 0 \qquad V' = 12.5 \text{ kN}$$

Using these results, we can now proceed to analyze each part of the frame. The analysis starts with the *corner* segment *OGR*, Fig. 7–14c. The three unknowns O_y, R_x, and R_y have been calculated using the equations of equilibrium. With these results segment *OJM* is analyzed next, Fig. 7–14d; then segment *JA*, Fig. 7–14e; *RPS*, Fig. 7–14f; *PMKN*, Fig. 7–14g; and *KB*, Fig. 7–14h. Complete this example and analyze segments *SIQ*, then *QNL*, and finally *LC*, and show that $C_x = 12.5$ kN, $C_y = 15.625$ kN, and $M_C = 37.5$ kN·m. Also, use the results and show that the moment diagram for *DMENF* is given in Fig. 7–14i.

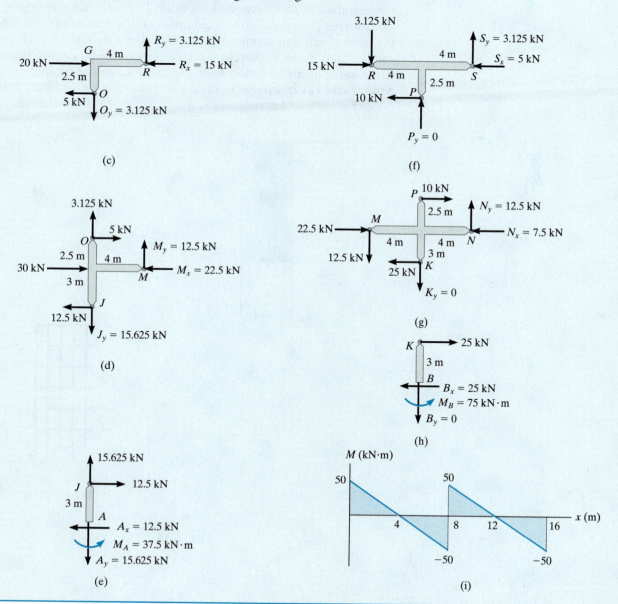

(c)

(f)

(d)

(g)

(h)

(e)

(i)

7.6   Lateral Loads on Building Frames: Cantilever Method

The cantilever method is based on the same action as a long cantilevered beam subjected to a transverse load. It may be recalled from mechanics of materials that such a loading causes a bending stress in the beam that varies linearly from the beam's neutral axis, Fig. 7–15a. In a similar manner, the lateral loads on a frame tend to tip the frame over, or cause a rotation of the frame about a "neutral axis" lying in a horizontal plane that passes through the columns between each floor. To counteract this tipping, the axial forces (or stress) in the columns will be tensile on one side of the neutral axis and compressive on the other side, Fig. 7–15b. Like the cantilevered beam, it therefore seems reasonable to assume this axial stress has a linear variation from the centroid of the column areas or neutral axis. *The **cantilever method** is therefore appropriate if the frame is tall and slender, or has columns with different cross-sectional areas.*

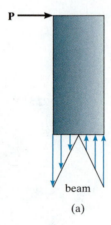

beam

(a)

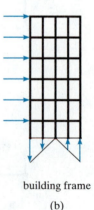

building frame

(b)

Fig. 7–15

In summary, using the cantilever method, the following assumptions apply to a fixed-supported frame.

1. A hinge is placed at the center of each girder, since this is assumed to be a point of zero moment.

2. A hinge is placed at the center of each column, since this is assumed to be a point of zero moment.

3. The axial *stress* in a column is proportional to its distance from the centroid of the cross-sectional areas of the columns at a given floor level. Since stress equals force per area, then in the special case of the *columns having equal cross-sectional areas,* the *force* in a column is also proportional to its distance from the centroid of the column areas.

These three assumptions reduce the frame to one that is both stable and statically determinate.

The following examples illustrate how to apply the cantilever method to analyze a building bent.

The building framework has rigid connections. A lateral-load analysis can be performed (approximately) by using the cantilever method of analysis.

EXAMPLE 7.7

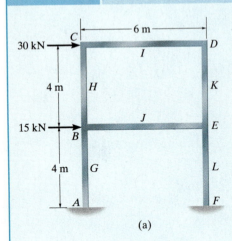

(a)

Fig. 7–16

Determine (approximately) the reactions at the base of the columns of the frame shown in Fig. 7–16a. The columns are assumed to have equal cross-sectional areas. Use the cantilever method of analysis.

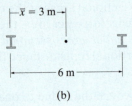

(b)

SOLUTION

First hinges are placed at the midpoints of the columns and girders. The locations of these points are indicated by the letters G through L in Fig. 7–16a. The centroid of the columns' cross-sectional areas can be determined by inspection, Fig. 7–16b, or analytically as follows:

$$\bar{x} = \frac{\Sigma \tilde{x} A}{\Sigma A} = \frac{0(A) + 6(A)}{A + A} = 3 \text{ m}$$

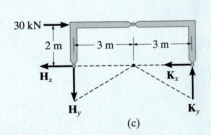

(c)

The axial *stress* in each column is thus proportional to its distance from this point. Here the columns have the same cross-sectional area, and so the force in each column is also proportional to its distance from the centroid. Hence, a section through the hinges H and K at the top story yields the free-body diagram shown in Fig. 7–16c. Note how the column to the left of the centroid must be subjected to tension and the one on the right is subjected to compression. This is necessary in order to counteract the tipping caused by the 30-kN force. Summing moments about the neutral axis, we have

$$\zeta + \Sigma M = 0; \qquad -30(2) + 3 H_y + 3 K_y = 0$$

The unknowns can be related by proportional triangles, Fig. 7–16c, that is,

$$\frac{H_y}{3} = \frac{K_y}{3} \qquad \text{or} \qquad H_y = K_y$$

Thus,

$$H_y = K_y = 10 \text{ kN}$$

In a similar manner, using a section of the frame through the hinges at G and L, Fig. 7–16d, we have

$$\zeta + \Sigma M = 0; \qquad -30(6) - 15(2) + 3G_y + 3L_y = 0$$

Since $G_y/3 = L_y/3$ or $G_y = L_y$, then

$$G_y = L_y = 35 \text{ kN}$$

Each part of the frame can now be analyzed using the above results. As in Examples 7.5 and 7.6, we begin at the upper corner where the applied loading occurs, i.e., segment HCI, Fig. 7–16a. Applying the three equations of equilibrium, $\Sigma M_I = 0$, $\Sigma F_x = 0$, $\Sigma F_y = 0$, yields the results for H_x, I_x, and I_y, respectively, shown on the free-body diagram in Fig. 7–16e. Using these results, segment IDK is analyzed next, Fig. 7–16f; followed by HJG, Fig. 7–16g; then KJL, Fig. 7–16h; and finally the bottom portions of the columns, Fig. 7–16i and Fig. 7–16j. The moment diagrams for each girder are shown in Fig. 7–16k.

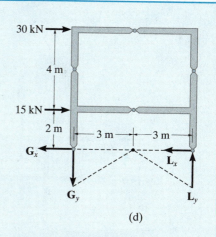

(d)

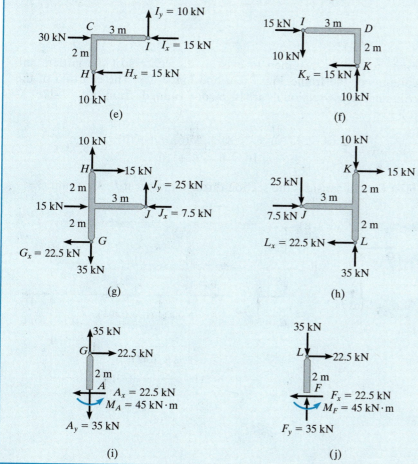

(e)

(f)

(g)

(h)

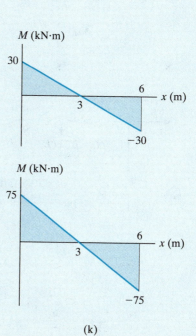

(i)

(j)

(k)

EXAMPLE 7.8

Show how to determine (approximately) the reactions at the base of the columns of the frame shown in Fig. 7–17a. The columns have the cross-sectional areas shown in Fig. 7–17b. Use the cantilever method of analysis.

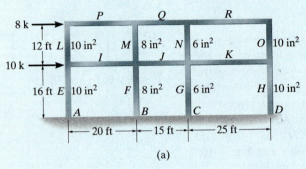

(a)

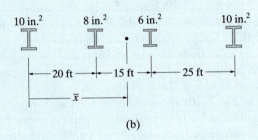

(b)

Fig. 7–17

SOLUTION

First, hinges are assumed to exist at the centers of the girders and columns of the frame, Fig. 7–17d and Fig. 7–17e. The centroid of the columns' cross-sectional areas is determined from Fig. 7–17b as follows:

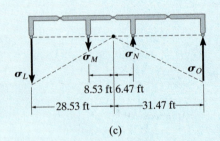

(c)

$$\bar{x} = \frac{\Sigma \tilde{x} A}{\Sigma A} = \frac{0(10) + 20(8) + 35(6) + 60(10)}{10 + 8 + 6 + 10} = 28.53 \text{ ft}$$

First we will consider the section through hinges at L, M, N and O.

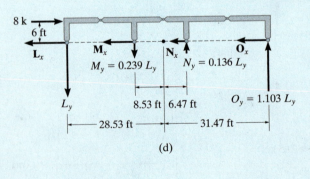

(d)

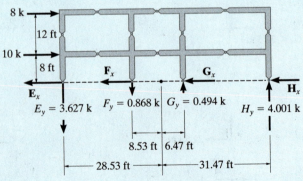

(e)

In this problem the columns have *different* cross-sectional areas, so we must consider the *axial stress* in each column to be proportional to its distance from the neutral axis, located at $\bar{x} = 28.53$ ft.

We can relate the column stresses by proportional triangles, Fig. 7–17c. Expressing the relations in terms of the force in each column, since $\sigma = F/A$, we have

$$\sigma_M = \frac{8.53 \text{ ft}}{28.53 \text{ ft}} \sigma_L; \qquad \frac{M_y}{8 \text{ in}^2} = \frac{8.53}{28.53}\left(\frac{L_y}{10 \text{ in}^2}\right) \qquad M_y = 0.239 L_y$$

$$\sigma_N = \frac{6.47 \text{ ft}}{28.53 \text{ ft}} \sigma_L; \qquad \frac{N_y}{6 \text{ in}^2} = \frac{6.47}{28.53}\left(\frac{L_y}{10 \text{ in}^2}\right) \qquad N_y = 0.136 L_y$$

$$\sigma_O = \frac{31.47 \text{ ft}}{28.53 \text{ ft}} \sigma_L; \qquad \frac{O_y}{10 \text{ in}^2} = \frac{31.47}{28.53}\left(\frac{L_y}{10 \text{ in}^2}\right) \qquad O_y = 1.103 L_y$$

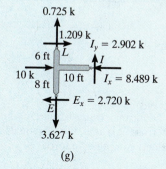

(f)

Now that each force is related to L_y, the free-body diagram is shown in Fig. 7–17d.

Note how the columns to the left of the centroid are subjected to tension and those on the right are subjected to compression. Why? Summing moments about the neutral axis, we have

$$\zeta + \Sigma M = 0; \qquad -8 \text{ k } (6 \text{ ft}) + L_y(28.53 \text{ ft}) + (0.239 L_y)(8.53 \text{ ft})$$
$$+ (0.136 L_y)(6.47 \text{ ft}) + (1.103 L_y)(31.47 \text{ ft}) = 0$$

Solving,

$$L_y = 0.725 \text{ k} \qquad M_y = 0.174 \text{ k} \qquad N_y = 0.0987 \text{ k} \qquad O_y = 0.800 \text{ k}$$

Using this same method, show that one obtains the results in Fig. 7–17e for the columns at $E, F, G,$ and H.

We can now proceed to analyze each part of the frame. As in the previous examples, we begin with the upper corner segment LP, Fig. 7–17f. Using the calculated results, segment LEI is analyzed next, Fig. 7–17g, followed by segment EA, Fig. 7–17h. One can continue to analyze the other segments in sequence, i.e., PQM, then $MJFI$, then FB, and so on.

(g)

(h)

7

PROBLEMS

Sec. 7.1–7.2

7–1. Determine (approximately) the force in each member of the truss. Assume the diagonals can support either a tensile or a compressive force.

7–2. Solve Prob. 7–1 assuming that the diagonals cannot support a compressive force.

7–5. Determine (approximately) the force in each member of the truss. Assume the diagonals can support either a tensile or a compressive force.

7–6. Solve Prob. 7–5 assuming that the diagonals cannot support a compressive force.

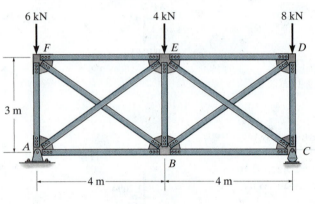

Probs. 7–1/2

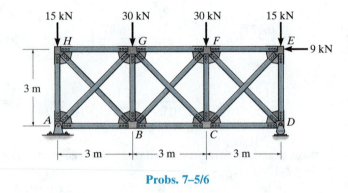

Probs. 7–5/6

7–3. Determine (approximately) the force in each member of the truss. Assume the diagonals can support either a tensile or a compressive force.

*7–4. Determine (approximately) the force in each member of the truss. Assume the diagonals cannot support a compressive force.

7–7. Determine (approximately) the force in each member of the truss. Assume the diagonals can support either a tensile or compressive force.

*7–8. Solve Prob. 7–7 assuming that the diagonals cannot support a compressive force.

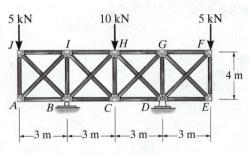

Probs. 7–3/4

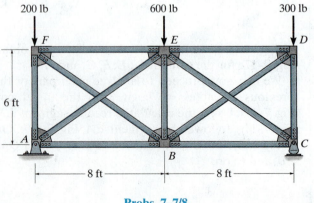

Probs. 7–7/8

7–9.   Determine (approximately) the force in each member of the truss. Assume the diagonals can support both tensile and compressive forces.

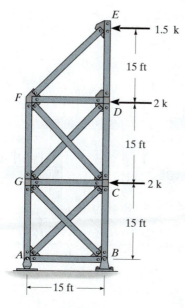

Prob. 7–9

7–11.   Determine (approximately) the force in each member of the truss. Assume the diagonals can support either a tensile or compressive force.

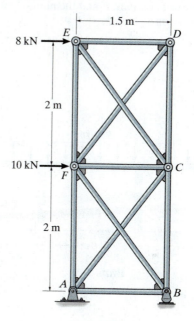

Prob. 7–11

7–10.   Determine (approximately) the force in each member of the truss. Assume the four diagonals cannot support a compressive force.

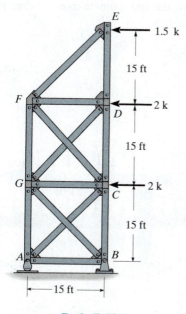

Prob. 7–10

*7–12.**   Determine (approximately) the force in each member of the truss. Assume the diagonals cannot support a compressive force.

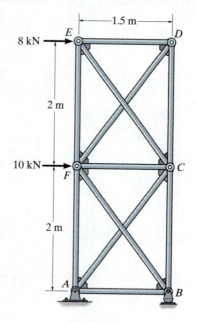

Prob. 7–12

7

Sec. 7.3

7–13. Determine (approximately) the internal moment that member *EF* exerts on joint *E* and the internal moment that member *FG* exerts on joint *F*.

7–15. Draw the approximate moment diagrams for each of the five girders.

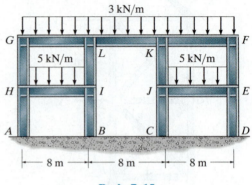

Prob. 7–15

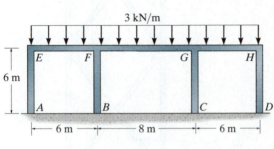

Prob. 7–13

7–14. Determine (approximately) the internal moments at joints *F* and *D*.

*7–16. Draw the approximate moment diagrams for each of the four girders.

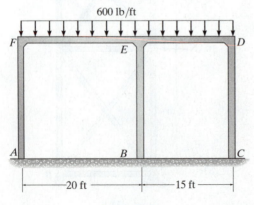

Prob. 7–14

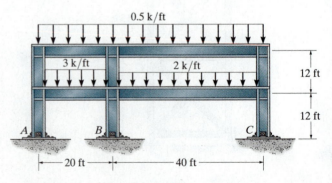

Prob. 7–16

7–17. Determine (approximately) the internal moments at joint *H* from *HG*, and at joint *J* from *JI* and *JK*.

7–19. Determine (approximately) the support reactions at *A* and *B* of the portal frame. Assume the supports are (a) pinned, and (b) fixed.

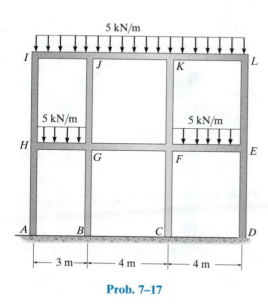

Prob. 7–17

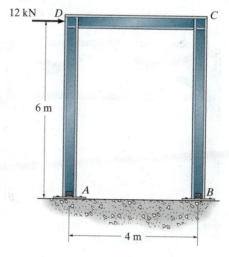

Prob. 7–19

7–18. Determine (approximately) the internal moments at joint *F* from *FG*, and at joint *E* from *ED* of the frame.

*7–20. Determine (approximately) the internal moment and shear at the ends of each member of the portal frame. Assume the supports at *A* and *D* are partially fixed, such that an inflection point is located at *h*/3 from the bottom of each column.

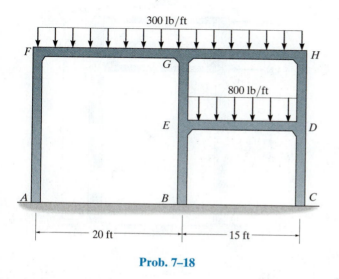

Prob. 7–18

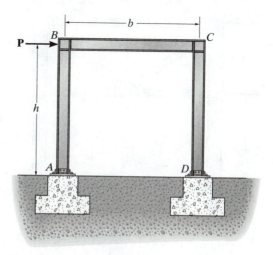

Prob. 7–20

7–21. Determine (approximately) the force in each truss member of the portal frame. Assume all members of the truss to be pin connected at their ends.

7–22. Solve Prob. 7–21 if the supports at A and B are fixed instead of pinned.

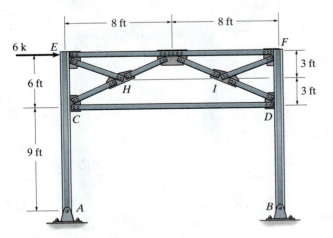

Probs. 7–21/22

7–23. Determine (approximately) the force in each truss member of the portal frame. Also find the reactions at the fixed column supports A and B. Assume all members of the truss to be pin connected at their ends.

*7–24. Solve Prob. 7–23 if the supports at A and B are pinned instead of fixed.

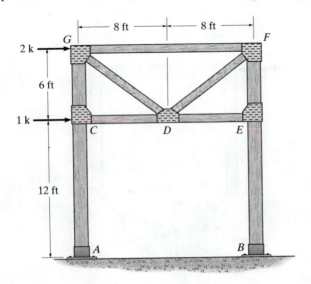

Probs. 7–23/24

7–25. Draw (approximately) the moment diagram for column AGF of the portal. Assume all truss members and the columns to be pin connected at their ends. Also determine the force in all the truss members.

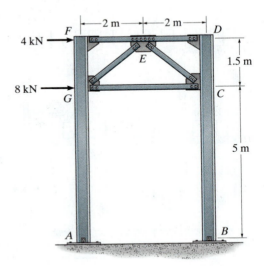

Prob. 7–25

7–26. Draw (approximately) the moment diagram for column AGF of the portal. Assume all the members of the truss to be pin connected at their ends. The columns are fixed at A and B. Also determine the force in all the truss members.

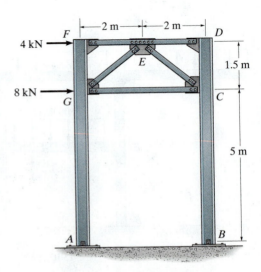

Prob. 7–26

7–27. Determine (approximately) the force in each truss member of the portal frame. Also, find the reactions at the column supports A and B. Assume all members of the truss and the columns to be pin connected at their ends.

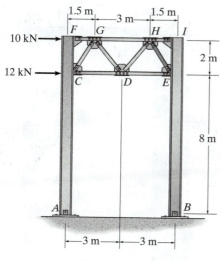

Prob. 7–27

7–30. Draw (approximately) the moment diagram for column ACD of the portal. Assume all truss members and the columns to be pin connected at their ends. Also determine the force in members FG, FH, and EH.

7–31. Solve Prob. 7–30 if the supports at A and B are fixed instead of pinned.

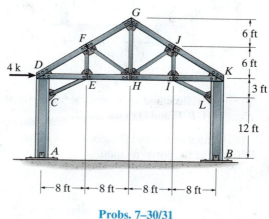

Probs. 7–30/31

*7–28. Determine (approximately) the force in members GF, GK, and JK of the portal frame. Also find the reactions at the fixed column supports A and B. Assume all members of the truss to be connected at their ends.

7–29. Solve Prob. 7–28 if the supports at A and B are pin connected instead of fixed.

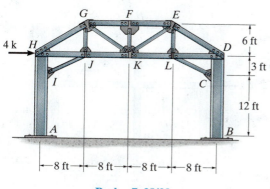

Probs. 7–28/29

*7–32. Draw (approximately) the moment diagram for column AJI of the portal. Assume all truss members and the columns to be pin connected at their ends. Also determine the force in members HG, HL, and KL.

7–33. Solve Prob. 7–32 if the supports at A and B are fixed instead of pinned.

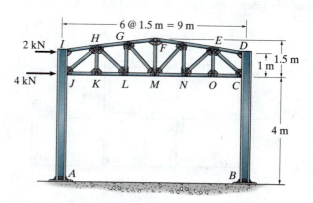

Probs. 7–32/33

7

Sec. 7.5–7.6

7–34. Use the portal method of analysis and draw the moment diagram for girder *FED*.

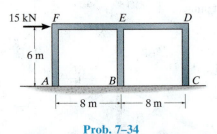

Prob. 7–34

7–35. Use the portal method and determine (approximately) the reactions at *A*, *B*, *C*, and *D* of the frame.

*7–36. Draw (approximately) the moment diagram for the girder *EFGH*. Use the portal method.

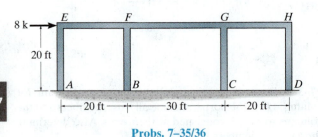

Probs. 7–35/36

7–37. Draw the moment diagram for girder *EFGH*. Use the portal method of analysis.

7–38. Solve Prob. 7–37 using the cantilever method of analysis. Each column has the same cross-sectional area.

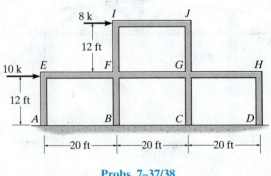

Probs. 7–37/38

7–39. Use the portal method of analysis and draw the moment diagram for column *AFE*.

*7–40. Solve Prob. 7–39 using the cantilever method of analysis. All the columns have the same cross-sectional area.

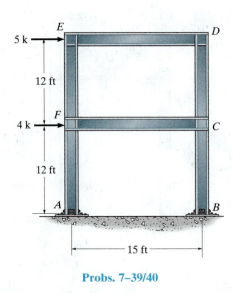

Probs. 7–39/40

7–41. Use the portal method and determine (approximately) the reactions at *A*.

7–42. Use the cantilever method and determine (approximately) the reactions at *A*. All of the columns have the same cross-sectional area.

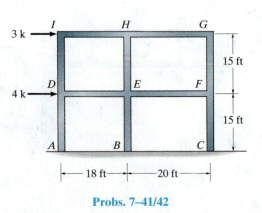

Probs. 7–41/42

7–43. Draw (approximately) the moment diagram for girder *PQRST* and column *BGLQ* of the building frame. Use the portal method.

*7–44. Draw (approximately) the moment diagram for girder *PQRST* and column *BGLQ* of the building frame. All columns have the same cross-sectional area. Use the cantilever method.

7–45. Draw the moment diagram for girder *IJKL* of the building frame. Use the portal method of analysis.

7–46. Solve Prob. 7–45 using the cantilever method of analysis. Each column has the cross-sectional area indicated.

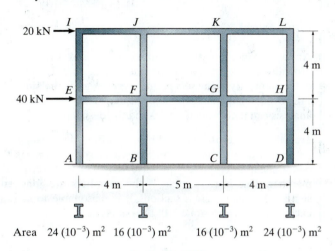

Probs. 7–45/46

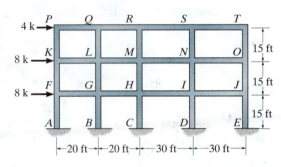

Probs. 7–43/44

PROJECT PROBLEMS

7–1P. The railroad trestle bridge shown in the photo is supported by reinforced concrete bents. Assume the two simply supported side girders, track bed, and two rails have a weight of 0.5 k/ft and the load imposed by a train is 7.2 k/ft (see Fig. 1–11). Each girder is 20 ft long. Apply the load over the entire bridge and determine the compressive force in the columns of each bent. For the analysis assume all joints are pin connected and neglect the weight of the bent. Are these realistic assumptions?

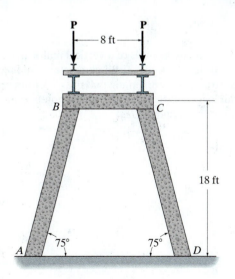

Project Prob. 7–1P

CHAPTER REVIEW

An approximate structural analysis is used to reduce a statically indeterminate structure to one that is statically determinate. By doing so a preliminary design of the members can be made, and once complete, the more exact indeterminate analysis can then be performed and the design refined.

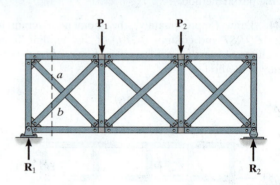

Trusses having cross-diagonal bracing within their panels can be analyzed by assuming the tension diagonal supports the panel shear and the compressive diagonal is a zero-force member. This is reasonable if the members are long and slender. For larger cross sections, it is reasonable to assume each diagonal carries one-half the panel shear.

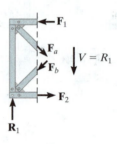

The approximate analysis of a vertical uniform load acting on a girder of length L of a fixed-connected building frame can be approximated by assuming that the girder does not support an axial load, and there are inflection points (hinges) located $0.1L$ from the supports.

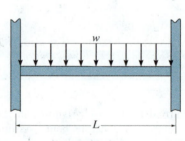

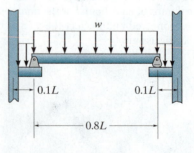

7

Portal frames having fixed supports are approximately analyzed by assuming there are hinges at the midpoint of each column height, measured to the bottom of the truss bracing. Also, for these, and pin-supported frames, each column is assumed to support half the shear load on the frame.

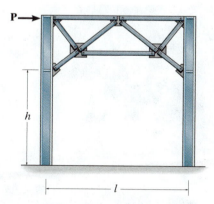

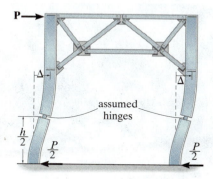

For fixed-connected building frames subjected to lateral loads, we can assume there are hinges at the centers of the columns and girders. If the frame has a low elevation, shear resistance is important and so we can use the portal method, where the interior columns at any floor level carry twice the shear as that of the exterior columns. For tall slender frames, the cantilever method can be used, where the axial stress in a column is proportional to its distance from the centroid of the cross-sectional area of all the columns at a given floor level.

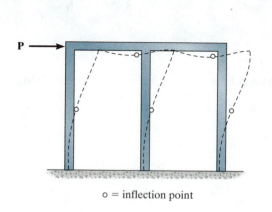

o = inflection point

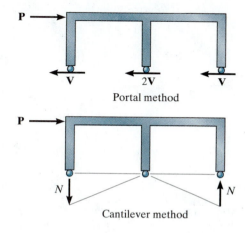

Portal method

Cantilever method

Chapter 8

© Lev Kropotov/Shutterstock

The deflection of the overhanging segments of this bridge deck must be carefully monitored while it is under construction.

Deflections

In this chapter we will show how to determine the elastic deflections of a beam using the method of double integration and two important geometrical methods, namely, the moment-area theorems and the conjugate-beam method. Double integration is used to obtain equations which define the slope and the elastic curve. The geometric methods provide a way to obtain the slope and deflection at specific points on the beam. Each of these methods has particular advantages or disadvantages, which will be discussed when each method is presented.

8.1   Deflection Diagrams and the Elastic Curve

Deflections of structures can occur from various sources, such as loads, temperature, fabrication errors, or settlement. In design, deflections must be limited in order to provide integrity and stability of roofs, and prevent cracking of attached brittle materials such as concrete, plaster or glass. Furthermore, a structure must not vibrate or deflect severely in order to "appear" safe for its occupants. More important, though, deflections at specified points in a structure must be determined if one is to analyze statically indeterminate structures.

The deflections to be considered throughout this text apply only to structures having *linear elastic material response*. Under this condition, a structure subjected to a load will return to its original undeformed position after the load is removed. The deflection of a structure is caused by its internal loadings such as normal force, shear force, or bending moment. For *beams* and *frames,* however, the greatest deflections are most often caused by *internal bending*, whereas *internal axial forces* cause the deflections of a *truss*.

TABLE 8.1

(1)

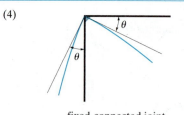

$\Delta = 0$
roller or rocker

(2)

$\Delta = 0$
pin

(3)

$\Delta = 0$
$\theta = 0$
fixed support

(4)

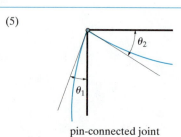

fixed-connected joint

(5)

θ_2

θ_1

pin-connected joint

Before the slope or displacement of a point on a beam or frame is determined, it is often helpful to sketch the deflected shape of the structure when it is loaded in order to partially check the results. This *deflection diagram* represents the *elastic curve* or locus of points which defines the displaced position of the centroid of the cross section along the members. For most problems the elastic curve can be sketched without much difficulty. When doing so, however, it is necessary to know the restrictions as to slope or displacement that often occur at a support or a connection. With reference to Table 8.1, supports that *resist a force,* such as a pin, *restrict displacement;* and those that *resist moment,* such as a fixed wall, *restrict rotation.* Note also that deflection of frame members that are fixed connected (4) causes the joint to rotate the connected members by the same amount θ. On the other hand, if a pin connection is used at the joint, the members will each have a *different slope* or rotation at the pin, since the pin cannot support a moment (5).

The two-member frames support both the dead load of the roof and a live snow loading. The frame can be considered pinned at the wall, fixed at the ground, and having a fixed-connected joint.

If the elastic curve seems difficult to establish, it is suggested that the moment diagram for the beam or frame be drawn first. By our sign convention for moments established in Chapter 4, a *positive moment* tends to bend a beam or horizontal member *concave upward*, Fig. 8–1. Likewise, a *negative moment* tends to bend the beam or member *concave downward*, Fig. 8–2. Therefore, *if the shape of the moment diagram is known, it will be easy to construct the elastic curve and vice versa.* For example, consider the beam in Fig. 8–3 with its associated moment diagram. Due to the pin-and-roller support, the displacement at A and D must be zero. Within the region of negative moment, the elastic curve is concave downward; and within the region of positive moment, the elastic curve is concave upward. In particular, there must be an *inflection point* at the point where the curve changes from concave down to concave up, since this is a point of zero moment. Using these same principles, note how the elastic curve for the beam in Fig. 8–4 was drawn based on its moment diagram. In particular, realize that the positive moment reaction from the wall keeps the initial slope of the beam horizontal.

positive moment,
concave upward

Fig. 8–1

negative moment,
concave downward

Fig. 8–2

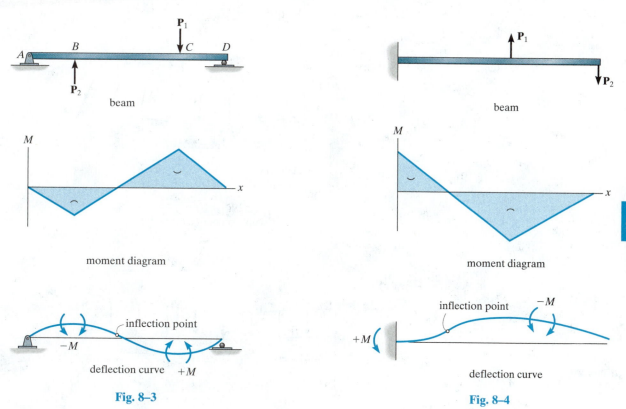

beam

moment diagram

deflection curve

Fig. 8–3

beam

moment diagram

deflection curve

Fig. 8–4

Being able to draw the deflection curve also helps engineers in locating the steel needed to reinforce a concrete beam, column, or wall. Concrete is rather weak in tension, so regions of a concrete structural member where tensile stresses are developed are reinforced with steel bars, called **reinforcing rods**. These rods prevent or control any cracking that may occur within these regions. Examples of the required placement of reinforcing steel in two beams is shown in Fig. 8–5. Notice how each member deflects under the load, how the internal moment acts, and the placement of the steel needed to resist the tensile stress caused by the bending.

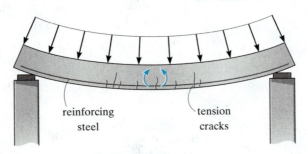

reinforcing   tension
steel   cracks

simply supported beam

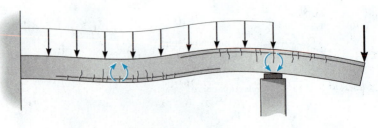

overhang beam

Fig. 8–5

EXAMPLE 8.1

Draw the deflected shape of each of the beams shown in Fig. 8–6.

SOLUTION

In Fig. 8–6a the roller at A allows free rotation with no deflection while the fixed wall at B prevents both rotation and deflection. The deflected shape is shown by the bold line. In Fig. 8–6b, no rotation or deflection can occur at A and B. In Fig. 8–6c, the couple moment will rotate end A. This will cause deflections at both ends of the beam since no deflection is possible at B and C. Notice that segment CD remains undeformed (a straight line) since no internal load acts within it. In Fig. 8–6d, the pin (internal hinge) at B allows free rotation, and so the slope of the deflection curve will suddenly change at this point while the beam is constrained by its supports. In Fig. 8–6e, the compound beam deflects as shown. The slope abruptly changes on each side of the pin at B. Finally, in Fig. 8–6f, span BC will deflect concave upwards due to the load. Since the beam is continuous, the end spans will deflect concave downwards.

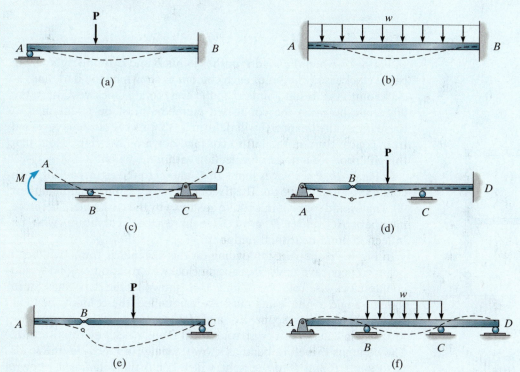

Fig. 8–6

EXAMPLE | 8.2

Draw the deflected shapes of each of the frames shown in Fig. 8–7.

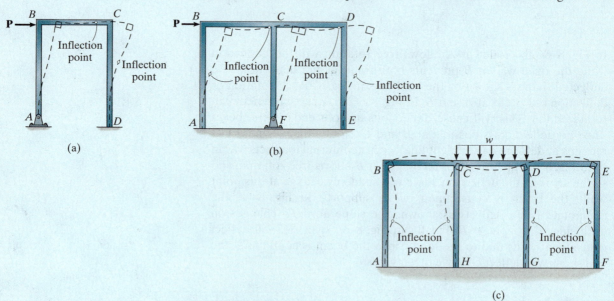

(a)

(b)

(c)

SOLUTION

In Fig. 8–7a, when the load **P** pushes joints B and C to the right, it will cause clockwise rotation of each column as shown by the dark lines at each joint. As a result, joints B and C must rotate clockwise. Since the 90° angle between the connected members must be maintained at these joints, the beam BC will deform so that its curvature is reversed from concave up on the left to concave down on the right. Note that this will produce a point of inflection within the beam.

In Fig. 8–7b, **P** displaces joints B, C, and D to the right, causing each column to bend as shown. The fixed joints B, C, D must maintain their 90° angles, and so the joints rotate as shown by the dark lines. Therefore members BC and CD must have a reversed curvature, with an inflection point near their midpoint.

In Fig. 8–7c, the vertical loading on this symmetric frame will bend beam CD concave upwards, causing clockwise rotation of joint C and counterclockwise rotation of joint D as shown by the dark lines. Since the 90° angle at the joints must be maintained, the columns bend as shown. This causes spans BC and DE to be concave downwards, resulting in counterclockwise rotation at B and clockwise rotation at E. The columns therefore bend as shown. Finally, in Fig. 8–7d, the loads push joints B and C to the right, which bends the columns as shown. The fixed joint B maintains its 90° angle; however, no restriction on the relative rotation between the members at C is possible since the joint is a pin. Consequently, member CD does not have a reverse curvature.

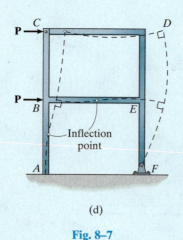

(d)

Fig. 8–7

8.2   Elastic-Beam Theory

In this section we will develop two important differential equations that relate the internal moment in a beam to the displacement and slope of its elastic curve. These equations form the basis for the deflection methods presented in this chapter, and for this reason the assumptions and limitations used in their development should be fully understood.

To derive these relationships, we will limit the analysis to the most common case of an initially straight beam that is elastically deformed by loads applied perpendicular to the beam's x axis and lying in the $x–v$ plane of symmetry for the beam's cross-sectional area, Fig. 8–8a. Due to the loading, the deformation of the beam is caused by both the internal shear force and bending moment. If the beam has a length that is much greater than its depth, the greatest deformation will be caused by bending, and therefore we will direct our attention to its effects. Deflections caused by shear will be discussed later in the chapter.

When the internal moment M deforms the element of the beam, each cross section remains plane and the angle between them becomes $d\theta$, Fig. 8–8b. The arc dx that represents a portion of the elastic curve intersects the neutral axis for each cross section. The *radius of curvature* for this arc is defined as the distance ρ, which is measured from the *center of curvature O'* to dx. Any arc on the element other than dx is subjected to a normal strain. For example, the strain in arc ds, located at a position y from the neutral axis, is $\epsilon = (ds' - ds)/ds$. However, $ds = dx = \rho \, d\theta$ and $ds' = (\rho - y) \, d\theta$, and so

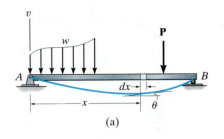

(a)

$$\epsilon = \frac{(\rho - y)\, d\theta - \rho\, d\theta}{\rho\, d\theta} \qquad \text{or} \qquad \frac{1}{\rho} = -\frac{\epsilon}{y}$$

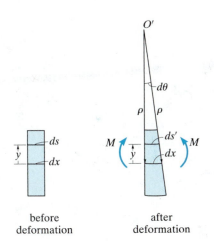

before deformation   after deformation

(b)

Fig. 8–8

If the material is homogeneous and behaves in a linear elastic manner, then Hooke's law applies, $\epsilon = \sigma/E$. Also, since the flexure formula applies, $\sigma = -My/I$. Combining these equations and substituting into the above equation, we have

$$\frac{1}{\rho} = \frac{M}{EI} \tag{8–1}$$

Here

ρ = the radius of curvature at a specific point on the elastic curve ($1/\rho$ is referred to as the *curvature*)

M = the internal moment in the beam at the point where ρ is to be determined

E = the material's modulus of elasticity

I = the beam's moment of inertia computed about the neutral axis

The product EI in this equation is referred to as the *flexural rigidity*, and it is always a positive quantity. Since $dx = \rho \, d\theta$, then from Eq. 8–1,

$$d\theta = \frac{M}{EI} dx \qquad (8\text{–}2)$$

If we choose the v axis positive upward, Fig. 8–8a, and if we can express the curvature $(1/\rho)$ in terms of x and v, we can then determine the elastic curve for the beam. In most calculus books it is shown that this curvature relationship is

$$\frac{1}{\rho} = \frac{d^2v/dx^2}{\left[1 + (dv/dx)^2\right]^{3/2}}$$

Therefore,

$$\frac{M}{EI} = \frac{d^2v/dx^2}{\left[1 + (dv/dx)^2\right]^{3/2}} \qquad (8\text{–}3)$$

This equation represents a nonlinear second-order differential equation. Its solution, $v = f(x)$, gives the exact shape of the elastic curve—assuming, of course, that beam deflections occur only due to bending. In order to facilitate the solution of a greater number of problems, Eq. 8–3 will be modified by making an important simplification. Since the slope of the elastic curve for most structures is very small, we will use small deflection theory and assume $dv/dx \approx 0$. Consequently its square will be negligible compared to unity and therefore Eq. 8–3 reduces to

$$\boxed{\frac{d^2v}{dx^2} = \frac{M}{EI}} \qquad (8\text{–}4)$$

It should also be pointed out that by assuming $dv/dx \approx 0$, the original length of the beam's axis x and the *arc* of its elastic curve will be approximately the same. In other words, ds in Fig. 8–8b is approximately equal to dx, since

$$ds = \sqrt{dx^2 + dv^2} = \sqrt{1 + (dv/dx)^2} \, dx \approx dx$$

This result implies that points on the elastic curve will only be displaced vertically and not horizontally.

Tabulated Results.
In the next section we will show how to apply Eq. 8–4 to find the slope of a beam and the equation of its elastic curve. The results from such an analysis for some common beam loadings often encountered in structural analysis are given in the table on the inside front cover of this book. Also listed are the slope and displacement at critical points on the beam. Obviously, no single table can account for the many different cases of loading and geometry that are encountered in practice. When a table is not available or is incomplete, the displacement or slope of a specific point on a beam or frame can be determined by using the double integration method or one of the other methods discussed in this and the next chapter.

8.3   The Double Integration Method

Once M is expressed as a function of position x, then successive integrations of Eq. 8–4 will yield the beam's slope, $\theta \approx \tan \theta = dv/dx = \int (M/EI) \, dx$ (Eq. 8–2), and the equation of the elastic curve, $v = f(x) = \int \left(\int (M/EI) \, dx \right) dx$, respectively. For each integration it is necessary to introduce a "constant of integration" and then solve for the constants to obtain a unique solution for a particular problem. Recall from Sec. 4.2 that if the loading on a beam is discontinuous— that is, it consists of a series of several distributed and concentrated loads—then several functions must be written for the internal moment, each valid within the region between the discontinuities. For example, consider the beam shown in Fig. 8–9. The internal moment in regions $AB, BC,$ and CD must be written in terms of the $x_1, x_2,$ and x_3 coordinates. Once these functions are integrated through the application of Eq. 8–4 and the constants of integration determined, the functions will give the slope and deflection (elastic curve) for each region of the beam for which they are valid. With these functions we can then determine the slope and deflection at *any point* along the beam.

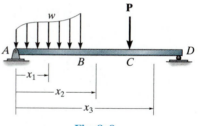

Fig. 8–9

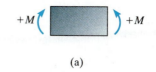

(a)

Sign Convention.

When applying Eq. 8–4, it is important to use the proper sign for M as established by the sign convention that was used in the derivation of this equation, Fig. 8–10a. Furthermore, recall that positive deflection, v, is upward, and as a result, the positive slope angle θ will be measured counterclockwise from the x axis. The reason for this is shown in Fig. 8–10b. Here, positive increases dx and dv in x and v create an increase $d\theta$ that is counterclockwise. Also, since the slope angle θ will be very small, its value in radians can be determined directly from $\theta \approx \tan \theta = dv/dx$.

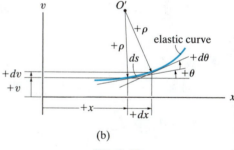

(b)

Fig. 8–10

Boundary and Continuity Conditions.

The constants of integration are determined by evaluating the functions for slope or displacement at a particular point on the beam where the value of the function is known. These values are called *boundary conditions*. For example, if the beam is supported by a roller or pin, then it is required that the displacement be zero at these points. Also, at a fixed support the slope and displacement are both zero.

If a single x coordinate cannot be used to express the equation for the beam's slope or the elastic curve, then continuity conditions must be used to evaluate some of the integration constants. Consider the beam in Fig. 8–11. Here the x_1 and x_2 coordinates are valid only within the regions AB and BC, respectively. Once the functions for the slope and deflection are obtained, they must give the same values for the slope and deflection at point B, $x_1 = x_2 = a$, so that the elastic curve is physically continuous. Expressed mathematically, this requires $\theta_1(a) = \theta_2(a)$ and $v_1(a) = v_2(a)$. These equations can be used to determine two constants of integration.

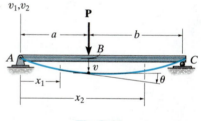

Fig. 8–11

8

Procedure for Analysis

The following procedure provides a method for determining the slope and deflection of a beam (or shaft) using the method of double integration. It should be realized that this method is suitable only for *elastic deflections* for which the beam's slope is very small. Furthermore, the method considers *only deflections due to bending*. Additional deflection due to shear generally represents only a few percent of the bending deflection, and so it is usually neglected in engineering practice.

Elastic Curve

- Draw an exaggerated view of the beam's elastic curve. Recall that points of zero slope and zero displacement occur at a fixed support, and zero displacement occurs at pin and roller supports.

- Establish the x and v coordinate axes. The x axis must be parallel to the undeflected beam and its origin at the left side of the beam, with a positive direction to the right.

- If several discontinuous loads are present, establish x coordinates that are valid for each region of the beam between the discontinuities.

- In all cases, the associated positive v axis should be directed upward.

Load or Moment Function

- For each region in which there is an x coordinate, express the internal moment M as a function of x.

- *Always* assume that M acts in the positive direction when applying the equation of moment equilibrium to determine $M = f(x)$.

Slope and Elastic Curve

- Provided EI is constant, apply the moment equation $EI\,d^2v/dx^2 = M(x)$, which requires two integrations. For each integration it is important to include a constant of integration. The constants are determined using the boundary conditions for the supports and the continuity conditions that apply to slope and displacement at points where two functions meet.

- Once the integration constants are determined and substituted back into the slope and deflection equations, the slope and displacement at *specific points* on the elastic curve can be determined. The numerical values obtained can be checked graphically by comparing them with the sketch of the elastic curve.

- *Positive* values for *slope* are *counterclockwise* and *positive displacement is upward*.

EXAMPLE 8.3

Each simply supported floor joist shown in the photo is subjected to a uniform design loading of 4 kN/m, Fig. 8–12a. Determine the maximum deflection of the joist. EI is constant.

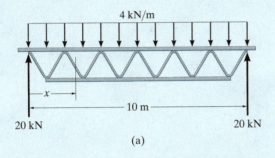

4 kN/m

10 m

20 kN   20 kN

(a)

Elastic Curve. Due to symmetry, the joist's maximum deflection will occur at its center. Only a single x coordinate is needed to determine the internal moment.

Moment Function. From the free-body diagram, Fig. 8–12b, we have

$$M = 20x - 4x\left(\frac{x}{2}\right) = 20x - 2x^2$$

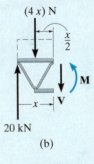

$(4x)$ N

$\frac{x}{2}$

M

x V

20 kN

(b)

Fig. 8–12

Slope and Elastic Curve. Applying Eq. 8–4 and integrating twice gives

$$EI\frac{d^2v}{dx^2} = 20x - 2x^2$$

$$EI\frac{dv}{dx} = 10x^2 - 0.6667x^3 + C_1$$

$$EIv = 3.333x^3 - 0.1667x^4 + C_1x + C_2$$

Here $v = 0$ at $x = 0$ so that $C_2 = 0$, and $v = 0$ at $x = 10$, so that $C_1 = -166.7$. The equation of the elastic curve is therefore

$$EIv = 3.333x^3 - 0.1667x^4 - 166.7x$$

At $x = 5$ m, note that $dv/dx = 0$. The maximum deflection is therefore

$$v_{max} = -\frac{521}{EI}$$ *Ans.*

EXAMPLE 8.4

The cantilevered beam shown in Fig. 8–13a is subjected to a couple moment M_0 at its end. Determine the equation of the elastic curve. EI is constant.

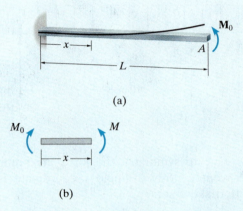

(a)

(b)

Fig. 8–13

SOLUTION

Elastic Curve. The load tends to deflect the beam as shown in Fig. 8–13a. By inspection, the internal moment can be represented throughout the beam using a single x coordinate.

Moment Function. From the free-body diagram, with **M** acting in the *positive direction*, Fig. 8–13b, we have

$$M = M_0$$

Slope and Elastic Curve. Applying Eq. 8–4 and integrating twice yields

$$EI\frac{d^2v}{dx^2} = M_0 \tag{1}$$

$$EI\frac{dv}{dx} = M_0 x + C_1 \tag{2}$$

$$EIv = \frac{M_0 x^2}{2} + C_1 x + C_2 \tag{3}$$

Using the boundary conditions $dv/dx = 0$ at $x = 0$ and $v = 0$ at $x = 0$, then $C_1 = C_2 = 0$. Substituting these results into Eqs. (2) and (3) with $\theta = dv/dx$, we get

$$\theta = \frac{M_0 x}{EI}$$

$$v = \frac{M_0 x^2}{2EI} \qquad\qquad Ans.$$

Maximum slope and displacement occur at A $(x = L)$, for which

$$\theta_A = \frac{M_0 L}{EI} \qquad\qquad (4)$$

$$v_A = \frac{M_0 L^2}{2EI} \qquad\qquad (5)$$

The *positive* result for θ_A indicates *counterclockwise* rotation and the *positive* result for v_A indicates that v_A is *upward*. This agrees with the results sketched in Fig. 8–13a.

In order to obtain some idea as to the actual *magnitude* of the slope and displacement at the end A, consider the beam in Fig. 8–13a to have a length of 12 ft, support a couple moment of 15 k · ft, and be made of steel having $E_{st} = 29(10^3)$ ksi. If this beam were designed *without* a factor of safety by assuming the allowable normal stress is equal to the yield stress $\sigma_{allow} = 36$ ksi, then a $W6 \times 9$ would be found to be adequate ($I = 16.4$ in.^{4}). From Eqs. (4) and (5) we get

$$\theta_A = \frac{15 \text{ k} \cdot \text{ft}(12 \text{ in./ft})(12 \text{ ft})(12 \text{ in./ft})}{29(10^3) \text{ k/in}^2(16.4 \text{ in}^4)} = 0.0545 \text{ rad}$$

$$v_A = \frac{15 \text{ k} \cdot \text{ft}(12 \text{ in./ft})(12 \text{ ft})^2(12 \text{ in./1 ft})^2}{2(29(10^3) \text{ k/in}^2)(16.4 \text{ in}^4)} = 3.92 \text{ in.}$$

Since $\theta_A^2 = 0.00297$ rad$^2 \ll 1$, this justifies the use of Eq. 8–4, rather than applying the more exact Eq. 8–3, for determining the deflection of the beam. Also, since this numerical application is for a *cantilevered beam,* we have obtained *larger values* for maximum θ and v than would have been obtained if the beam were supported using pins, rollers, or other supports.

EXAMPLE 8.5

The beam in Fig. 8–14a is subjected to a load **P** at its end. Determine the displacement at *C*. *EI* is constant.

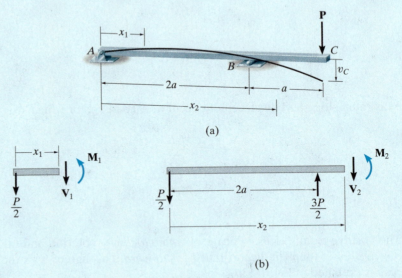

(a)

(b)

Fig. 8–14

SOLUTION

Elastic Curve.   The beam deflects into the shape shown in Fig. 8–14a. Due to the loading, two *x* coordinates must be considered.

Moment Functions.   Using the free-body diagrams shown in Fig. 8–14b, we have

$$M_1 = -\frac{P}{2}x_1 \qquad 0 \le x_1 \le 2a$$

$$M_2 = -\frac{P}{2}x_2 + \frac{3P}{2}(x_2 - 2a)$$

$$= Px_2 - 3Pa \qquad 2a \le x_2 \le 3a$$

Slope and Elastic Curve.   Applying Eq. 8–4,

for x_1,
$$EI\frac{d^2v_1}{dx_1^2} = -\frac{P}{2}x_1$$

$$EI\frac{dv_1}{dx_1} = -\frac{P}{4}x_1^2 + C_1 \tag{1}$$

$$EIv_1 = -\frac{P}{12}x_1^3 + C_1x_1 + C_2 \tag{2}$$

For x_2,
$$EI\frac{d^2v_2}{dx_2^2} = Px_2 - 3Pa$$

$$EI\frac{dv_2}{dx_2} = \frac{P}{2}x_2^2 - 3Pax_2 + C_3 \tag{3}$$

$$EIv_2 = \frac{P}{6}x_2^3 - \frac{3}{2}Pax_2^2 + C_3x_2 + C_4 \tag{4}$$

The *four* constants of integration are determined using *three* boundary conditions, namely, $v_1 = 0$ at $x_1 = 0$, $v_1 = 0$ at $x_1 = 2a$, and $v_2 = 0$ at $x_2 = 2a$, and *one* continuity equation. Here the continuity of slope at the roller requires $dv_1/dx_1 = dv_2/dx_2$ at $x_1 = x_2 = 2a$. (Note that continuity of displacement at B has been indirectly considered in the boundary conditions, since $v_1 = v_2 = 0$ at $x_1 = x_2 = 2a$.) Applying these four conditions yields

$v_1 = 0$ at $x_1 = 0$; $0 = 0 + 0 + C_2$

$v_1 = 0$ at $x_1 = 2a$; $0 = -\dfrac{P}{12}(2a)^3 + C_1(2a) + C_2$

$v_2 = 0$ at $x_2 = 2a$; $0 = \dfrac{P}{6}(2a)^3 - \dfrac{3}{2}Pa(2a)^2 + C_3(2a) + C_4$

$\dfrac{dv_1(2a)}{dx_1} = \dfrac{dv_2(2a)}{dx_2}$; $-\dfrac{P}{4}(2a)^2 + C_1 = \dfrac{P}{2}(2a)^2 - 3Pa(2a) + C_3$

Solving, we obtain

$$C_1 = \frac{Pa^2}{3} \qquad C_2 = 0 \qquad C_3 = \frac{10}{3}Pa^2 \qquad C_4 = -2Pa^3$$

Substituting C_3 and C_4 into Eq. (4) gives

$$v_2 = \frac{P}{6EI}x_2^3 - \frac{3Pa}{2EI}x_2^2 + \frac{10Pa^2}{3EI}x_2 - \frac{2Pa^3}{EI}$$

The displacement at C is determined by setting $x_2 = 3a$. We get

$$v_C = -\frac{Pa^3}{EI} \qquad\qquad \textit{Ans.}$$

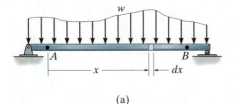

(a)

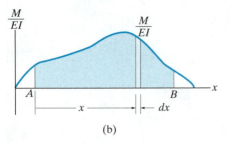

(b)

Fig. 8–15

8.4　Moment-Area Theorems

The initial ideas for the two moment-area theorems were developed by Otto Mohr and later stated formally by Charles E. Greene in 1873. These theorems provide a semigraphical technique for determining the slope and deflection at a specific point on a beam due to bending. They are particularly advantageous when used to solve problems involving beams subjected to a series of concentrated loadings or having segments with different moments of inertia.

To develop the theorems, reference is made to the beam in Fig. 8–15a. If we draw the moment diagram for the beam and then divide it by the flexural rigidity, EI, the "M/EI diagram" shown in Fig. 8–15b results. By Eq. 8–2,

$$d\theta = \left(\frac{M}{EI}\right) dx$$

Thus it can be seen that the change $d\theta$ in the slope of the tangents on either side of the element dx is equal to the lighter-shaded *area* under the M/EI diagram. Integrating from point A on the elastic curve to point B, Fig. 8–15c, we have

$$\theta_{B/A} = \int_A^B \frac{M}{EI} dx \tag{8–5}$$

This equation forms the basis for the first moment-area theorem.

> **Theorem 1: The change in slope between any two points on the elastic curve equals the area of the M/EI diagram between these two points.**

The notation $\theta_{B/A}$ is referred to as the angle of the tangent at B measured with respect to the tangent at A, Fig. 8–15c. From the proof it should be evident that this angle is measured *counterclockwise* from tangent A to tangent B if the area of the M/EI diagram is *positive,* Fig. 8–15b. Conversely, if this area is *negative,* or below the x axis, the angle $\theta_{B/A}$ is measured *clockwise* from tangent A to tangent B. Furthermore, from the dimensions of Eq. 8–5, $\theta_{B/A}$ is measured in radians.

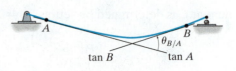

elastic curve
(c)

The second moment-area theorem is based on the relative deviation of *tangents* to the elastic curve. Shown in Fig. 8–15d is a greatly exaggerated view of the *vertical deviation dt* of the tangents on each side of the differential element *dx*. This deviation is measured along a vertical line passing through point *A*. Since the slope of the elastic curve and its deflection are assumed to be very small, it is satisfactory to approximate the length of each tangent line by *x* and the arc *ds'* by *dt*. Using the circular-arc formula $s = \theta r$, where *r* is of length *x*, we can write $dt = x\, d\theta$. Using Eq. 8–2, $d\theta = (M/EI)\, dx$, the vertical deviation of the tangent at *A* with respect to the tangent at *B* can be found by integration, in which case

$$t_{A/B} = \int_A^B x \frac{M}{EI} dx \qquad (8\text{–}6)$$

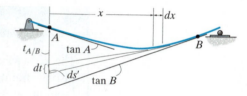

elastic curve
(d)

Recall from statics that the centroid of an area is determined from $\bar{x} \int dA = \int x\, dA$. Since $\int M/EI\, dx$ represents an area of the M/EI diagram, we can also write

$$t_{A/B} = \bar{x} \int_A^B \frac{M}{EI} dx \qquad (8\text{–}7)$$

Here $\bar{x}$ is the distance from the vertical axis through *A* to the *centroid* of the area between *A* and *B*, Fig. 8–15e.

The second moment-area theorem can now be stated as follows:

> **Theorem 2: The vertical deviation of the tangent at a point (*A*) on the elastic curve with respect to the tangent extended from another point (*B*) equals the "moment" of the area under the M/EI diagram between the two points (*A* and *B*). This moment is computed about point *A* (the point on the elastic curve), where the deviation $t_{A/B}$ is to be determined.**

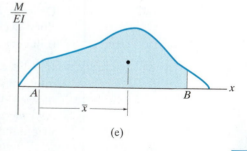

(e)

Provided the moment of a *positive* M/EI area from *A* to *B* is determined, as in Fig. 8–15e, it indicates that the tangent at point *A* is *above* the tangent to the curve extended from point *B*, Fig. 8–15f. Similarly, *negative* M/EI areas indicate that the tangent at *A* is *below* the tangent extended from *B*. Note that in general $t_{A/B}$ is not equal to $t_{B/A}$, which is shown in Fig. 8–15f. Specifically, the moment of the area under the M/EI diagram between *A* and *B* is determined about point *A* to find $t_{A/B}$, and it is determined about point *B* to find $t_{B/A}$.

It is important to realize that the moment-area theorems can only be used to determine the angles or deviations between two tangents on the beam's elastic curve. In general, they *do not* give a direct solution for the slope or displacement at a point on the beam. These unknowns must first be related to the angles or vertical deviations of tangents at points on the elastic curve. Usually the tangents at the supports are drawn in this regard since these points do not undergo displacement and/or have zero slope. Specific cases for establishing these geometric relationships are given in the example problems.

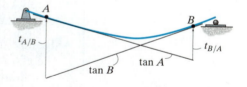

elastic curve
(f)

8

Procedure for Analysis

The following procedure provides a method that may be used to determine the displacement and slope at a point on the elastic curve of a beam using the moment-area theorems.

M/EI Diagram

- Determine the support reactions and draw the beam's M/EI diagram.
- If the beam is loaded with concentrated forces, the M/EI diagram will consist of a series of straight line segments, and the areas and their moments required for the moment-area theorems will be relatively easy to compute.
- If the loading consists of a series of concentrated forces and distributed loads, it may be simpler to compute the required M/EI areas and their moments by drawing the M/EI diagram in parts, using the method of superposition as discussed in Sec. 4.5. In any case, the M/EI diagram will consist of parabolic or perhaps higher-order curves, and it is suggested that the table on the inside back cover be used to locate the area and centroid under each curve.

Elastic Curve

- Draw an exaggerated view of the beam's elastic curve. Recall that points of zero slope occur at fixed supports and zero displacement occurs at all fixed, pin, and roller supports.

- If it becomes difficult to draw the general shape of the elastic curve, use the moment (or M/EI) diagram. Realize that when the beam is subjected to a *positive moment* the beam bends *concave up,* whereas *negative moment* bends the beam *concave down.* Furthermore, an inflection point or change in curvature occurs where the moment in the beam (or M/EI) is zero.

- The displacement and slope to be determined should be indicated on the curve. Since the moment-area theorems apply only between two tangents, attention should be given as to which tangents should be constructed so that the angles or deviations between them will lead to the solution of the problem. In this regard, *the tangents at the points of unknown slope and displacement and at the supports should be considered,* since the beam usually has zero displacement and/or zero slope at the supports.

Moment-Area Theorems

- Apply Theorem 1 to determine the angle between two tangents, and Theorem 2 to determine vertical deviations between these tangents.

- Realize that Theorem 2 in general *will not* yield the displacement of a point on the elastic curve. When applied properly, it will only give the vertical distance or deviation of a tangent at point A on the elastic curve from the tangent at B.

- After applying either Theorem 1 or Theorem 2, the algebraic sign of the answer can be verified from the angle or deviation as indicated on the elastic curve.

EXAMPLE 8.6

Determine the slope at points B and C of the beam shown in Fig. 8–16a. Take $E = 29(10^3)$ ksi and $I = 600\,in^4$.

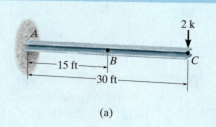

(a)

SOLUTION

M/EI Diagram. This diagram is shown in Fig. 8–16b. It is easier to solve the problem in terms of EI and substitute the numerical data as a last step.

Elastic Curve. The 2-k load causes the beam to deflect as shown in Fig. 8–16c. (The beam is deflected concave down, since M/EI is negative.) Here the tangent at A (the support) is *always horizontal*. The tangents at B and C are also indicated. We are required to find θ_B and θ_C. By the construction, the angle between tan A and tan B, that is, $\theta_{B/A}$, is equivalent to θ_B.

$$\theta_B = \theta_{B/A}$$

Also,

$$\theta_C = \theta_{C/A}$$

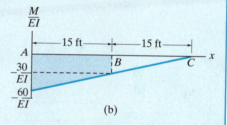

(b)

Moment-Area Theorem. Applying Theorem 1, $\theta_{B/A}$ is equal to the area under the M/EI diagram between points A and B; that is,

$$\theta_B = \theta_{B/A} = -\left(\frac{30\,k \cdot ft}{EI}\right)(15\,ft) - \frac{1}{2}\left(\frac{60\,k \cdot ft}{EI} - \frac{30\,k \cdot ft}{EI}\right)(15\,ft)$$

$$= -\frac{675\,k \cdot ft^2}{EI}$$

Substituting numerical data for E and I, and converting feet to inches, we have

$$\theta_B = \frac{-675\,k \cdot ft^2(144\,in^2/1\,ft^2)}{29(10^3)\,k/in^2(600\,in^4)}$$

$$= -0.00559\,rad \qquad \textit{Ans.}$$

The *negative sign* indicates that the angle is measured clockwise from A, Fig. 8–16c.

In a similar manner, the area under the M/EI diagram between points A and C equals $\theta_{C/A}$. We have

$$\theta_C = \theta_{C/A} = \frac{1}{2}\left(-\frac{60\,k \cdot ft}{EI}\right)(30\,ft) = -\frac{900\,k \cdot ft^2}{EI}$$

Substituting numerical values for EI, we have

$$\theta_C = \frac{-900\,k \cdot ft^2(144\,in^2/ft^2)}{29(10^3)\,k/in^2(600\,in^4)}$$

$$= -0.00745\,rad \qquad \textit{Ans.}$$

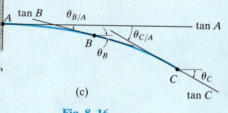

(c)

Fig. 8–16

8

EXAMPLE   8.7

$I_{AB} = 8(10^6) \text{ mm}^4$ $I_{BC} = 4(10^6) \text{ mm}^4$
—— 4 m —— —— 3 m ——

(a)

Determine the deflection at points B and C of the beam shown in Fig. 8–17a. Values for the moment of inertia of each segment are indicated in the figure. Take $E = 200$ GPa.

SOLUTION

M/EI Diagram.   By inspection, the moment diagram for the beam is a rectangle. Here we will construct the M/EI diagram relative to I_{BC}, realizing that $I_{AB} = 2I_{BC}$. Fig. 8–17b. Numerical data for EI_{BC} will be substituted as a last step.

Elastic Curve.   The couple moment at C causes the beam to deflect as shown in Fig. 8–17c. The tangents at A (the support), B, and C are indicated. We are required to find Δ_B and Δ_C. These displacements can be related *directly* to the deviations between the tangents, so that from the construction Δ_B is equal to the deviation of tan B relative to tan A; that is,

$$\Delta_B = t_{B/A}$$

Also,

$$\Delta_C = t_{C/A}$$

Moment-Area Theorem.   Applying Theorem 2, $t_{B/A}$ is equal to the moment of the area under the M/EI_{BC} diagram between A and B computed about point B, since this is the point where the tangential deviation is to be determined. Hence, from Fig. 8–17b,

$$\Delta_B = t_{B/A} = \left[\frac{250 \text{ N} \cdot \text{m}}{EI_{BC}} (4 \text{ m}) \right] (2 \text{ m}) = \frac{2000 \text{ N} \cdot \text{m}^3}{EI_{BC}}$$

Substituting the numerical data yields

$$\Delta_B = \frac{2000 \text{ N} \cdot \text{m}^3}{\left[200(10^9) \text{ N/m}^2 \right] \left[4(10^6) \text{ mm}^4 (1 \text{ m}^4/(10^3)^4 \text{ mm}^4) \right]}$$

$$= 0.0025 \text{ m} = 2.5 \text{ mm}. \qquad \textit{Ans.}$$

Likewise, for $t_{C/A}$ we must compute the moment of the entire M/EI_{BC} diagram from A to C about point C. We have

$$\Delta_C = t_{C/A} = \left[\frac{250 \text{ N} \cdot \text{m}}{EI_{BC}} (4 \text{ m}) \right] (5 \text{ m}) + \left[\frac{500 \text{ N} \cdot \text{m}}{EI_{BC}} (3 \text{ m}) \right] (1.5 \text{ m})$$

$$= \frac{7250 \text{ N} \cdot \text{m}^3}{EI_{BC}} = \frac{7250 \text{ N} \cdot \text{m}^3}{\left[200(10^9) \text{ N/m}^2 \right] \left[4(10^6)(10^{-12}) \text{ m}^4 \right]}$$

$$= 0.00906 \text{ m} = 9.06 \text{ mm} \qquad \textit{Ans.}$$

Since both answers are *positive*, they indicate that points B and C lie *above* the tangent at A.

$\dfrac{M}{EI_{BC}}$

$\dfrac{250}{EI_{BC}}$ $\dfrac{500}{EI_{BC}}$

A —2 m— B C → x
—— 4 m —— —— 3 m ——

(b)

A B C tan C
tan B $\Delta_C = t_{C/A}$
$\Delta_B = t_{B/A}$ tan A

(c)

Fig. 8–17

EXAMPLE 8.8

Determine the slope at point C of the beam in Fig. 8–18a. $E = 200$ GPa, $I = 6(10^6)$ mm^4.

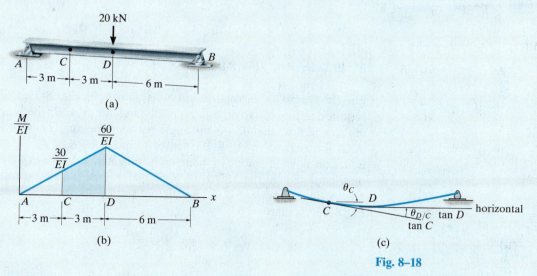

Fig. 8–18

SOLUTION

M/EI Diagram. Fig. 8–18b.

Elastic Curve. Since the loading is applied symmetrically to the beam, the elastic curve is symmetric, as shown in Fig. 8–18c. We are required to find θ_C. This can easily be done, realizing that the tangent at D is *horizontal*, and therefore, by the construction, the angle $\theta_{D/C}$ between tan C and tan D is equal to θ_C; that is,

$$\theta_C = \theta_{D/C}$$

Moment-Area Theorem. Using Theorem 1, $\theta_{D/C}$ is equal to the shaded area under the M/EI diagram between points C and D. We have

$$\theta_C = \theta_{D/C} = 3 \text{ m}\left(\frac{30 \text{ kN} \cdot \text{m}}{EI}\right) + \frac{1}{2}(3 \text{ m})\left(\frac{60 \text{ kN} \cdot \text{m}}{EI} - \frac{30 \text{ kN} \cdot \text{m}}{EI}\right)$$

$$= \frac{135 \text{ kN} \cdot \text{m}^2}{EI}$$

Thus,

$$\theta_C = \frac{135 \text{ kN} \cdot \text{m}^2}{\left[200(10^6) \text{ kN/m}^2\right]\left[6(10^6)(10^{-12}) \text{ m}^4\right]} = 0.112 \text{ rad} \quad Ans.$$

8

EXAMPLE **8.9**

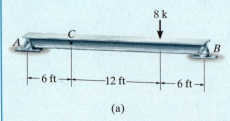

(a)

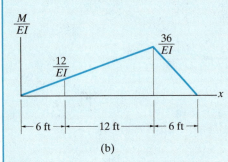

(b)

(c)

Fig. 8–19

Determine the slope at point C of the beam in Fig. 8–19a. $E = 29(10^3)$ ksi, $I = 600$ in^4.

SOLUTION

M/EI Diagram.   Fig. 8–19b.

Elastic Curve.   The elastic curve is shown in Fig. 8–19c. We are required to find θ_C. To do this, establish tangents at A, B (the supports), and C and note that $\theta_{C/A}$ is the angle between the tangents at A and C. Also, the angle ϕ in Fig. 8–19c can be found using $\phi = t_{B/A}/L_{AB}$. This equation is valid since $t_{B/A}$ is actually very small, so that $t_{B/A}$ can be approximated by the length of a circular arc defined by a radius of $L_{AB} = 24$ ft and sweep of ϕ. (Recall that $s = \theta r$.) From the geometry of Fig. 8–19c, we have

$$\theta_C = \phi - \theta_{C/A} = \frac{t_{B/A}}{24} - \theta_{C/A} \qquad (1)$$

Moment-Area Theorems.   Using Theorem 1, $\theta_{C/A}$ is equivalent to the area under the M/EI diagram between points A and C; that is,

$$\theta_{C/A} = \frac{1}{2}(6 \text{ ft})\left(\frac{12 \text{ k} \cdot \text{ft}}{EI}\right) = \frac{36 \text{ k} \cdot \text{ft}^2}{EI}$$

Applying Theorem 2, $t_{B/A}$ is equivalent to the moment of the area under the M/EI diagram between B and A about point B, since this is the point where the tangential deviation is to be determined. We have

$$t_{B/A} = \left[6 \text{ ft} + \frac{1}{3}(18 \text{ ft})\right]\left[\frac{1}{2}(18 \text{ ft})\left(\frac{36 \text{ k} \cdot \text{ft}}{EI}\right)\right]$$
$$+ \frac{2}{3}(6 \text{ ft})\left[\frac{1}{2}(6 \text{ ft})\left(\frac{36 \text{ k} \cdot \text{ft}}{EI}\right)\right]$$
$$= \frac{4320 \text{ k} \cdot \text{ft}^3}{EI}$$

Substituting these results into Eq. 1, we have

$$\theta_C = \frac{4320 \text{ k} \cdot \text{ft}^3}{(24 \text{ ft}) EI} - \frac{36 \text{ k} \cdot \text{ft}^2}{EI} = \frac{144 \text{ k} \cdot \text{ft}^2}{EI}$$

so that

$$\theta_C = \frac{144 \text{ k} \cdot \text{ft}^2}{29(10^3) \text{ k/in}^2(144 \text{ in}^2/\text{ft}^2)\, 600 \text{ in}^4(1 \text{ ft}^4/(12)^4 \text{ in}^4)}$$
$$= 0.00119 \text{ rad} \qquad\qquad \textit{Ans.}$$

EXAMPLE 8.10

Determine the deflection at C of the beam shown in Fig. 8–20a. Take $E = 29(10^3)$ ksi, $I = 21$ in^4.

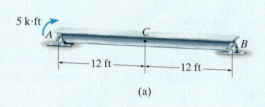

(a)

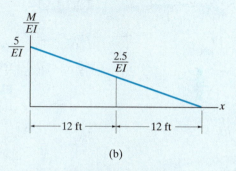

(b)

SOLUTION

***M/EI* Diagram.**   Fig. 8–20b.

Elastic Curve.   Here we are required to find Δ_C, Fig. 8–20c. This is not necessarily the maximum deflection of the beam, since the loading and hence the elastic curve are *not symmetric*. Also indicated in Fig. 8–20c are the tangents at A, B (the supports), and C. If $t_{A/B}$ is determined, then Δ' can be found from proportional triangles, that is, $\Delta'/12 = t_{A/B}/24$ or $\Delta' = t_{A/B}/2$. From the construction in Fig. 8–20c, we have

$$\Delta_C = \frac{t_{A/B}}{2} - t_{C/B} \qquad (1)$$

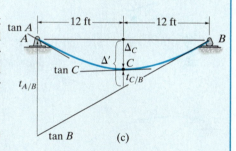

(c)

Fig. 8–20

Moment-Area Theorem.   We will apply Theorem 2 to determine $t_{A/B}$ and $t_{C/B}$. Here $t_{A/B}$ is the moment of the M/EI diagram between A and B about point A,

$$t_{A/B} = \left[\frac{1}{3}(24 \text{ ft})\right]\left[\frac{1}{2}(24 \text{ ft})\left(\frac{5 \text{ k} \cdot \text{ft}}{EI}\right)\right] = \frac{480 \text{ k} \cdot \text{ft}^3}{EI}$$

and $t_{C/B}$ is the moment of the M/EI diagram between C and B about C.

$$t_{C/B} = \left[\frac{1}{3}(12 \text{ ft})\right]\left[\frac{1}{2}(12 \text{ ft})\left(\frac{2.5 \text{ k} \cdot \text{ft}}{EI}\right)\right] = \frac{60 \text{ k} \cdot \text{ft}^3}{EI}$$

Substituting these results into Eq. (1) yields

$$\Delta_C = \frac{1}{2}\left(\frac{480 \text{ k} \cdot \text{ft}^3}{EI}\right) - \frac{60 \text{ k} \cdot \text{ft}^3}{EI} = \frac{180 \text{ k} \cdot \text{ft}^3}{EI}$$

Working in units of kips and inches, we have

$$\Delta_C = \frac{180 \text{ k} \cdot \text{ft}^3(1728 \text{ in}^3/\text{ft}^3)}{29(10^3) \text{ k/in}^2(21 \text{ in}^4)}$$

$$= 0.511 \text{ in.} \qquad \qquad Ans.$$

8

EXAMPLE 8.11

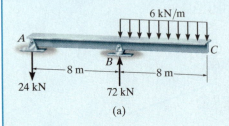

6 kN/m

A

B

C

—8 m— —8 m—

24 kN   72 kN

(a)

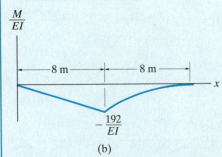

$\frac{M}{EI}$

—8 m— —8 m—

x

$-\frac{192}{EI}$

(b)

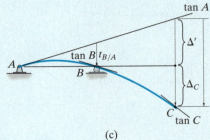

tan A

Δ'

tan B $t_{B/A}$

A

B

Δ_C

C tan C

(c)

Fig. 8–21

Determine the deflection at point C of the beam shown in Fig. 8–21a. $E = 200$ GPa, $I = 250(10^6)$ mm^4.

SOLUTION

M/EI Diagram. As shown in Fig. 8–21b, this diagram consists of a triangular and a parabolic segment.

Elastic Curve. The loading causes the beam to deform as shown in Fig. 8–21c. We are required to find Δ_C. By constructing tangents at A, B (the supports), and C, it is seen that $\Delta_C = t_{C/A} - \Delta'$. However, Δ' can be related to $t_{B/A}$ by proportional triangles, that is, $\Delta'/16 = t_{B/A}/8$ or $\Delta' = 2t_{B/A}$. Hence

$$\Delta_C = t_{C/A} - 2t_{B/A} \qquad (1)$$

Moment-Area Theorem. We will apply Theorem 2 to determine $t_{C/A}$ and $t_{B/A}$. Using the table on the inside back cover for the parabolic segment and considering the moment of the M/EI diagram between A and C about point C, we have

$$t_{C/A} = \left[\frac{3}{4}(8\text{ m})\right]\left[\frac{1}{3}(8\text{ m})\left(-\frac{192\text{ kN}\cdot\text{m}}{EI}\right)\right]$$
$$+ \left[\frac{1}{3}(8\text{ m}) + 8\text{ m}\right]\left[\frac{1}{2}(8\text{ m})\left(-\frac{192\text{ kN}\cdot\text{m}}{EI}\right)\right]$$
$$= -\frac{11\,264\text{ kN}\cdot\text{m}^3}{EI}$$

The moment of the M/EI diagram between A and B about point B gives

$$t_{B/A} = \left[\frac{1}{3}(8\text{ m})\right]\left[\frac{1}{2}(8\text{ m})\left(-\frac{192\text{ kN}\cdot\text{m}}{EI}\right)\right] = -\frac{2048\text{ kN}\cdot\text{m}^3}{EI}$$

Why are these terms negative? Substituting the results into Eq. (1) yields

$$\Delta_C = -\frac{11\,264\text{ kN}\cdot\text{m}^3}{EI} - 2\left(-\frac{2048\text{ kN}\cdot\text{m}^3}{EI}\right)$$
$$= -\frac{7168\text{ kN}\cdot\text{m}^3}{EI}$$

Thus,

$$\Delta_C = \frac{-7168\text{ kN}\cdot\text{m}^3}{\left[200(10^6)\text{ kN/m}^2\right]\left[250(10^6)(10^{-12})\text{ m}^4\right]}$$
$$= -0.143\text{ m} \qquad \textit{Ans.}$$

EXAMPLE 8.12

Determine the slope at the roller B of the double overhang beam shown in Fig. 8–22a. Take $E = 200$ GPa, $I = 18(10^6)$ mm^4.

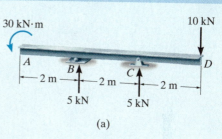

(a)

SOLUTION

M/EI Diagram.   The M/EI diagram can be simplified by drawing it in parts as discussed in Sec. 4.5. Here we will consider the M/EI diagrams for the three loadings each acting on a cantilever beam fixed at D, Fig. 8–22b. (The 10-kN load is not considered since it produces no moment about D.)

Elastic Curve.   If tangents are drawn at B and C, Fig. 8–22c, the slope B can be determined by finding $t_{C/B}$, and for small angles,

$$\theta_B = \frac{t_{C/B}}{2\ \text{m}} \qquad (1)$$

Moment Area Theorem.   To determine $t_{C/B}$ we apply the moment area theorem by finding the moment of the M/EI diagram between BC about point C. This only involves the shaded area under two of the diagrams in Fig. 8–22b. Thus,

$$t_{C/B} = (1\ \text{m})\left[(2\ \text{m})\left(\frac{-30\ \text{kN}\cdot\text{m}}{EI}\right)\right] + \left(\frac{2\ \text{m}}{3}\right)\left[\frac{1}{2}(2\ \text{m})\left(\frac{10\ \text{kN}\cdot\text{m}}{EI}\right)\right]$$

$$= -\frac{53.33\ \text{kN}\cdot\text{m}^3}{EI}$$

Substituting the positive value into Eq. (1),

$$\theta_B = \frac{53.33\ \text{kN}\cdot\text{m}^3}{(2\ \text{m})\left[200(10^6)\ \text{kN/m}^3\right]\left[18(10^6)(10^{-12})\ \text{m}^4\right]}$$

$$= 0.00741\ \text{rad} \qquad\qquad\qquad Ans.$$

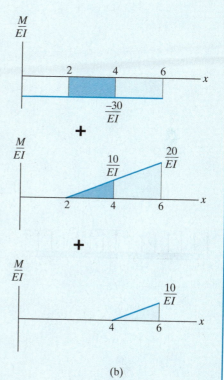

(b)

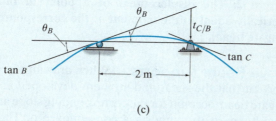

(c)

Fig. 8–22

8

8.5   Conjugate-Beam Method

The conjugate-beam method was developed by H. Müller-Breslau in 1865. Essentially, it requires the same amount of computation as the moment-area theorems to determine a beam's slope or deflection at a specific point. However, this method relies only on the principles of statics, and hence its application will be more familiar.

The basis for the method comes from the *similarity* of Eq. 4–1 and 4–2 to Eq. 8–2 and Eq. 8–4. To show this similarity, we can write these equations as follows:

$$\frac{dV}{dx} = w \qquad\qquad\qquad \frac{d^2M}{dx^2} = w$$

$$\frac{d\theta}{dx} = \frac{M}{EI} \qquad\qquad\qquad \frac{d^2v}{dx^2} = \frac{M}{EI}$$

Or integrating,

$$V = \int w\, dx \qquad\qquad M = \int\left[\int w\, dx\right] dx$$

$$\updownarrow \qquad \updownarrow \qquad\qquad \updownarrow \qquad \updownarrow$$

$$\theta = \int\left(\frac{M}{EI}\right) dx \qquad\qquad v = \int\left[\int\left(\frac{M}{EI}\right) dx\right] dx$$

Here the *shear V* compares with the *slope θ*, the *moment M* compares with the *displacement v*, and the *external load w* compares with the M/EI diagram. To make use of this comparison we will now consider a beam having the same length as the real beam, but referred to here as the "conjugate beam," Fig. 8–23. The conjugate beam is "loaded" with the M/EI diagram derived from the load w on the real beam. From the above comparisons, we can state two theorems related to the conjugate beam, namely,

> **Theorem 1: The slope at a point in the real beam is numerically equal to the shear at the corresponding point in the conjugate beam.**

> **Theorem 2: The displacement of a point in the real beam is numerically equal to the moment at the corresponding point in the conjugate beam.**

Conjugate-Beam Supports.

When drawing the conjugate beam it is important that the shear and moment developed at the supports of the conjugate beam account for the corresponding slope and displacement of the real beam at its supports, a consequence of Theorems 1 and 2.

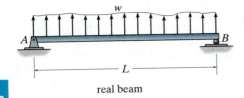

real beam

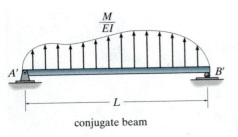

conjugate beam

Fig. 8–23

For example, as shown in Table 8.2, a pin or roller support at the end of the real beam provides *zero displacement,* but the beam has a nonzero slope. Consequently, from Theorems 1 and 2, the conjugate beam must be supported by a pin or roller, since this support has *zero moment* but has a shear or end reaction. When the real beam is fixed supported (3), both the slope and displacement at the support are zero. Here the conjugate beam has a free end, since at this end there is zero shear and zero moment. Corresponding real and conjugate-beam supports for other cases are listed in the table. Examples of real and conjugate beams are shown in Fig. 8–24. Note that, as a rule, neglecting axial force, statically determinate real beams have statically determinate conjugate beams; and statically indeterminate real beams, as in the last case in Fig. 8–24, become unstable conjugate beams. Although this occurs, the M/EI loading will provide the necessary "equilibrium" to hold the conjugate beam stable.

TABLE 8.2

	Real Beam		Conjugate Beam	
1)	θ $\Delta = 0$	pin	V $M = 0$	pin
2)	θ $\Delta = 0$	roller	V $M = 0$	roller
3)	$\theta = 0$ $\Delta = 0$	fixed	$V = 0$ $M = 0$	free
4)	θ Δ	free	V M	fixed
5)	θ $\Delta = 0$	internal pin	V $M = 0$	hinge
6)	θ $\Delta = 0$	internal roller	V $M = 0$	hinge
7)	θ Δ	hinge	V M	internal roller

8

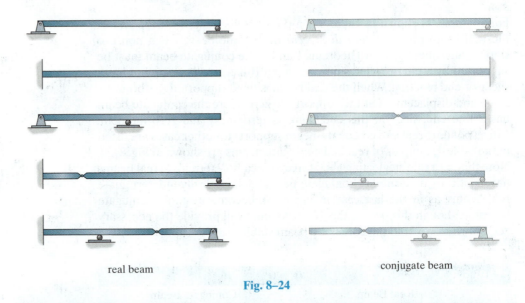

real beam   conjugate beam

Fig. 8–24

Procedure for Analysis

The following procedure provides a method that may be used to determine the displacement and slope at a point on the elastic curve of a beam using the conjugate-beam method.

Conjugate Beam

- Draw the conjugate beam for the real beam. This beam has the same length as the real beam and has corresponding supports as listed in Table 8.2.

- In general, if the real support allows a *slope*, the conjugate support must develop a *shear;* and if the real support allows a *displacement,* the conjugate support must develop a *moment.*

- The conjugate beam is loaded with the real beam's M/EI diagram. This loading is assumed to be *distributed* over the conjugate beam and is directed *upward* when M/EI is *positive* and *downward* when M/EI is *negative.* In other words, the loading always acts *away* from the beam.

Equilibrium

- Using the equations of equilibrium, determine the reactions at the conjugate beam's supports.

- Section the conjugate beam at the point where the slope θ and displacement Δ of the real beam are to be determined. At the section show the unknown shear V' and moment M' acting in their positive sense.

- Determine the shear and moment using the equations of equilibrium. V' and M' equal θ and Δ, respectively, for the real beam. In particular, if these values are *positive,* the *slope* is *counterclockwise* and the *displacement* is *upward.*

EXAMPLE 8.13

Determine the slope and deflection at point B of the steel beam shown in Fig. 8–25a. The reactions have been computed. $E = 29(10^3)$ ksi, $I = 800$ in^4.

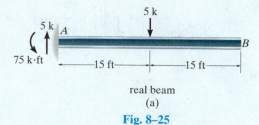

real beam
(a)

Fig. 8–25

SOLUTION

Conjugate Beam. The conjugate beam is shown in Fig. 8–25b. The supports at A' and B' correspond to supports A and B on the real beam, Table 8.2. It is important to understand why this is so. The M/EI diagram is *negative*, so the distributed load acts *downward*, i.e., away from the beam.

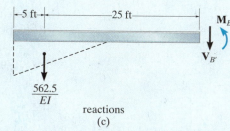

conjugate beam
(b)

Equilibrium. Since θ_B and Δ_B are to be determined, we must compute $V_{B'}$ and $M_{B'}$ in the conjugate beam, Fig. 8–25c.

$$+\uparrow \Sigma F_y = 0; \qquad -\frac{562.5 \text{ k} \cdot \text{ft}^2}{EI} - V_{B'} = 0$$

$$\theta_B = V_{B'} = -\frac{562.5 \text{ k} \cdot \text{ft}^2}{EI}$$

$$= \frac{-562.5 \text{ k} \cdot \text{ft}^2}{29(10^3) \text{ k/in}^2 (144 \text{ in}^2/\text{ft}^2) 800 \text{ in}^4 (1 \text{ ft}^4/(12)^4 \text{ in}^4)}$$

$$= -0.00349 \text{ rad} \qquad \qquad Ans.$$

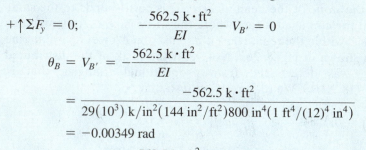

reactions
(c)

$$\zeta + \Sigma M_{B'} = 0; \qquad \frac{562.5 \text{ k} \cdot \text{ft}^2}{EI}(25 \text{ ft}) + M_{B'} = 0$$

$$\Delta_B = M_{B'} = -\frac{14\,062.5 \text{ k} \cdot \text{ft}^3}{EI}$$

$$= \frac{-14\,062.5 \text{ k} \cdot \text{ft}^3}{29(10^3)(144) \text{ k/ft}^2 \left[800/(12)^4\right] \text{ft}^4}$$

$$= -0.0873 \text{ ft} = -1.05 \text{ in.} \qquad \qquad Ans.$$

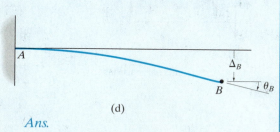

(d)

The negative signs indicate the slope of the beam is measured clockwise and the displacement is downward, Fig. 8–25d.

EXAMPLE | 8.14

Determine the maximum deflection of the steel beam shown in Fig. 8–26a. The reactions have been computed. $E = 200$ GPa, $I = 60(10^6)$ mm^4.

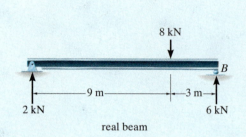

real beam

(a)

Fig. 8–26

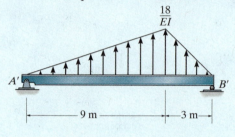

conjugate beam

(b)

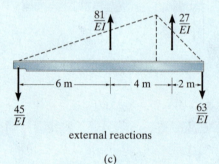

external reactions

(c)

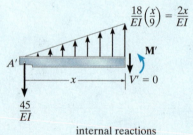

internal reactions

(d)

SOLUTION

Conjugate Beam. The conjugate beam loaded with the M/EI diagram is shown in Fig. 8–26b. Since the M/EI diagram is positive, the distributed load acts upward (away from the beam).

Equilibrium. The external reactions on the conjugate beam are determined first and are indicated on the free-body diagram in Fig. 8–26c. *Maximum deflection* of the real beam occurs at the point where the *slope* of the beam is *zero*. This corresponds to the same point in the conjugate beam where the *shear* is *zero*. Assuming this point acts within the region $0 \le x \le 9$ m from A', we can isolate the section shown in Fig. 8–26d. Note that the peak of the distributed loading was determined from proportional triangles, that is, $w/x = (18/EI)/9$. We require $V' = 0$ so that

$$+\uparrow \Sigma F_y = 0; \qquad -\frac{45}{EI} + \frac{1}{2}\left(\frac{2x}{EI}\right)x = 0$$

$$x = 6.71 \text{ m} \qquad (0 \le x \le 9 \text{ m}) \text{ OK}$$

Using this value for x, the maximum deflection in the real beam corresponds to the moment M'. Hence,

$$\zeta + \Sigma M = 0; \qquad \frac{45}{EI}(6.71) - \left[\frac{1}{2}\left(\frac{2(6.71)}{EI}\right)6.71\right]\frac{1}{3}(6.71) + M' = 0$$

$$\Delta_{max} = M' = -\frac{201.2 \text{ kN} \cdot \text{m}^3}{EI}$$

$$= \frac{-201.2 \text{ kN} \cdot \text{m}^3}{\left[200(10^6) \text{ kN/m}^2\right]\left[60(10^6) \text{ mm}^4(1 \text{ m}^4/(10^3)^4 \text{ mm}^4)\right]}$$

$$= -0.0168 \text{ m} = -16.8 \text{ mm} \qquad \qquad Ans.$$

The negative sign indicates the deflection is downward.

EXAMPLE | 8.15

The girder in Fig. 8–27a is made from a continuous beam and reinforced at its center with cover plates where its moment of inertia is larger. The 12-ft end segments have a moment of inertia of $I = 450$ in⁴, and the center portion has a moment of inertia of $I' = 900$ in⁴. Determine the deflection at the center C. Take $E = 29(10^3)$ ksi. The reactions have been calculated.

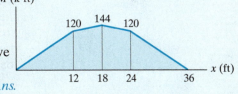

real beam

(a)

Fig. 8–27

SOLUTION

Conjugate Beam. The moment diagram for the beam is determined first, Fig. 8–27b. Since $I' = 2I$, for simplicity, we can express the load on the conjugate beam in terms of the constant EI, as shown in Fig. 8–27c.

Equilibrium. The reactions on the conjugate beam can be calculated by the symmetry of the loading or using the equations of equilibrium. The results are shown in Fig. 8–27d. Since the deflection at C is to be determined, we must compute the internal moment at C'. Using the method of sections, segment $A'C'$ is isolated and the resultants of the distributed loads and their locations are determined, Fig. 8–27e. Thus,

$$\zeta + \Sigma M_{C'} = 0; \quad \frac{1116}{EI}(18) - \frac{720}{EI}(10) - \frac{360}{EI}(3) - \frac{36}{EI}(2) + M_{C'} = 0$$

$$M_{C'} = -\frac{11\ 736\ \text{k}\cdot\text{ft}^3}{EI}$$

Substituting the numerical data for EI and converting units, we have

$$\Delta_C = M_{C'} = -\frac{11\ 736\ \text{k}\cdot\text{ft}^3(1728\ \text{in}^3/\text{ft}^3)}{29(10^3)\ \text{k/in}^2(450\ \text{in}^4)} = -1.55\ \text{in.} \quad Ans.$$

The negative sign indicates that the deflection is downward.

M (k·ft)

moment diagram

(b)

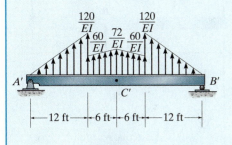

conjugate beam

(c)

external reactions

(d)

internal reactions

(e)

8

EXAMPLE   8.16

Determine the displacement of the pin at B and the slope of each beam segment connected to the pin for the compound beam shown in Fig. 8–28a. $E = 29(10^3)$ ksi, $I = 30$ in^4.

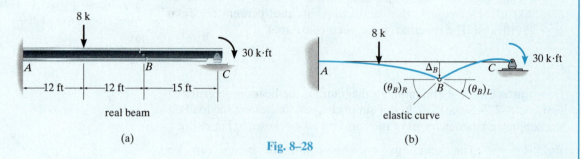

real beam

(a)

Fig. 8–28

elastic curve

(b)

SOLUTION

Conjugate Beam.   The elastic curve for the beam is shown in Fig. 8–28b in order to identify the unknown displacement Δ_B and the slopes $(\theta_B)_L$ and $(\theta_B)_R$ to the left and right of the pin. Using Table 8.2, the conjugate beam is shown in Fig. 8–28c. For simplicity in calculation, the M/EI diagram has been drawn in *parts* using the principle of superposition as described in Sec. 4.5. Here the beam is cantilevered from the left support, A. The moment diagrams for the 8-k load, the reactive force $C_y = 2$ k, and the 30-k·ft loading are given. Notice that negative regions of this diagram develop a downward distributed load and positive regions have a distributed load that acts upward.

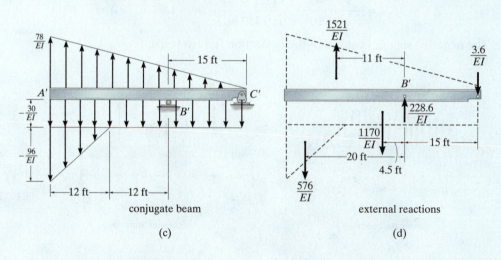

conjugate beam

(c)

external reactions

(d)

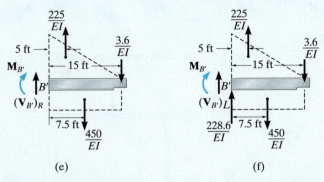

(e)   (f)

Equilibrium. The external reactions at B' and C' are calculated first and the results are indicated in Fig. 8–28d. In order to determine $(\theta_B)_R$, the conjugate beam is sectioned just to the *right* of B' and the shear force $(V_{B'})_R$ is computed, Fig. 8–28e. Thus,

$$+\uparrow\Sigma F_y = 0; \qquad (V_{B'})_R + \frac{225}{EI} - \frac{450}{EI} - \frac{3.6}{EI} = 0$$

$$(\theta_B)_R = (V_{B'})_R = \frac{228.6 \text{ k} \cdot \text{ft}^2}{EI}$$

$$= \frac{228.6 \text{ k} \cdot \text{ft}^2}{\left[29(10^3)(144) \text{ k/ft}^2\right]\left[30/(12)^4\right]\text{ft}^4}$$

$$= 0.0378 \text{ rad} \qquad\qquad Ans.$$

The internal moment at B' yields the displacement of the pin. Thus,

$$\zeta +\Sigma M_{B'} = 0; \qquad -M_{B'} + \frac{225}{EI}(5) - \frac{450}{EI}(7.5) - \frac{3.6}{EI}(15) = 0$$

$$\Delta_B = M_{B'} = -\frac{2304 \text{ k} \cdot \text{ft}^3}{EI}$$

$$= \frac{-2304 \text{ k} \cdot \text{ft}^3}{\left[29(10^3)(144) \text{ k/ft}^2\right]\left[30/(12)^4\right]\text{ft}^4}$$

$$= -0.381 \text{ ft} = -4.58 \text{ in.} \qquad Ans.$$

The slope $(\theta_B)_L$ can be found from a section of beam just to the *left* of B', Fig. 8–28f. Thus,

$$+\uparrow\Sigma F_y = 0; \qquad (V_{B'})_L + \frac{228.6}{EI} + \frac{225}{EI} - \frac{450}{EI} - \frac{3.6}{EI} = 0$$

$$(\theta_B)_L = (V_{B'})_L = 0 \qquad\qquad Ans.$$

Obviously, $\Delta_B = M_{B'}$ for this segment is the *same* as previously calculated, since the moment arms are only slightly different in Figs. 8–28e and 8–28f.

8

FUNDAMENTAL PROBLEMS

F8–1. Draw the deflected shape of each beam.

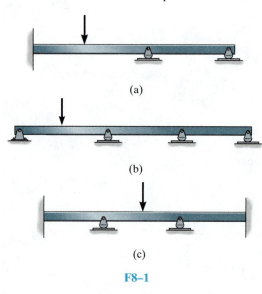

(a)

(b)

(c)

F8–1

F8–2. Draw the deflected shape of each beam.

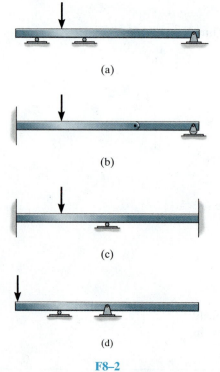

(a)

(b)

(c)

(d)

F8–2

F8–3. Draw the deflected shape of each frame.

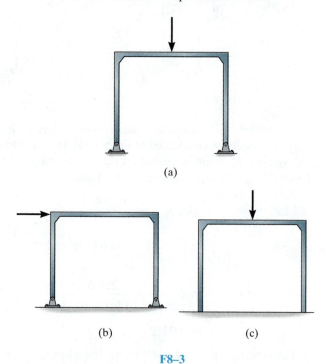

(a)

(b)   (c)

F8–3

F8–4. Draw the deflected shape of each frame.

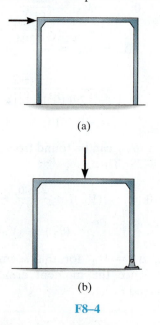

(a)

(b)

F8–4

F8–5. Draw the deflected shape of each frame.

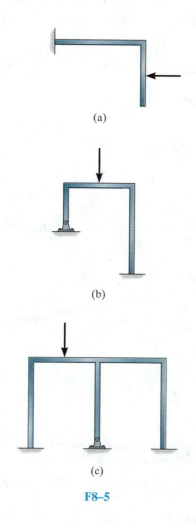

(a)

(b)

(c)

F8–5

F8–6. Determine the equation of the elastic curve for the beam using the x coordinate that is valid for $0 < x < L$. EI is constant.

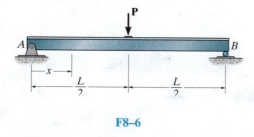

F8–6

F8–7. Determine the equation of the elastic curve for the beam using the x coordinate that is valid for $0 < x < L$. EI is constant.

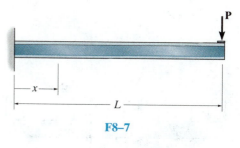

F8–7

F8–8. Determine the equation of the elastic curve for the beam using the x coordinate that is valid for $0 < x < L$. EI is constant.

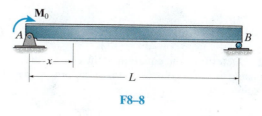

F8–8

F8–9. Determine the equation of the elastic curve for the beam using the x coordinate that is valid for $0 < x < L$. EI is constant.

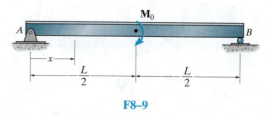

F8–9

8

F8–10. Determine the equation of the elastic curve for the beam using the x coordinate that is valid for $0 < x < L$. EI is constant.

F8–12. Use the moment-area theorems and determine the slope at A and deflection at A. EI is constant.

F8–13. Solve Prob. F8–12 using the conjugate beam method.

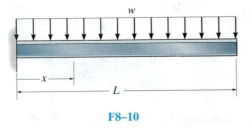

F8–10

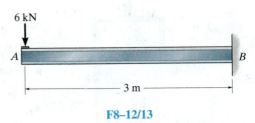

F8–12/13

F8–11. Determine the equation of the elastic curve for the beam using the x coordinate that is valid for $0 < x < L$. EI is constant.

F8–14. Use the moment-area theorems and determine the slope at B and deflection at B. EI is constant.

F8–15. Solve Prob. F8–14 using the conjugate beam method.

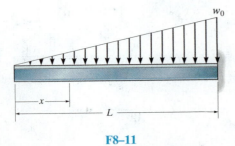

F8–11

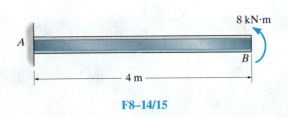

F8–14/15

F8–16. Use the moment-area theorems and determine the slope at *A* and displacement at *C*. *EI* is constant.

F8–17. Solve Prob. F8–16 using the conjugate beam method.

F8–20. Use the moment-area theorems and determine the slope at *A* and displacement at *C*. *EI* is constant.

F8–21. Solve Prob. F8–20 using the conjugate beam method.

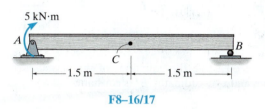

F8–16/17

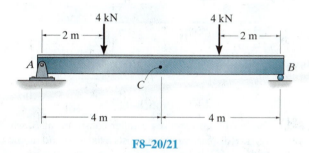

F8–20/21

F8–18. Use the moment-area theorems and determine the slope at *A* and displacement at *C*. *EI* is constant.

F8–19. Solve Prob. F8–18 using the conjugate beam method.

F8–22. Use the moment-area theorems and determine the slope at *B* and displacement at *B*. *EI* is constant.

F8–23. Solve Prob. F8–22 using the conjugate beam method.

8

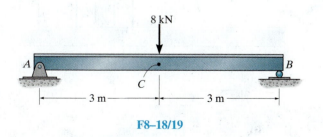

F8–18/19

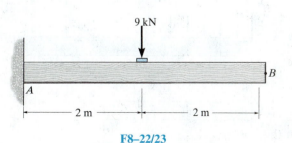

F8–22/23

PROBLEMS

Sec. 8.1–8.3

8–1. Determine the equations of the elastic curve for the beam using the x_1 and x_2 coordinates. Specify the slope at A and the maximum deflection. EI is constant.

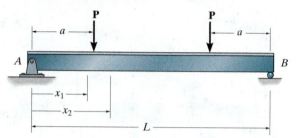

Prob. 8–1

8–2. The bar is supported by a roller constraint at B, which allows vertical displacement but resists axial load and moment. If the bar is subjected to the loading shown, determine the slope at A and the deflection at C. EI is constant.

8–3. Determine the deflection at B of the bar in Prob. 8–2.

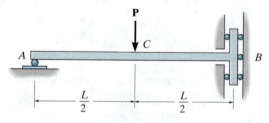

Probs. 8–2/3

*8–4. Determine the equations of the elastic curve using the coordinates x_1 and x_2 and specify the slope and deflection at B. EI is constant.

8–5. Determine the equations of the elastic curve using the coordinates x_1 and x_3 and specify the slope and deflection at point B. EI is constant.

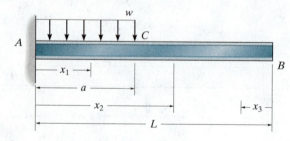

Probs. 8–4/5

8–6. Determine the equations of the elastic curve for the beam using the x coordinate. Specify the slope at A and the maximum deflection of the beam. EI is constant.

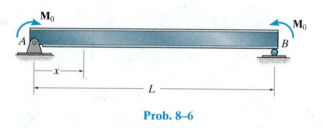

Prob. 8–6

8–7. Determine the equations of the elastic curve using the x_1 and x_2 coordinates. EI is constant.

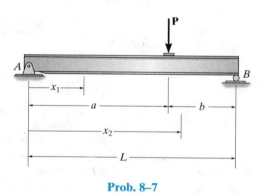

Prob. 8–7

*8–8. Determine the equations of the elastic curve using the coordinates x_1 and x_2 and specify the slope at C and displacement at B. EI is constant.

8–9. Determine the equations of the elastic curve using the coordinates x_1 and x_3 and specify the slope at B and deflection at C. EI is constant.

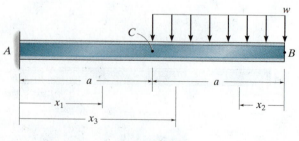

Probs. 8–8/9

Sec. 8.4–8.5

8–10. Determine the slope at B and the maximum displacement of the beam. Use the moment-area theorems. Take $E = 29(10^3)$ ksi, $I = 500$ in^4.

8–11. Solve Prob. 8–10 using the conjugate-beam method.

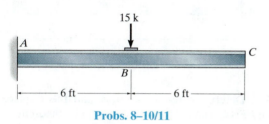

15 k

6 ft   6 ft

Probs. 8–10/11

*8–12. Use the moment-area theorems and determine the slope at A and displacement at C. EI is constant.

8–13. Solve Prob. 8–12 using the conjugate beam method.

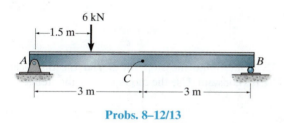

6 kN

—1.5 m—

3 m   3 m

Probs. 8–12/13

8–14. Determine the value of a so that the slope at A is equal to zero. EI is constant. Use the moment-area theorems.

8–15. Solve Prob. 8–14 using the conjugate-beam method.

*8–16. Determine the value of a so that the displacement at C is equal to zero. EI is constant. Use the moment-area theorems.

8–17. Solve Prob. 8–16 using the conjugate-beam method.

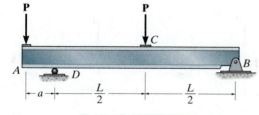

P   P
           C
A   D   B
|—a—| $\frac{L}{2}$ $\frac{L}{2}$

Probs. 8–14/15/16/17

8–18. Determine the slope at D and the displacement at the end C of the beam. EI is constant. Use the moment-area theorems.

8–19. Solve Prob. 8–18 using the conjugate-beam method.

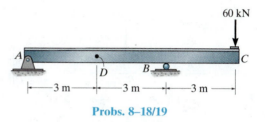

60 kN

A   D   B   C

3 m   3 m   3 m

Probs. 8–18/19

*8–20. Determine the slope and the displacement at the end C of the beam. $E = 200$ GPa, $I = 70(10^6)$ mm^4. Use the moment-area theorems.

8–21. Solve Prob. 8–20 using the conjugate-beam method.

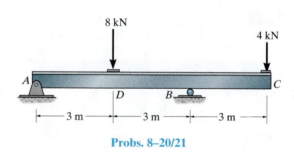

8 kN
                                    4 kN
A   D   B   C

3 m   3 m   3 m

Probs. 8–20/21

8–22. Determine the displacement and slope at C. EI is constant. Use the moment-area theorems.

8–23. Solve Prob. 8–22 using the conjugate-beam method.

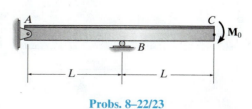

A   C
                       M_0
         B

L   L

Probs. 8–22/23

8

*8–24. Determine the slope at B and the maximum displacement of the beam. Use the moment-area theorems. Take $E = 29(10^3)$ ksi, $I = 500$ in^4.

8–25. Solve Prob. 8–24 using the conjugate-beam method.

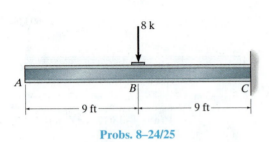

Probs. 8–24/25

8–26. Determine the slope and displacement at C. EI is constant. Use the moment-area theorems.

8–27. Solve Prob. 8–26 using the conjugate-beam method.

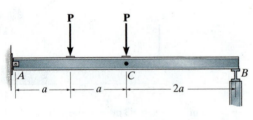

Probs. 8–26/27

*8–28. Determine the slope and the displacement at C. EI is constant. Use the moment-area theorems.

8–29. Solve Prob. 8–28 using the conjugate-beam method.

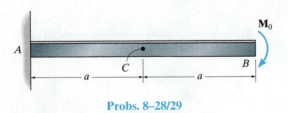

Probs. 8–28/29

8–30. Determine the displacement at B. EI is constant. Use the conjugate-beam method.

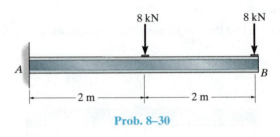

Prob. 8–30

8–31. Determine the slope at A and the displacement at D. EI is constant. Use the moment-area theorems.

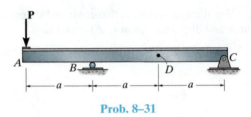

Prob. 8–31

*8–32. Determine the slope at B and the displacement at C. EI is constant. Use the moment-area theorems.

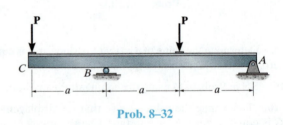

Prob. 8–32

8–33. Determine the slope at B and the displacement at C. EI is constant. Use the conjugate-beam method.

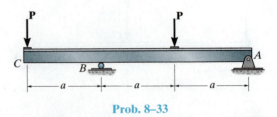

Prob. 8–33

8–34. Determine the displacement at the center B of the beam and the slope at A. EI is constant. Use the moment-area theorems.

8–35. Solve Prob. 8–34 using the conjugate-beam method.

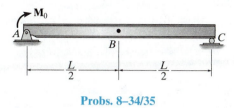

Probs. 8–34/35

8–37. Determine the displacement at C and the slope at B. EI is constant. Use the conjugate-beam method.

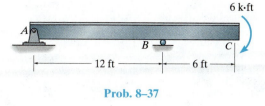

Prob. 8–37

*8–36. Determine the slope to the left and right of B and the displacement at D. EI is constant. Use the moment-area theorems.

8–38. Determine the displacement at C and the slope at D. Assume A is a fixed support, B is a pin, and D is a roller. Use the conjugate-beam method.

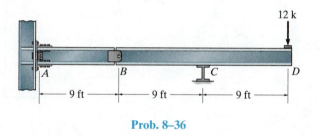

Prob. 8–36

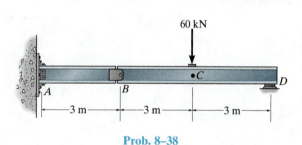

Prob. 8–38

8

CHAPTER REVIEW

The deflection of a member (or structure) can always be established provided the moment diagram is known, because positive moment will tend to bend the member concave upwards, and negative moment will tend to bend the member concave downwards. Likewise, the general shape of the moment diagram can be determined if the deflection curve is known.

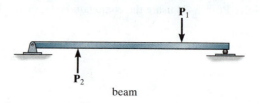

beam

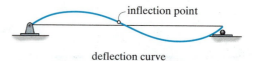

deflection curve

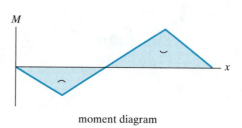

moment diagram

Deflection of a beam due to bending can be determined by using double integration of the equation.

$$\frac{d^2v}{dx^2} = \frac{M}{EI}$$

Here the internal moment M must be expressed as a function of the x coordinates that extend across the beam. The constants of integration are obtained from the boundary conditions, such as zero deflection at a pin or roller support and zero deflection and slope at a fixed support. If several x coordinates are necessary, then the continuity of slope and deflection must be considered, where at $x_1 = x_2 = a$, $\theta_1(a) = \theta_2(a)$ and $v_1(a) = v_2(a)$.

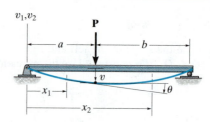

If the moment diagram has a simple shape, the moment-area theorems or the conjugate beam method can be used to determine the deflection and slope at a point on the beam.

The moment-area theorems consider the angles and vertical deviation between the tangents at two points A and B on the elastic curve. The change in slope is found from the area under the M/EI diagram between the two points, and the deviation is determined from the moment of the M/EI diagram area about the point where the deviation occurs.

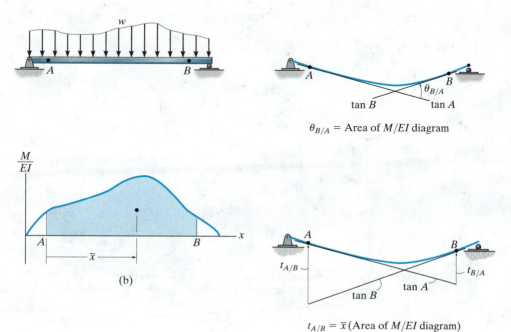

$$\theta_{B/A} = \text{Area of } M/EI \text{ diagram}$$

$$t_{A/B} = \bar{x} \,(\text{Area of } M/EI \text{ diagram})$$

(b)

The conjugate beam method is very methodical and requires application of the principles of statics. Quite simply, one establishes the conjugate beam using Table 8.2, then considers the loading as the M/EI diagram. The slope (deflection) at a point on the real beam is then equal to the shear (moment) at the same point on the conjugate beam.

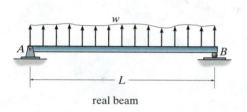

real beam

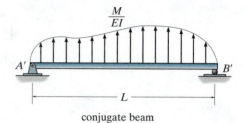

conjugate beam

8

Chapter 9

© Il Fede/Fotolia

The deflection of the ends of this arch as it is being constructed can be determined using the energy methods.

Deflections Using Energy Methods

In this chapter, we will show how to apply energy methods to solve problems involving slope and deflection. The chapter begins with a discussion of work and strain energy, followed by a development of the principle of work and energy. The method of virtual work and Castigliano's theorem are then developed, and these methods are used to determine the displacements at points on trusses, beams, and frames.

9.1 External Work and Strain Energy

The semigraphical methods presented in the previous chapters are very effective for finding the displacements and slopes at points in *beams* subjected to rather simple loadings. For more complicated loadings or for structures such as trusses and frames, it is suggested that energy methods be used for the computations. Most energy methods are based on the *conservation of energy principle*, which states that the work done by all the external forces acting on a structure, U_e, is transformed into internal work or strain energy, U_i, which is developed when the structure deforms. If the material's elastic limit is not exceeded, the *elastic strain energy* will return the structure to its undeformed state when the loads are removed. The conservation of energy principle can be stated mathematically as

$$U_e = U_i \qquad\qquad (9\text{–}1)$$

Before developing any of the energy methods based on this principle, however, we will first determine the external work and strain energy caused by a force and a moment. The formulations to be presented will provide a basis for understanding the work and energy methods that follow.

External Work—Force. When a force **F** undergoes a displacement dx in the *same direction* as the force, the work done is $dU_e = F\,dx$. If the total displacement is x, the work becomes

$$U_e = \int_0^x F\,dx \qquad (9\text{--}2)$$

Consider now the effect caused by an axial force applied to the end of a bar as shown in Fig. 9–1a. As the magnitude of **F** is *gradually* increased from zero to some limiting value $F = P$, the final elongation of the bar becomes Δ. If the material has a linear elastic response, then $F = (P/\Delta)x$. Substituting into Eq. 9–2, and integrating from 0 to Δ, we get

$$U_e = \tfrac{1}{2}P\Delta \qquad (9\text{--}3)$$

which represents the shaded *triangular area* in Fig. 9–1a.

We may also conclude from this that as a force is gradually applied to the bar, and its magnitude builds linearly from zero to some value P, the work done is equal to the *average force magnitude* $(P/2)$ times the displacement (Δ).

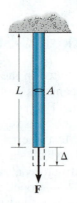

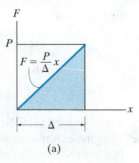

(a)

Fig. 9–1

Suppose now that **P** is already applied to the bar and that *another force* **F'** is now applied, so the bar deflects further by an amount Δ', Fig. 9–1b. The work done by **P** (not **F'**) when the bar undergoes the further deflection Δ' is then

$$U_e' = P\Delta' \qquad (9\text{–}4)$$

Here the work represents the shaded *rectangular area* in Fig. 9–1b. In this case **P** does not change its magnitude since Δ' is caused only by **F'**. Therefore, work is simply the force magnitude (P) times the displacement Δ').

In summary, then, when a force **P** is applied to the bar, followed by application of the force **F'**, the total work done by both forces is represented by the triangular area ACE in Fig. 9–1b. The triangular area ABG represents the work of **P** that is caused by its displacement Δ, the triangular area BCD represents the work of **F'** since this force causes a displacement Δ', and lastly, the shaded rectangular area $BDEG$ represents the additional work done by **P** when displaced Δ' as caused by **F'**.

External Work—Moment.

The work of a moment is defined by the product of the magnitude of the moment **M** and the angle $d\theta$ through which it rotates, that is, $dU_e = M\,d\theta$, Fig. 9–2. If the total angle of rotation is θ radians, the work becomes

$$U_e = \int_0^\theta M\,d\theta \qquad (9\text{–}5)$$

As in the case of force, if the moment is applied *gradually* to a structure having linear elastic response from zero to M, the work is then

$$U_e = \tfrac{1}{2}M\theta \qquad (9\text{–}6)$$

However, if the moment is already applied to the structure and other loadings further distort the structure by an amount θ', then **M** rotates θ', and the work is

$$U_e' = M\theta' \qquad (9\text{–}7)$$

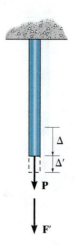

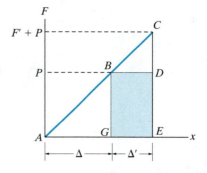

(b)

Fig. 9–1

9

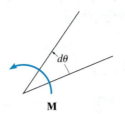

Fig. 9–2

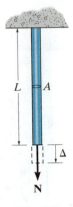

L A

Δ

N

Fig. 9–3

Strain Energy—Axial Force.

When an axial force N is applied gradually to the bar in Fig. 9–3, it will strain the material such that the *external work* done by N will be converted into *strain energy*, which is stored in the bar (Eq. 9–1). Provided the material is *linearly elastic*, Hooke's law is valid, $\sigma = E\epsilon$, and if the bar has a constant cross-sectional area A and length L, the normal stress is $\sigma = N/A$ and the final strain is $\epsilon = \Delta/L$. Consequently, $N/A = E(\Delta/L)$, and the final deflection is

$$\Delta = \frac{NL}{AE} \tag{9–8}$$

Substituting into Eq. 9–3, with $P = N$, the strain energy in the bar is therefore

$$U_i = \frac{N^2 L}{2AE} \tag{9–9}$$

Strain Energy—Bending.

Consider the beam shown in Fig. 9–4a, which is distorted by the *gradually* applied loading P and w. These loads create an internal moment M in the beam at a section located a distance x from the left support. The resulting rotation of the differential element dx, Fig. 9–4b, can be found from Eq. 8–2, that is, $d\theta = (M/EI)\,dx$. Consequently, the strain energy, or work stored in the element, is determined from Eq. 9–6 since the internal moment is gradually developed. Hence,

$$dU_i = \frac{M^2\,dx}{2EI} \tag{9–10}$$

The strain energy for the beam is determined by integrating this result over the beam's entire length L. The result is

$$U_i = \int_0^L \frac{M^2\,dx}{2EI} \tag{9–11}$$

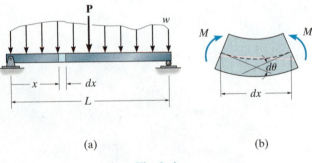

(a)   (b)

Fig. 9–4

9.2 Principle of Work and Energy

Now that the work and strain energy for a force and a moment have been formulated, we will illustrate how the conservation of energy or the principle of work and energy can be applied to determine the displacement at a point on a structure. To do this, consider finding the displacement Δ at the point where the force **P** is applied to the cantilever beam in Fig. 9–5. From Eq. 9–3, the external work is $U_e = \frac{1}{2}P\Delta$. To obtain the resulting strain energy, we must first determine the internal moment as a function of position x in the beam and then apply Eq. 9–11. In this case $M = -Px$, so that

$$U_i = \int_0^L \frac{M^2\,dx}{2EI} = \int_0^L \frac{(-Px)^2\,dx}{2EI} = \frac{1}{6}\frac{P^2L^3}{EI}$$

Equating the external work to internal strain energy and solving for the unknown displacement Δ, we have

$$U_e = U_i$$

$$\frac{1}{2}P\Delta = \frac{1}{6}\frac{P^2L^3}{EI}$$

$$\Delta = \frac{PL^3}{3EI}$$

   Although the solution here is quite direct, application of this method is limited to only a few select problems. It will be noted that only *one load* may be applied to the structure, since if more than one load were applied, there would be an unknown displacement under each load, and yet it is possible to write only *one* "work" equation for the beam. Furthermore, *only the displacement under the force can be obtained*, since the external work depends upon both the force and its corresponding displacement. One way to circumvent these limitations is to use the method of virtual work or Castigliano's theorem, both of which are explained in the following sections.

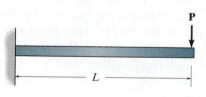
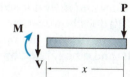

Fig. 9–5

9.3   Principle of Virtual Work

The principle of virtual work was developed by John Bernoulli in 1717 and is sometimes referred to as the unit-load method. It provides a general means of obtaining the displacement and slope at a specific point on a structure, be it a beam, frame, or truss.

Before developing the principle of virtual work, it is necessary to make some general statements regarding the principle of work and energy, which was discussed in the previous section. If we take a deformable structure of any shape or size and apply a series of *external loads* **P** to it, it will cause *internal* loads **u** at points throughout the structure. *It is necessary that the external and internal loads be related by the equations of equilibrium.* As a consequence of these loadings, external displacements Δ will occur at the **P** loads and internal displacements δ will occur at each point of internal load **u**. In general, *these displacements do not have to be elastic*, and they may not be related to the loads; however, *the external and internal displacements must be related by the compatibility of the displacements.* In other words, if the external displacements are known, the corresponding internal displacements are uniquely defined. In general, then, the principle of work and energy states:

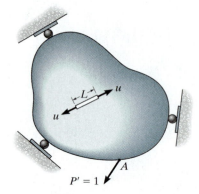

Apply virtual load $P' = 1$

(a)

$$\underset{\substack{\text{Work of}\\\text{External Loads}}}{\Sigma P \Delta} \quad = \quad \underset{\substack{\text{Work of}\\\text{Internal Loads}}}{\Sigma u \delta} \qquad (9\text{--}12)$$

Based on this concept, the principle of virtual work will now be developed. To do this, we will consider the structure (or body) to be of arbitrary shape as shown in Fig. 9–6b.* Suppose it is necessary to determine the displacement Δ of point A on the body caused by the "real loads" **P**$_1$, **P**$_2$, and **P**$_3$. It is to be understood that these loads cause no movement of the supports; in general, however, they can strain the material *beyond the elastic limit.* Since no external load acts on the body at A and in the direction of Δ, the displacement Δ can be determined by *first* placing on the body a ***"virtual" load*** such that this force **P'** acts in the *same direction* as Δ, Fig. 9–6a. For convenience, which will be apparent later, we will choose **P'** to have a "unit" magnitude, that is, $P' = 1$. The term "virtual" is used to describe the load, since *it is imaginary and does not actually exist as part of the real loading.* The unit load (**P'**) does, however, create an internal virtual load **u** in a representative element or fiber of the body, as shown in Fig. 9–6a. Here it is required that **P'** and **u** be related by the equations of equilibrium.†

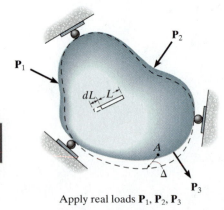

Apply real loads **P**$_1$, **P**$_2$, **P**$_3$

(b)

Fig. 9–6

---

*This arbitrary shape will later represent a specific truss, beam, or frame.
†Although these loads will cause virtual displacements, we will not be concerned with their magnitudes.

Once the virtual loadings are applied, *then* the body is subjected to the *real loads* $\mathbf{P_1}$, $\mathbf{P_2}$, and $\mathbf{P_3}$, Fig. 9–6b. Point A will be displaced an amount Δ, causing the element to deform an amount dL. As a result, the external virtual force $\mathbf{P'}$ and internal virtual load $\mathbf{u}$ "ride along" by Δ and dL, respectively, and therefore perform *external virtual work* of $1 \cdot \Delta$ on the body and internal virtual work of $u \cdot dL$ on the element. Realizing that the external virtual work is equal to the internal virtual work done on all the elements of the body, we can write the virtual-work equation as

$$\overset{\displaystyle \text{virtual loadings}}{\underset{\displaystyle \text{real displacements}}{1 \cdot \Delta = \Sigma u \cdot dL}} \qquad\qquad (9\text{–}13)$$

where

$P' = 1 = $ external virtual unit load acting in the direction of Δ.

$u = $ internal virtual load acting on the element in the direction of dL.

$\Delta = $ external displacement caused by the real loads.

$dL = $ internal deformation of the element caused by the real loads.

By choosing $P' = 1$, it can be seen that the solution for Δ follows directly, since $(1)\Delta = \Sigma u\, dL$.

In a similar manner, if the rotational displacement or slope of the tangent at a point on a structure is to be determined, a virtual *couple moment* $\mathbf{M'}$ having a "unit" magnitude is applied at the point. As a consequence, this couple moment causes a virtual load $\mathbf{u_\theta}$ in one of the elements of the body. Assuming that the real loads deform the element an amount dL, the rotation θ can be found from the virtual-work equation

$$\overset{\displaystyle \text{virtual loadings}}{\underset{\displaystyle \text{real displacements}}{1 \cdot \theta = \Sigma u_\theta \cdot dL}} \qquad\qquad (9\text{–}14)$$

where

$M' = 1 = $ external virtual unit couple moment acting in the direction of θ.

$u_\theta = $ internal virtual load acting on an element in the direction of dL.

$\theta = $ external rotational displacement or slope in radians caused by the real loads.

$dL = $ internal deformation of the element caused by the real loads.

This method for applying the principle of virtual work is often referred to as the ***method of virtual forces***, since a virtual force is applied resulting in the calculation of a *real displacement*. The equation of virtual work in this case represents a *compatibility requirement* for the structure. Although not important here, realize that we can also apply the principle of virtual work as a ***method of virtual displacements***. In this case virtual displacements are imposed on the structure while the structure is subjected to *real loadings*. This method can be used to determine a force on or in a structure,* so that the equation of virtual work is then expressed as an *equilibrium requirement*.

9.4   Method of Virtual Work: Trusses

We can use the method of virtual work to determine the displacement of a truss joint when the truss is subjected to an external loading, temperature change, or fabrication errors. Each of these situations will now be discussed.

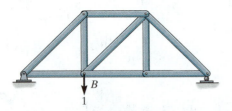

B

1

Apply virtual unit load to B

(a)

External Loading.   For the purpose of explanation let us consider the vertical displacement Δ of joint B of the truss in Fig. 9–7a. Here a typical element of the truss would be one of its *members* having a length L, Fig. 9–7b. If the applied loadings $\mathbf{P}_1$ and $\mathbf{P}_2$ cause a *linear elastic material response*, then this element deforms an amount $\Delta L = NL/AE$, where N is the normal or axial force in the member, caused by the loads. Applying Eq. 9–13, the virtual-work equation for the truss is therefore

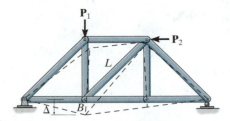

$\mathbf{P}_1$

L

$\mathbf{P}_2$

Δ B

Apply real loads $\mathbf{P}_1$, $\mathbf{P}_2$

(b)

Fig. 9–7

$$1 \cdot \Delta = \sum \frac{nNL}{AE} \tag{9–15}$$

where

$1 =$ external virtual unit load acting on the truss joint in the stated direction of Δ.

$n =$ internal virtual normal force in a truss member caused by the external virtual unit load.

$\Delta =$ external joint displacement caused by the real loads on the truss.

$N =$ internal normal force in a truss member caused by the real loads.

$L =$ length of a member.

$A =$ cross-sectional area of a member.

$E =$ modulus of elasticity of a member.

   The formulation of this equation follows naturally from the development in Sec. 9.3. Here the external virtual unit load creates internal virtual forces **n** in each of the truss members. The real loads then cause the truss joint to be displaced Δ in the same direction as the virtual unit load, and each member is displaced NL/AE in the same direction as its respective **n** force. Consequently, the external virtual work $1 \cdot \Delta$ equals the internal virtual work or the internal (virtual) strain energy stored in *all* the truss members, that is, $\Sigma nNL/AE$.

*It was used in this manner in Sec. 6.3 with reference to the Müller-Breslau principle.

Temperature. In some cases, truss members may change their length due to temperature. If α is the coefficient of thermal expansion for a member and ΔT is the change in its temperature, the change in length of a member is $\Delta L = \alpha \, \Delta T \, L$. Hence, we can determine the displacement of a selected truss joint due to this temperature change from Eq. 9–13, written as

$$1 \cdot \Delta = \Sigma n \alpha \, \Delta T \, L \qquad (9\text{–}16)$$

where

$1 = $ external virtual unit load acting on the truss joint in the stated direction of Δ.

$n = $ internal virtual normal force in a truss member caused by the external virtual unit load.

$\Delta = $ external joint displacement caused by the temperature change.

$\alpha = $ coefficient of thermal expansion of member.

$\Delta T = $ change in temperature of member.

$L = $ length of member.

Fabrication Errors and Camber. Occasionally, errors in fabricating the lengths of the members of a truss may occur. Also, in some cases truss members must be made slightly longer or shorter in order to give the truss a camber. Camber is often built into a bridge truss so that the bottom cord will curve upward by an amount equivalent to the downward deflection of the cord when subjected to the bridge's full dead weight. If a truss member is shorter or longer than intended, the displacement of a truss joint from its expected position can be determined from direct application of Eq. 9–13, written as

$$1 \cdot \Delta = \Sigma n \, \Delta L \qquad (9\text{–}17)$$

where

$1 = $ external virtual unit load acting on the truss joint in the stated direction of Δ.

$n = $ internal virtual normal force in a truss member caused by the external virtual unit load.

$\Delta = $ external joint displacement caused by the fabrication errors.

$\Delta L = $ difference in length of the member from its intended size as caused by a fabrication error.

A combination of the right sides of Eqs. 9–15 through 9–17 will be necessary if both external loads act on the truss and some of the members undergo a thermal change or have been fabricated with the wrong dimensions.

Procedure for Analysis

The following procedure may be used to determine a specific displacement of any joint on a truss using the method of virtual work.

Virtual Forces n

- Place the unit load on the truss at the joint where the desired displacement is to be determined. The load should be in the same direction as the specified displacement, e.g., horizontal or vertical.

- With the unit load so placed, and all the real loads *removed* from the truss, use the method of joints or the method of sections and calculate the internal **n** force in each truss member. Assume that tensile forces are positive and compressive forces are negative.

Real Forces N

- Use the method of sections or the method of joints to determine the **N** force in each member. These forces are caused only by the real loads acting on the truss. Again, assume tensile forces are positive and compressive forces are negative.

Virtual-Work Equation

- Apply the equation of virtual work, to determine the desired displacement. It is important to retain the algebraic sign for each of the corresponding **n** and **N** forces when substituting these terms into the equation.

- If the resultant sum $\Sigma nNL/AE$ is positive, the displacement Δ is in the same direction as the unit load. If a negative value results, Δ is opposite to the unit load.

- When applying $1 \cdot \Delta = \Sigma n\alpha \, \Delta TL$, realize that if any of the members undergoes an *increase in temperature*, ΔT will be *positive*, whereas a *decrease in temperature* results in a *negative* value for ΔT.

- For $1 \cdot \Delta = \Sigma n \, \Delta L$, when a fabrication error *increases the length* of a member, ΔL is *positive*, whereas a *decrease in length* is *negative*.

- When applying any formula, attention should be paid to the units of each numerical quantity. In particular, the virtual unit load can either be assigned no units, or have any arbitrary unit (lb, kip, N, etc.), since the **n** forces will have these *same units*, and as a result the units for both the virtual unit load and the **n** forces will cancel from both sides of the equation.

EXAMPLE | 9.1

Determine the vertical displacement of joint C of the steel truss shown in Fig. 9–8a. The cross-sectional area of each member is $A = 0.5$ in^2 and $E = 29(10^3)$ ksi.

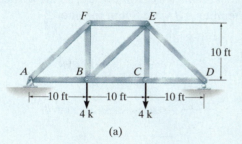

(a)

SOLUTION

Virtual Forces n. Only a vertical 1-k load is placed at joint C, and the force in each member is calculated using the method of joints. The results are shown in Fig. 9–8b. Positive numbers indicate tensile forces and negative numbers indicate compressive forces.

Real Forces N. The real forces in the members are calculated using the method of joints. The results are shown in Fig. 9–8c.

Virtual-Work Equation. Arranging the data in tabular form, we have

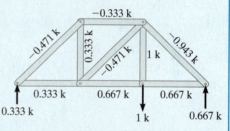

virtual forces **n**

(b)

Member	n (k)	N (k)	L (ft)	nNL (k$^2 \cdot$ ft)
AB	0.333	4	10	13.33
BC	0.667	4	10	26.67
CD	0.667	4	10	26.67
DE	−0.943	−5.66	14.14	75.42
FE	−0.333	−4	10	13.33
EB	−0.471	0	14.14	0
BF	0.333	4	10	13.33
AF	−0.471	−5.66	14.14	37.71
CE	1	4	10	40
				$\Sigma 246.47$

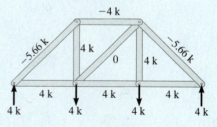

real forces **N**

(c)

Fig. 9–8

Thus

$$1 \text{ k} \cdot \Delta_{C_v} = \sum \frac{nNL}{AE} = \frac{246.47 \text{ k}^2 \cdot \text{ft}}{AE}$$

Converting the units of member length to inches and substituting the numerical values for A and E, we have

$$1 \text{ k} \cdot \Delta_{C_v} = \frac{(246.47 \text{ k}^2 \cdot \text{ft})(12 \text{ in./ft})}{(0.5 \text{ in}^2)(29(10^3) \text{ k/in}^2)}$$

$$\Delta_{C_v} = 0.204 \text{ in.} \qquad\qquad Ans.$$

9

EXAMPLE 9.2

The cross-sectional area of each member of the truss shown in Fig. 9–9a is $A = 400$ mm^2 and $E = 200$ GPa. (a) Determine the vertical displacement of joint C if a 4-kN force is applied to the truss at C. (b) If no loads act on the truss, what would be the vertical displacement of joint C if member AB were 5 mm too short?

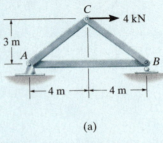

(a)

Fig. 9–9

SOLUTION

Part (a)

Virtual Forces n.   Since the *vertical displacement* of joint C is to be determined, a virtual force of 1 kN is applied at C in the vertical direction. The units of this force are the *same* as those of the real loading. The support reactions at A and B are calculated and the **n** force in each member is determined by the method of joints as shown on the free-body diagrams of joints A and B, Fig. 9–9b.

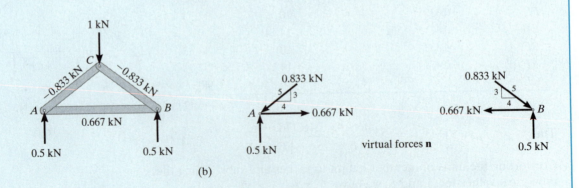

(b)

Real Forces N.   The joint analysis of A and B when the real load of 4 kN is applied to the truss is given in Fig. 9–9c.

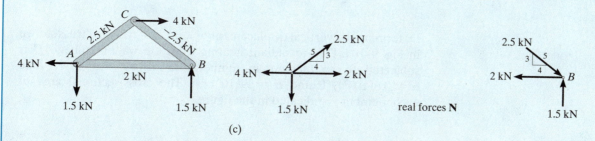

(c)

Virtual-Work Equation.   Since AE is constant, each of the terms nNL can be arranged in tabular form and computed. Here positive numbers indicate tensile forces and negative numbers indicate compressive forces.

Member	n (kN)	N (kN)	L (m)	$n\,NL$ (kN$^2 \cdot$ m)
AB	0.667	2	8	10.67
AC	-0.833	2.5	5	-10.41
CB	-0.833	-2.5	5	10.41
				$\Sigma\,10.67$

Thus,

$$1\ \text{kN} \cdot \Delta_{C_v} = \sum \frac{nNL}{AE} = \frac{10.67\ \text{kN}^2 \cdot \text{m}}{AE}$$

Substituting the values $A = 400\ \text{mm}^2 = 400(10^{-6})\ \text{m}^2$, $E = 200\ \text{GPa} = 200(10^6)\ \text{kN/m}^2$, we have

$$1\ \text{kN} \cdot \Delta_{C_v} = \frac{10.67\ \text{kN}^2 \cdot \text{m}}{400(10^{-6})\ \text{m}^2 (200(10^6)\ \text{kN/m}^2)}$$

$$\Delta_{C_v} = 0.000133\ \text{m} = 0.133\ \text{mm} \qquad \textit{Ans.}$$

Part (b).   Here we must apply Eq. 9–17. Since the vertical displacement of C is to be determined, we can use the results of Fig. 9–9b. Only member AB undergoes a change in length, namely, of $\Delta L = -0.005$ m.
Thus,

$$1 \cdot \Delta = \Sigma n\,\Delta L$$
$$1\ \text{kN} \cdot \Delta_{C_v} = (0.667\ \text{kN})(-0.005\ \text{m})$$
$$\Delta_{C_v} = -0.00333\ \text{m} = -3.33\ \text{mm} \qquad \textit{Ans.}$$

The negative sign indicates joint C is displaced *upward,* opposite to the 1-kN vertical load. Note that if the 4-kN load and fabrication error are both accounted for, the resultant displacement is then $\Delta_{C_v} = 0.133 - 3.33 = -3.20$ mm (upward).

EXAMPLE 9.3

Determine the vertical displacement of joint C of the steel truss shown in Fig. 9–10a. Due to radiant heating from the wall, member AD is subjected to an *increase* in temperature of $\Delta T = +120°F$. Take $\alpha = 0.6(10^{-5})/°F$ and $E = 29(10^3)$ ksi. The cross-sectional area of each member is indicated in the figure.

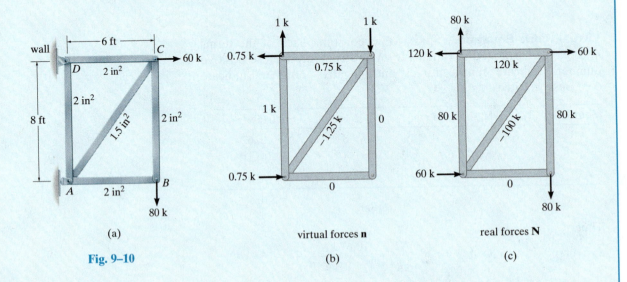

(a)

Fig. 9–10

virtual forces **n**

(b)

real forces **N**

(c)

SOLUTION

Virtual Forces n.　A *vertical* 1-k load is applied to the truss at joint C, and the forces in the members are computed, Fig. 9–10b.

Real Forces N.　Since the **n** forces in members AB and BC are *zero*, the **N** forces in these members do *not* have to be computed. Why? For completion, though, the entire real-force analysis is shown in Fig. 9–10c.

Virtual-Work Equation.　Both loads and temperature affect the deformation; therefore, Eqs. 9–15 and 9–16 are combined. Working in units of kips and inches, we have

$$1 \cdot \Delta_{C_v} = \sum \frac{nNL}{AE} + \sum n\alpha \, \Delta T \, L$$

$$= \frac{(0.75)(120)(6)(12)}{2\left[29(10^3)\right]} + \frac{(1)(80)(8)(12)}{2\left[29(10^3)\right]}$$

$$+ \frac{(-1.25)(-100)(10)(12)}{1.5\left[29(10^3)\right]} + (1)\left[0.6(10^{-5})\right](120)(8)(12)$$

$$\Delta_{C_v} = 0.658 \text{ in.} \qquad\qquad Ans.$$

9.5 Castigliano's Theorem

In 1879 Alberto Castigliano, an Italian railroad engineer, published a book in which he outlined a method for determining the deflection or slope at a point in a structure, be it a truss, beam, or frame. This method, which is referred to as *Castigliano's second theorem*, or the *method of least work*, applies only to structures that have constant temperature, unyielding supports, and *linear elastic* material response. If the displacement of a point is to be determined, the theorem states that it is equal to the first partial derivative of the strain energy in the structure with respect to a force acting at the point and in the direction of displacement. In a similar manner, the slope at a point in a structure is equal to the first partial derivative of the strain energy in the structure with respect to a couple moment acting at the point and in the direction of rotation.

To derive Castigliano's second theorem, consider a body (structure) of any arbitrary shape which is subjected to a series of n forces $P_1, P_2, \ldots, P_n$. Since the external work done by these loads is equal to the internal strain energy stored in the body, we can write

$$U_i = U_e$$

The external work is a function of the external loads $(U_e = \Sigma \int P \, dx)$. Thus,

$$U_i = U_e = f(P_1, P_2, \ldots, P_n)$$

Now, if any one of the forces, say P_i, is increased by a differential amount dP_i, the internal work is also increased such that the new strain energy becomes

$$U_i + dU_i = U_i + \frac{\partial U_i}{\partial P_i} dP_i \qquad (9\text{–}18)$$

This value, however, should not depend on the sequence in which the n forces are applied to the body. For example, if we apply dP_i to the body *first*, then this will cause the body to be displaced a differential amount $d\Delta_i$ in the direction of dP_i. By Eq. 9–3 $\left(U_e = \frac{1}{2} P\Delta\right)$, the increment of strain energy would be $\frac{1}{2} dP_i \, d\Delta_i$. This quantity, however, is a second-order differential and may be neglected. Further application of the loads $P_1, P_2, \ldots, P_n$, which displace the body $\Delta_1, \Delta_2, \ldots, \Delta_n$, yields the strain energy.

$$U_i + dU_i = U_i + dP_i \Delta_i \qquad (9\text{–}19)$$

Here, as before, U_i is the internal strain energy in the body, caused by the loads $P_1, P_2, \ldots, P_n$, and $dU_i = dP_i \Delta_i$ is the *additional* strain energy caused by dP_i. (Eq. 9–4, $U_e = P\Delta'$.)

In summary, then, Eq. 9–18 represents the strain energy in the body determined by first applying the loads $P_1, P_2, \ldots, P_n$, *then* dP_i, and Eq. 9–19 represents the strain energy determined by first applying dP_i and *then* the loads $P_1, P_2, \ldots, P_n$. Since these two equations must be equal, we require

$$\Delta_i = \frac{\partial U_i}{\partial P_i} \qquad (9\text{-}20)$$

which proves the theorem; i.e., the displacement Δ_i in the direction of P_i is equal to the first partial derivative of the strain energy with respect to P_i.*

It should be noted that Eq. 9–20 is a statement regarding the *structure's compatibility*. Also, the above derivation requires that *only conservative forces* be considered for the analysis. These forces do work that is independent of the path and therefore create no energy loss. Since forces causing a linear elastic response are conservative, the theorem is restricted to *linear elastic behavior* of the material. This is unlike the method of virtual force discussed in the previous section, which applied to *both* elastic and inelastic behavior.

9.6   Castigliano's Theorem for Trusses

The strain energy for a member of a truss is given by Eq. 9–9, $U_i = N^2 L / 2AE$. Substituting this equation into Eq. 9–20 and omitting the subscript i, we have

$$\Delta = \frac{\partial}{\partial P} \sum \frac{N^2 L}{2AE}$$

It is generally easier to perform the differentiation prior to summation. In the general case L, A, and E are constant for a given member, and therefore we may write

$$\boxed{\Delta = \sum N \left(\frac{\partial N}{\partial P} \right) \frac{L}{AE}} \qquad (9\text{-}21)$$

where

$\Delta =$ external joint displacement of the truss.
$P =$ external force applied to the truss joint in the direction of Δ.
$N =$ internal force in a member caused by *both* the force P and the loads on the truss.
$L =$ length of a member.
$A =$ cross-sectional area of a member.
$E =$ modulus of elasticity of a member.

---

*Castigliano's first theorem is similar to his second theorem; however, it relates the load P_i to the partial derivative of the strain energy with respect to the corresponding displacement, that is, $P_i = \partial U_i / \partial \Delta_i$. The proof is similar to that given above and, like the method of virtual displacement, Castigliano's first theorem applies to both elastic and inelastic material behavior. This theorem is another way of expressing the *equilibrium requirements* for a structure, and since it has very limited use in structural analysis, it will not be discussed in this book.

This equation is similar to that used for the method of virtual work, Eq. 9–15 ($1 \cdot \Delta = \Sigma nNL/AE$), except n is replaced by $\partial N/\partial P$. Notice that in order to determine this partial derivative it will be necessary to treat P as a *variable* (not a specific numerical quantity), and furthermore, each member force N must be expressed as a function of P. As a result, computing $\partial N/\partial P$ generally requires slightly more calculation than that required to compute each n force directly. These terms will of course be the same, since n or $\partial N/\partial P$ is simply the change of the internal member force with respect to the load P, or the change in member force per unit load.

Procedure for Analysis

The following procedure provides a method that may be used to determine the displacement of any joint of a truss using Castigliano's theorem.

External Force P

- Place a force P on the truss at the joint where the desired displacement is to be determined. This force is assumed to have a *variable magnitude* in order to obtain the change $\partial N/\partial P$. Be sure **P** is directed along the line of action of the displacement.

Internal Forces N

- Determine the force N in each member caused by both the real (numerical) loads and the variable force P. Assume tensile forces are positive and compressive forces are negative.

- Compute the respective partial derivative $\partial N/\partial P$ for each member.

- After N and $\partial N/\partial P$ have been determined, assign P its numerical value if it has replaced a real force on the truss. Otherwise, set P equal to zero.

Castigliano's Theorem

- Apply Castigliano's theorem to determine the desired displacement Δ. It is important to retain the algebraic signs for corresponding values of N and $\partial N/\partial P$ when substituting these terms into the equation.

- If the resultant sum $\Sigma N(\partial N/\partial P)L/AE$ is positive, Δ is in the same direction as P. If a negative value results, Δ is opposite to P.

EXAMPLE   9.4

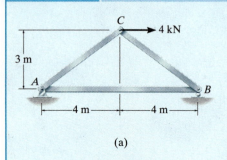

(a)

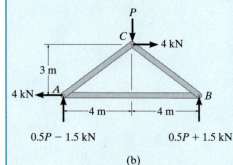

(b)

$N_{AC} = 0.833P - 2.5$ kN

4 kN

$N_{AB} = 0.667P + 2$ kN

$0.5P - 1.5$ kN

$N_{BC} = 0.833P + 2.5$ kN

$N_{AB} = 0.667P + 2$ kN

$0.5P + 1.5$ kN

(c)

Fig. 9–11

Determine the vertical displacement of joint C of the truss shown in Fig. 9–11a. The cross-sectional area of each member is $A = 400$ mm^2 and $E = 200$ GPa.

SOLUTION

External Force P.   A vertical force $\mathbf{P}$ is applied to the truss at joint C, since this is where the vertical displacement is to be determined, Fig. 9–11b.

Internal Forces N.   The reactions at the truss supports at A and B are determined and the results are shown in Fig. 9–11b. Using the method of joints, the N forces in each member are determined, Fig. 9–11c.* For convenience, these results along with the partial derivatives $\partial N / \partial P$ are listed in tabular form as follows:

Member	N	$\dfrac{\partial N}{\partial P}$	$N(P = 0)$	L	$N\left(\dfrac{\partial N}{\partial P}\right)L$
AB	$0.667P + 2$	0.667	2	8	10.67
AC	$-(0.833P - 2.5)$	-0.833	2.5	5	-10.42
BC	$-(0.833P + 2.5)$	-0.833	-2.5	5	10.42

$$\Sigma = 10.67 \text{ kN} \cdot \text{m}$$

Since P does not actually exist as a real load on the truss, we require $P = 0$ in the table above.

Castigliano's Theorem.   Applying Eq. 9–21, we have

$$\Delta_{C_v} = \Sigma N\left(\frac{\partial N}{\partial P}\right)\frac{L}{AE} = \frac{10.67 \text{ kN} \cdot \text{m}}{AE}$$

Substituting $A = 400$ mm$^2 = 400(10^{-6})$ m^2, $E = 200$ GPa $= 200(10^9)$ Pa, and converting the units of N from kN to N, we have

$$\Delta_{C_v} = \frac{10.67(10^3) \text{ N} \cdot \text{m}}{400(10^{-6}) \text{ m}^2 (200(10^9) \text{ N/m}^2)} = 0.000133 \text{ m} = 0.133 \text{ mm} \quad \textit{Ans.}$$

This solution should be compared with the virtual-work method of Example 9.2.

---

*It may be more convenient to analyze the truss with just the 4-kN load on it, then analyze the truss with the P load on it. The results can then be added together to give the N forces.

EXAMPLE | 9.5

Determine the horizontal displacement of joint D of the truss shown in Fig. 9–12a. Take $E = 29(10^3)$ ksi. The cross-sectional area of each member is indicated in the figure.

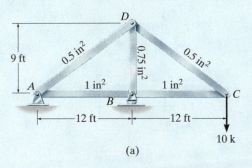

(a)

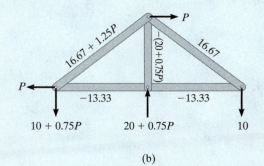

(b)

Fig. 9–12

SOLUTION

External Force P.   Since the horizontal displacement of D is to be determined, a horizontal variable force P is applied to joint D, Fig. 9–12b.

Internal Forces N.   Using the method of joints, the force N in each member is computed.* Again, when applying Eq. 9–21, we set $P = 0$ since this force does not actually exist on the truss. The results are shown in Fig. 9–12b. Arranging the data in tabular form, we have

Member	N	$\dfrac{\partial N}{\partial P}$	$N(P = 0)$	L	$N\left(\dfrac{\partial N}{\partial P}\right)L$
AB	−13.33	0	−13.33	12	0
BC	−13.33	0	−13.33	12	0
CD	16.67	0	16.67	15	0
DA	16.67 + 1.25P	1.25	16.67	15	312.50
BD	−(20 + 0.75P)	−0.75	− 20	9	135.00

Castigliano's Theorem.   Applying Eq. 9–21, we have

$$\Delta_{D_h} = \sum N\left(\frac{\partial N}{\partial P}\right)\frac{L}{AE} = 0 + 0 + 0 + \frac{312.50 \text{ k} \cdot \text{ft}(12 \text{ in.}/\text{ft})}{(0.5 \text{ in}^2)\left[29(10^3) \text{ k}/\text{in}^2\right]} + \frac{135.00 \text{ k} \cdot \text{ft}(12 \text{ in.}/\text{ft})}{(0.75 \text{ in}^2)\left[29(10^3) \text{ k}/\text{in}^2\right]}$$

$$= 0.333 \text{ in.} \qquad\qquad Ans.$$

*As in the preceding example, it may be preferable to perform a separate analysis of the truss loaded with 10 k and loaded with P and then superimpose the results.

EXAMPLE 9.6

Determine the vertical displacement of joint C of the truss shown in Fig. 9–13a. Assume that $A = 0.5$ in^2 and $E = 29(10^3)$ ksi.

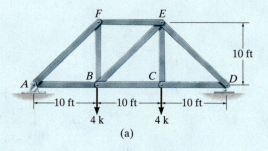

(a)

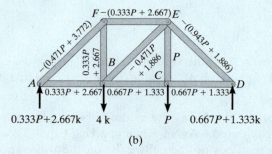

(b)

Fig. 9–13

SOLUTION

External Force P. The 4-k force at C is replaced with a *variable force P* at joint C, Fig. 9–13b.

Internal Forces N. The method of joints is used to determine the force N in each member of the truss. The results are summarized in Fig. 9–13b. Here $P = 4$ k when we apply Eq. 9–21. The required data can be arranged in tabulated form as follows:

Member	N	$\dfrac{\partial N}{\partial P}$	$N\,(P = 4\,k)$	L	$N\left(\dfrac{\partial N}{\partial P}\right)L$
AB	$0.333P + 2.667$	0.333	4	10	13.33
BC	$0.667P + 1.333$	0.667	4	10	26.67
CD	$0.667P + 1.333$	0.667	4	10	26.67
DE	$-(0.943P + 1.886)$	-0.943	-5.66	14.14	75.42
EF	$-(0.333P + 2.667)$	-0.333	-4	10	13.33
FA	$-(0.471P + 3.771)$	-0.471	-5.66	14.14	37.71
BF	$0.333P + 2.667$	0.333	4	10	13.33
BE	$-0.471P - 1.886$	-0.471	0	14.14	0
CE	P	1	4	10	40

$$\Sigma = 246.47 \text{ k} \cdot \text{ft}$$

Castigliano's Theorem.   Substituting the data into Eq. 9–21, we have

$$\Delta_{C_v} = \sum N\left(\frac{\partial N}{\partial P}\right)\frac{L}{AE} = \frac{246.47 \text{ k} \cdot \text{ft}}{AE}$$

Converting the units of member length to inches and substituting the numerical value for AE, we have

$$\Delta_{C_v} = \frac{(246.47 \text{ k} \cdot \text{ft})(12 \text{ in./ft})}{(0.5 \text{ in}^2)(29(10^3) \text{ k/in}^2)} = 0.204 \text{ in.} \qquad \textit{Ans.}$$

The similarity between this solution and that of the virtual-work method, Example 9.1, should be noted.

9

Apply virtual unit load to point A

(a)

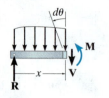

Apply real load w

(b)

Fig. 9–14

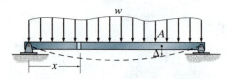

Apply virtual unit couple moment to point A

(a)

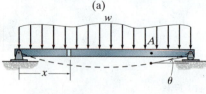

Apply real load w

Fig. 9–15

9.7 Method of Virtual Work: Beams and Frames

The method of virtual work can also be applied to deflection problems involving beams and frames. Since strains due to *bending* are the *primary cause* of beam or frame deflections, we will discuss their effects first. Deflections due to shear, axial and torsional loadings, and temperature will be considered in Sec. 9.8.

The principle of virtual work, or more exactly, the method of virtual force, may be formulated for beam and frame deflections by considering the beam shown in Fig. 9–14b. Here the displacement Δ of point A is to be determined. To compute Δ a virtual unit load acting in the direction of Δ is placed on the beam at A, and the *internal virtual moment* $\mathbf{m}$ is determined by the method of sections at an arbitrary location x from the left support, Fig. 9–14a. When the real loads act on the beam, Fig. 9–14b, point A is displaced Δ. Provided these loads cause *linear elastic material response*, then from Eq. 8–2, the element dx deforms or rotates $d\theta = (M/EI)\,dx$.* Here M is the internal moment at x caused by the real loads. Consequently, the *external virtual work* done by the unit load is $1 \cdot \Delta$, and the *internal virtual work* done by the moment $\mathbf{m}$ is $m\,d\theta = m(M/EI)\,dx$. Summing the effects on all the elements dx along the beam requires an integration, and therefore Eq. 9–13 becomes

$$1 \cdot \Delta = \int_0^L \frac{mM}{EI}\,dx \qquad (9\text{–}22)$$

where

$1 =$ external virtual unit load acting on the beam or frame in the direction of Δ.

$m =$ internal virtual moment in the beam or frame, expressed as a function of x and caused by the external virtual unit load.

$\Delta =$ external displacement of the point caused by the real loads acting on the beam or frame.

$M =$ internal moment in the beam or frame, expressed as a function of x and caused by the real loads.

$E =$ modulus of elasticity of the material.

$I =$ moment of inertia of cross-sectional area, computed about the neutral axis.

In a similar manner, if the tangent rotation or slope angle θ at a point A on the beam's elastic curve is to be determined, Fig. 9–15, a unit couple moment is first applied at the point, and the corresponding internal moments m_θ have to be determined. Since the work of the unit couple is $1 \cdot \theta$, then

$$1 \cdot \theta = \int_0^L \frac{m_\theta M}{EI}\,dx \qquad (9\text{–}23)$$

*Recall that if the material is strained beyond its elastic limit, the principle of virtual work can still be applied, although in this case a nonlinear or plastic analysis must be used.

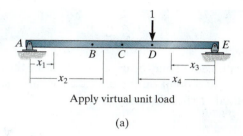

Apply virtual unit load

(a)

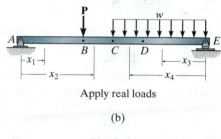

Apply real loads

(b)

Fig. 9–16

When applying Eqs. 9–22 and 9–23, it is important to realize that the definite integrals on the right side actually represent the amount of virtual strain energy that is *stored* in the beam. If concentrated forces or couple moments act on the beam or the distributed load is discontinuous, a single integration cannot be performed across the beam's entire length. Instead, separate x coordinates will have to be chosen within regions that have no discontinuity of loading. Also, it is not necessary that each x have the same origin; however, the x selected for determining the real moment M in a particular region must be the *same* x as that selected for determining the virtual moment m or m_θ within the same region. For example, consider the beam shown in Fig. 9–16. In order to determine the displacement of D, four regions of the beam must be considered, and therefore four integrals having the form $\int (mM/EI)\, dx$ must be evaluated. We can use x_1 to determine the strain energy in region AB, x_2 for region BC, x_3 for region DE, and x_4 for region DC. In any case, each x coordinate should be selected so that both M and m (or m_θ) can be easily formulated.

Integration Using Tables.   When the structure is subjected to a relatively simple loading, and yet the solution for a displacement requires several integrations, a *tabular method* may be used to perform these integrations. To do so the moment diagrams for each member are drawn first for both the real and virtual loadings. By matching these diagrams for m and M with those given in the table on the inside front cover, the integral $\int mM\, dx$ can be determined from the appropriate formula. Examples 9.8 and 9.10 illustrate the application of this method.

Procedure for Analysis

The following procedure may be used to determine the displacement and/or the slope at a point on the elastic curve of a beam or frame using the method of virtual work.

Virtual Moments m or m

- Place a *unit load* on the beam or frame at the point and in the direction of the desired *displacement*.

- If the *slope* is to be determined, place a *unit couple moment* at the point.

- Establish appropriate x coordinates that are valid within regions of the beam or frame where there is no discontinuity of real or virtual load.

- With the virtual load in place, and all the real loads *removed* from the beam or frame, calculate the internal moment m or m_θ as a function of each x coordinate.

- Assume m or m_θ acts in the conventional positive direction as indicated in Fig. 4–1.

Real Moments

- Using the *same x* coordinates as those established for m or m_θ, determine the internal moments M caused only by the real loads.

- Since m or m_θ was assumed to act in the conventional positive direction, *it is important that positive M acts in this same direction.* This is necessary since positive or negative internal work depends upon the directional sense of load (defined by $\pm m$ or $\pm m_\theta$) and displacement (defined by $\pm M\, dx/EI$).

Virtual-Work Equation

- Apply the equation of virtual work to determine the desired displacement Δ or rotation θ. It is important to retain the algebraic sign of each integral calculated within its specified region.

- If the algebraic sum of all the integrals for the entire beam or frame is positive, Δ or θ is in the same direction as the virtual unit load or unit couple moment, respectively. If a negative value results, the direction of Δ or θ is opposite to that of the unit load or unit couple moment.

9

EXAMPLE | 9.7

Determine the displacement of point B of the steel beam shown in Fig. 9–17a. Take $E = 200$ GPa, $I = 500(10^6)$ mm^4.

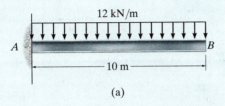

12 kN/m

A B

10 m

(a)

SOLUTION

Virtual Moment m. The vertical displacement of point B is obtained by placing a virtual unit load of 1 kN at B, Fig. 9–17b. By inspection there are no discontinuities of loading on the beam for *both* the real and virtual loads. Thus, a *single x* coordinate can be used to determine the virtual strain energy. This coordinate will be selected with its origin at B, since then the reactions at A do not have to be determined in order to find the internal moments m and M. Using the method of sections, the internal moment m is formulated as shown in Fig. 9–17b.

Real Moment M. Using the *same x* coordinate, the internal moment M is formulated as shown in Fig. 9–17c.

Virtual-Work Equation. The vertical displacement of B is thus

$$1 \text{ kN} \cdot \Delta_B = \int_0^L \frac{mM}{EI} dx = \int_0^{10} \frac{(-1x)(-6x^2)\, dx}{EI}$$

$$1 \text{ kN} \cdot \Delta_B = \frac{15(10^3) \text{ kN}^2 \cdot \text{m}^3}{EI}$$

or

$$\Delta_B = \frac{15(10^3) \text{ kN} \cdot \text{m}^3}{200(10^6) \text{ kN/m}^2 (500(10^6) \text{ mm}^4)(10^{-12} \text{ m}^4/\text{mm}^4)}$$

$$= 0.150 \text{ m} = 150 \text{ mm} \qquad \textit{Ans.}$$

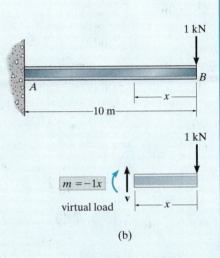

1 kN

A B

x

10 m

1 kN

$m = -1x$ $\overset{\curvearrowleft}{\underset{\mathbf{V}}{}}$

virtual load x

(b)

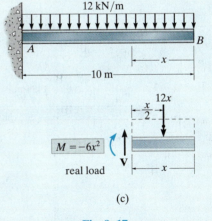

12 kN/m

A B

x

10 m

12x

$\frac{x}{2}$

$M = -6x^2$ $\overset{\curvearrowleft}{\underset{\mathbf{V}}{}}$

real load x

(c)

Fig. 9–17

9

EXAMPLE 9.8

Determine the slope θ at point B of the steel beam shown in Fig. 9–18a. Take $E = 200$ GPa, $I = 60(10^6)$ mm^4.

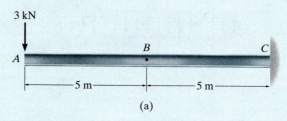

(a)

Fig. 9–18

SOLUTION

Virtual Moment m.   The slope at B is determined by placing a virtual unit couple moment of 1 kN·m at B, Fig. 9–18b. Here two x coordinates must be selected in order to determine the total virtual strain energy in the beam. Coordinate x_1 accounts for the strain energy within segment AB and coordinate x_2 accounts for that in segment BC. The internal moments m_θ within each of these segments are computed using the method of sections as shown in Fig. 9–18b.

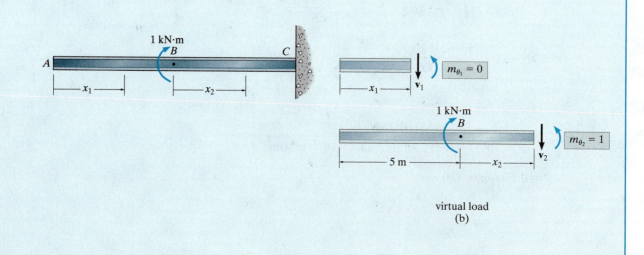

virtual load
(b)

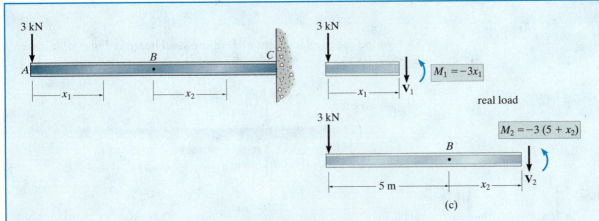

(c)

Real Moments M. Using the *same* coordinates x_1 and x_2, the internal moments M are computed as shown in Fig. 9–18c.

Virtual-Work Equation. The slope at B is thus given by

$$1 \cdot \theta_B = \int_0^L \frac{m_\theta M}{EI} dx$$

$$= \int_0^5 \frac{(0)(-3x_1)\, dx_1}{EI} + \int_0^5 \frac{(1)[-3(5 + x_2)]\, dx_2}{EI}$$

$$\theta_B = \frac{-112.5 \text{ kN} \cdot \text{m}^2}{EI} \qquad (1)$$

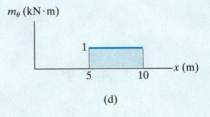

(d)

We can also evaluate the integrals $\int m_\theta M\, dx$ graphically, using the table given on the inside front cover of the book. To do so it is first necessary to draw the moment diagrams for the beams in Figs. 9–18b and 9–18c. These are shown in Figs. 9–18d and 9–18e, respectively. Since there is no moment m for $0 \le x < 5$ m, we use only the shaded rectangular and trapezoidal areas to evaluate the integral. Finding these shapes in the appropriate row and column of the table, we have

$$\int_5^{10} m_\theta M\, dx = \tfrac{1}{2} m_\theta(M_1 + M_2)L = \tfrac{1}{2}(1)(-15 - 30)5$$

$$= -112.5 \text{ kN}^2 \cdot \text{m}^3$$

This is the same value as that determined in Eq. 1. Thus,

$$(1 \text{ kN} \cdot \text{m}) \cdot \theta_B = \frac{-112.5 \text{ kN}^2 \cdot \text{m}^3}{200(10^6) \text{ kN/m}^2 [\, 60(10^6) \text{ mm}^4\,](10^{-12} \text{ m}^4/\text{mm}^4)}$$

$$\theta_B = -0.00938 \text{ rad} \qquad \qquad Ans.$$

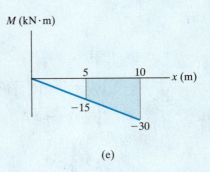

(e)

The *negative* sign indicates θ_B is *opposite* to the direction of the virtual couple moment shown in Fig. 9–18b.

EXAMPLE 9.9

Determine the displacement at D of the steel beam in Fig. 9–19a. Take $E = 29(10^3)$ ksi, $I = 800$ in^4.

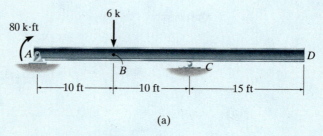

(a)

Fig. 9–19

SOLUTION

Virtual Moments m.   The beam is subjected to a virtual unit load at D as shown in Fig. 9–19b. By inspection, *three coordinates*, such as x_1, x_2, and x_3, must be used to cover all the regions of the beam. Notice that these coordinates cover regions where no discontinuities in either real or virtual load occur. The internal moments m have been computed in Fig. 9–19b using the method of sections.

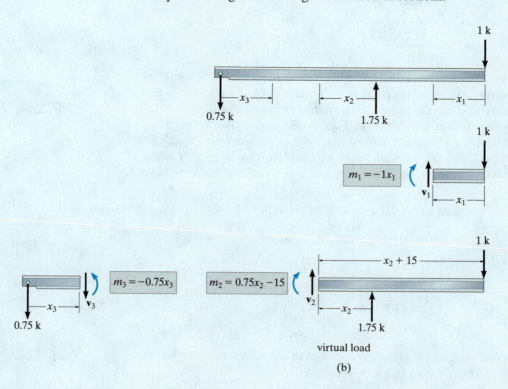

virtual load

(b)

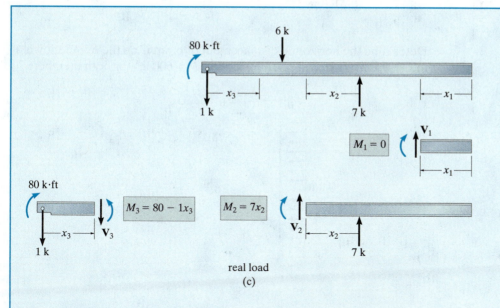

real load
(c)

Real Moments M.   The reactions on the beam are computed first; then, using the *same x* coordinates as those used for *m*, the internal moments *M* are determined as shown in Fig. 9–19c.

Virtual-Work Equation.   Applying the equation of virtual work to the beam using the data in Figs. 9–19b and 9–19c, we have

$$1 \cdot \Delta_D = \int_0^L \frac{mM}{EI}\, dx$$

$$= \int_0^{15} \frac{(-1x_1)(0)\, dx_1}{EI} + \int_0^{10} \frac{(0.75x_2 - 15)(7x_2)\, dx_2}{EI}$$

$$+ \int_0^{10} \frac{(-0.75x_3)(80 - 1x_3)\, dx_3}{EI}$$

$$\Delta_D = \frac{0}{EI} - \frac{3500}{EI} - \frac{2750}{EI} = -\frac{6250 \text{ k} \cdot \text{ft}^3}{EI}$$

or

$$\Delta_D = \frac{-6250 \text{ k} \cdot \text{ft}^3 (12)^3 \text{in}^3/\text{ft}^3}{29(10^3) \text{ k}/\text{in}^2 (800 \text{ in}^4)}$$

$$= -0.466 \text{ in.} \qquad \textit{Ans.}$$

The negative sign indicates the displacement is upward, opposite to the downward unit load, Fig. 9–19b. Also note that m_1 did not actually have to be calculated since $M_1 = 0$.

EXAMPLE 9.10

Determine the horizontal displacement of point C on the frame shown in Fig. 9–20a. Take $E = 29(10^3)$ ksi and $I = 600$ in^4 for both members.

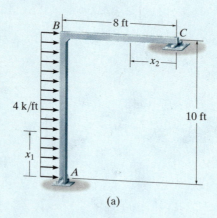

(a)

Fig. 9–20

SOLUTION

Virtual Moments m.   For convenience, the coordinates x_1 and x_2 in Fig. 9–20a will be used. A *horizontal* unit load is applied at C, Fig. 9–20b. Why? The support reactions and internal virtual moments are computed as shown.

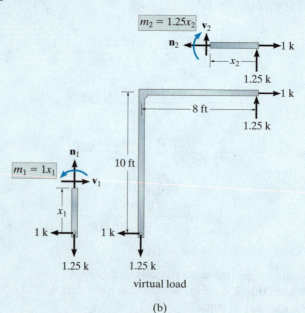

(b)

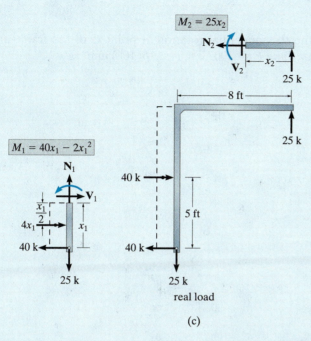

$$M_2 = 25x_2$$

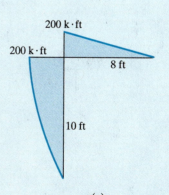

real load

(c)

Real Moments M. In a similar manner the support reactions and real moments are computed as shown in Fig. 9–20c.

Virtual-Work Equation. Using the data in Figs. 9–20b and 9–20c, we have

$$1 \cdot \Delta_{C_h} = \int_0^L \frac{mM}{EI}dx = \int_0^{10} \frac{(1x_1)(40x_1 - 2x_1^2)dx_1}{EI} + \int_0^8 \frac{(1.25x_2)(25x_2)dx_2}{EI}$$

$$\Delta_{C_h} = \frac{8333.3}{EI} + \frac{5333.3}{EI} = \frac{13\,666.7\,\text{k}\cdot\text{ft}^3}{EI} \qquad (1)$$

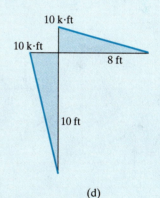

(d)

If desired, the integrals $\int mM/dx$ can also be evaluated graphically using the table on the inside front cover. The moment diagrams for the frame in Figs. 9–20b and 9–20c are shown in Figs. 9–20d and 9–20e, respectively. Thus, using the formulas for similar shapes in the table yields

$$\int mM\, dx = \tfrac{5}{12}(10)(200)(10) + \tfrac{1}{3}(10)(200)(8)$$

$$= 8333.3 + 5333.3 = 13\,666.7\,\text{k}^2\cdot\text{ft}^3$$

This is the same as that calculated in Eq. 1. Thus

$$\Delta_{C_h} = \frac{13\,666.7\,\text{k}\cdot\text{ft}^2}{\left[29(10^3)\,\text{k/in}^2((12)^2\,\text{in}^2/\text{ft}^2)\right]\left[600\,\text{in}\,(\text{ft}^4/(12)^4\,\text{in}^4)\right]}$$

$$= 0.113\,\text{ft} = 1.36\,\text{in.} \qquad \textit{Ans.}$$

(e)

EXAMPLE 9.11

Determine the tangential rotation at point C of the frame shown in Fig. 9–21a. Take $E = 200$ GPa, $I = 15(10^6)$ mm^4.

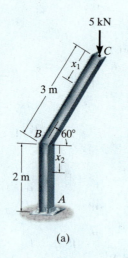

(a)

Fig. 9–21

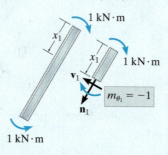

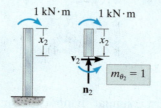

virtual load

(b)

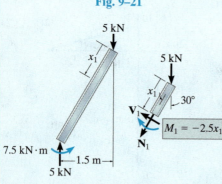

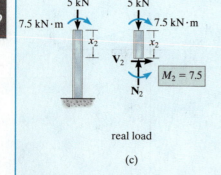

real load

(c)

SOLUTION

Virtual Moments m. The coordinates x_1 and x_2 shown in Fig. 9–21a will be used. A unit couple moment is applied at C and the internal moments m_θ are calculated, Fig. 9–21b.

Real Moments M. In a similar manner, the real moments M are calculated as shown in Fig. 9–21c.

Virtual-Work Equation. Using the data in Figs. 9–21b and 9–21c, we have

$$1 \cdot \theta_C = \int_0^L \frac{m_\theta M}{EI} dx = \int_0^3 \frac{(-1)(-2.5x_1)\, dx_1}{EI} + \int_0^2 \frac{(1)(7.5)\, dx_2}{EI}$$

$$\theta_C = \frac{11.25}{EI} + \frac{15}{EI} = \frac{26.25 \text{ kN} \cdot \text{m}^2}{EI}$$

or

$$\theta_C = \frac{26.25 \text{ kN} \cdot \text{m}^2}{200(10^6) \text{ kN/m}^2 \big[15(10^6) \text{ mm}^4\big](10^{-12} \text{ m}^4/\text{mm}^4)}$$

$$= 0.00875 \text{ rad} \qquad\qquad\qquad Ans.$$

9.8   Virtual Strain Energy Caused by Axial Load, Shear, Torsion, and Temperature

Although deflections of beams and frames are caused primarily by bending strain energy, in some structures the additional strain energy of axial load, shear, torsion, and perhaps temperature may become important. Each of these effects will now be considered.

Axial Load.   Frame members can be subjected to axial loads, and the virtual strain energy caused by these loadings has been established in Sec. 9.4. For members having a constant cross-sectional area, we have

$$U_n = \frac{nNL}{AE} \tag{9-24}$$

where
n = internal virtual axial load caused by the external virtual unit load.
N = internal axial force in the member caused by the real loads.
E = modulus of elasticity for the material.
A = cross-sectional area of the member.
L = member's length.

Shear.   In order to determine the virtual strain energy in a beam due to shear, we will consider the beam element dx shown in Fig. 9–22. The shearing distortion dy of the element as caused by the *real loads* is $dy = \gamma\, dx$. If the shearing strain γ is caused by *linear elastic material response*, then Hooke's law applies, $\gamma = \tau/G$. Therefore, $dy = (\tau/G)\, dx$. We can express the shear stress as $\tau = K(V/A)$, where K is a *form factor* that depends upon the shape of the beam's cross-sectional area A. Hence, we can write $dy = K(V/GA)\, dx$. The internal virtual work done by a virtual shear force v, acting on the element *while* it is deformed dy, is therefore $dU_s = v\, dy = v(KV/GA)\, dx$. For the entire beam, the virtual strain energy is determined by integration.

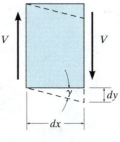

Fig. 9–22

$$U_s = \int_0^L K\!\left(\frac{vV}{GA}\right) dx \tag{9-25}$$

where
v = internal virtual shear in the member, expressed as a function of x and caused by the external virtual unit load.
V = internal shear in the member, expressed as a function of x and caused by the real loads.
A = cross-sectional area of the member.
K = form factor for the cross-sectional area:
$\quad K = 1.2$ for rectangular cross sections.
$\quad K = 10/9$ for circular cross sections.
$\quad K \approx 1$ for wide-flange and I-beams, where A is the area of the web.
G = shear modulus of elasticity for the material.

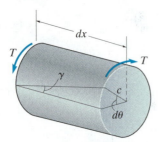

Fig. 9–23

Torsion.   Often three-dimensional frameworks are subjected to torsional loadings. If the member has a *circular* cross-sectional area, no warping of its cross section will occur when it is loaded. As a result, the virtual strain energy in the member can easily be derived. To do so consider an element dx of the member that is subjected to an applied torque **T**, Fig. 9–23. This torque causes a shear strain of $\gamma = (c\,d\theta)/dx$. Provided *linear elastic material response* occurs, then $\gamma = \tau/G$, where $\tau = Tc/J$. Thus, the angle of twist $d\theta = (\gamma\,dx)/c = (\tau/Gc)\,dx = (T/GJ)\,dx$. If a virtual unit load is applied to the structure that causes an internal virtual torque **t** in the member, then after applying the real loads, the virtual strain energy in the member of length dx will be $dU_t = t\,d\theta = tT\,dx/GJ$. Integrating over the length L of the member yields

$$U_t = \frac{tTL}{GJ} \qquad (9\text{–}26)$$

where

$t =$ internal virtual torque caused by the external virtual unit load.

$T =$ internal torque in the member caused by the real loads.

$G =$ shear modulus of elasticity for the material.

$J =$ polar moment of inertia for the cross section, $J = \pi c^4/2$, where c is the radius of the cross-sectional area.

$L =$ member's length.

The virtual strain energy due to torsion for members having noncircular cross-sectional areas is determined using a more rigorous analysis than that presented here.

Temperature.   In Sec. 9.4 we considered the effect of a *uniform temperature change* ΔT on a truss member and indicated that the member will elongate or shorten by an amount $\Delta L = \alpha\,\Delta TL$. In some cases, however, a structural member can be subjected to a *temperature difference across its depth*, as in the case of the beam shown in Fig. 9–24a. If this occurs, it is possible to determine the displacement of points along the elastic curve of the beam by using the principle of virtual work. To do so we must first compute the amount of *rotation* of a differential element dx of the beam as caused by the thermal gradient that acts over the beam's cross section. For the sake of discussion we will choose the most common case of a beam having a neutral axis located at the mid-depth (c) of the beam. If we plot the temperature profile, Fig. 9–24b, it will be noted that the mean temperature is $T_m = (T_1 + T_2)/2$. If $T_1 > T_2$, the temperature difference at the top of the element causes strain elongation, while that at the bottom causes strain contraction. In both cases the difference in temperature is $\Delta T_m = T_1 - T_m = T_m - T_2$.

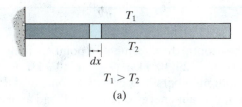

T_1

T_2

dx

$T_1 > T_2$

(a)

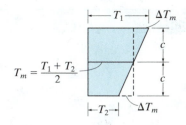

T_1 ΔT_m

c

$T_m = \dfrac{T_1 + T_2}{2}$

c

T_2 ΔT_m

temperature profile

(b)

Since the thermal change of length at the top and bottom is $\delta x = \alpha \, \Delta T_m \, dx$, Fig. 9–24c, then the rotation of the element is

$$d\theta = \frac{\alpha \, \Delta T_m \, dx}{c}$$

If we apply a virtual unit load at a point on the beam where a displacement is to be determined, or apply a virtual unit couple moment at a point where a rotational displacement of the tangent is to be determined, then this loading creates a virtual moment **m** in the beam at the point where the element dx is located. When the temperature gradient is imposed, the virtual strain energy in the beam is then

$$U_{\text{temp}} = \int_0^L \frac{m\alpha \, \Delta T_m \, dx}{c} \qquad (9\text{–}27)$$

where
 m = internal virtual moment in the beam expressed as a function of x and caused by the external virtual unit load or unit couple moment.
 α = coefficient of thermal expansion.
 ΔT_m = temperature difference between the mean temperature and the temperature at the top or bottom of the beam.
 c = mid-depth of the beam.

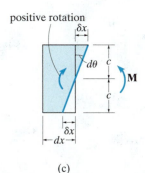

positive rotation

δx

$d\theta$ c

M

c

δx

dx

(c)

Fig. 9–24

Unless otherwise stated, *this text will consider only beam and frame deflections due to bending*. In general, though, beam and frame members may be subjected to several of the other loadings discussed in this section. However, as previously mentioned, the additional deflections caused by shear and axial force alter the deflection of beams by only a few percent and are therefore generally ignored for even "small" two- or three-member frame analysis of one-story height. If these and the other effects of torsion and temperature are to be considered for the analysis, then one simply adds their virtual strain energy as defined by Eqs. 9–24 through 9–27 to the equation of virtual work defined by Eq. 9–22 or Eq. 9–23. The following examples illustrate application of these equations.

9

EXAMPLE 9.12

Determine the horizontal displacement of point C on the frame shown in Fig. 9–25a. Take $E = 29(10^3)$ ksi, $G = 12(10^3)$ ksi, $I = 600$ in⁴, and $A = 80$ in² for both members. The cross-sectional area is rectangular. Include the internal strain energy due to axial load and shear.

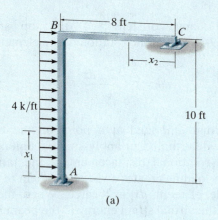

(a)

Fig. 9–25

SOLUTION

Here we must apply a horizontal unit load at C. The necessary free-body diagrams for the real and virtual loadings are shown in Figs. 9–25b and 9–25c.

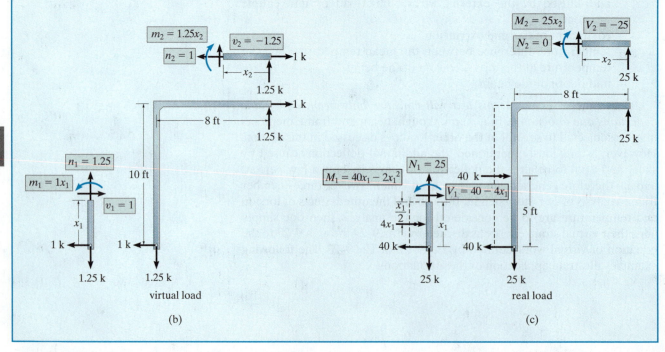

(b)   (c)

Bending.   The virtual strain energy due to bending has been determined in Example 9.10. There it was shown that

$$U_b = \int_0^L \frac{mM\,dx}{EI} = \frac{13\,666.7\ \text{k}^2\cdot\text{ft}^3}{EI} = \frac{13\,666.7\ \text{k}^2\cdot\text{ft}^3\,(12^3\ \text{in}^3/1\ \text{ft}^3)}{[29(10^3)\ \text{k/in}^2](600\ \text{in}^4)} = 1.357\ \text{in.}\cdot\text{k}$$

Axial load.   From the data in Fig. 9–25b and 9–25c, we have

$$U_a = \sum \frac{nNL}{AE}$$

$$= \frac{1.25\ \text{k}(25\ \text{k})(120\ \text{in.})}{80\ \text{in}^2\big[29(10^3)\ \text{k/in}^2\big]} + \frac{1\ \text{k}(0)(96\ \text{in.})}{80\ \text{in}^2\big[29(10^3)\ \text{k/in}^2\big]}$$

$$= 0.001616\ \text{in.}\cdot\text{k}$$

Shear.   Applying Eq. 9–25 with $K = 1.2$ for rectangular cross sections, and using the shear functions shown in Fig. 9–25b and 9–25c, we have

$$U_s = \int_0^L K\left(\frac{vV}{GA}\right) dx$$

$$= \int_0^{10} \frac{1.2(1)(40 - 4x_1)\,dx_1}{GA} + \int_0^8 \frac{1.2(-1.25)(-25)\,dx_2}{GA}$$

$$= \frac{540\ \text{k}^2\cdot\text{ft}(12\ \text{in.}/\text{ft})}{\big[12(10^3)\ \text{k/in}^2\big](80\ \text{in}^2)} = 0.00675\ \text{in.}\cdot\text{k}$$

Applying the equation of virtual work, we have

$$1\ \text{k}\cdot\Delta_{C_h} = 1.357\ \text{in.}\cdot\text{k} + 0.001616\ \text{in.}\cdot\text{k} + 0.00675\ \text{in.}\cdot\text{k}$$

$$\Delta_{C_h} = 1.37\ \text{in.} \qquad\qquad\qquad \textit{Ans.}$$

Including the effects of shear and axial load contributed only a 0.6% increase in the answer to that determined only from bending.

EXAMPLE   9.13

The beam shown in Fig. 9–26a is used in a building subjected to two different thermal environments. If the temperature at the top surface of the beam is 80°F and at the bottom surface is 160°F, determine the vertical deflection of the beam at its midpoint due to the temperature gradient. Take $\alpha = 6.5(10^{-6})/°F$.

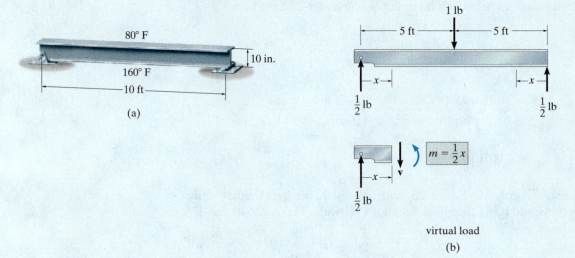

Fig. 9–26

SOLUTION

Since the deflection at the center of the beam is to be determined, a virtual unit load is placed there and the internal virtual moment in the beam is calculated, Fig. 9–26b.

The mean temperature at the center of the beam is $(160° + 80°)/2 = 120°F$, so that for application of Eq. 9–27, $\Delta T_m = 120°F - 80°F = 40°F$. Also, $c = 10$ in./2 $= 5$ in. Applying the principle of virtual work, we have

$$1 \text{ lb} \cdot \Delta_{C_v} = \int_0^L \frac{m\alpha\,\Delta T_m\,dx}{c}$$

$$= 2\int_0^{60 \text{ in.}} \frac{\left(\tfrac{1}{2}x\right)6.5(10^{-6})/°F(40°F)}{5 \text{ in.}}\,dx$$

$$\Delta_{C_v} = 0.0936 \text{ in.} \qquad\qquad Ans.$$

The result indicates a very negligible deflection.

9.9 Castigliano's Theorem for Beams and Frames

The internal bending strain energy for a beam or frame is given by Eq. 9–11 ($U_i = \int M^2\, dx/2EI$). Substituting this equation into Eq. 9–20 ($\Delta_i = \partial U_i/\partial P_i$) and omitting the subscript i, we have

$$\Delta = \frac{\partial}{\partial P}\int_0^L \frac{M^2\, dx}{2EI}$$

Rather than squaring the expression for internal moment M, integrating, and then taking the partial derivative, it is generally easier to differentiate prior to integration. Provided E and I are constant, we have

$$\Delta = \int_0^L M\left(\frac{\partial M}{\partial P}\right)\frac{dx}{EI} \qquad (9\text{–}28)$$

where
$\Delta = $ external displacement of the point caused by the real loads acting on the beam or frame.
$P = $ external force applied to the beam or frame in the direction of Δ.
$M = $ internal moment in the beam or frame, expressed as a function of x and caused by both the force P and the real loads on the beam.
$E = $ modulus of elasticity of beam material.
$I = $ moment of inertia of cross-sectional area computed about the neutral axis.

If the slope θ at a point is to be determined, we must find the partial derivative of the internal moment M with respect to an *external couple moment* M' acting at the point, i.e.,

$$\theta = \int_0^L M\left(\frac{\partial M}{\partial M'}\right)\frac{dx}{EI} \qquad (9\text{–}29)$$

The above equations are similar to those used for the method of virtual work, Eqs. 9–22 and 9–23, except $\partial M/\partial P$ and $\partial M/\partial M'$ replace m and m_θ, respectively. As in the case for trusses, slightly more calculation is generally required to determine the partial derivatives and apply Castigliano's theorem rather than use the method of virtual work. Also, recall that this theorem applies only to material having a linear elastic response. If a more complete accountability of strain energy in the structure is desired, the strain energy due to shear, axial force, and torsion must be included. The derivations for shear and torsion follow the same development as Eqs. 9–25 and 9–26. The strain energies and their derivatives are, respectively,

$$U_s = K \int_0^L \frac{V^2\, dx}{2AG} \qquad \frac{\partial U_s}{\partial P} = \int_0^L \frac{V}{AG}\left(\frac{\partial V}{\partial P}\right) dx$$

$$U_t = \int_0^L \frac{T^2\, dx}{2JG} \qquad \frac{\partial U_t}{\partial P} = \int_0^L \frac{T}{JG}\left(\frac{\partial T}{\partial P}\right) dx$$

These effects, however, will not be included in the analysis of the problems in this text since beam and frame deflections are caused mainly by bending strain energy. Larger frames, or those with unusual geometry, can be analyzed by computer, where these effects can readily be incorporated into the analysis.

Procedure for Analysis

The following procedure provides a method that may be used to determine the deflection and/or slope at a point in a beam or frame using Castigliano's theorem.

External Force P or Couple Moment M'

- Place a force **P** on the beam or frame at the point and in the direction of the desired displacement.
- If the slope is to be determined, place a couple moment **M'** at the point.
- It is assumed that both P and M' have a *variable magnitude* in order to obtain the changes $\partial M/\partial P$ or $\partial M/\partial M'$.

Internal Moments M

- Establish appropriate x coordinates that are valid within regions of the beam or frame where there is no discontinuity of force, distributed load, or couple moment.
- Calculate the internal moment M as a function of P or M' and each x coordinate. Also, compute the partial derivative $\partial M/\partial P$ or $\partial M/\partial M'$ for each coordinate x.
- After M and $\partial M/\partial P$ or $\partial M/\partial M'$ have been determined, assign P or M' its numerical value if it has replaced a real force or couple moment. Otherwise, set P or M' equal to zero.

Castigliano's Theorem

- Apply Eq. 9–28 or 9–29 to determine the desired displacement Δ or slope θ. It is important to retain the algebraic signs for corresponding values of M and $\partial M/\partial P$ or $\partial M/\partial M'$.
- If the resultant sum of all the definite integrals is positive, Δ or θ is in the same direction as **P** or **M'**.

EXAMPLE 9.14

Determine the displacement of point B of the beam shown in Fig. 9–27a. Take $E = 200$ GPa, $I = 500(10^6)$ mm^4.

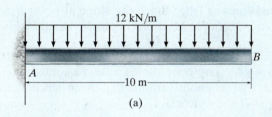

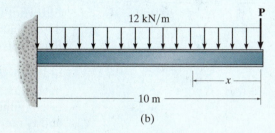

(a)

(b)

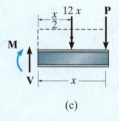

(c)

Fig. 9–27

SOLUTION

External Force P.   A vertical force $\mathbf{P}$ is placed on the beam at B as shown in Fig. 9–27b.

Internal Moments M.   A single x coordinate is needed for the solution, since there are no discontinuities of loading between A and B. Using the method of sections, Fig. 9–27c, we have

$$\zeta + \Sigma M = 0; \qquad -M - (12x)\left(\frac{x}{2}\right) - Px = 0$$

$$M = -6x^2 - Px \qquad \frac{\partial M}{\partial P} = -x$$

Setting $P = 0$, its actual value, yields

$$M = -6x^2 \qquad \frac{\partial M}{\partial P} = -x$$

Castigliano's Theorem.   Applying Eq. 9–28, we have

$$\Delta_B = \int_0^L M\left(\frac{\partial M}{\partial P}\right)\frac{dx}{EI} = \int_0^{10} \frac{(-6x^2)(-x)\,dx}{EI} = \frac{15(10^3)\;\text{kN} \cdot \text{m}^3}{EI}$$

or

$$\Delta_B = \frac{15(10^3)\;\text{kN} \cdot \text{m}^3}{200(10^6)\;\text{kN/m}^2\left[500(10^6)\;\text{mm}^4\right](10^{-12}\;\text{m}^4/\text{mm}^4)}$$

$$= 0.150\;\text{m} = 150\;\text{mm} \qquad\qquad Ans.$$

The similarity between this solution and that of the virtual-work method, Example 9.7, should be noted.

EXAMPLE | 9.15

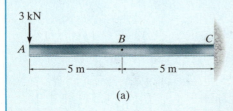

(a)

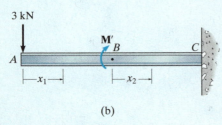

(b)

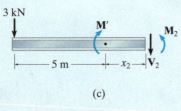

(c)

Fig. 9–28

Determine the slope at point B of the beam shown in Fig. 9–28a. Take $E = 200$ GPa, $I = 60(10^6)$ mm^4.

SOLUTION

External Couple Moment M′. Since the slope at point B is to be determined, an external couple $\mathbf{M'}$ is placed on the beam at this point, Fig. 9–28b.

Internal Moments M. Two coordinates, x_1 and x_2, must be used to determine the internal moments within the beam since there is a discontinuity, M', at B. As shown in Fig. 9–28b, x_1 ranges from A to B and x_2 ranges from B to C. Using the method of sections, Fig. 9–28c, the internal moments and the partial derivatives are computed as follows:

For x_1:

$$\zeta + \Sigma M = 0; \qquad M_1 + 3x_1 = 0$$
$$M_1 = -3x_1$$
$$\frac{\partial M_1}{\partial M'} = 0$$

For x_2:

$$\zeta + \Sigma M = 0; \qquad M_2 - M' + 3(5 + x_2) = 0$$
$$M_2 = M' - 3(5 + x_2)$$
$$\frac{\partial M_2}{\partial M'} = 1$$

Castigliano's Theorem. Setting $M' = 0$, its actual value, and applying Eq. 9–29, we have

$$\theta_B = \int_0^L M \left(\frac{\partial M}{\partial M'} \right) \frac{dx}{EI}$$

$$= \int_0^5 \frac{(-3x_1)(0)\, dx_1}{EI} + \int_0^5 \frac{-3(5 + x_2)(1)\, dx_2}{EI} = -\frac{112.5\ \text{kN} \cdot \text{m}^2}{EI}$$

or

$$\theta_B = \frac{-112.5\ \text{kN} \cdot \text{m}^2}{200(10^6)\ \text{kN/m}^2 \left[60(10^6)\ \text{mm}^4 \right] (10^{-12}\ \text{m}^4/\text{mm}^4)}$$

$$= -0.00938\ \text{rad} \qquad\qquad Ans.$$

The negative sign indicates that θ_B is opposite to the direction of the couple moment $\mathbf{M'}$. Note the similarity between this solution and that of Example 9.8.

EXAMPLE 9.16

Determine the vertical displacement of point C of the beam shown in Fig. 9–29a. Take $E = 200$ GPa, $I = 150(10^6)$ mm^4.

SOLUTION

External Force P. A vertical force $\mathbf{P}$ is applied at point C, Fig. 9–29b. Later this force will be set equal to a fixed value of 20 kN.

Internal Moments M. In this case two x coordinates are needed for the integration, Fig. 9–29b, since the load is discontinuous at C. Using the method of sections, Fig. 9–29c, we have

For x_1:

$$\zeta + \Sigma M = 0; \qquad -(24 + 0.5P)x_1 + 8x_1\left(\frac{x_1}{2}\right) + M_1 = 0$$

$$M_1 = (24 + 0.5P)x_1 - 4x_1^2$$

$$\frac{\partial M_1}{\partial P} = 0.5x_1$$

For x_2:

$$\zeta + \Sigma M = 0; \qquad -M_2 + (8 + 0.5P)x_2 = 0$$

$$M_2 = (8 + 0.5P)x_2$$

$$\frac{\partial M_2}{\partial P} = 0.5x_2$$

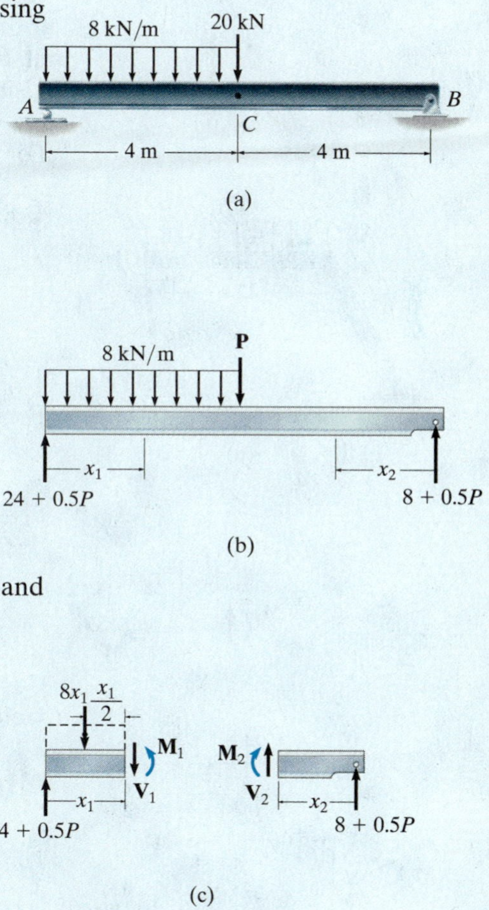

Castigliano's Theorem. Setting $P = 20$ kN, its actual value, and applying Eq. 9–28 yields

$$\Delta_{C_v} = \int_0^L M\left(\frac{\partial M}{\partial P}\right)\frac{dx}{EI}$$

$$= \int_0^4 \frac{(34x_1 - 4x_1^2)(0.5x_1)\,dx_1}{EI} + \int_0^4 \frac{(18x_2)(0.5x_2)\,dx_2}{EI}$$

$$= \frac{234.7 \text{ kN} \cdot \text{m}^3}{EI} + \frac{192 \text{ kN} \cdot \text{m}^3}{EI} = \frac{426.7 \text{ kN} \cdot \text{m}^3}{EI}$$

or

$$\Delta_{C_v} = \frac{426.7 \text{ kN} \cdot \text{m}^3}{200(10^6) \text{ kN/m}^2 \left[150(10^6) \text{ mm}^4\right](10^{-12} \text{ m}^4/\text{mm}^4)}$$

$$= 0.0142 \text{ m} = 14.2 \text{ mm} \qquad\qquad \textit{Ans.}$$

(a)

(b)

(c)

Fig. 9–29

EXAMPLE 9.17

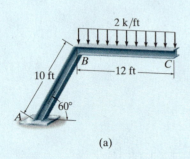

(a)

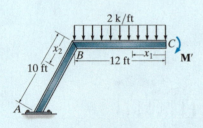

(b)

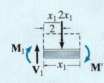

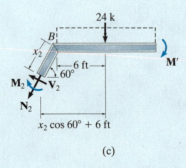

(c)

Fig. 9–30

Determine the slope at point C of the two-member frame shown in Fig. 9–30a. The support at A is fixed. Take $E = 29(10^3)$ ksi, $I = 600$ in^4.

SOLUTION

External Couple Moment M'.   A variable moment **M'** is applied to the frame at point C, since the slope at this point is to be determined, Fig. 9–30b. Later this moment will be set equal to zero.

Internal Moments M.   Due to the discontinuity of internal loading at B, two coordinates, x_1 and x_2, are chosen as shown in Fig. 9–30b. Using the method of sections, Fig. 9–30c, we have

For x_1:

$$\zeta + \Sigma M = 0; \qquad -M_1 - 2x_1\left(\frac{x_1}{2}\right) - M' = 0$$

$$M_1 = -\left(x_1^2 + M'\right)$$

$$\frac{\partial M_1}{\partial M'} = -1$$

For x_2:

$$\zeta + \Sigma M = 0; \qquad -M_2 - 24(x_2\cos 60° + 6) - M' = 0$$

$$M_2 = -24(x_2\cos 60° + 6) - M'$$

$$\frac{\partial M_2}{\partial M'} = -1$$

Castigliano's Theorem.   Setting $M' = 0$ and applying Eq. 9–29 yields

$$\theta_C = \int_0^L M\left(\frac{\partial M}{\partial M'}\right)\frac{dx}{EI}$$

$$= \int_0^{12}\frac{\left(-x_1^2\right)(-1)\,dx_1}{EI} + \int_0^{10}\frac{-24(x_2\cos 60° + 6)(-1)\,dx_2}{EI}$$

$$= \frac{576\text{ k}\cdot\text{ft}^2}{EI} + \frac{2040\text{ k}\cdot\text{ft}^2}{EI} = \frac{2616\text{ k}\cdot\text{ft}^2}{EI}$$

$$\theta_C = \frac{2616\text{ k}\cdot\text{ft}^2(144\text{ in}^2/\text{ft}^2)}{29(10^3)\text{ k/in}^2(600\text{ in}^4)} = 0.0216\text{ rad} \qquad \textit{Ans.}$$

FUNDAMENTAL PROBLEMS

F9–1. Determine the vertical displacement of joint *B*. *AE* is constant. Use the principle of virtual work.

F9–2. Solve Prob. F9–2 using Castigliano's theorem.

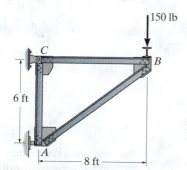

F9–1/2

F9–3. Determine the horizontal displacement of joint *A*. *AE* is constant. Use the principle of virtual work.

F9–4. Solve Prob. F9–3 using Castigliano's theorem.

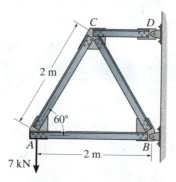

F9–3/4

F9–5. Determine the horizontal displacement of joint *D*. *AE* is constant. Use the principle of virtual work.

F9–6. Solve Prob. F9–5 using Castigliano's theorem.

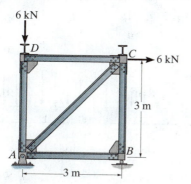

F9–5/6

F9–7. Determine the vertical displacement of joint *D*. *AE* is constant. Use the principle of virtual work.

F9–8. Solve Prob. F9–7 using Castigliano's theorem.

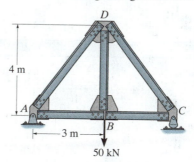

F9–7/8

F9–9. Determine the vertical displacement of joint *B*. *AE* is constant. Use the principle of virtual work.

F9–10. Solve Prob. F9–9 using Castigliano's theorem.

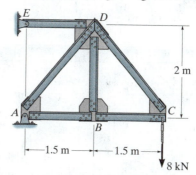

F9–9/10

F9–11. Determine the vertical displacement of joint *C*. *AE* is constant. Use the principle of virtual work.

F9–12. Solve Prob. F9–11 using Castigliano's theorem.

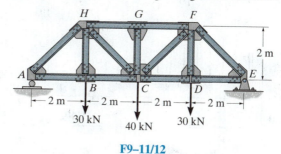

F9–11/12

F9–13. Determine the slope and displacement at point *A*. *EI* is constant. Use the principle of virtual work.

F9–14. Solve Prob. F9–13 using Castigliano's theorem.

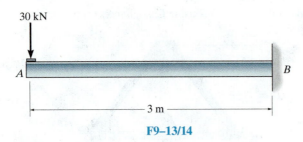

F9–13/14

F9–15. Determine the slope and displacement at point *A*. *EI* is constant. Use the principle of virtual work.

F9–16. Solve Prob. F9–15 using Castigliano's theorem.

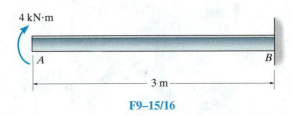

F9–15/16

F9–17. Determine the slope and displacement at point *B*. *EI* is constant. Use the principle of virtual work.

F9–18. Solve Prob. F9–17 using Castigliano's theorem.

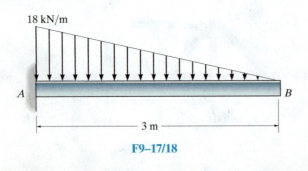

F9–17/18

F9–19. Determine the slope at *A* and displacement at point *C*. *EI* is constant. Use the principle of virtual work.

F9–20. Solve Prob. F9–19 using Castigliano's theorem.

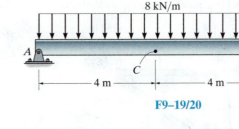

F9–19/20

F9–21. Determine the slope and displacement at point *C*. *EI* is constant. Use the principle of virtual work.

F9–22. Solve Prob. F9–21 using Castigliano's theorem.

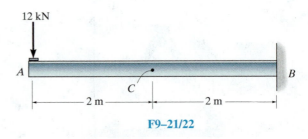

F9–21/22

F9–23. Determine the displacement at point *C*. *EI* is constant. Use the principle of virtual work.

F9–24. Solve Prob. F9–23 using Castigliano's theorem.

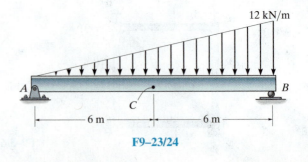

F9–23/24

PROBLEMS

Sec. 9.1–9.6

9–1. Determine the horizontal displacement of joint C. Assume the members are pin connected at their end points. Take $A = 200$ mm^2, and $E = 200$ GPa for each member. Use the method of virtual work.

9–2. Solve Prob. 9–1 using Castigliano's theorem.

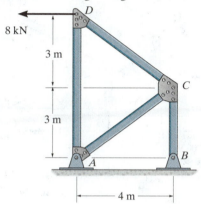

Probs. 9–1/2

9–3. Use the method of virtual work and determine the vertical displacement of point A. Each steel member has a cross-sectional area of 400 mm^2, $E = 200$ GPa.

*9–4. Solve Prob. 9–3 using Castigliano's theorem.

9–5. Use the method of virtual work and determine the vertical displacement of point B. Each A-36 steel member has a cross-sectional area of 400 mm^2, $E = 200$ GPa.

9–6. Solve Prob. 9–5 using Castigliano's theorem.

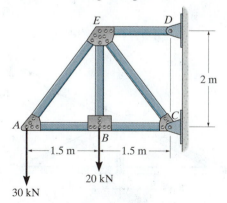

Probs. 9–3/4/5/6

9–7. Determine the vertical displacement of joint A. Each bar is made of steel and has the cross-sectional area shown. Take $E = 29(10^3)$ ksi. Use the method of virtual work.

*9–8. Solve Prob. 9–7 using Castigliano's theorem.

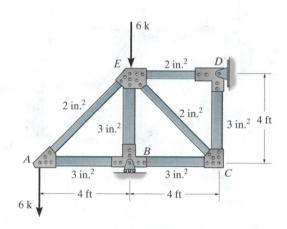

Probs. 9–7/8

9–9. Use the method of virtual work and determine the vertical displacement of joint H. Each steel member has a cross-sectional area of 4.5 in^2.

9–10. Solve Prob. 9–9 using Castigliano's theorem.

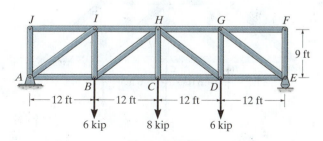

Probs. 9–9/10

9

9–11. Determine the vertical displacement of joint A of the truss. Each member has a cross-sectional area of $A = 300 \text{ mm}^2$. $E = 200$ GPa. Use the method of virtual work.

*9–12. Solve Prob. 9–11 using Castigliano's theorem.

9–15. Determine the vertical displacement of joint A. Assume the members are pin connected at their end points. Take $A = 2 \text{ in}^2$ and $E = 29 (10^3)$ for each member. Use the method of virtual work.

*9–16. Solve Prob. 9–15 using Castigliano's theorem.

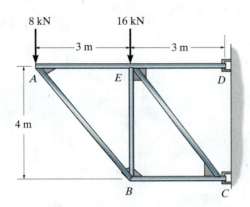

Probs. 9–11/12

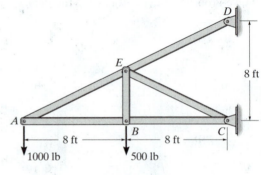

Probs. 9–15/16

9–13. Determine the vertical displacement of joint C. Assume the members are pin connected at their end points. AE is constant. Use the method of virtual work.

9–14. Solve Prob. 9–13 using Castigliano's theorem.

9–17. Determine the vertical displacement of joint A if members AB and BC experience a temperature increase of $\Delta T = 200°\text{F}$. Take $A = 2 \text{ in}^2$ and $E = 29(10^3)$ ksi. Also, $\alpha = 6.60 (10^{-6})/°\text{F}$.

9–18. Determine the vertical displacement of joint A if member AE is fabricated 0.5 in. too short.

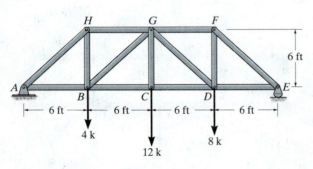

Probs. 9–13/14

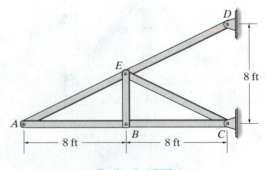

Probs. 9–17/18

Sec. 9.7–9.9

9–19. Determine the displacement of point C and the slope at point B. EI is constant. Use the principle of virtual work.

*9–20. Solve Prob. 9–19 using Castigliano's theorem.

9–23. Determine the displacement at point D. Use the principle of virtual work. EI is constant.

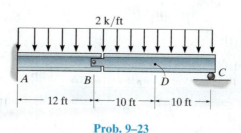

Prob. 9–23

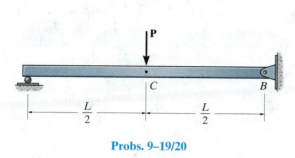

Probs. 9–19/20

9–21. Determine the slope and displacement at point C. Use the principle of virtual work. EI is constant.

9–22. Solve Prob. 9–21 using Castigliano's theorem.

*9–24. Determine the slope and displacement at the end C of the beam $E = 200$ GPa, $I = 70(10^6)$ mm^4. Use the method of virtual work.

9–25. Solve Prob. 9–24 using Castigliano's theorem.

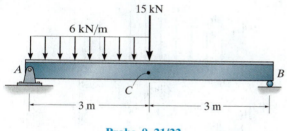

Probs. 9–21/22

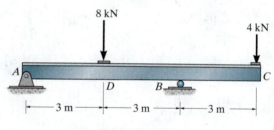

Probs. 9–24/25

9–26.   Determine the displacement and slope at point C of the cantilever beam. The moment of inertia of each segment is indicated in the figure. Take $E = 29(10^3)$ ksi. Use the principle of virtual work.

9–27.   Solve Prob. 9–26 using Castigliano's theorem.

9–30.   Determine the slope and displacement at point C. EI is constant. Use the method of virtual work.

9–31.   Solve Prob. 9–30 using Castigliano's theorem.

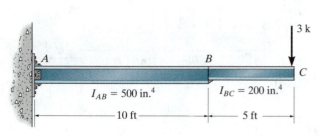

Probs. 9–26/27

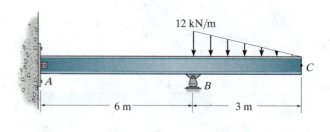

Probs. 9–30/31

*****9–28.**   Determine the displacement and slope at point C of the cantilever beam. The moment of inertia of each segment is indicated in the figure. Take $E = 29(10^3)$ ksi. Use the principle of virtual work.

9–29.   Solve Prob. 9–28 using Castigliano's theorem.

*****9–32.**   Determine the displacement of point C. Use the method of virtual work. EI is constant.

9–33.   Solve Prob. 9–32 using Castigliano's theorem.

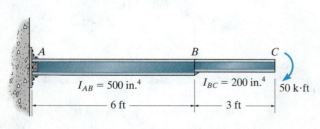

Probs. 9–28/29

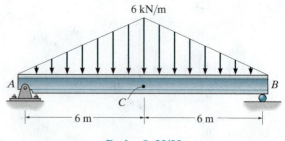

Probs. 9–32/33

9

9–34. Determine the slope and displacement at point B. Assume the support at A is a pin and C is a roller. Take $E = 29(10^3)$ ksi, $I = 300$ in^4. Use the method of virtual work.

9–35. Solve Prob. 9–34 using Castigliano's theorem.

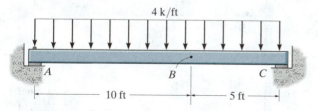

4 k/ft

10 ft — 5 ft

Probs. 9–34/35

*9–36. Determine the slope and displacement at point B. Assume the support at A is a pin and C is a roller. Account for the additional strain energy due to shear. Take $E = 29(10^3)$ ksi, $I = 300$ in^4, $G = 12(10^3)$ ksi, and assume AB has a cross-sectional area of $A = 7.50$ in^2. Use the method of virtual work.

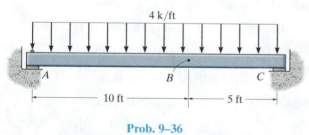

4 k/ft

10 ft — 5 ft

Prob. 9–36

9–37. Determine the displacement of point C. Use the method of virtual work. EI is constant.

9–38. Solve Prob. 9–37 using Castigliano's theorem.

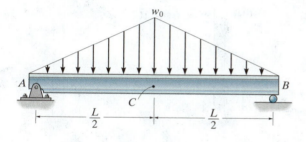

w_0

$\frac{L}{2}$ $\frac{L}{2}$

Probs. 9–37/38

9–39. Use the method of virtual work and determine the displacement of point C of the beam made from steel. $E = 29(10^3)$ ksi, $I = 245$ in^4.

*9–40. Use the method of virtual work and determine the slope at A of the beam made from steel. $E = 29(10^3)$ ksi, $I = 245$ in^4.

9–41. Solve Prob. 9–40 using Castigliano's theorem.

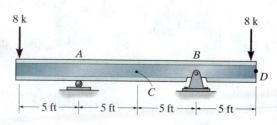

8 k   8 k

5 ft — 5 ft — 5 ft — 5 ft

Probs. 9–39/40/41

9–42. Determine the displacement at point D. Use the principle of virtual work. EI is constant.

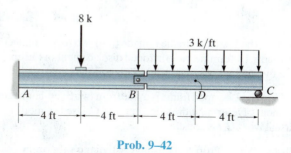

8 k   3 k/ft

4 ft — 4 ft — 4 ft — 4 ft

Prob. 9–42

9–43. Determine the displacement at point D. Use Castigliano's theorem. EI is constant.

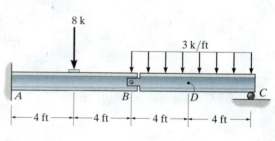

8 k   3 k/ft

4 ft — 4 ft — 4 ft — 4 ft

Prob. 9–43

*9–44. Determine the horizontal displacement at A. Take $E = 29(10^3)$ ksi. The moment of inertia of each segment of the frame is indicated in the figure. Assume D is a pin support. Use the method of virtual work.

9–45. Solve Prob. 9–44 using Castigliano's theorem.

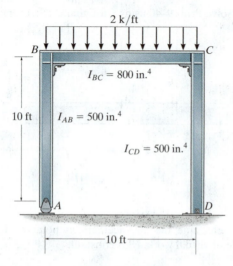

2 k/ft

$I_{BC} = 800$ in.^{4}

$I_{AB} = 500$ in.^{4}

$I_{CD} = 500$ in.^{4}

10 ft

10 ft

Probs. 9–44/45

9–46. The L-shaped frame is made from two fixed-connected segments. Determine the horizontal displacement of the end C. Use the method of virtual work. EI is constant.

9–47. The L-shaped frame is made from two fixed-connected segments. Determine the slope at point C. Use the method of virtual work. EI is constant.

*9–48. Solve Prob. 9–47 using Castigliano's theorem.

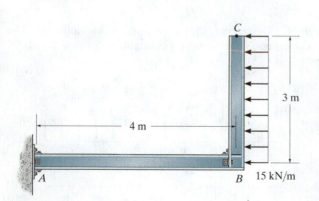

C
3 m
4 m
A   B   15 kN/m

Probs. 9–46/47/48

9–49. Determine the vertical displacement of joint D. Use the method of virtual work. Take $E = 29(10^3)$ ksi. Assume the members are pin connected at their ends.

9–50. Solve Prob. 9–49 using Castigliano's theorem.

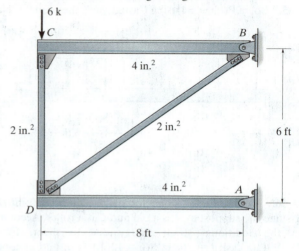

6 k

C   B
4 in.^{2}
2 in.^{2} 2 in.^{2} 6 ft
4 in.^{2} A
D
8 ft

Probs. 9–49/50

9–51. Determine the horizontal displacement at C. Take $E = 29(10^3)$ ksi, $I = 150$ in^4 for each member. Use the method of virtual work.

*9–52. Solve Prob. 9–51 using Castigliano's theorem.

9–53. Determine the horizontal displacement of the rocker at B. Take $E = 29(10^3)$ ksi, $I = 150$ in^4 for each member. Use the method of virtual work.

9–54. Solve Prob. 9–53 using Castigliano's theorem.

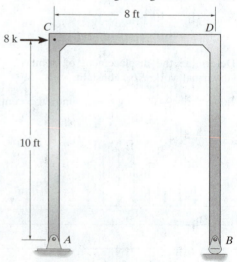

8 ft
C   D
8 k
10 ft
A   B

Probs. 9–51/52/53/54

9–55. Determine the vertical displacement of point *C*. *EI* is constant. Use the method of virtual work.

*9–56. Solve Prob. 9–55 using Castigliano's theorem.

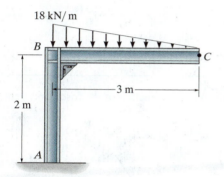

Probs. 9–55/56

9–57. Determine the slope at *A* and the vertical displacement at *B*. Use the method of virtual work.

9–58. Solve Prob. 9–57 using Castigliano's theorem.

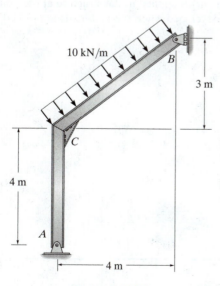

Probs. 9–57/58

9–59. The bent rod has a radius of 0.75 in. Determine the displacement at *C* in the direction of the 150-lb force. Include the effects of bending, axial, shear, and torsional strain energy. Use the method of virtual work. Take $E = 29 (10^3)$ ksi, $G = 11(10^3)$ ksi.

*9–60. Solve Prob. 9–59 using Castigliano's theorem.

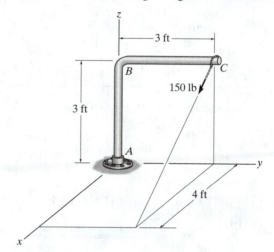

Probs. 9–59/60

9–61. Determine the vertical displacement of joint *C*. Assume the supports at *A* and *E* are pins and joint *C* is pin connected. *EI* is constant. Use the method of virtual work.

9–62. Solve Prob. 9–61 using Castigliano's theorem.

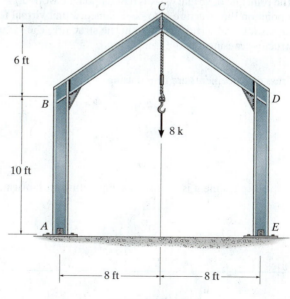

Probs. 9–61/62

9

CHAPTER REVIEW

All energy methods are based on the conservation of energy principle, which states that the work done by all external forces acting on the structure, U_e, is transformed into internal work or strain energy. U_i, developed in the members when the structure deforms.

$$U_e = U_i$$

A force (moment) does work U when it undergoes a displacement (rotation) in the direction of the force (moment).

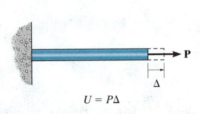

$$U = P\Delta$$

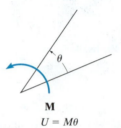

$$U = M\theta$$

The principle of virtual work is based upon the work done by a "virtual" or imaginary unit force. If the deflection (rotation) at a point on the structure is to be obtained, a unit virtual force (couple moment) is applied to the structure at the point. This causes internal virtual loadings in the structure. The virtual work is then developed when the real loads are placed on the structure causing it to deform.

Truss displacements are found using

$$1 \cdot \Delta = \sum \frac{nNL}{AE}$$

If the displacement is caused by temperature, or fabrication errors, then

$$1 \cdot \Delta = \sum n\alpha \Delta TL \qquad\qquad 1 \cdot \Delta = \sum n \Delta L$$

For beams and frames, the displacement (rotation) is defined from

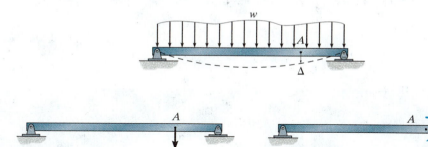

$$1 \cdot \Delta = \int_0^L \frac{mM}{EI} \, dx$$

$$1 \cdot \theta = \int_0^L \frac{m_\theta M}{EI} \, dx$$

Castigliano's second theorem, called the method of least work, can be used to determine the deflections in structures that respond elastically. It states that the displacement (rotation) at a point on a structure is equal to the first partial derivative of the strain energy in the structure with respect to a force P (couple moment M') acting at the point and in the direction of the displacement (rotation). For a truss

$$\Delta = \sum N \left(\frac{\partial N}{\partial P} \right) \frac{L}{AE}$$

For beams and frames

$$\Delta = \int_0^L M \left(\frac{\partial M}{\partial P} \right) \frac{dx}{EI}$$

$$\theta = \int_0^L M \left(\frac{\partial M}{\partial M'} \right) \frac{dx}{EI}$$

9

Chapter 10

© Tony Freeman/Science Source

The fixed-connected joints of this concrete framework make this a statically indeterminate structure.

Analysis of Statically Indeterminate Structures by the Force Method

In this chapter we will apply the *force* or *flexibility* method to analyze statically indeterminate trusses, beams, and frames. At the end of the chapter we will present a method for drawing the influence line for a statically indeterminate beam or frame.

10.1 Statically Indeterminate Structures

Recall from Sec. 2.4 that a structure of any type is classified as **statically indeterminate** when the number of unknown reactions or internal forces exceeds the number of equilibrium equations available for its analysis. In this section we will discuss the merits of using indeterminate structures and two fundamental ways in which they may be analyzed. Realize that most of the structures designed today are statically indeterminate. This indeterminacy may arise as a result of added supports or members, or by the general form of the structure. For example, reinforced concrete buildings are almost always statically indeterminate since the columns and beams are poured as continuous members through the joints and over supports.

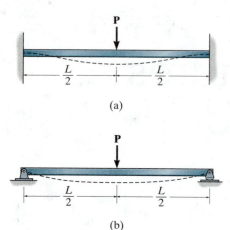

(a)

(b)

Fig. 10–1

Advantages and Disadvantages.

Although the analysis of a statically indeterminate structure is more involved than that of a statically determinate one, there are usually several very important reasons for choosing this type of structure for design. Most important, for a given loading the maximum stress and deflection of an indeterminate structure are generally *smaller* than those of its statically determinate counterpart. For example, the statically indeterminate, fixed-supported beam in Fig. 10–1a will be subjected to a maximum moment of $M_{max} = PL/8$, whereas the same beam, when simply supported, Fig. 10–1b, will be subjected to twice the moment, that is, $M_{max} = PL/4$. As a result, the fixed-supported beam has one fourth the deflection and one half the stress at its center of the one that is simply supported.

Another important reason for selecting a statically indeterminate structure is because it has a tendency to redistribute its load to its redundant supports in cases where faulty design or overloading occurs. In these cases, the structure maintains its stability and collapse is prevented. This is particularly important when *sudden* lateral loads, such as wind or earthquake, are imposed on the structure. To illustrate, consider again the fixed-end beam in Fig. 10–1a. As **P** is increased, the beam's material at the walls and at the center of the beam begins to *yield* and forms localized "plastic hinges," which causes the beam to deflect as if it were hinged or pin connected at these points. Although the deflection becomes large, the walls will develop horizontal force and moment reactions that will hold the beam and thus prevent it from totally collapsing. In the case of the simply supported beam, Fig. 10–1b, an excessive load **P** will cause the "plastic hinge" to form only at the center of the beam, and due to the large vertical deflection, the supports will not develop the horizontal force and moment reactions that may be necessary to prevent total collapse.

Although statically indeterminate structures can support a loading with thinner members and with increased stability compared to their statically determinate counterparts, there are cases when these advantages may instead become disadvantages. The cost savings in material must be compared with the added cost necessary to fabricate the structure, since oftentimes it becomes more costly to construct the supports and joints of an indeterminate structure compared to one that is determinate. More important, though, because statically indeterminate structures have redundant support reactions, one has to be very careful to prevent differential displacement of the supports, since this effect will introduce internal stress in the structure. For example, if the wall at one end of the fixed-end beam in Fig. 10–1a were to settle, stress would be developed in the beam because of this "forced" deformation. On the other hand, if the beam were simply supported or statically determinate, Fig. 10–1b, then any settlement of its end would not cause the beam to deform, and therefore no stress would be developed in the beam. In general, then, any deformation, such as that caused by relative support displacement, or changes in member lengths caused by temperature or fabrication errors, will introduce additional stresses in the structure, which must be considered when designing indeterminate structures.

Methods of Analysis.

When analyzing any indeterminate structure, it is necessary to satisfy *equilibrium*, *compatibility*, and *force-displacement* requirements for the structure. *Equilibrium* is satisfied when the reactive forces hold the structure at rest, and *compatibility* is satisfied when the various segments of the structure fit together without intentional breaks or overlaps. The *force-displacement* requirements depend upon the way the structure's material responds to loads. In this text we have assumed this to be a linear elastic response. In general there are two different ways to satisfy these three requirements. For a statically indeterminate structure, they are the **force** or **flexibility method**, and the **displacement** or **stiffness method**.

Force Method.

The force method was originally developed by James Clerk Maxwell in 1864 and later refined by Otto Mohr and Heinrich Müller-Breslau. This method was one of the first available for the analysis of statically indeterminate structures. Since compatibility forms the basis for this method, it has sometimes been referred to as the *compatibility method* or the *method of consistent displacements*. This method consists of writing equations that satisfy the *compatibility* and *force-displacement* requirements for the structure in order to determine the redundant *forces*. Once these forces have been determined, the remaining reactive forces on the structure are determined by satisfying the equilibrium requirements. The fundamental principles involved in applying this method are easy to understand and develop, and they will be discussed in this chapter.

Displacement Method.

The displacement method of analysis is based on first writing force-displacement relations for the members and then satisfying the *equilibrium requirements* for the structure. In this case the *unknowns* in the equations are *displacements*. Once the displacements are obtained, the forces are determined from the compatibility and force-displacement equations. We will study some of the classical techniques used to apply the displacement method in Chapters 11 and 12. Since almost all present day computer software for structural analysis is developed using this method we will present a matrix formulation of the displacement method in Chapters 14, 15, and 16.

Each of these two methods of analysis, which are outlined in Fig. 10–2, has particular advantages and disadvantages, depending upon the geometry of the structure and its degree of indeterminacy. A discussion of the usefulness of each method will be given after each has been presented.

	Unknowns	Equations Used for Solution	Coefficients of the Unknowns
Force Method	Forces	Compatibility and Force Displacement	Flexibility Coefficients
Displacement Method	Displacements	Equilibrium and Force Displacement	Stiffness Coefficients

Fig. 10–2

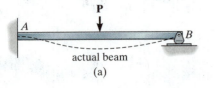

actual beam

(a)

$\parallel$

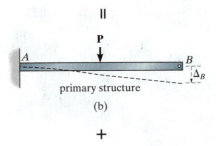

primary structure

(b)

$+$

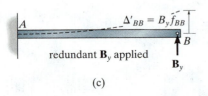

redundant $\mathbf{B}_y$ applied

(c)

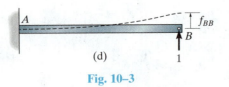

(d)

Fig. 10–3

10.2   Force Method of Analysis: General Procedure

Perhaps the best way to illustrate the principles involved in the force method of analysis is to consider the beam shown in Fig. 10–3a. If its free-body diagram were drawn, there would be four unknown support reactions; and since three equilibrium equations are available for solution, the beam is indeterminate to the first degree. Consequently, one additional equation is necessary for solution. To obtain this equation, we will use the principle of superposition and consider the *compatibility of displacement* at one of the supports. This is done by choosing one of the support reactions as "redundant" and temporarily removing its effect on the beam so that the beam then becomes statically determinate and stable. This beam is referred to as the **primary structure**. Here we will remove the restraining action of the rocker at B. As a result, the load $\mathbf{P}$ will cause B to be displaced downward by an amount Δ_B as shown in Fig. 10–3b. By superposition, however, the unknown reaction at B, i.e., $\mathbf{B}_y$, causes the beam at B to be displaced Δ'_{BB} upward, Fig. 10–3c. Here the first letter in this double-subscript notation refers to the point (B) where the deflection is specified, and the second letter refers to the point (B) where the unknown reaction acts. Assuming positive displacements act upward, then from Figs. 10–3a through 10–3c we can write the necessary compatibility equation at the rocker as

$$(+\uparrow) \qquad\qquad 0 = -\Delta_B + \Delta'_{BB}$$

Let us now denote the displacement at B caused by a *unit load* acting in the direction of $\mathbf{B}_y$ as the *linear flexibility coefficient* f_{BB}, Fig. 10–3d. Using the same scheme for this double-subscript notation as above, f_{BB} is the deflection at B caused by a unit load at B. Since the material behaves in a linear-elastic manner, a force of $\mathbf{B}_y$ acting at B, instead of the unit load, will cause a proportionate increase in f_{BB}. Thus we can write

$$\Delta'_{BB} = B_y f_{BB}$$

When written in this format, it can be seen that the linear flexibility coefficient f_{BB} is a *measure of the deflection per unit force,* and so its units are m/N, ft/lb, etc. The compatibility equation above can therefore be written in terms of the unknown B_y as

$$0 = -\Delta_B + B_y f_{BB}$$

10

Using the methods of Chapter 8 or 9, or the deflection table on the inside front cover of the book, the appropriate load-displacement relations for the deflection Δ_B, Fig. 10–3b, and the flexibility coefficient f_{BB}, Fig. 10–3d, can be obtained and the solution for B_y determined, that is, $B_y = \Delta_B/f_{BB}$. Once this is accomplished, the three reactions at the wall A can then be found from the equations of equilibrium.

As stated previously, the choice of the redundant is *arbitrary*. For example, the moment at A, Fig. 10–4a, can be determined *directly* by removing the capacity of the beam to support a moment at A, that is, by replacing the fixed support by a pin. As shown in Fig. 10–4b, the rotation at A caused by the load $\mathbf{P}$ is θ_A, and the rotation at A caused by the redundant $\mathbf{M}_A$ at A is θ'_{AA}, Fig. 10–4c. If we denote an *angular flexibility coefficient* α_{AA} as the angular displacement at A caused by a unit couple moment applied to A, Fig. 10–4d, then

$$\theta'_{AA} = M_A \alpha_{AA}$$

Thus, the angular flexibility coefficient measures the angular displacement per unit couple moment, and therefore it has units of rad/N·m or rad/lb·ft, etc. The compatibility equation for rotation at A therefore requires

$$(\zeta+) \qquad\qquad 0 = \theta_A + M_A \alpha_{AA}$$

In this case, $M_A = -\theta_A/\alpha_{AA}$, a negative value, which simply means that $\mathbf{M}_A$ acts in the opposite direction to the unit couple moment.

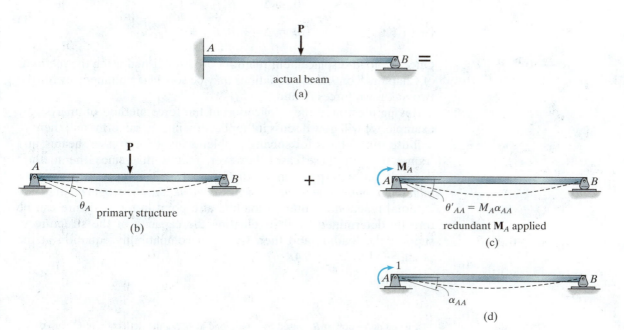

actual beam
(a)

primary structure
(b)

$\theta'_{AA} = M_A \alpha_{AA}$
redundant $\mathbf{M}_A$ applied
(c)

α_{AA}
(d)

Fig. 10–4

10

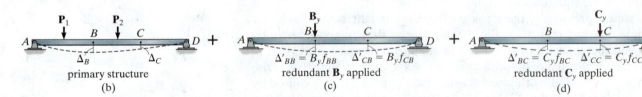

actual beam
(a)

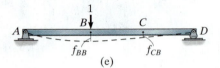

primary structure
(b)

redundant B_y applied
(c)

redundant C_y applied
(d)

Fig. 10–5

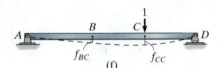

(e)

(f)

A third example that illustrates application of the force method is given in Fig. 10–5a. Here the beam is indeterminate to the second degree and therefore two compatibility equations will be necessary for the solution. We will choose the vertical forces at the roller supports, B and C, as redundants. The resultant statically determinate beam deflects as shown in Fig. 10–5b when the redundants are removed. Each redundant force, which is *assumed* to act downward, deflects this beam as shown in Fig. 10–5c and 10–5d, respectively. Here the flexibility coefficients*f_{BB} and f_{CB} are found from a unit load acting at B, Fig. 10–5e; and f_{CC} and f_{BC} are found from a unit load acting at C, Fig. 10–5f. By superposition, the compatibility equations for the deflection at B and C, respectively, are

$$(+\downarrow) \qquad\qquad 0 = \Delta_B + B_y f_{BB} + C_y f_{BC}$$

$$(+\downarrow) \qquad\qquad 0 = \Delta_C + B_y f_{CB} + C_y f_{CC}$$

$$(10\text{–}1)$$

Once the load-displacement relations are established using the methods of Chapters 8 or 9, these equations may be solved simultaneously for the two unknown forces B_y and C_y.

Having illustrated the application of the force method of analysis by example, we will now discuss its application in general terms and then we will use it as a basis for solving problems involving trusses, beams, and frames. For all these cases, however, realize that since the method depends on superposition of displacements, it is necessary that *the material remain linear elastic when loaded.* Also, recognize that *any* external reaction or internal loading at a point in the structure can be directly determined by first releasing the capacity of the structure to support the loading and then writing a compatibility equation at the point. See Example 10.4.

*f_{BB} is the deflection at B caused by a unit load at B; f_{CB} the deflection at C caused by a unit load at B.

Procedure for Analysis

The following procedure provides a general method for determining the reactions or internal loadings of statically indeterminate structures using the force or flexibility method of analysis.

Principle of Superposition

Determine the number of degrees n to which the structure is indeterminate. Then specify the n unknown redundant forces or moments that must be removed from the structure in order to make it statically determinate and stable. Using the principle of superposition, draw the statically indeterminate structure and show it to be equal to a series of corresponding statically *determinate* structures. The primary structure supports the same external loads as the statically indeterminate structure, and each of the other structures added to the primary structure shows the structure loaded with a separate redundant force or moment. Also, sketch the elastic curve on each structure and indicate symbolically the displacement or rotation at the point of each redundant force or moment.

Compatibility Equations

Write a compatibility equation for the displacement or rotation at each point where there is a redundant force or moment. These equations should be expressed in terms of the unknown redundants and their corresponding flexibility coefficients obtained from unit loads or unit couple moments that are collinear with the redundant forces or moments.

Determine all the deflections and flexibility coefficients using the table on the inside front cover or the methods of Chapter 8 or 9.* Substitute these load-displacement relations into the compatibility equations and solve for the unknown redundants. In particular, if a numerical value for a redundant is negative, it indicates the redundant acts opposite to its corresponding unit force or unit couple moment.

Equilibrium Equations

Draw a free-body diagram of the structure. Since the redundant forces and/or moments have been calculated, the remaining unknown reactions can be determined from the equations of equilibrium.

It should be realized that once all the support reactions have been obtained, the shear and moment diagrams can then be drawn, and the deflection at any point on the structure can be determined using the same methods outlined previously for statically determinate structures.

10

*It is suggested that if the M/EI diagram for a beam consists of simple segments, the moment-area theorems or the conjugate-beam method be used. Beams with complicated M/EI diagrams, that is, those with many curved segments (parabolic, cubic, etc.), can be readily analyzed using the method of virtual work or by Castigliano's second theorem.

10.3   Maxwell's Theorem of Reciprocal Displacements; Betti's Law

When Maxwell developed the force method of analysis, he also published a theorem that relates the flexibility coefficients of any two points on an elastic structure—be it a truss, a beam, or a frame. This theorem is referred to as the theorem of reciprocal displacements and may be stated as follows: *The displacement of a point B on a structure due to a unit load acting at point A is equal to the displacement of point A when the unit load is acting at point B, that is,* $f_{BA} = f_{AB}$.

Proof of this theorem is easily demonstrated using the principle of virtual work. For example, consider the beam in Fig. 10–6. When a real unit load acts at A, assume that the internal moments in the beam are represented by m_A. To determine the flexibility coefficient at B, that is, f_{BA}, a virtual unit load is placed at B, Fig. 10–7, and the internal moments m_B are computed. Then applying Eq. 9–18 yields

$$f_{BA} = \int \frac{m_B m_A}{EI} dx$$

Likewise, if the flexibility coefficient f_{AB} is to be determined when a real unit load acts at B, Fig. 10–7, then m_B represents the internal moments in the beam due to a real unit load. Furthermore, m_A represents the internal moments due to a virtual unit load at A, Fig. 10–6. Hence,

$$f_{AB} = \int \frac{m_A m_B}{EI} dx$$

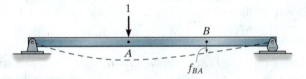

Fig. 10–6

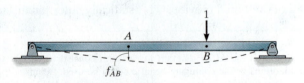

Fig. 10–7

Both integrals obviously give the same result, which proves the theorem. The theorem also applies for reciprocal rotations, and may be stated as follows: *The rotation at point B on a structure due to a unit couple moment acting at point A is equal to the rotation at point A when the unit couple moment is acting at point B.* Furthermore, using a unit force and unit couple moment, applied at separate points on the structure, we may also state: *The rotation in radians at point B on a structure due to a unit load acting at point A is equal to the displacement at point A when a unit couple moment is acting at point B.*

As a consequence of this theorem, some work can be saved when applying the force method to problems that are statically indeterminate to the second degree or higher. For example, only one of the two flexibility coefficients f_{BC} or f_{CB} has to be calculated in Eqs. 10–1, since $f_{BC} = f_{CB}$. Furthermore, the theorem of reciprocal displacements has applications in structural model analysis and for constructing influence lines using the Müller-Breslau principle (see Sec. 10.10).

When the theorem of reciprocal displacements is formalized in a more general sense, it is referred to as **Betti's law**. Briefly stated: The virtual work δU_{AB} done by a system of forces $\Sigma \mathbf{P}_B$ that undergo a displacement caused by a system of forces $\Sigma \mathbf{P}_A$ is equal to the virtual work δU_{BA} caused by the forces $\Sigma \mathbf{P}_A$ when the structure deforms due to the system of forces $\Sigma \mathbf{P}_B$. In other words, $\delta U_{AB} = \delta U_{BA}$. The proof of this statement is similar to that given above for the reciprocal-displacement theorem.

10.4 Force Method of Analysis: Beams

The force method applied to beams was outlined in Sec. 10.2. Using the Procedure for Analysis also given in Sec. 10.2, we will present several examples that illustrate the application of this technique. Before we do, however, it is worth mentioning a case where engineers have designed a structure that has the advantages of being statically indeterminate, yet the analysis of it is reduced to one that is statically determinate. It is a

These bridge girders are statically indeterminate, since they are continuous over their piers.

trussed cantilever bridge, such as the one shown in the photo. It consists of two cantilevered sections AB and CD, and a center suspended span BC that was floated out and lifted in place. This span was pinned at B and suspended from C by a primary vertical member CE. The advantages for using this design can be explained by considering the three simple beams resting on supports at, A, B, C, and D, as shown in Fig. 10–8a. When a uniform distributed loading is applied to each beam it produces a maximum moment of $0.125\,wL^2$. If instead a *continuous beam* is used, Fig. 10–8b, then a statically indeterminate analysis would produce a moment diagram that has a maximum moment in the beam of $0.10\,wL^2$. Although this is a 25% reduction in the maximum moment, unfortunately any slight settlement of one of the bridge piers would introduce larger reactions at the supports, and also larger moments in the beam. To get around this disadvantage, pins can be introduced in the span at the points E and F of zero moment, Fig. 10–8c. The beam then becomes statically determinate and yet continuity of the span is maintained. In this case any settlement of a support would not affect the reactions. The cantilevered bridge span works on the same principle. Here one side of the span, AB in the photo, is supported by a pin and the other side by a vertical member in order to allow free horizontal movement, to accommodate the traction force (friction) of vehicles on the deck and its thermal expansion and contraction.

Cantilever bridge.

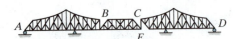

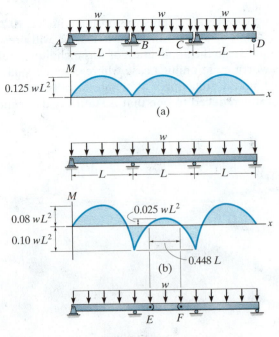

Fig. 10–8

EXAMPLE 10.1

Determine the reaction at the roller support B of the beam shown in Fig. 10–9a. EI is constant.

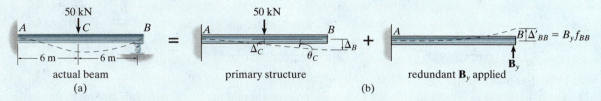

actual beam
(a)

primary structure
(b)

redundant B_y applied

Fig. 10–9

SOLUTION

Principle of Superposition. By inspection, the beam is statically indeterminate to the first degree. The redundant will be taken as B_y so that this force can be determined *directly*. Fig. 10–9b shows application of the principle of superposition. Notice that removal of the redundant requires that the roller support or the constraining action of the beam in the direction of B_y be removed. Here we have assumed that B_y acts upward on the beam.

Compatibility Equation. Taking positive displacement as upward, Fig. 10–9b, we have

$(+\uparrow)$ $\qquad\qquad 0 = -\Delta_B + B_y f_{BB}$ $\qquad\qquad$ (1)

The terms Δ_B and f_{BB} are easily obtained using the table on the inside front cover. In particular, note that $\Delta_B = \Delta_C + \theta_C(6\text{ m})$. Thus,

$$\Delta_B = \frac{P(L/2)^3}{3EI} + \frac{P(L/2)^2}{2EI}\left(\frac{L}{2}\right)$$

$$= \frac{(50\text{ kN})(6\text{ m})^3}{3EI} + \frac{(50\text{ kN})(6\text{ m})^2}{2EI}(6\text{ m}) = \frac{9000\text{ kN}\cdot\text{m}^3}{EI}\downarrow$$

$$f_{BB} = \frac{PL^3}{3EI} = \frac{1(12\text{ m})^3}{3EI} = \frac{576\text{ m}^3}{EI}\uparrow$$

Substituting these results into Eq. (1) yields

$(+\uparrow)$ $\qquad 0 = -\dfrac{9000}{EI} + B_y\left(\dfrac{576}{EI}\right)$ $\qquad B_y = 15.6\text{ kN}$ $\qquad$ *Ans.*

If this reaction is placed on the free-body diagram of the beam, the reactions at A can be obtained from the three equations of equilibrium, Fig. 10–9c.

Having determined all the reactions, the moment diagram can be constructed as shown in Fig. 10–9d.

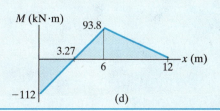

(c)

(d)

10

EXAMPLE   10.2

Draw the shear and moment diagrams for the beam shown in Fig. 10–10a. The support at B settles 1.5 in. Take $E = 29(10^3)$ ksi, $I = 750$ in^4.

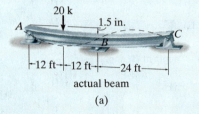

20 k   1.5 in.

A B C

$-$12 ft$-$12 ft$-$24 ft$-$

actual beam

(a)

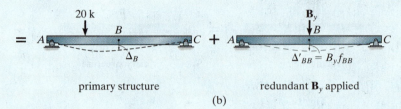

$=$ primary structure $+$ redundant $\mathbf{B}_y$ applied

(b)

Fig. 10–10

SOLUTION

Principle of Superposition.   By inspection, the beam is indeterminate to the first degree. The center support B will be chosen as the redundant, so that the roller at B is removed, Fig. 10–10b. Here $\mathbf{B}_y$ is assumed to act downward on the beam.

Compatibility Equation.   With reference to point B in Fig. 10–10b, using units of inches, we require

$$(+\downarrow) \qquad\qquad 1.5 \text{ in.} = \Delta_B + B_y f_{BB} \qquad\qquad (1)$$

We will use the table on the inside front cover. Note that for Δ_B the equation for the deflection curve requires $0 < x < a$. Since $x = 24$ ft, then $a = 36$ ft. Thus,

$$\Delta_B = \frac{Pbx}{6LEI}(L^2 - b^2 - x^2) = \frac{20(12)(24)}{6(48)EI}\left[(48)^2 - (12)^2 - (24)^2\right]$$

$$= \frac{31,680 \text{ k} \cdot \text{ft}^3}{EI}$$

$$f_{BB} = \frac{PL^3}{48EI} = \frac{1(48)^3}{48\,EI} = \frac{2304 \text{ ft}^3}{EI}$$

Substituting these values into Eq. (1), we get

$$1.5 \text{ in.} \, (29(10^3)\,\text{k/in}^2)(750 \text{ in}^4)$$

$$= 31,680 \text{ k} \cdot \text{ft}^3(12\,\text{in./ft})^3 + B_y(2304 \text{ ft}^3)(12\,\text{in./ft})^3$$

$$B_y = -5.56 \text{ k}$$

The negative sign indicates that $\mathbf{B}_y$ acts *upward* on the beam.

Equilibrium Equations. From the free-body diagram shown in Fig. 10–10c we have

$\zeta + \Sigma M_A = 0;$ $\qquad -20(12) + 5.56(24) + C_y(48) = 0$

$\qquad\qquad\qquad C_y = 2.22\ k$

$+\uparrow \Sigma F_y = 0;$ $\qquad A_y - 20 + 5.56 + 2.22 = 0$

$\qquad\qquad\qquad A_y = 12.22\ k$

Using these results, verify the shear and moment diagrams shown in Fig. 10–10d.

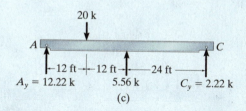

(c)

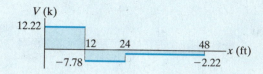

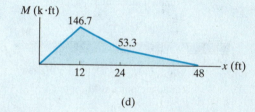

(d)

EXAMPLE 10.3

Draw the shear and moment diagrams for the beam shown in Fig. 10–11a. *EI* is constant. Neglect the effects of axial load.

SOLUTION

Principle of Superposition.   Since axial load is neglected, the beam is indeterminate to the second degree. The two end moments at *A* and *B* will be considered as the redundants. The beam's capacity to resist these moments is removed by placing a pin at *A* and a rocker at *B*. The principle of superposition applied to the beam is shown in Fig. 10–11*b*.

Compatibility Equations.   Reference to points *A* and *B*, Fig. 10–11*b*, requires

$(\circlearrowleft+)$ $\qquad\qquad$ $0 = \theta_A + M_A\alpha_{AA} + M_B\alpha_{AB}$ $\qquad\qquad$ (1)

$(\circlearrowleft+)$ $\qquad\qquad$ $0 = \theta_B + M_A\alpha_{BA} + M_B\alpha_{BB}$ $\qquad\qquad$ (2)

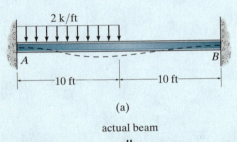

(a)

actual beam

||

primary structure

+

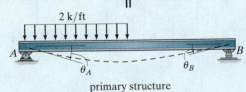

$\theta'_{AA} = M_A\alpha_{AA}$ $\qquad$ $\theta'_{BA} = M_A\alpha_{BA}$

redundant moment **M**$_A$ applied

+

$\theta'_{AB} = M_B\alpha_{AB}$ $\qquad$ $\theta'_{BB} = M_B\alpha_{BB}$

redundant moment **M**$_B$ applied

(b)

Fig. 10–11

The required slopes and angular flexibility coefficients can be determined using the table on the inside front cover. We have

$$\theta_A = \frac{3wL^3}{128EI} = \frac{3(2)(20)^3}{128EI} = \frac{375}{EI}$$

$$\theta_B = \frac{7wL^3}{384EI} = \frac{7(2)(20)^3}{384EI} = \frac{291.7}{EI}$$

$$\alpha_{AA} = \frac{ML}{3EI} = \frac{1(20)}{3EI} = \frac{6.67}{EI}$$

$$\alpha_{BB} = \frac{ML}{3EI} = \frac{1(20)}{3EI} = \frac{6.67}{EI}$$

$$\alpha_{AB} = \frac{ML}{6EI} = \frac{1(20)}{6EI} = \frac{3.33}{EI}$$

Note that $\alpha_{BA} = \alpha_{AB}$, a consequence of Maxwell's theorem of reciprocal displacements.

Substituting the data into Eqs. (1) and (2) yields

$$0 = \frac{375}{EI} + M_A\left(\frac{6.67}{EI}\right) + M_B\left(\frac{3.33}{EI}\right)$$

$$0 = \frac{291.7}{EI} + M_A\left(\frac{3.33}{EI}\right) + M_B\left(\frac{6.67}{EI}\right)$$

Canceling EI and solving these equations simultaneously, we have

$$M_A = -45.8 \text{ k} \cdot \text{ft} \qquad M_B = -20.8 \text{ k} \cdot \text{ft}$$

Using these results, the end shears are calculated, Fig. 10–11c, and the shear and moment diagrams plotted.

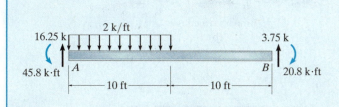

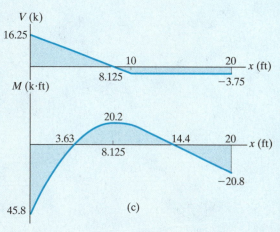

(c)

10

EXAMPLE | **10.4**

Determine the reactions at the supports for the beam shown in Fig. 10–12a. *EI* is constant.

SOLUTION

Principle of Superposition. By inspection, the beam is indeterminate to the first degree. Here, for the sake of illustration, we will choose the internal moment at support *B* as the redundant. Consequently, the beam is cut open and end pins or an internal hinge are placed at *B* in order to release *only* the capacity of the beam to resist moment at this point, Fig. 10–12b. The internal moment at *B* is applied to the beam in Fig. 10–12c.

Compatibility Equations. From Fig. 10–12a we require the relative rotation of one end of one beam with respect to the end of the other beam to be zero, that is,

$$(\zeta+) \qquad\qquad \theta_B + M_B \alpha_{BB} = 0$$

where

$$\theta_B = \theta_B' + \theta_B''$$

and

$$\alpha_{BB} = \alpha_{BB}' + \alpha_{BB}''$$

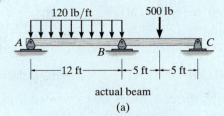

120 lb/ft   500 lb

A   *B*   *C*

|——12 ft——|—5 ft—|—5 ft—|

actual beam

(a)

||

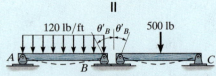

120 lb/ft θ_B' θ_B' 500 lb

A   *B*   *C*

primary structure

(b)

+

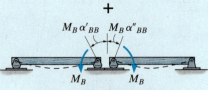

$M_B\,\alpha_{BB}'$ $M_B\,\alpha_{BB}''$

M_B M_B

redundant $\mathbf{M}_B$ applied

(c)

Fig. 10–12

10

The slopes and angular flexibility coefficients can be determined from the table on the inside front cover, that is,

$$\theta_B' = \frac{wL^3}{24EI} = \frac{120(12)^3}{24EI} = \frac{8640 \text{ lb} \cdot \text{ft}^2}{EI}$$

$$\theta_B'' = \frac{PL^2}{16EI} = \frac{500(10)^2}{16EI} = \frac{3125 \text{ lb} \cdot \text{ft}^2}{EI}$$

$$\alpha_{BB}' = \frac{ML}{3EI} = \frac{1(12)}{3EI} = \frac{4 \text{ ft}}{EI}$$

$$\alpha_{BB}'' = \frac{ML}{3EI} = \frac{1(10)}{3EI} = \frac{3.33 \text{ ft}}{EI}$$

Thus

$$\frac{8640 \text{ lb} \cdot \text{ft}^2}{EI} + \frac{3125 \text{ lb} \cdot \text{ft}^2}{EI} + M_B\left(\frac{4 \text{ ft}}{EI} + \frac{3.33 \text{ ft}}{EI}\right) = 0$$

$$M_B = -1604 \text{ lb} \cdot \text{ft}$$

The negative sign indicates M_B acts in the opposite direction to that shown in Fig. 10–12c. Using this result, the reactions at the supports are calculated as shown in Fig. 10–12d. Furthermore, the shear and moment diagrams are shown in Fig. 10–12e.

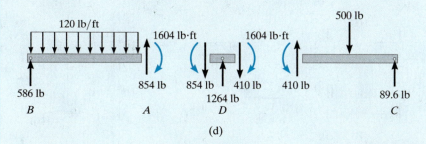

(d)

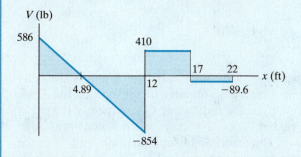
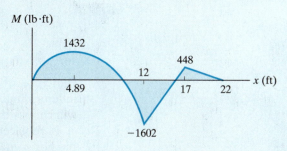

(e)

10

10.5   Force Method of Analysis: Frames

The force method is very useful for solving problems involving statically indeterminate frames that have a single story and unusual geometry, such as gabled frames. Problems involving multistory frames, or those with a high degree of indeterminacy, are best solved using the slope-deflection, moment-distribution, or the stiffness method discussed in later chapters.

The following examples illustrate the application of the force method using the procedure for analysis outlined in Sec. 10.2.

EXAMPLE | 10.5

The saddle bent shown in the photo is used to support the bridge deck. Assuming EI is constant, a drawing of it along with the dimensions and loading is shown in Fig. 10–13a. Determine the horizontal support reaction at A.

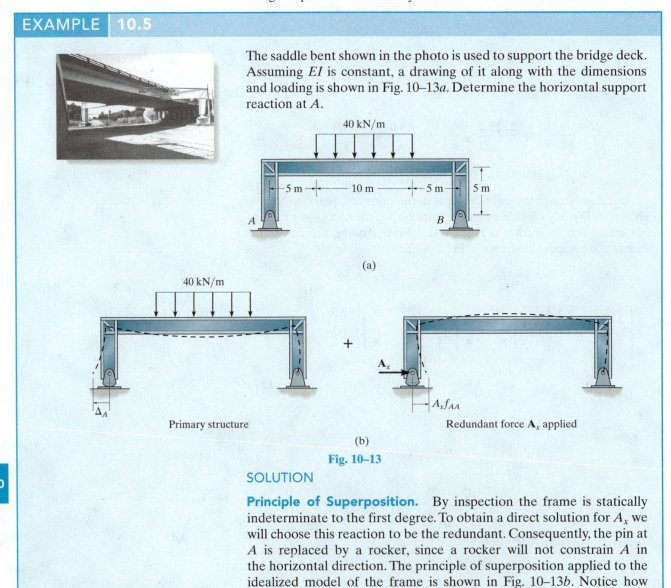

Fig. 10–13

SOLUTION

Principle of Superposition. By inspection the frame is statically indeterminate to the first degree. To obtain a direct solution for A_x we will choose this reaction to be the redundant. Consequently, the pin at A is replaced by a rocker, since a rocker will not constrain A in the horizontal direction. The principle of superposition applied to the idealized model of the frame is shown in Fig. 10–13b. Notice how the frame deflects in each case.

Compatibility Equation. Reference to point A in Fig. 10–13b requires

$(\xrightarrow{+})$ $\qquad\qquad 0 = \Delta_A + A_x f_{AA}$ $\qquad\qquad$ (1)

The terms Δ_A and f_{AA} will be determined using the method of virtual work. Because of symmetry of geometry *and* loading we need only three x coordinates. These and the internal moments are shown in Figs. 10–13c and 10–13d. It is important that each x coordinate be the *same* for both the real and virtual loadings. Also, the positive directions for **M** and **m** must be the *same*.

For Δ_A we require application of real loads, Fig. 10–13c, and a virtual unit load at A, Fig. 10–13d. Thus,

$$\Delta_A = \int_0^L \frac{Mm}{EI}dx = 2\int_0^5 \frac{(0)(1x_1)dx_1}{EI} + 2\int_0^5 \frac{(200x_2)(-5)dx_2}{EI}$$

$$+ 2\int_0^5 \frac{(1000 + 200x_3 - 20x_3^2)(-5)dx_3}{EI}$$

$$= 0 - \frac{25\,000}{EI} - \frac{66\,666.7}{EI} = -\frac{91\,666.7}{EI}$$

For f_{AA} we require application of a real unit load and a virtual unit load acting at A, Fig. 10–13d. Thus,

$$f_{AA} = \int_0^L \frac{mm}{EI}dx = 2\int_0^5 \frac{(1x_1)^2dx_1}{EI} + 2\int_0^5 (5)^2dx_2 + 2\int_0^5 (5)^2d$$

$$= \frac{583.33}{EI}$$

Substituting the results into Eq. (1) and solving yields

$$0 = \frac{-91\,666.7}{EI} + A_x\left(\frac{583.33}{EI}\right)$$

$$A_x = 157 \text{ kN} \qquad\qquad\qquad\qquad Ans.$$

Draw the free-body diagram for the bent and show that $A_y = B_y = 200$ kN, and $B_x = A_x = 157$ kN.

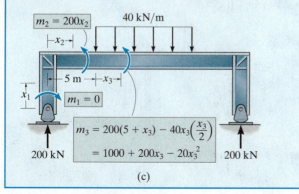

(c)

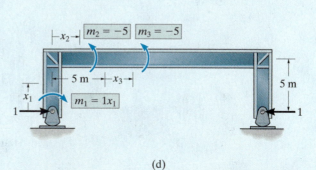

(d)

10

EXAMPLE 10.6

Determine the moment at the fixed support A for the frame shown in Fig. 10–14a. The support at B is a rocker. EI is constant.

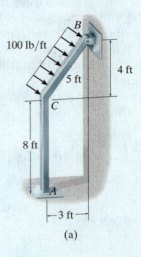

100 lb/ft

B

5 ft

4 ft

C

8 ft

A

3 ft

(a)

Fig. 10–14

SOLUTION

Principle of Superposition.   The frame is indeterminate to the first degree. A direct solution for $\mathbf{M}_A$ can be obtained by choosing this as the redundant. Thus the capacity of the frame to support a moment at A is removed and therefore a pin is used at A for support. The principle of superposition applied to the frame is shown in Fig. 10–14b.

Compatibility Equation.   Reference to point A in Fig. 10–14b requires

$(\zeta+)$ $0 = \theta_A + M_A \alpha_{AA}$ (1)

As in the preceding example, θ_A and α_{AA} will be computed using the method of virtual work. The frame's x coordinates and internal moments are shown in Figs. 10–14c and 10–14d.

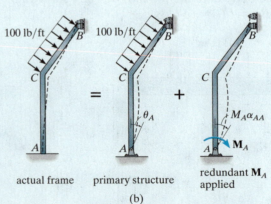

100 lb/ft B 100 lb/ft B B

C C C

$=$ $+$

θ_A $M_A\alpha_{AA}$

A A A $\mathbf{M}_A$

actual frame   primary structure   redundant $\mathbf{M}_A$
                                              applied

(b)

For θ_A we require application of the real loads, Fig. 10–14c, and a virtual unit couple moment at A, Fig. 10–14d. Thus,

$$\theta_A = \sum \int_0^L \frac{Mm_\theta \, dx}{EI}$$

$$= \int_0^8 \frac{(29.17x_1)(1 - 0.0833x_1) \, dx}{EI}$$

$$+ \int_0^5 \frac{\left(296.7x_2 - 50x_2^2\right)(0.0667x_2)dx_2}{EI}$$

$$= \frac{518.5}{EI} + \frac{303.2}{EI} = \frac{821.8}{EI}$$

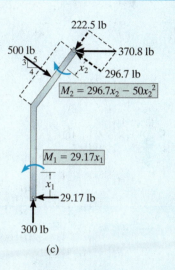

(c)

For α_{AA} we require application of a real unit couple moment and a virtual unit couple moment acting at A, Fig. 10–14d. Thus,

$$\alpha_{AA} = \sum \int_0^L \frac{m_\theta m_\theta}{EI} dx$$

$$= \int_0^8 \frac{(1 - 0.0833x_1)^2 \, dx_1}{EI} + \int_0^5 \frac{(0.0667x_2)^2 \, dx_2}{EI}$$

$$= \frac{3.85}{EI} + \frac{0.185}{EI} = \frac{4.04}{EI}$$

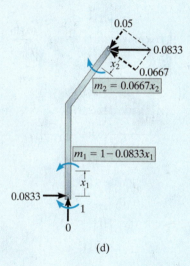

(d)

Substituting these results into Eq. (1) and solving yields

$$0 = \frac{821.8}{EI} + M_A\left(\frac{4.04}{EI}\right) \qquad M_A = -204 \text{ lb} \cdot \text{ft} \qquad \textit{Ans.}$$

10

The negative sign indicates $\mathbf{M}_A$ acts in the opposite direction to that shown in Fig. 10–14b.

10.6 Force Method of Analysis: Trusses

The degree of indeterminacy of a truss can usually be determined by inspection; however, if this becomes difficult, use Eq. 3–1, $b + r > 2j$. Here the unknowns are represented by the number of bar forces (b) plus the support reactions (r), and the number of available equilibrium equations is $2j$ since two equations can be written for each of the (j) joints.

The force method is quite suitable for analyzing trusses that are statically indeterminate to the first or second degree. The following examples illustrate application of this method using the procedure for analysis outlined in Sec. 10.2.

EXAMPLE | **10.7**

Determine the force in member AC of the truss shown in Fig. 10–15a. AE is the same for all the members.

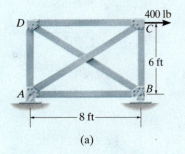

(a)

Fig. 10–15

SOLUTION

Principle of Superposition.   By inspection the truss is indeterminate to the first degree.* Since the force in member AC is to be determined, member AC will be chosen as the redundant. This requires "cutting" this member so that it cannot sustain a force, thereby making the truss statically determinate and stable. The principle of superposition applied to the truss is shown in Fig. 10–15b.

Compatibility Equation.   With reference to member AC in Fig. 10–15b, we require the relative displacement Δ_{AC}, which occurs at the ends of the cut member AC due to the 400-lb load, plus the relative displacement $F_{AC}f_{ACAC}$ caused by the redundant force acting alone, to be equal to zero, that is,

$$0 = \Delta_{AC} + F_{AC}f_{ACAC} \tag{1}$$

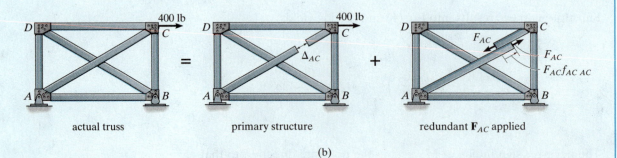

actual truss   primary structure   redundant $\mathbf{F}_{AC}$ applied

(b)

*Applying Eq. 3–1, $b + r > 2j$ or $6 + 3 > 2(4)$, $9 > 8$, $9 - 8 = $ 1st degree.

Here the flexibility coefficient f_{ACAC} represents the relative displacement of the cut ends of member AC caused by a "real" unit load acting at the cut ends of member AC. This term, f_{ACAC}, and Δ_{AC} will be computed using the method of virtual work. The force analysis, using the method of joints, is summarized in Figs. 10–15c and 10–15d.

For Δ_{AC} we require application of the real load of 400 lb, Fig. 10–15c, and a virtual unit force acting at the cut ends of member AC, Fig. 10–15d. Thus,

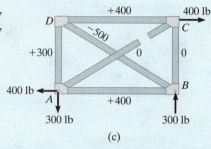

(c)

$$\Delta_{AC} = \sum \frac{nNL}{AE}$$

$$= 2\left[\frac{(-0.8)(400)(8)}{AE}\right] + \frac{(-0.6)(0)(6)}{AE} + \frac{(-0.6)(300)(6)}{AE}$$

$$+ \frac{(1)(-500)(10)}{AE} + \frac{(1)(0)(10)}{AE}$$

$$= -\frac{11\ 200}{AE}$$

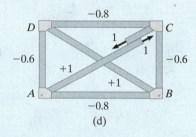

(d)

For f_{ACAC} we require application of real unit forces and virtual unit forces acting on the cut ends of member AC, Fig. 10–15d. Thus,

$$f_{ACAC} = \sum \frac{n^2 L}{AE}$$

$$= 2\left[\frac{(-0.8)^2(8)}{AE}\right] + 2\left[\frac{(-0.6)^2(6)}{AE}\right] + 2\left[\frac{(1)^2 10}{AE}\right]$$

$$= \frac{34.56}{AE}$$

Substituting the data into Eq. (1) and solving yields

$$0 = -\frac{11\ 200}{AE} + \frac{34.56}{AE} F_{AC}$$

$$F_{AC} = 324\ \text{lb (T)} \qquad\qquad Ans.$$

Since the numerical result is positive, AC is subjected to tension as assumed, Fig. 10–15b. Using this result, the forces in the other members can be found by equilibrium, using the method of joints.

10

EXAMPLE   10.8

Determine the force in each member of the truss shown in Fig. 10–16a if the turnbuckle on member AC is used to shorten the member by 0.5 in. Each bar has a cross-sectional area of 0.2 in^2, and $E = 29(10^6)$ psi.

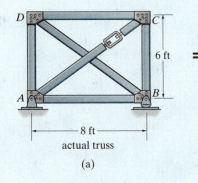

6 ft

8 ft

actual truss

(a)

=

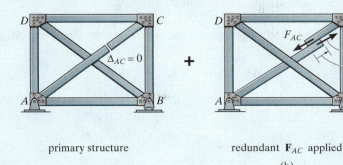

$\Delta_{AC} = 0$

primary structure

+

F_{AC}

$-F_{AC}$
$F_{AC}f_{ACAC}$

redundant $\mathbf{F}_{AC}$ applied

(b)

Fig. 10–16

SOLUTION

Principle of Superposition. This truss has the same geometry as that in Example 10.7. Since AC has been shortened, we will choose it as the redundant, Fig. 10–16b.

Compatibility Equation. Since no external loads act on the primary structure (truss), there will be no relative displacement between the ends of the sectioned member caused by load; that is, $\Delta_{AC} = 0$. The flexibility coefficient f_{ACAC} has been determined in Example 10.7, so

$$f_{ACAC} = \frac{34.56}{AE}$$

Assuming the amount by which the bar is shortened is positive, the compatibility equation for the bar is therefore

$$0.5 \text{ in.} = 0 + \frac{34.56}{AE}F_{AC}$$

Realizing that f_{ACAC} is a measure of displacement per unit force, we have

$$0.5 \text{ in.} = 0 + \frac{34.56 \text{ ft}(12 \text{ in./ft})}{(0.2 \text{ in}^2)\left[29(10^6) \text{ lb/in}^2\right]}F_{AC}$$

Thus,

$$F_{AC} = 6993 \text{ lb} = 6.99 \text{ k (T)} \qquad \textit{Ans.}$$

Since no external forces act on the truss, the external reactions are zero. Therefore, using F_{AC} and analyzing the truss by the method of joints yields the results shown in Fig. 10–16c.

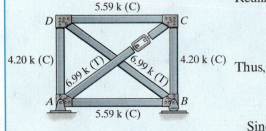

5.59 k (C)

4.20 k (C)

6.99 k (T)

6.99 k (T)

4.20 k (C)

5.59 k (C)

(c)

10

10.7 Composite Structures

Composite structures are composed of some members subjected only to axial force, while other members are subjected to bending. If the structure is statically indeterminate, the force method can conveniently be used for its analysis. The following example illustrates the procedure.

EXAMPLE 10.9

The simply supported queen-post trussed beam shown in the photo is to be designed to support a uniform load of 2 kN/m. The dimensions of the structure are shown in Fig. 10–17a. Determine the force developed in member *CE*. Neglect the thickness of the beam and assume the truss members are pin connected to the beam. Also, neglect the effect of axial compression and shear in the beam. The cross-sectional area of each strut is 400 mm², and for the beam $I = 20(10^6)$ mm⁴. Take $E = 200$ GPa.

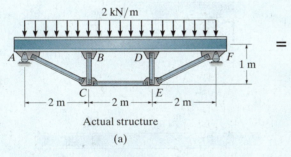

Actual structure
(a)

Fig. 10–17

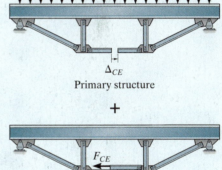

Primary structure

Redundant $\mathbf{F}_{CE}$ applied
(b)

SOLUTION

Principle of Superposition. If the force in one of the truss members is known, then the force in all the other members, as well as in the beam, can be determined by statics. Hence, the structure is indeterminate to the first degree. For solution the force in member *CE* is chosen as the redundant. This member is therefore sectioned to eliminate its capacity to sustain a force. The principle of superposition applied to the structure is shown in Fig. 10–17b.

Compatibility Equation. With reference to the relative displacement of the cut ends of member *CE*, Fig. 10–17b, we require

$$0 = \Delta_{CE} + F_{CE} f_{CE\ CE} \qquad (1)$$

The method of virtual work will be used to find Δ_{CE} and $f_{CE\,CE}$. The necessary force analysis is shown in Figs. 10–17c and 10–17d.

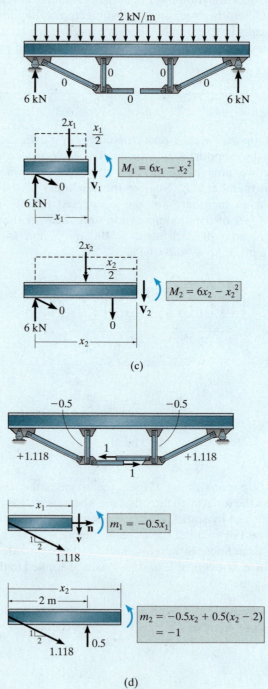

$$M_1 = 6x_1 - x_1^2$$

$$M_2 = 6x_2 - x_2^2$$

(c)

$$m_1 = -0.5x_1$$

$$m_2 = -0.5x_2 + 0.5(x_2 - 2)$$
$$= -1$$

(d)

For Δ_{CE} we require application of the real loads, Fig. 10–17c, and a virtual unit load applied to the cut ends of member CE, Fig. 10–17d. Here we will use symmetry of *both* loading and geometry, and only consider the bending strain energy in the beam and, of course, the axial strain energy in the truss members. Thus,

$$\Delta_{CE} = \int_0^L \frac{Mm}{EI}dx + \sum \frac{nNL}{AE} = 2\int_0^2 \frac{(6x_1 - x_1^2)(-0.5x_1)dx_1}{EI}$$

$$+ 2\int_2^3 \frac{(6x_2 - x_2^2)(-1)dx_2}{EI} + 2\left(\frac{(1.118)(0)(\sqrt{5})}{AE}\right)$$

$$+ 2\left(\frac{(-0.5)(0)(1)}{AE}\right) + \left(\frac{1(0)2}{AE}\right)$$

$$= -\frac{12}{EI} - \frac{17.33}{EI} + 0 + 0 + 0$$

$$= \frac{-29.33(10^3)}{200(10^9)(20)(10^{-6})} = -7.333(10^{-3})\,\text{m}$$

For $f_{CE\,CE}$ we require application of a real unit load and a virtual unit load at the cut ends of member CE, Fig. 10–17d. Thus,

$$f_{CE\,CE} = \int_0^L \frac{m^2dx}{EI} + \sum \frac{n^2L}{AE} = 2\int_0^2 \frac{(-0.5x_1)^2dx_1}{EI} + 2\int_2^3 \frac{(-1)^2dx_2}{EI}$$

$$+ 2\left(\frac{(1.118)^2(\sqrt{5})}{AE}\right) + 2\left(\frac{(-0.5)^2(1)}{AE}\right) + \left(\frac{(1)^2(2)}{AE}\right)$$

$$= \frac{1.3333}{EI} + \frac{2}{EI} + \frac{5.590}{AE} + \frac{0.5}{AE} + \frac{2}{AE}$$

$$= \frac{3.333(10^3)}{200(10^9)(20)(10^{-6})} + \frac{8.090(10^3)}{400(10^{-6})(200(10^9))}$$

$$= 0.9345(10^{-3})\,\text{m/kN}$$

Substituting the data into Eq. (1) yields

$$0 = -7.333(10^{-3})\,\text{m} + F_{CE}(0.9345(10^{-3})\,\text{m/kN})$$

$$F_{CE} = 7.85\,\text{kN} \qquad\qquad \textit{Ans.}$$

10

10.8   Additional Remarks on the Force Method of Analysis

Now that the basic ideas regarding the force method have been developed, we will proceed to generalize its application and discuss its usefulness.

When computing the flexibility coefficients, f_{ij} (or α_{ij}), for the structure, it will be noticed that they depend only on the material and geometrical properties of the members and *not* on the loading of the primary structure. Hence these values, once determined, can be used to compute the reactions for any loading.

For a structure having n redundant reactions, $\mathbf{R}_n$, we can write n compatibility equations, namely:

$$\Delta_1 + f_{11}R_1 + f_{12}R_2 + \cdots + f_{1n}R_n = 0$$
$$\Delta_2 + f_{21}R_1 + f_{22}R_2 + \cdots + f_{2n}R_n = 0$$
$$\vdots$$
$$\Delta_n + f_{n1}R_1 + f_{n2}R_2 + \cdots + f_{nn}R_n = 0$$

Here the displacements, $\Delta_1, \ldots, \Delta_n$, are caused by *both* the *real loads* on the primary structure and by *support settlement* or *dimensional changes* due to temperature differences or fabrication errors in the members. To simplify computation for structures having a large degree of indeterminacy, the above equations can be recast into a matrix form,

$$\begin{bmatrix} f_{11} & f_{12} & \cdots & f_{1n} \\ f_{21} & f_{22} & \cdots & f_{2n} \\ & & \vdots & \\ f_{n1} & f_{n2} & \cdots & f_{nn} \end{bmatrix} \begin{bmatrix} R_1 \\ R_2 \\ \vdots \\ R_n \end{bmatrix} = - \begin{bmatrix} \Delta_1 \\ \Delta_2 \\ \vdots \\ \Delta_n \end{bmatrix} \qquad (10\text{--}2)$$

or simply

$$\mathbf{fR} = -\mathbf{\Delta}$$

In particular, note that $f_{ij} = f_{ji}$ ($f_{12} = f_{21}$, etc.), a consequence of Maxwell's theorem of reciprocal displacements (or Betti's law). Hence the *flexibility matrix* will be *symmetric,* and this feature is beneficial when solving large sets of linear equations, as in the case of a highly indeterminate structure.

Throughout this chapter we have determined the flexibility coefficients using the method of virtual work as it applies to the *entire structure*. It is possible, however, to obtain these coefficients for *each member* of the structure, and then, using transformation equations, to obtain their values for the entire structure. This approach is covered in books devoted to matrix analysis of structures, and will not be covered in this text.*

---

*See, for example, H. C. Martin, *Introduction to Matrix Methods of Structural Analysis*, McGraw-Hill, New York.

Although the details for applying the force method of analysis using computer methods will also be omitted here, we can make some general observations and comments that apply when using this method to solve problems that are highly indeterminate and thus involve large sets of equations. In this regard, numerical accuracy for the solution is improved if the flexibility coefficients located near the main diagonal of the **f** matrix are larger than those located off the diagonal. To achieve this, some thought should be given to selection of the primary structure. To facilitate computations of $\mathbf{f}_{ij}$, it is also desirable to choose the primary structure so that it is somewhat symmetric. This will tend to yield some flexibility coefficients that are similar or may be zero. Lastly, the deflected shape of the primary structure should be *similar* to that of the actual structure. If this occurs, then the redundants will induce only *small* corrections to the primary structure, which results in a more accurate solution of Eq. 10–2.

10.9 Symmetric Structures

A structural analysis of any highly indeterminate structure, or for that matter, even a statically determinate structure, can be simplified provided the designer or analyst can recognize those structures that are symmetric and support either symmetric or antisymmetric loadings. In a general sense, a structure can be classified as being *symmetric* provided half of it develops the same internal loadings and deflections as its mirror image reflected about its central axis. Normally symmetry requires the material composition, geometry, supports, and loading to be the same on each side of the structure. However, this does not always have to be the case. Notice that for horizontal stability a pin is required to support the beam and truss in Figs. 10–18a and 10–18b. Here the horizontal reaction at the pin is zero, and so both of these structures will deflect and produce the same internal loading as their reflected counterpart. As a result, they can be classified as being symmetric. Realize that this would not be the case for the frame, Fig. 10–18c, if the fixed support at A was replaced by a pin, since then the deflected shape and internal loadings would not be the same on its left and right sides.

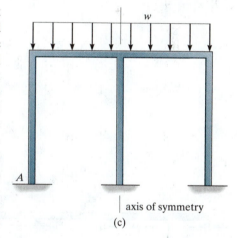

axis of symmetry

(c)

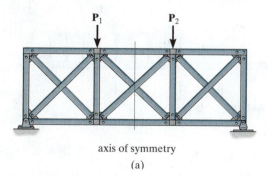

axis of symmetry

(a)

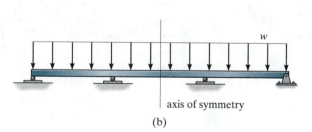

axis of symmetry

(b)

Fig. 10–18

Sometimes a symmetric structure supports an antisymmetric loading, that is, the loading on its reflected side has the opposite direction, such as shown by the two examples in Fig. 10–19. Provided the structure is symmetric and its loading is either symmetric or antisymmetric, then a structural analysis will only have to be performed on half the members of the structure since the same (symmetric) or opposite (antisymmetric) results will be produced on the other half. If a structure is symmetric and its applied loading is unsymmetrical, then it is possible to transform this loading into symmetric and antisymmetric components. To do this, *the loading is first divided in half, then it is reflected to the other side of the structure and both symmetric and antisymmetric components are produced.* For example, the loading on the beam in Fig. 10–20a is divided by two and reflected about the beam's axis of symmetry. From this, the symmetric and antisymmetric components of the load are produced as shown in Fig. 10–20b. When added together these components produce the original loading. A separate structural analysis can now be performed using the symmetric and antisymmetric loading components and the results superimposed to obtain the actual behavior of the structure.

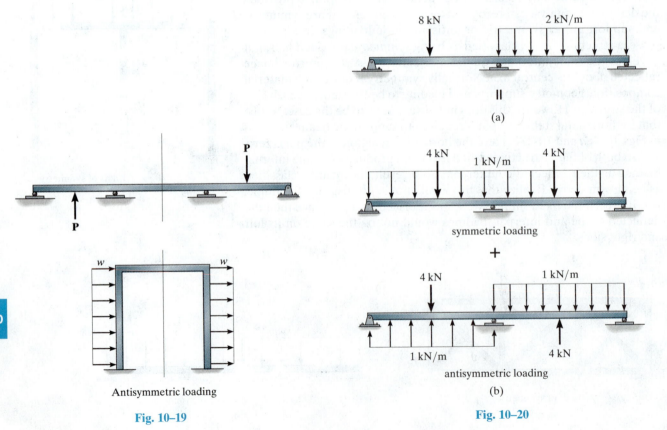

Antisymmetric loading

Fig. 10–19

Fig. 10–20

10.10   Influence Lines for Statically Indeterminate Beams

In Sec. 6.3 we discussed the use of the Müller-Breslau principle for drawing the influence line for the reaction, shear, and moment at a point in a statically determinate beam. In this section we will extend this method and apply it to statically indeterminate beams.

Recall that, for a beam, *the **Müller-Breslau principle** states that the influence line for a function (reaction, shear, or moment) is to the same scale as the deflected shape of the beam when the beam is acted upon by the function.* To draw the deflected shape properly, the capacity of the beam to resist the applied function must be *removed* so the beam can deflect when the function is applied. For *statically determinate beams,* the deflected shapes (or the influence lines) will be a series of *straight line segments.* For *statically indeterminate beams, curves* will result. Construction of each of the three types of influence lines (reaction, shear, and moment) will now be discussed for a statically indeterminate beam. In each case we will illustrate the validity of the Müller-Breslau principle using Maxwell's theorem of reciprocal displacements.

Reaction at A.

To determine the influence line for the reaction at A in Fig. 10–21a, a unit load is placed on the beam at successive points, and at each point the reaction at A must be determined. A plot of these results yields the influence line. For example, when the load is at point D, Fig. 10–21a, the reaction at A, which represents the ordinate of the influence line at D, can be determined by the force method. To do this, the principle of superposition is applied, as shown in Figs. 10–21a through 10–21c. The compatibility equation for point A is thus $0 = f_{AD} + A_y f_{AA}$ or $A_y = -f_{AD}/f_{AA}$; however, by Maxwell's theorem of reciprocal displacements $f_{AD} = -f_{DA}$, Fig. 10–21d, so that we can also compute A_y (or the ordinate of the influence line at D) using the equation

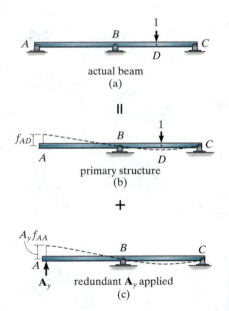

actual beam
(a)

$\parallel$

primary structure
(b)

$+$

redundant $\mathbf{A}_y$ applied
(c)

$$A_y = \left(\frac{1}{f_{AA}}\right) f_{DA}$$

By comparison, the Müller-Breslau principle requires removal of the support at A and application of a vertical unit load. The resulting deflection curve, Fig. 10–21d, is to some scale the shape of the influence line for A_y. From the equation above, however, it is seen that the scale factor is $1/f_{AA}$.

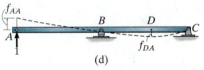

(d)

Fig. 10–21

Shear at E.

If the influence line for the shear at point E of the beam in Fig. 10–22a is to be determined, then by the Müller-Breslau principle the beam is imagined cut open at this point and a *sliding device* is inserted at E, Fig. 10–22b. This device will transmit a moment and normal force but no shear. When the beam deflects due to positive unit shear loads acting at E, the slope on each side of the guide remains the same, and the deflection curve represents to some scale the influence line for the shear at E, Fig. 10–22c. Had the basic method for establishing the influence line for the shear at E been applied, it would then be necessary to apply a unit load at each point D and compute the internal shear at E, Fig. 10–22a. This value, V_E, would represent the ordinate of the influence line at D. Using the force method and Maxwell's theorem of reciprocal displacements, as in the previous case, it can be shown that

$$V_E = \left(\frac{1}{f_{EE}} \right) f_{DE}$$

This again establishes the validity of the Müller-Breslau principle, namely, a positive unit shear load applied to the beam at E, Fig. 10–22c, will cause the beam to deflect into the *shape* of the influence line for the shear at E. Here the scale factor is $(1/f_{EE})$.

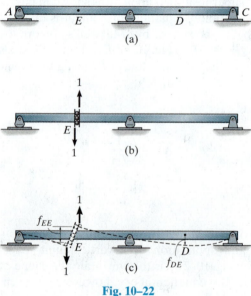

(a)

(b)

(c)

Fig. 10–22

Moment at E. The influence line for the moment at E in Fig. 10–23a can be determined by placing a *pin* or *hinge* at E, since this connection transmits normal and shear forces but cannot resist a moment, Fig. 10–23b. Applying a positive unit couple moment, the beam then deflects to the dashed position in Fig. 10–23c, which yields to some scale the influence line, again a consequence of the Müller-Breslau principle. Using the force method and Maxwell's reciprocal theorem, we can show that

$$M_E = \left(\frac{1}{\alpha_{EE}}\right) f_{DE}$$

The scale factor here is $(1/\alpha_{EE})$.

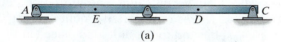

(a)

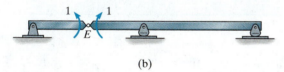

(b)

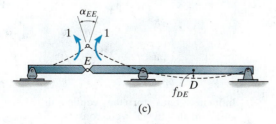

(c)

Fig. 10–23

10

Procedure for Analysis

The following procedure provides a method for establishing the influence line for the reaction, shear, or moment at a point on a beam using the Müller-Breslau technique.

Qualitative Influence Line

At the point on the beam for which the influence line is to be determined, place a connection that will remove the capacity of the beam to support the function of the influence line. If the function is a vertical *reaction*, use a vertical *roller guide*; if the function is *shear*, use a *sliding device*; or if the function is *moment*, use a *pin* or *hinge*. Place a unit load at the connection acting on the beam in the "positive direction" of the function. Draw the deflection curve for the beam. This curve represents to some scale the shape of the influence line for the beam.

Quantitative Influence Line

If numerical values of the influence line are to be determined, compute the *displacement* of successive points along the beam when the beam is subjected to the unit load placed at the connection mentioned above. Divide each value of displacement by the displacement determined at the point where the unit load acts. By applying this scalar factor, the resulting values are the ordinates of the influence line.

Influence lines for the continuous girder of this trestle were constructed in order to properly design the girder.

10.11   Qualitative Influence Lines for Frames

The Müller-Breslau principle provides a quick method and is of great value for establishing the general shape of the influence line for building frames. Once the influence-line *shape* is known, one can immediately specify the *location* of the live loads so as to create the greatest influence of the function (reaction, shear, or moment) in the frame. For example, the shape of the influence line for the *positive* moment at the center I of girder FG of the frame in Fig. 10–24a is shown by the dashed lines. Thus, uniform loads would be placed only on girders AB, CD, and FG in order to create the largest positive moment at I. With the frame loaded in this manner, Fig. 10–24b, an indeterminate analysis of the frame could then be performed to determine the critical moment at I.

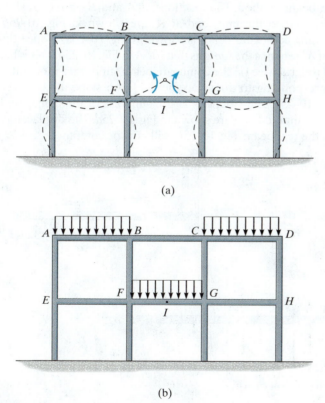

(a)

(b)

Fig. 10–24

EXAMPLE 10.10

Draw the influence line for the vertical reaction at A for the beam in Fig. 10–25a. EI is constant. Plot numerical values every 6 ft.

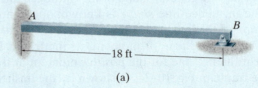

(a)

SOLUTION

The capacity of the beam to resist the reaction $\mathbf{A}_y$ is removed. This is done using a vertical roller device shown in Fig. 10–25b. Applying a vertical unit load at A yields the shape of the influence line shown in Fig. 10–25c.

In order to determine ordinates of the influence line we will use the conjugate-beam method. The reactions at A and B on the "real beam," when subjected to the unit load at A, are shown in Fig. 10–25b. The corresponding conjugate beam is shown in Fig. 10–25d. Notice that the support at A' remains the *same* as that for A in Fig. 10–25b. This is because a vertical roller device on the conjugate beam supports a moment but no shear, corresponding to a displacement but no slope at A on the real beam, Fig. 10–25c. The reactions at the supports of the conjugate beam have been computed and are shown in Fig. 10–25d. The displacements of points on the real beam, Fig. 10–25b, will now be computed.

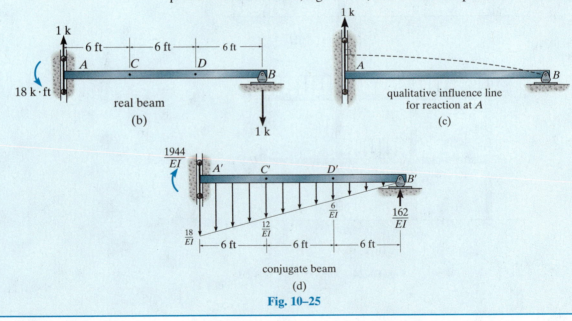

real beam
(b)

qualitative influence line
for reaction at A
(c)

conjugate beam
(d)

Fig. 10–25

For B', since no moment exists on the conjugate beam at B', Fig. 10–25d, then

$$\Delta_B = M_{B'} = 0$$

For D', Fig. 10–25e:

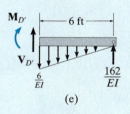

(e)

$$\Sigma M_{D'} = 0; \qquad \Delta_D = M_{D'} = \frac{162}{EI}(6) - \frac{1}{2}\left(\frac{6}{EI}\right)(6)(2) = \frac{936}{EI}$$

For C', Fig. 10–25f:

$$\Sigma M_{C'} = 0; \qquad \Delta_C = M_{C'} = \frac{162}{EI}(12) - \frac{1}{2}\left(\frac{12}{EI}\right)(12)(4) = \frac{1656}{EI}$$

For A', Fig. 10–25d:

$$\Delta_A = M_{A'} = \frac{1944}{EI}$$

Since a vertical 1-k load acting at A on the beam in Fig. 10–25a will cause a vertical reaction at A of 1 k, the displacement at A, $\Delta_A = 1944/EI$, should correspond to a numerical value of 1 for the influence-line ordinate at A. Thus, dividing the other computed displacements by this factor, we obtain

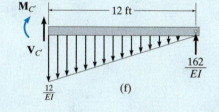

(f)

x	A_y
A	1
C	0.852
D	0.481
B	0

A plot of these values yields the influence line shown in Fig. 10–25g.

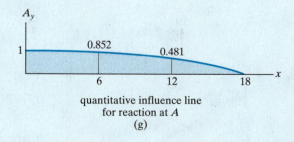

quantitative influence line
for reaction at A
(g)

EXAMPLE │ 10.11

Draw the influence line for the shear at D for the beam in Fig. 10–26a. EI is constant. Plot numerical values every 9 ft.

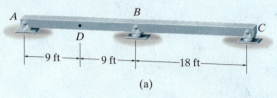

(a)

Fig. 10–26

SOLUTION

The capacity of the beam to resist shear at D is removed. This is done using the roller device shown in Fig. 10–26b. Applying a positive unit shear at D yields the shape of the influence line shown in Fig. 10–26c.

The support reactions at A, B, and C on the "real beam" when subjected to the unit shear at D are shown in Fig. 10–26b. The corresponding conjugate beam is shown in Fig. 10–26d. Here an external couple moment $\mathbf{M}_{D'}$ must be applied at D' in order to cause a different *internal moment* just to the left and just to the right of D'. These internal moments correspond to the displacements just to the left and just to the right of D on the real beam, Fig. 10–26c. The reactions at the supports A', B', C' and the external moment $\mathbf{M}_{D'}$ on the conjugate beam have been computed and are shown in Fig. 10–26e. As an exercise verify the calculations.

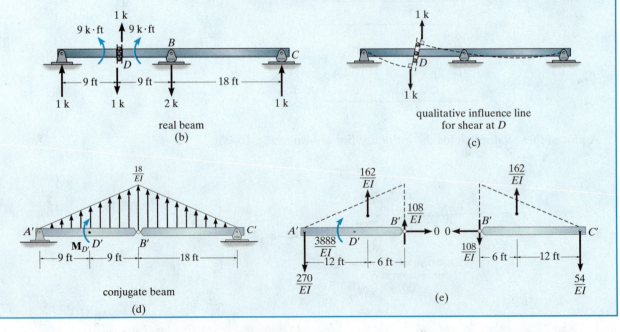

Since there is a *discontinuity* of moment at D', the internal moment just to the left and right of D' will be computed. Just to the left of D', Fig. 10–26f, we have

$$\Sigma M_{D'_L} = 0; \qquad \Delta_{D_L} = M_{D'_L} = \frac{40.5}{EI}(3) - \frac{270}{EI}(9) = -\frac{2308.5}{EI}$$

Just to the right of D', Fig. 10–26g, we have

$$\Sigma M_{D'_R} = 0; \qquad \Delta_{D_R} = M_{D'_R} = \frac{40.5}{EI}(3) - \frac{270}{EI}(9) + \frac{3888}{EI} = \frac{1579.5}{EI}$$

From Fig. 10–26e,

$$\Delta_A = M_{A'} = 0 \qquad \Delta_B = M_{B'} = 0 \qquad \Delta_C = M_{C'} = 0$$

For point E, Fig. 10–26b, using the method of sections at the corresponding point E' on the conjugate beam, Fig. 10–26h, we have

$$\Sigma M_{E'} = 0; \qquad \Delta_E = M_{E'} = \frac{40.5}{EI}(3) - \frac{54}{EI}(9) = -\frac{364.5}{EI}$$

The ordinates of the influence line are obtained by dividing each of the above values by the scale factor $M_{D'} = 3888/EI$. We have

x	V_D
A	0
D_L	−0.594
D_R	0.406
B	0
E	−0.0938
C	0

A plot of these values yields the influence line shown in Fig. 10–26i.

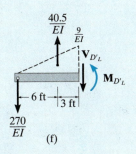

(f)

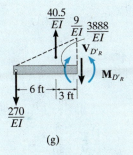

(g)

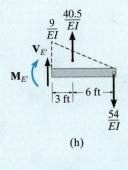

(h)

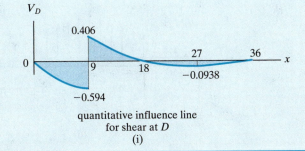

quantitative influence line
for shear at D
(i)

10

EXAMPLE   10.12

Draw the influence line for the moment at D for the beam in Fig. 10–27a. EI is constant. Plot numerical values every 9 ft.

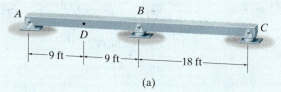

(a)

Fig. 10–27

SOLUTION

A hinge is inserted at D in order to remove the capacity of the beam to resist moment at this point, Fig. 10–27b. Applying positive unit couple moments at D yields the influence line shown in Fig. 10–27c. The reactions at A, B, and C on the "real beam" when subjected to the unit couple moments at D are shown in Fig. 10–27b. The corresponding conjugate beam and its reactions are shown in Fig. 10–27d. It is suggested that the reactions be verified in both cases. From Fig. 10–27d, note that

$$\Delta_A = M_{A'} = 0 \qquad \Delta_B = M_{B'} = 0 \qquad \Delta_C = M_{C'} = 0$$

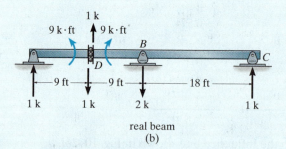

real beam
(b)

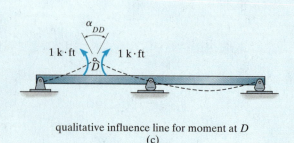

qualitative influence line for moment at D
(c)

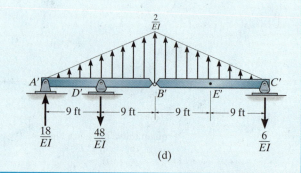

(d)

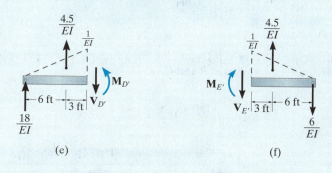

(e)   (f)

For point D', Fig. 10–27e:

$$\Sigma M_{D'} = 0; \qquad \Delta_D = M_{D'} = \frac{4.5}{EI}(3) + \frac{18}{EI}(9) = \frac{175.5}{EI}$$

For point E', Fig. 10–27f:

$$\Sigma M_{E'} = 0; \qquad \Delta_E = M_{E'} = \frac{4.5}{EI}(3) - \frac{6}{EI}(9) = -\frac{40.5}{EI}$$

The angular displacement α_{DD} at D of the "real beam" in Fig. 10–27c is defined by the reaction at D' on the conjugate beam. This factor, $D'_y = 48/EI$, is divided into the above values to give the ordinates of the influence line, that is,

x	M_D
A	0
D	3.656
B	0
E	−0.844
C	0

A plot of these values yields the influence line shown in Fig. 10–27g.

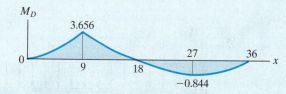

quantitative influence line
for moment at D
(g)

FUNDAMENTAL PROBLEMS

F10–1. Determine the reactions at the fixed support at A and the roller at B. EI is constant.

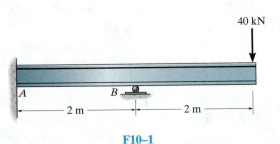

40 kN

A B

2 m   2 m

F10–1

F10–2. Determine the reactions at the fixed supports at A and the roller at B. EI is constant.

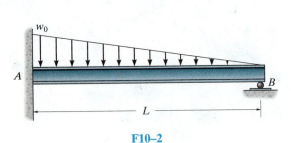

w_0

A B

L

F10–2

F10–3. Determine the reactions at the fixed support at A and the roller at B. Support B settles 5 mm. Take $E = 200$ GPa and $I = 300(10^6)$ mm^4.

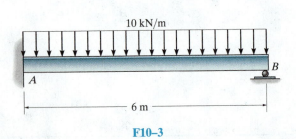

10 kN/m

A B

6 m

F10–3

F10–4. Determine the reactions at the pin at A and the rollers at B and C.

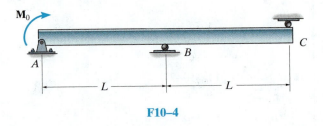

M_0

A B C

L L

F10–4

F10–5. Determine the reactions at the pin A and the rollers at B and C on the beam. EI is constant.

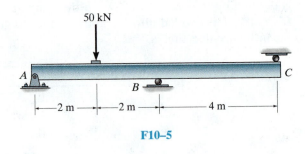

50 kN

A B C

2 m   2 m   4 m

F10–5

F10–6. Determine the reactions at the pin at A and the rollers at B and C on the beam. Support B settles 5 mm. Take $E = 200$ GPa, $I = 300(10^6)$ mm^4.

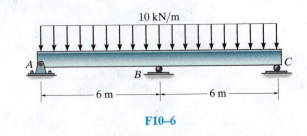

10 kN/m

A B C

6 m   6 m

F10–6

PROBLEMS

Sec. 10.1–10.4

10–1. Determine the reactions at the supports then draw the moment diagram. Assume the support at B is a roller.

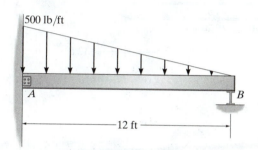

500 lb/ft

A B

—— 12 ft ——

Prob. 10–1

10–2. Determine the reactions at the supports $A, B,$ and $C,$ then draw the shear and moment diagrams. EI is constant.

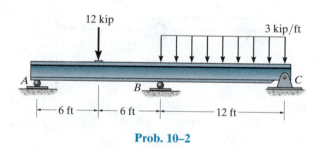

12 kip   3 kip/ft

A B C

— 6 ft — 6 ft — 12 ft —

Prob. 10–2

10–3. Determine the reactions at the supports, then draw the moment diagram. Assume the support at B is a roller. EI is constant.

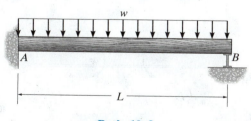

w

A B

—— L ——

Prob. 10–3

*10–4. Determine the reactions at the supports $A, B,$ and C; then draw the shear and moment diagram. EI is constant.

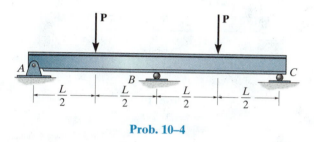

P   P

A B C

$\frac{L}{2}$ $\frac{L}{2}$ $\frac{L}{2}$ $\frac{L}{2}$

Prob. 10–4

10–5. Determine the reactions at the supports, then draw the moment diagram. Assume the support at B is a roller. EI is constant.

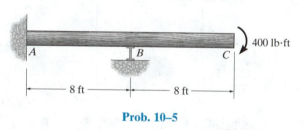

A B C 400 lb·ft

—— 8 ft —— 8 ft ——

Prob. 10–5

10–6. Determine the reactions at the supports, then draw the moment diagram. Assume B and C are rollers and A is pinned. The support at B settles downward 0.25 ft. Take $E = 29(10^3)$ ksi, $I = 500$ in⁴.

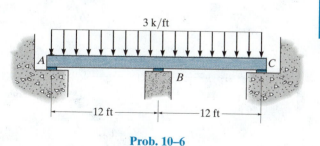

3 k/ft

A B C

—— 12 ft —— 12 ft ——

Prob. 10–6

10

10–7. Determine the value of a so that the maximum positive moment has the same magnitude as the maximum negative moment. EI is constant.

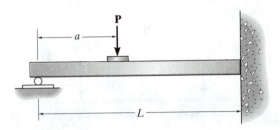

Prob. 10–7

10–10. Determine the reactions at the supports. The moment of inertia for each segment is shown in the figure. Assume the support at B is a roller. Take $E = 29(10^3)$ ksi.

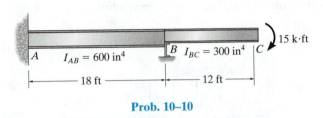

A $I_{AB} = 600$ in^4 B $I_{BC} = 300$ in^4 C 15 k·ft

18 ft — 12 ft

Prob. 10–10

*10–8.** Draw the moment diagram. Assume A and C are rollers and B is pinned. EI is constant.

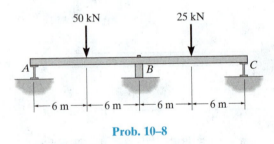

50 kN   25 kN

A B C

6 m — 6 m — 6 m — 6 m

Prob. 10–8

10–11. If the pin support B settles 0.5 in., determine the maximum moment developed in the beam. Take $E = 29(10^3)$ ksi, $I = 300$ in^4.

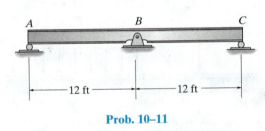

A B C

12 ft — 12 ft

Prob. 10–11

10–9. Draw the moment diagram for the beam. Assume the support at A is fixed and B and C are rollers. EI is constant. Use the three-moment equation.

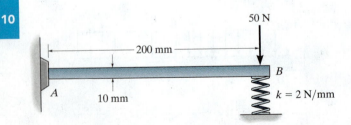

50 N

200 mm

B

A 10 mm $k = 2$ N/mm

Prob. 10–9

*10–12.** Determine the moment reactions at the supports A and B, then draw the moment diagram. EI is constant.

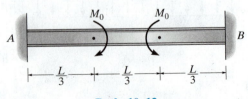

M_0 M_0

A B

$\frac{L}{3}$ $\frac{L}{3}$ $\frac{L}{3}$

Prob. 10–12

Sec. 10.5

10–13. Determine the reactions at the supports. Assume A and C are pins and the joint at B is fixed connected. EI is constant.

10–15. Determine the reactions at the supports, then draw the moment diagram for each member. EI is constant.

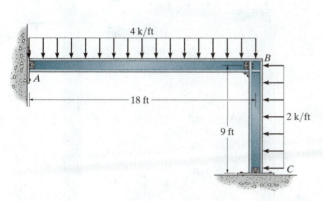

Prob. 10–13

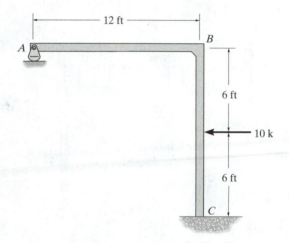

Prob. 10–15

10–14. Determine the reactions at the supports, then draw the moment diagrams for each member. EI is constant.

*10–16. Determine the reactions at the supports. Assume A is fixed connected. E is constant.

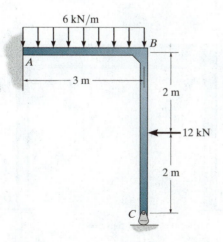

Prob. 10–14

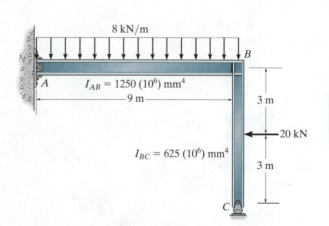

Prob. 10–16

10

10–17. Determine the reactions at the supports, then draw the moment diagram for each member. *EI* is constant.

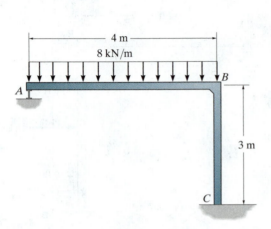

Prob. 10–17

10–18. Determine the reactions at the supports. Assume *A* is a fixed and the joint at *B* is fixed connected. *EI* is constant.

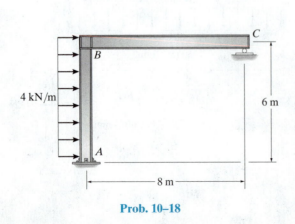

Prob. 10–18

10–19. Determine the reactions at the supports. *E* is constant.

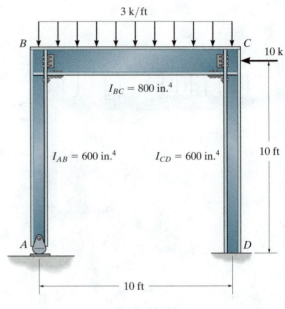

Prob. 10–19

*10–20. Determine the reactions at the supports, then draw the moment diagram for each member. *EI* is constant.

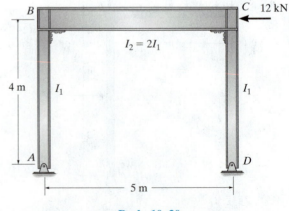

Prob. 10–20

10–21. Determine the reactions at the supports. Assume A and D are pins. EI is constant.

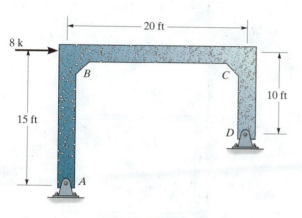

Prob. 10–21

10–23. Determine the reactions at the supports. Assume A and B are pins. EI is constant.

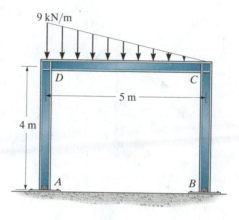

Prob. 10–23

10–22. Determine the reactions at the supports, then draw the moment diagrams for each member. Assume A and B are pins and the joint at C is fixed connected. EI is constant.

*10–24. Two boards each having the same EI and length L are crossed perpendicular to each other as shown. Determine the vertical reactions at the supports. Assume the boards just touch each other before the load **P** is applied.

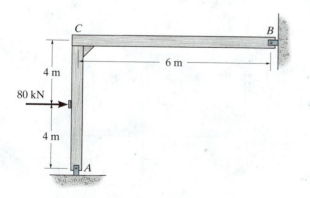

Prob. 10–22

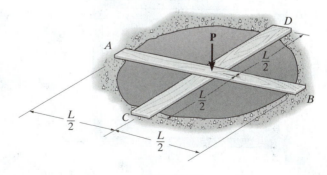

Prob. 10–24

10

Sec. 10.6

10–25. Determine the force in each member. AE is constant.

10–27. Determine the force in member AC of the truss. AE is constant.

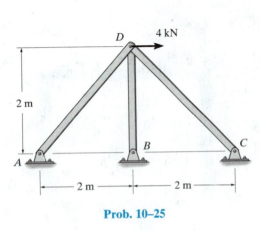

Prob. 10–25

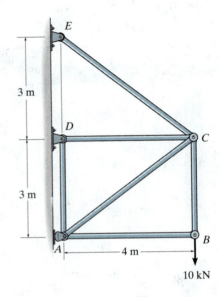

Prob. 10–27

10–26. Determine the force in each member of the truss. The cross-sectional area of each member is indicated in the figure. $E = 29(10^3)$ ksi. Assume the members are pin connected at their ends.

*10–28. Determine the force in member BE of the pin-connected truss. AE is constant.

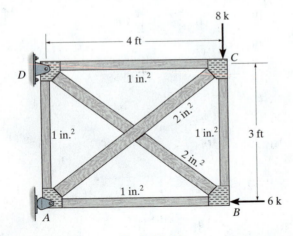

Prob. 10–26

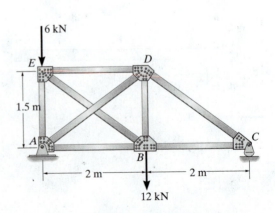

Prob. 10–28

10

10–29. Determine the force in member AD of the pin-connected truss. AE is constant.

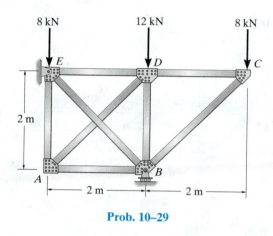

8 kN   12 kN   8 kN

Prob. 10–29

10–30. Determine the force in member BD. AE is constant.

10–31. Determine the force in member BC. AE is constant.

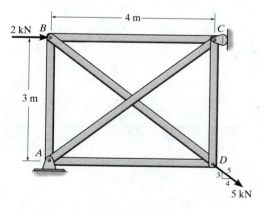

Probs. 10–30/31

*10–32.** Determine the force in each member of the truss. AE is constant.

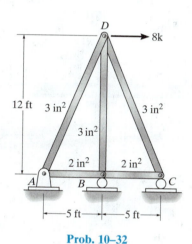

Prob. 10–32

10–33. Determine the force in member GB of the truss. AE is constant.

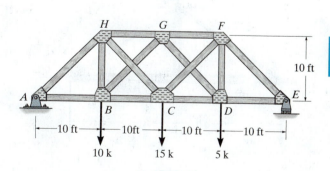

Prob. 10–33

10

Sec. 10.7

10–34. Determine the reactions at the supports, then draw the moment diagram. Each spring is originally unstretched and has a stiffness $k = 12\,EI/L^3$. EI is constant.

*10–36. The trussed beam supports the uniform distributed loading. If all the truss members have a cross-sectional area of 1.25 in², determine the force in member BC. Neglect both the depth and axial compression in the beam. Take $E = 29(10^3)$ ksi for all members. Also, for the beam $I_{AD} = 750$ in⁴. Assume A is a pin and D is a rocker.

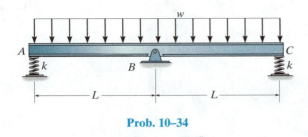

Prob. 10–34

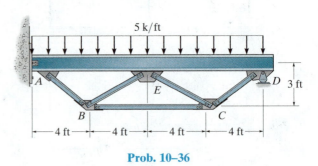

Prob. 10–36

10–35. The trussed beam supports the two loads of 3 k. If the two cables have a cross sectional area of 0.5 in² and the strut CF has a cross sectional area of 3 in², determine the force in the strut. Neglect both the depth and axial compression in the beam. Take $E = 29(10^3)$ ksi for all members. Also, $I_{AE} = 450$ in⁴.

10–37. The king-post trussed beam supports a concentrated force of 40 k at its center. Determine the force in each of the three struts. The struts each have a cross-sectional area of 2 in². Assume they are pin connected at their end points. Neglect both the depth of the beam and the effect of axial compression in the beam. Take $E = 29(10^3)$ ksi for both the beam and struts. Also, $I_{AB} = 400$ in⁴.

10–38. Determine the maximum moment in the beam in Prob. 10–37.

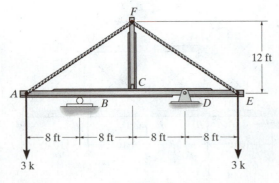

Prob. 10–35

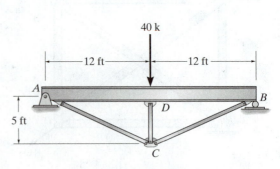

Probs. 10–37/38

10–39. Determine the reactions at the fixed support D. EI is constant for both beams.

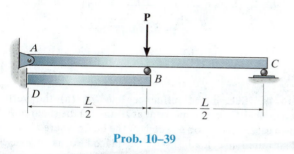

Prob. 10–39

10–41. The compound beam segments meet in the center using a smooth contact (roller). Determine the reactions at the fixed supports A and B when the load **P** is applied. EI is constant.

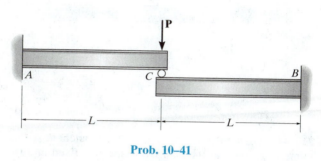

Prob. 10–41

*10–40. The beam AB has a moment of inertia $I = 800(10^6)$mm^4, and rests on the smooth supports at its ends. A 50 mm in diameter post CD is placed at the center of the beam. If the temperature of the rod is increased by 100 °C, determine the force developed in the rod. The beam and rod are both made of steel for which $E = 200$ GPa and $\alpha = 11.7(10^{-6})/$°C.

10–42. The structural assembly supports the loading shown. Draw the moment diagrams for each of the beams. Take $I = 100(10^6)$ mm^4 for the beams and $A = 200$ mm^2 for the tie rod. All members are made of steel for which $E = 200$ GPa.

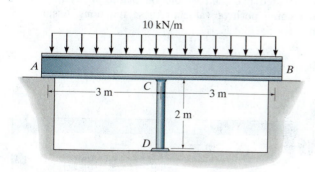

Prob. 10–40

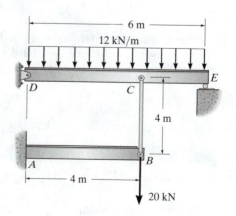

Prob. 10–42

10

Sec. 10.10–10.11

10–43. Draw the influence line for the reaction at C. Plot numerical values at the peaks. Assume A is a pin and B and C are rollers. EI is constant.

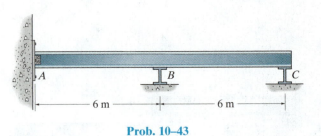

Prob. 10–43

*10–44. Draw the influence line for the moment at A. Plot numerical values at the peaks. Assume A is fixed and the support at B is a roller. EI is constant.

10–45. Draw the influence line for the vertical reaction at B. Plot numerical values at the peaks. Assume A is fixed and the support at B is a roller. EI is constant.

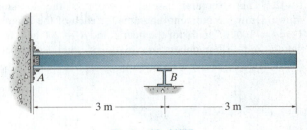

Probs. 10–44/45

10–46. Draw the influence line for the shear at C. Plot numerical values every 1.5 m. Assume A is fixed and the support at B is a roller. EI is constant.

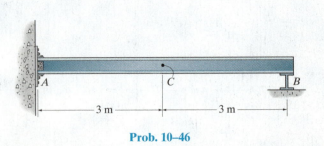

Prob. 10–46

10–47. Draw the influence line for the reaction at C. Plot the numerical values every 5 ft. EI is constant.

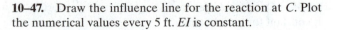

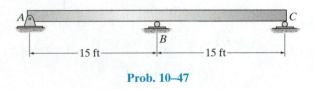

Prob. 10–47

*10–48. Sketch the influence line for (a) the vertical reaction at C, (b) the moment at B, and (c) the shear at E. In each case, indicate on a sketch of the beam where a uniform distributed live load should be placed so as to cause a maximum positive value of these functions. Assume the beam is fixed at F.

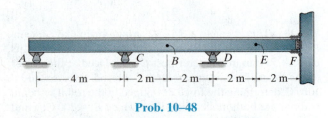

Prob. 10–48

10–49. Sketch the influence line for (a) the vertical reaction at C, (b) the moment at B, and (c) the shear at E. In each case, indicate on a sketch of the beam where a uniform distributed live load should be placed so as to cause a maximum positive value of these functions. Assume the beam is fixed at F.

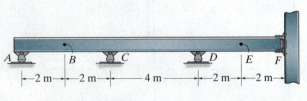

Prob. 10–49

10–50. Use the Müller-Breslau principle to sketch the general shape of the influence line for the moment at A.

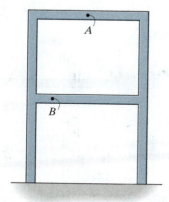

Prob. 10–50

*10–52. Use the Müller-Breslau principle to sketch the general shape of the influence line for the moment at A.

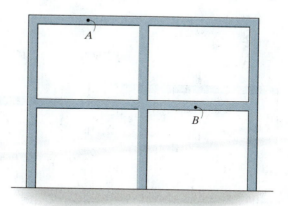

Prob. 10–52

10–51. Use the Müller-Breslau principle to sketch the general shape of the influence line for the moment at A and the shear at B.

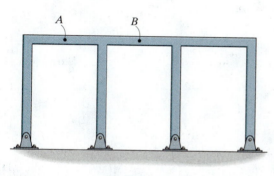

Prob. 10–51

10–53. Use the Müller-Breslau principle to sketch the general shape of the influence line for the moment at A and the shear at B.

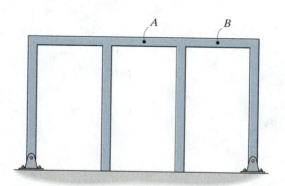

Prob. 10–53

10

CHAPTER REVIEW

The analysis of a statically indeterminate structure requires satisfying equilibrium, compatibility, and the force-displacement relationships for the structure. A force method of analysis consists of writing equations that satisfy compatibility and the force-displacement requirements, which then gives a direct solution for the redundant reactions. Once obtained, the remaining reactions are found from the equilibrium equations.

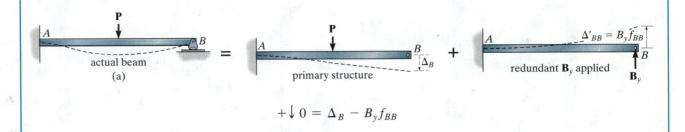

actual beam
(a)

primary structure

redundant $\mathbf{B}_y$ applied

$$+\downarrow 0 = \Delta_B - B_y f_{BB}$$

Simplification of the force method is possible, using Maxwell's theorem of reciprocal displacements, which states that the displacement of a point B on a structure due to a unit load acting at point A, f_{BA}, is equal to the displacement of point A when the load acts at B, f_{AB}.

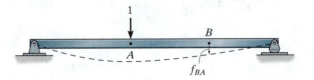

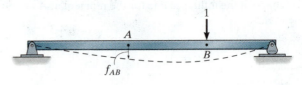

$$f_{BA} = f_{AB}$$

10

The analysis of a statically indeterminate structure can be simplified if the structure has symmetry of material, geometry, and loading about its central axis. In particular, structures having an asymmetric loading can be replaced with a superposition of a symmetric and antisymmetric load.

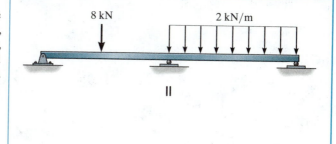

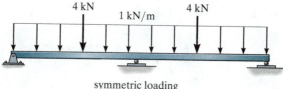

symmetric loading

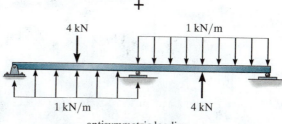

antisymmetric loading

Influence lines for statically indeterminate structures will consist of *curved lines*. They can be sketched using the Müller-Breslau principle, which states that the influence line shape for either the reaction, shear, or moment is to the same scale as the deflected shape of the structure when it is acted upon by the reaction, shear, or moment, respectively. By using Maxwell's theorem of reciprocal deflections, it is possible to obtain specific values of the ordinates of an influence line.

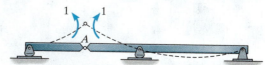

influence line shape for moment at *A*

10

Chapter 11

© Tim Gainey/Alamy

The members of this concrete building frame are all fixed connected, and so the frame is statically indeterminate.

Displacement Method of Analysis: Slope-Deflection Equations

In this chapter we will briefly outline the basic ideas for analyzing structures using the displacement method of analysis. Once these concepts have been presented, we will develop the general equations of slope deflection and then use them to analyze statically indeterminate beams and frames.

11.1 Displacement Method of Analysis: General Procedures

All structures must satisfy equilibrium, load-displacement, and compatibility of displacements requirements in order to ensure their safety. It was stated in Sec. 10.1 that there are two different ways to satisfy these requirements when analyzing a statically indeterminate structure. The force method of analysis, discussed in the previous chapter, is based on identifying the unknown redundant forces and then satisfying the structure's compatibility equations. This is done by expressing the displacements in terms of the loads by using the load-displacement relations. The solution of the resultant equations yields the redundant reactions, and then the equilibrium equations are used to determine the remaining reactions on the structure.

The *displacement method* works the opposite way. It first requires satisfying equilibrium equations for the structure. To do this the unknown displacements are written in terms of the loads by using the load-displacement relations, then these equations are solved for the displacements. Once the displacements are obtained, the unknown loads are determined from the compatibility equations using the load-displacement relations. Every displacement method follows this general procedure. In this chapter, the procedure will be generalized to produce the slope-deflection equations. In Chapter 12, the moment-distribution method will be developed. This method sidesteps the calculation of the displacements and instead makes it possible to apply a series of converging corrections that allow direct calculation of the end moments. Finally, in Chapters 14, 15, and 16, we will illustrate how to apply this method using matrix analysis, making it suitable for use on a computer.

In the discussion that follows we will show how to identify the unknown displacements in a structure and we will develop some of the important load-displacement relations for beam and frame members. The results will be used in the next section and in later chapters as the basis for applying the displacement method of analysis.

Degrees of Freedom.

When a structure is loaded, specified points on it, called *nodes*, will undergo unknown displacements. These displacements are referred to as the ***degrees of freedom*** for the structure, and in the displacement method of analysis it is important to specify these degrees of freedom since they become the unknowns when the method is applied. The number of these unknowns is referred to as the degree in which the structure is kinematically indeterminate.

To determine the kinematic indeterminacy we can imagine the structure to consist of a series of members connected to nodes, which are usually located at *joints, supports,* at the *ends* of a member, or where the members have a sudden *change in cross section.* In three dimensions, each node on a frame or beam can have at most three linear displacements and three rotational displacements; and in two dimensions, each node can have at most two linear displacements and one rotational displacement. Furthermore, nodal displacements may be restricted by the supports, or due to assumptions based on the behavior of the structure. For example, if the structure is a beam and only deformation due to bending is considered, then there can be no linear displacement along the axis of the beam since this displacement is caused by axial-force deformation.

To clarify these concepts we will consider some examples, beginning with the beam in Fig. 11–1a. Here any load **P** applied to the beam will cause node A only to rotate (neglecting axial deformation), while node B is completely restricted from moving. Hence the beam has only one unknown degree of freedom, θ_A, and is therefore kinematically indeterminate to the first degree. The beam in Fig. 11–1b has nodes at $A, B,$ and C, and so has four degrees of freedom, designated by the rotational displacements θ_A, θ_B, θ_C, and the vertical displacement Δ_C. It is kinematically indeterminate to the fourth degree. Consider now the frame in Fig. 11–1c. Again, if we neglect axial deformation of the members, an arbitrary loading **P** applied to the frame can cause nodes B and C to rotate, and these nodes can be displaced horizontally by an *equal* amount. The frame therefore has three degrees of freedom, θ_B, θ_C, Δ_B, and thus it is kinematically indeterminate to the third degree.

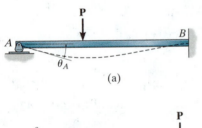

(a)

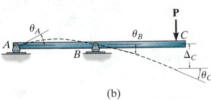

(b)

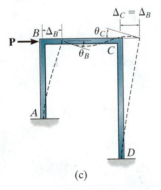

(c)

Fig. 11–1

11

In summary, specifying the kinematic indeterminacy or the number of unconstrained degrees of freedom for the structure is a necessary first step when applying a displacement method of analysis. It identifies the number of unknowns in the problem, based on the assumptions made regarding the deformation behavior of the structure. Furthermore, once these nodal displacements are known, the deformation of the structural members can be completely specified, and the loadings within the members obtained.

11.2 Slope-Deflection Equations

As indicated previously, the method of consistent displacements studied in Chapter 10 is called a force method of analysis, because it requires writing equations that relate the unknown forces or moments in a structure. Unfortunately, its use is limited to structures which are *not* highly indeterminate. This is because much work is required to set up the compatibility equations, and furthermore each equation written involves *all the unknowns*, making it difficult to solve the resulting set of equations unless a computer is available. By comparison, the slope-deflection method is not as involved. As we shall see, it requires less work both to write the necessary equations for the solution of a problem and to solve these equations for the unknown displacements and associated internal loads. Also, the method can be easily programmed on a computer and used to analyze a wide range of indeterminate structures.

The slope-deflection method was originally developed by Heinrich Manderla and Otto Mohr for the purpose of studying secondary stresses in trusses. Later, in 1915, G. A. Maney developed a refined version of this technique and applied it to the analysis of indeterminate beams and framed structures.

General Case.
The slope-deflection method is so named since it relates the unknown slopes and deflections to the applied load on a structure. In order to develop the general form of the slope-deflection equations, we will consider the typical span AB of a continuous beam as shown in Fig. 11–2, which is subjected to the arbitrary loading and has a constant EI. We wish to relate the beam's internal end moments M_{AB} and M_{BA} in terms of its three degrees of freedom, namely, its angular displacements θ_A and θ_B, and linear displacement Δ which could be caused by a relative settlement between the supports. Since we will be developing a formula, *moments* and *angular displacements* will be considered *positive* when they act *clockwise on the span*, as shown in Fig. 11–2. Furthermore, the *linear displacement* Δ is considered *positive* as shown, since this displacement causes the cord of the span and the span's cord angle ψ to rotate *clockwise*.

The slope-deflection equations can be obtained by using the principle of superposition by considering *separately* the moments developed at each support due to each of the displacements, θ_A, θ_B, and Δ, and then the loads.

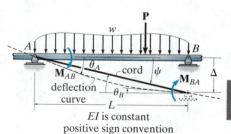

Fig. 11–2

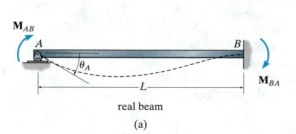

real beam

(a)

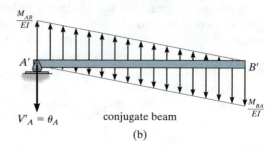

conjugate beam

(b)

Fig. 11–3

Angular Displacement at A, θ_A.

Consider node A of the member shown in Fig. 11–3a to rotate θ_A while its far-end node B is *held fixed*. To determine the moment M_{AB} needed to cause this displacement, we will use the conjugate-beam method. For this case the conjugate beam is shown in Fig. 11–3b. Notice that the end shear at A' acts downward on the beam, since θ_A is clockwise. The deflection of the "real beam" in Fig. 11–3a is to be zero at A and B, and therefore the corresponding sum of the *moments* at each end A' and B' of the conjugate beam must also be zero. This yields

$$\zeta + \Sigma M_{A'} = 0; \qquad \left[\frac{1}{2}\left(\frac{M_{AB}}{EI}\right)L\right]\frac{L}{3} - \left[\frac{1}{2}\left(\frac{M_{BA}}{EI}\right)L\right]\frac{2L}{3} = 0$$

$$\zeta + \Sigma M_{B'} = 0; \qquad \left[\frac{1}{2}\left(\frac{M_{BA}}{EI}\right)L\right]\frac{L}{3} - \left[\frac{1}{2}\left(\frac{M_{AB}}{EI}\right)L\right]\frac{2L}{3} + \theta_A L = 0$$

from which we obtain the following load-displacement relationships.

$$M_{AB} = \frac{4EI}{L}\theta_A \qquad\qquad (11\text{–}1)$$

$$M_{BA} = \frac{2EI}{L}\theta_A \qquad\qquad (11\text{–}2)$$

Angular Displacement at B, θ_B.

In a similar manner, if end B of the beam rotates to its final position θ_B, while end A is *held fixed*, Fig. 11–4, we can relate the applied moment M_{BA} to the angular displacement θ_B and the reaction moment M_{AB} at the wall. The results are

$$M_{BA} = \frac{4EI}{L}\theta_B \qquad\qquad (11\text{–}3)$$

$$M_{AB} = \frac{2EI}{L}\theta_B \qquad\qquad (11\text{–}4)$$

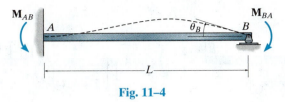

M_{AB} A θ_B B M_{BA} L

Fig. 11–4

Relative Linear Displacement, Δ.

If the far node B of the member is displaced relative to A, so that the cord of the member rotates clockwise (positive displacement) and yet both ends do not rotate, then equal but opposite moment and shear reactions are developed in the member, Fig. 11–5a. As before, the moment **M** can be related to the displacement Δ using the conjugate-beam method. In this case, the conjugate beam, Fig. 11–5b, is free at both ends, since the real beam (member) is fixed supported. However, due to the *displacement* of the real beam at B, the *moment* at the end B' of the conjugate beam must have a magnitude of Δ as indicated.* Summing moments about B', we have

$$\zeta + \Sigma M_{B'} = 0; \qquad \left[\frac{1}{2}\frac{M}{EI}(L)\left(\frac{2}{3}L\right)\right] - \left[\frac{1}{2}\frac{M}{EI}(L)\left(\frac{1}{3}L\right)\right] - \Delta = 0$$

$$\boxed{M_{AB} = M_{BA} = M = \frac{-6EI}{L^2}\Delta} \qquad (11\text{–}5)$$

By our sign convention, this induced moment is negative since for equilibrium it acts counterclockwise on the member.

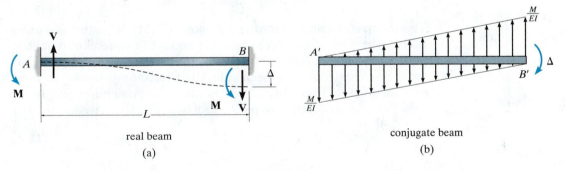

real beam

(a)

conjugate beam

(b)

Fig. 11–5

---

*The moment diagrams shown on the conjugate beam were determined by the method of superposition for a simply supported beam, as explained in Sec. 4.5.

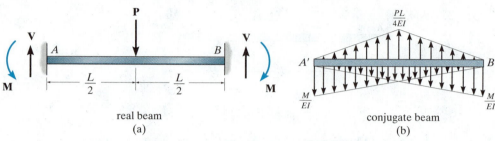

real beam
(a)

conjugate beam
(b)

Fig. 11–6

Fixed-End Moments.

In the previous cases we have considered relationships between the displacements and the necessary moments M_{AB} and M_{BA} acting at nodes A and B, respectively. In general, however, the linear or angular displacements of the nodes are caused by loadings acting on the *span* of the member, not by moments acting at its nodes. In order to develop the slope-deflection equations, we must transform these *span loadings* into equivalent moments acting at the nodes and then use the load-displacement relationships just derived. This is done simply by finding the reaction moment that each load develops at the nodes. For example, consider the fixed-supported member shown in Fig. 11–6a, which is subjected to a concentrated load **P** at its center. The conjugate beam for this case is shown in Fig. 11–6b. Since we require the slope at each end to be zero,

$$+\uparrow\Sigma F_y = 0; \qquad \left[\frac{1}{2}\left(\frac{PL}{4EI}\right)L\right] - 2\left[\frac{1}{2}\left(\frac{M}{EI}\right)L\right] = 0$$

$$M = \frac{PL}{8}$$

This moment is called a ***fixed-end moment*** (FEM). Note that according to our sign convention, it is negative at node A (counterclockwise) and positive at node B (clockwise). For convenience in solving problems, fixed-end moments have been calculated for other loadings and are tabulated on the inside back cover of the book. Assuming these FEMs have been determined for a specific problem (Fig. 11–7), we have

$$M_{AB} = (\text{FEM})_{AB} \qquad M_{BA} = (\text{FEM})_{BA} \qquad (11\text{–}6)$$

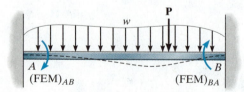

Fig. 11–7

Slope-Deflection Equation.

If the end moments due to each displacement (Eqs. 11–1 through 11–5) and the loading (Eq. 11–6) are added together, the resultant moments at the ends can be written as

$$M_{AB} = 2E\left(\frac{I}{L}\right)\left[2\theta_A + \theta_B - 3\left(\frac{\Delta}{L}\right)\right] + (FEM)_{AB}$$

$$M_{BA} = 2E\left(\frac{I}{L}\right)\left[2\theta_B + \theta_A - 3\left(\frac{\Delta}{L}\right)\right] + (FEM)_{BA}$$

$$(11-7)$$

This pedestrian bridge has a reinforced concrete deck. Since it extends over all its supports, it is indeterminate to the second degree. The slope deflection equations provide a convenient method for finding the internal moments in each span.

Since these two equations are similar, the result can be expressed as a single equation. Referring to one end of the span as the near end (N) and the other end as the far end (F), and letting the *member stiffness* be represented as $k = I/L$, and the *span's cord rotation* as ψ (psi) $= \Delta/L$, we can write

$$M_N = 2Ek(2\theta_N + \theta_F - 3\psi) + (FEM)_N$$

For Internal Span or End Span with Far End Fixed

$$(11-8)$$

where

M_N = internal moment in the near end of the span; this moment is *positive clockwise* when acting on the span.

E, k = modulus of elasticity of material and span stiffness $k = I/L$.

θ_N, θ_F = near- and far-end slopes or angular displacements of the span at the supports; the angles are measured in *radians* and are *positive clockwise*.

ψ = span rotation of its cord due to a linear displacement, that is, $\psi = \Delta/L$; this angle is measured in *radians* and is *positive clockwise*.

$(FEM)_N$ = fixed-end moment at the near-end support; the moment is *positive clockwise* when acting on the span; refer to the table on the inside back cover for various loading conditions.

From the derivation Eq. 11–8 is both a compatibility and load-displacement relationship found by considering only the effects of bending and neglecting axial and shear deformations. It is referred to as the general *slope-deflection equation*. When used for the solution of problems, this equation is applied *twice* for each member span (AB); that is, application is from A to B and from B to A for span AB in Fig. 11–2.

11

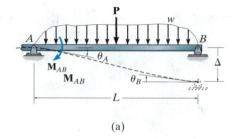

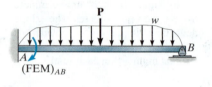

(a)

(b)

Fig. 11–8

Pin-Supported End Span.

Occasionally an end span of a beam or frame is supported by a pin or roller at its *far end*, Fig. 11–8a. When this occurs, the moment at the roller or pin must be zero; and provided the angular displacement θ_B at this support does not have to be determined, we can modify the general slope-deflection equation so that it has to be applied *only once* to the span rather than twice. To do this we will apply Eq. 11–8 or Eqs. 11–7 to each end of the beam in Fig. 11–8. This results in the following two equations:

$$M_N = 2Ek(2\theta_N + \theta_F - 3\psi) + (FEM)_N$$
$$0 = 2Ek(2\theta_F + \theta_N - 3\psi) + 0$$

(11–9)

Here the $(FEM)_F$ is equal to zero since the far end is pinned, Fig. 11–8b. Furthermore, the $(FEM)_N$ can be obtained, for example, using the table in the right-hand column on the inside back cover of this book. Multiplying the first equation by 2 and subtracting the second equation from it *eliminates* the unknown θ_F and yields

$$\boxed{M_N = 3Ek(\theta_N - \psi) + (FEM)_N}$$
Only for End Span with Far End Pinned or Roller Supported

(11–10)

Since the moment at the far end is zero, only *one* application of this equation is necessary for the end span. This simplifies the analysis since the general equation, Eq. 11–8, would require *two* applications for this span and therefore involve the (extra) unknown angular displacement θ_B (or θ_F) at the end support.

To summarize application of the slope-deflection equations, consider the continuous beam shown in Fig. 11–9, which has four degrees of freedom. Here Eq. 11–8 can be applied twice to each of the three spans, i.e., from A to B, B to A, B to C, C to B, C to D, and D to C. These equations would involve the four unknown rotations, θ_A, θ_B, θ_C, θ_D. Since the end moments at A and D are zero, however, it is not necessary to determine θ_A and θ_D. A shorter solution occurs if we apply Eq. 11–10 from B to A and C to D and then apply Eq. 11–8 from B to C and C to B. These four equations will involve only the unknown rotations θ_B and θ_C.

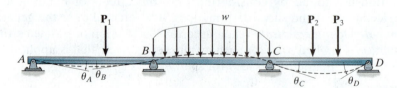

Fig. 11–9

11.3   Analysis of Beams

Procedure for Analysis

Degrees of Freedom

Label all the supports and joints (nodes) in order to identify the spans of the beam or frame between the nodes. By drawing the deflected shape of the structure, it will be possible to identify the number of degrees of freedom. Here each node can possibly have an angular displacement and a linear displacement. *Compatibility* at the nodes is maintained provided the members that are fixed connected to a node undergo the same displacements as the node. If these displacements are unknown, and in general they will be, then for convenience *assume* they act in the *positive direction* so as to cause *clockwise* rotation of a member or joint, Fig. 11–2.

Slope-Deflection Equations

The slope-deflection equations relate the unknown moments applied to the nodes to the displacements of the nodes for any span of the structure. If a load exists on the span, compute the FEMs using the table given on the inside back cover. Also, if a node has a linear displacement, Δ, compute $\psi = \Delta/L$ for the adjacent spans. Apply Eq. 11–8 to each end of the span, thereby generating *two* slope-deflection equations for each span. However, if a span at the *end* of a continuous beam or frame is pin supported, apply Eq. 11–10 only to the restrained end, thereby generating *one* slope-deflection equation for the span.

Equilibrium Equations

Write an equilibrium equation for each unknown degree of freedom for the structure. Each of these equations should be expressed in terms of unknown internal moments as specified by the slope-deflection equations. For beams and frames write the moment equation of equilibrium at each support, and for frames also write joint moment equations of equilibrium. If the frame sidesways or deflects horizontally, column shears should be related to the moments at the ends of the column. This is discussed in Sec. 11.5.

Substitute the slope-deflection equations into the equilibrium equations and solve for the unknown joint displacements. These results are then substituted into the slope-deflection equations to determine the internal moments at the ends of each member. If any of the results are *negative*, they indicate *counterclockwise* rotation; whereas *positive* moments and displacements are applied *clockwise*.

11

EXAMPLE 11.1

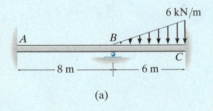

6 kN/m

A

B

C

8 m

6 m

(a)

Draw the shear and moment diagrams for the beam shown in Fig. 11–10a. *EI* is constant.

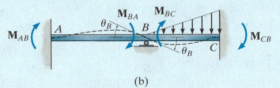

M_{AB} A θ_B M_{BA} M_{BC} B M_{CB} θ_B C

(b)

Fig. 11–10

SOLUTION

Slope-Deflection Equations.   Two spans must be considered in this problem. Since there is *no* span having the far end pinned or roller supported, Eq. 11–8 applies to the solution. Using the formulas for the FEMs tabulated for the triangular loading given on the inside back cover, we have

$$(FEM)_{BC} = -\frac{wL^2}{30} = -\frac{6(6)^2}{30} = -7.2 \text{ kN} \cdot \text{m}$$

$$(FEM)_{CB} = \frac{wL^2}{20} = \frac{6(6)^2}{20} = 10.8 \text{ kN} \cdot \text{m}$$

Note that $(FEM)_{BC}$ is negative since it acts counterclockwise *on the beam* at *B*. Also, $(FEM)_{AB} = (FEM)_{BA} = 0$ since there is no load on span *AB*.

In order to identify the unknowns, the elastic curve for the beam is shown in Fig. 11–10b. As indicated, there are four unknown internal moments. Only the slope at B, θ_B, is unknown. Since *A* and *C* are fixed supports, $\theta_A = \theta_C = 0$. Also, since the supports do not settle, nor are they displaced up or down, $\psi_{AB} = \psi_{BC} = 0$. For span *AB*, considering *A* to be the near end and *B* to be the far end, we have

$$M_N = 2E\left(\frac{I}{L}\right)(2\theta_N + \theta_F - 3\psi) + (FEM)_N$$

$$M_{AB} = 2E\left(\frac{I}{8}\right)[2(0) + \theta_B - 3(0)] + 0 = \frac{EI}{4}\theta_B \qquad (1)$$

Now, considering *B* to be the near end and *A* to be the far end, we have

$$M_{BA} = 2E\left(\frac{I}{8}\right)[2\theta_B + 0 - 3(0)] + 0 = \frac{EI}{2}\theta_B \qquad (2)$$

In a similar manner, for span *BC* we have

$$M_{BC} = 2E\left(\frac{I}{6}\right)[2\theta_B + 0 - 3(0)] - 7.2 = \frac{2EI}{3}\theta_B - 7.2 \qquad (3)$$

$$M_{CB} = 2E\left(\frac{I}{6}\right)[2(0) + \theta_B - 3(0)] + 10.8 = \frac{EI}{3}\theta_B + 10.8 \qquad (4)$$

11

Equilibrium Equations. The above four equations contain five unknowns. The necessary fifth equation comes from the condition of moment equilibrium at support B. The free-body diagram of a segment of the beam at B is shown in Fig. 11–10c. Here M_{BA} and M_{BC} are assumed to act in the positive direction to be consistent with the slope-deflection equations.* The beam shears contribute negligible moment about B since the segment is of differential length. Thus,

$$\zeta + \Sigma M_B = 0; \qquad M_{BA} + M_{BC} = 0 \qquad (5)$$

To solve, substitute Eqs. (2) and (3) into Eq. (5), which yields

$$\theta_B = \frac{6.17}{EI}$$

Resubstituting this value into Eqs. (1)–(4) yields

$$M_{AB} = 1.54 \text{ kN} \cdot \text{m}$$

$$M_{BA} = 3.09 \text{ kN} \cdot \text{m}$$

$$M_{BC} = -3.09 \text{ kN} \cdot \text{m}$$

$$M_{CB} = 12.86 \text{ kN} \cdot \text{m}$$

The negative value for M_{BC} indicates that this moment acts counterclockwise on the beam, not clockwise as shown in Fig. 11–10b.

Using these results, the shears at the end spans are determined from the equilibrium equations, Fig. 11–10d. The free-body diagram of the entire beam and the shear and moment diagrams are shown in Fig. 11–10e.

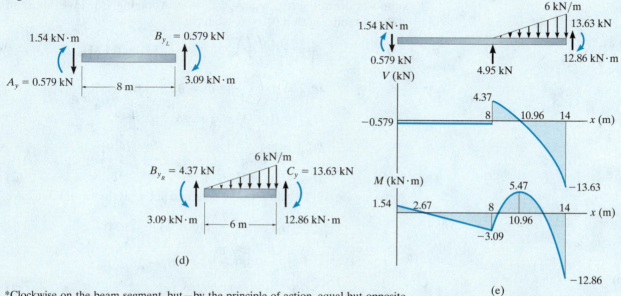

(c)

(d)

(e)

*Clockwise on the beam segment, but—by the principle of action, equal but opposite reaction—counterclockwise on the support.

EXAMPLE 11.2

Draw the shear and moment diagrams for the beam shown in Fig. 11–11a. EI is constant.

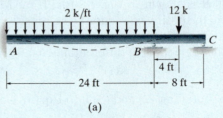

(a)

Fig. 11–11

SOLUTION

Slope-Deflection Equations.   Two spans must be considered in this problem. Equation 11–8 applies to span AB. We can use Eq. 11–10 for span BC since the *end C is on a roller*. Using the formulas for the FEMs tabulated on the inside back cover, we have

$$(\text{FEM})_{AB} = -\frac{wL^2}{12} = -\frac{1}{12}(2)(24)^2 = -96 \text{ k} \cdot \text{ft}$$

$$(\text{FEM})_{BA} = \frac{wL^2}{12} = \frac{1}{12}(2)(24)^2 = 96 \text{ k} \cdot \text{ft}$$

$$(\text{FEM})_{BC} = -\frac{3PL}{16} = -\frac{3(12)(8)}{16} = -18 \text{ k} \cdot \text{ft}$$

Note that $(\text{FEM})_{AB}$ and $(\text{FEM})_{BC}$ are negative since they act counterclockwise on the beam at A and B, respectively. Also, since the supports do not settle, $\psi_{AB} = \psi_{BC} = 0$. Applying Eq. 11–8 for span AB and realizing that $\theta_A = 0$, we have

$$M_N = 2E\left(\frac{I}{L}\right)(2\theta_N + \theta_F - 3\psi) + (\text{FEM})_N$$

$$M_{AB} = 2E\left(\frac{I}{24}\right)[2(0) + \theta_B - 3(0)] - 96$$

$$M_{AB} = 0.08333EI\theta_B - 96 \tag{1}$$

$$M_{BA} = 2E\left(\frac{I}{24}\right)[2\theta_B + 0 - 3(0)] + 96$$

$$M_{BA} = 0.1667EI\theta_B + 96 \tag{2}$$

Applying Eq. 11–10 with B as the near end and C as the far end, we have

$$M_N = 3E\left(\frac{I}{L}\right)(\theta_N - \psi) + (\text{FEM})_N$$

$$M_{BC} = 3E\left(\frac{I}{8}\right)(\theta_B - 0) - 18$$

$$M_{BC} = 0.375EI\theta_B - 18 \tag{3}$$

Remember that Eq. 11–10 is *not* applied from C (near end) to B (far end).

Equilibrium Equations.   The above three equations contain four unknowns. The necessary fourth equation comes from the conditions of equilibrium at the support B. The free-body diagram is shown in Fig. 11–11b. We have

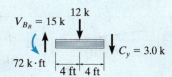

(b)

$$\zeta + \Sigma M_B = 0; \qquad M_{BA} + M_{BC} = 0 \qquad (4)$$

To solve, substitute Eqs. (2) and (3) into Eq. (4), which yields

$$\theta_B = -\frac{144.0}{EI}$$

Since θ_B is negative (counterclockwise) the elastic curve for the beam has been correctly drawn in Fig. 11–11a. Substituting θ_B into Eqs. (1)–(3), we get

$$M_{AB} = -108.0 \text{ k} \cdot \text{ft}$$

$$M_{BA} = 72.0 \text{ k} \cdot \text{ft}$$

$$M_{BC} = -72.0 \text{ k} \cdot \text{ft}$$

Using these data for the moments, the shear reactions at the ends of the beam spans have been determined in Fig. 11–11c. The shear and moment diagrams are plotted in Fig. 11–11d.

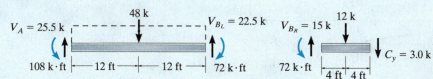

(c)

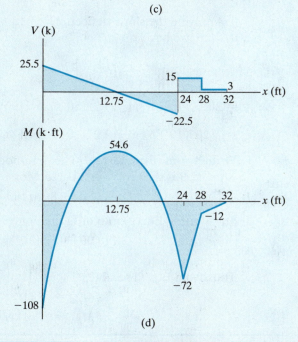

(d)

EXAMPLE 11.3

Determine the moment at A and B for the beam shown in Fig. 11–12a. The support at B is displaced (settles) 80 mm. Take $E = 200$ GPa, $I = 5(10^6)$ mm^4.

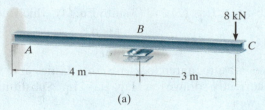

(a)

Fig. 11–12

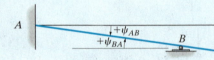

(b)

SOLUTION

Slope-Deflection Equations. Only one span (AB) must be considered in this problem since the moment $\mathbf{M}_{BC}$ due to the overhang can be calculated from statics. Since there is no loading on span AB, the FEMs are zero. As shown in Fig. 11–12b, the downward displacement (settlement) of B causes the cord for span AB to rotate clockwise. Thus,

$$\psi_{AB} = \psi_{BA} = \frac{0.08 \text{ m}}{4} = 0.02 \text{ rad}$$

The stiffness for AB is

$$k = \frac{I}{L} = \frac{5(10^6) \text{ mm}^4(10^{-12}) \text{ m}^4/\text{mm}^4}{4 \text{ m}} = 1.25(10^{-6}) \text{ m}^3$$

Applying the slope-deflection equation, Eq. 11–8, to span AB, with $\theta_A = 0$, we have

$$M_N = 2E\left(\frac{I}{L}\right)(2\theta_N + \theta_F - 3\psi) + (\text{FEM})_N$$

$$M_{AB} = 2(200(10^9) \text{ N/m}^2)\left[1.25(10^{-6}) \text{ m}^3\right][2(0) + \theta_B - 3(0.02)] + 0 \quad (1)$$

$$M_{BA} = 2(200(10^9) \text{ N/m}^2)\left[1.25(10^{-6}) \text{ m}^3\right][2\theta_B + 0 - 3(0.02)] + 0 \quad (2)$$

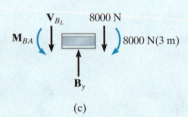

(c)

Equilibrium Equations. The free-body diagram of the beam at support B is shown in Fig. 11–12c. Moment equilibrium requires

$$\zeta + \Sigma M_B = 0; \qquad M_{BA} - 8000 \text{ N}(3 \text{ m}) = 0$$

Substituting Eq. (2) into this equation yields

$$1(10^6)\theta_B - 30(10^3) = 24(10^3)$$
$$\theta_B = 0.054 \text{ rad}$$

Thus, from Eqs. (1) and (2),

$$M_{AB} = -3.00 \text{ kN} \cdot \text{m}$$
$$M_{BA} = 24.0 \text{ kN} \cdot \text{m}$$

11

EXAMPLE 11.4

Determine the internal moments at the supports of the beam shown in Fig. 11–13a. The roller support at C is pushed downward 0.1 ft by the force $\mathbf{P}$. Take $E = 29(10^3)$ ksi, $I = 1500$ in^4.

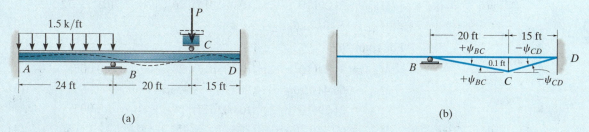

(a)   (b)

Fig. 11–13

SOLUTION

Slope-Deflection Equations. Three spans must be considered in this problem. Equation 11–8 applies since the end supports A and D are fixed. Also, only span AB has FEMs.

$$(\text{FEM})_{AB} = -\frac{wL^2}{12} = -\frac{1}{12}(1.5)(24)^2 = -72.0 \text{ k} \cdot \text{ft}$$

$$(\text{FEM})_{BA} = \frac{wL^2}{12} = \frac{1}{12}(1.5)(24)^2 = 72.0 \text{ k} \cdot \text{ft}$$

As shown in Fig. 11–13b, the displacement (or settlement) of the support C causes ψ_{BC} to be positive, since the cord for span BC rotates clockwise, and ψ_{CD} to be negative, since the cord for span CD rotates counterclockwise. Hence,

$$\psi_{BC} = \frac{0.1 \text{ ft}}{20 \text{ ft}} = 0.005 \text{ rad} \qquad \psi_{CD} = -\frac{0.1 \text{ ft}}{15 \text{ ft}} = -0.00667 \text{ rad}$$

Also, expressing the units for the stiffness in feet, we have

$$k_{AB} = \frac{1500}{24(12)^4} = 0.003014 \text{ ft}^3 \qquad k_{BC} = \frac{1500}{20(12)^4} = 0.003617 \text{ ft}^3$$

$$k_{CD} = \frac{1500}{15(12)^4} = 0.004823 \text{ ft}^3$$

Noting that $\theta_A = \theta_D = 0$ since A and D are fixed supports, and applying the slope-deflection Eq. 11–8 twice to each span, we have

11

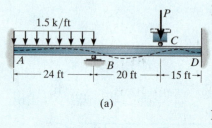

(a)

For span AB:

$$M_{AB} = 2[29(10^3)(12)^2](0.003014)[2(0) + \theta_B - 3(0)] - 72$$

$$M_{AB} = 25\ 173.6\theta_B - 72 \tag{1}$$

$$M_{BA} = 2[29(10^3)(12)^2](0.003014)[2\theta_B + 0 - 3(0)] + 72$$

$$M_{BA} = 50\ 347.2\theta_B + 72 \tag{2}$$

For span BC:

$$M_{BC} = 2[29(10^3)(12)^2](0.003617)[2\theta_B + \theta_C - 3(0.005)] + 0$$

$$M_{BC} = 60\ 416.7\theta_B + 30\ 208.3\theta_C - 453.1 \tag{3}$$

$$M_{CB} = 2[29(10^3)(12)^2](0.003617)[2\theta_C + \theta_B - 3(0.005)] + 0$$

$$M_{CB} = 60\ 416.7\theta_C + 30\ 208.3\theta_B - 453.1 \tag{4}$$

For span CD:

$$M_{CD} = 2[29(10^3)(12)^2](0.004823)[2\theta_C + 0 - 3(-0.00667)] + 0$$

$$M_{CD} = 80\ 555.6\theta_C + 0 + 805.6 \tag{5}$$

$$M_{DC} = 2[29(10^3)(12)^2](0.004823)[2(0) + \theta_C - 3(-0.00667)] + 0$$

$$M_{DC} = 40\ 277.8\theta_C + 805.6 \tag{6}$$

Equilibrium Equations.   These six equations contain eight unknowns. Writing the moment equilibrium equations for the supports at B and C, Fig. 10–13b, we have

$$\circlearrowleft + \Sigma M_B = 0; \qquad\qquad M_{BA} + M_{BC} = 0 \tag{7}$$

$$\circlearrowleft + \Sigma M_C = 0; \qquad\qquad M_{CB} + M_{CD} = 0 \tag{8}$$

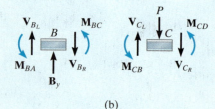

(b)

In order to solve, substitute Eqs. (2) and (3) into Eq. (7), and Eqs. (4) and (5) into Eq. (8). This yields

$$\theta_C + 3.667\theta_B = 0.01262$$

$$-\theta_C - 0.214\theta_B = 0.00250$$

Thus,

$$\theta_B = 0.00438 \text{ rad} \qquad \theta_C = -0.00344 \text{ rad}$$

The negative value for θ_C indicates counterclockwise rotation of the tangent at C, Fig. 11–13a. Substituting these values into Eqs. (1)–(6) yields

$$M_{AB} = 38.2 \text{ k} \cdot \text{ft} \qquad\qquad\qquad \textit{Ans.}$$

$$M_{BA} = 292 \text{ k} \cdot \text{ft} \qquad\qquad\qquad \textit{Ans.}$$

$$M_{BC} = -292 \text{ k} \cdot \text{ft} \qquad\qquad\qquad \textit{Ans.}$$

$$M_{CB} = -529 \text{ k} \cdot \text{ft} \qquad\qquad\qquad \textit{Ans.}$$

$$M_{CD} = 529 \text{ k} \cdot \text{ft} \qquad\qquad\qquad \textit{Ans.}$$

$$M_{DC} = 667 \text{ k} \cdot \text{ft} \qquad\qquad\qquad \textit{Ans.}$$

Apply these end moments to spans BC and CD and show that $V_{C_L} = 41.05$ k, $V_{C_R} = -79.73$ k and the force on the roller is $P = 121$ k.

11

11.4 Analysis of Frames: No Sidesway

A frame will not sidesway, or be displaced to the left or right, provided it is properly restrained. Examples are shown in Fig. 11–14. Also, no sidesway will occur in an unrestrained frame provided it is symmetric with respect to both loading and geometry, as shown in Fig. 11–15. For both cases the term ψ in the slope-deflection equations is equal to zero, since bending does not cause the joints to have a linear displacement.

The following examples illustrate application of the slope-deflection equations using the procedure for analysis outlined in Sec. 11.3 for these types of frames.

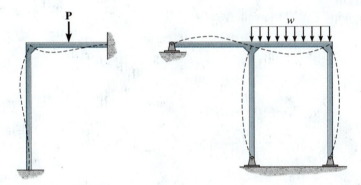

Fig. 11–14

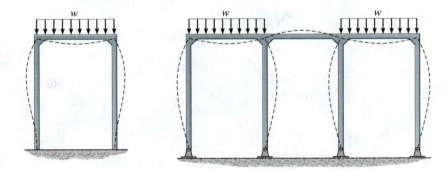

Fig. 11–15

11

EXAMPLE   11.5

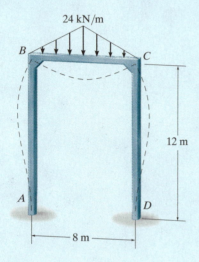

24 kN/m

12 m

8 m

(a)

Fig. 11–16

Determine the moments at each joint of the frame shown in Fig. 11–16a. EI is constant.

SOLUTION

Slope-Deflection Equations.   Three spans must be considered in this problem: AB, BC, and CD. Since the spans are fixed supported at A and D, Eq. 11–8 applies for the solution.
From the table on the inside back cover, the FEMs for BC are

$$(\text{FEM})_{BC} = -\frac{5wL^2}{96} = -\frac{5(24)(8)^2}{96} = -80 \text{ kN} \cdot \text{m}$$

$$(\text{FEM})_{CB} = \frac{5wL^2}{96} = \frac{5(24)(8)^2}{96} = 80 \text{ kN} \cdot \text{m}$$

Note that $\theta_A = \theta_D = 0$ and $\psi_{AB} = \psi_{BC} = \psi_{CD} = 0$, since no sidesway will occur.
Applying Eq. 11–8, we have

$$M_N = 2Ek(2\theta_N + \theta_F - 3\psi) + (\text{FEM})_N$$

$$M_{AB} = 2E\left(\frac{I}{12}\right)[2(0) + \theta_B - 3(0)] + 0$$

$$M_{AB} = 0.1667EI\theta_B \tag{1}$$

$$M_{BA} = 2E\left(\frac{I}{12}\right)[2\theta_B + 0 - 3(0)] + 0$$

$$M_{BA} = 0.333EI\theta_B \tag{2}$$

$$M_{BC} = 2E\left(\frac{I}{8}\right)[2\theta_B + \theta_C - 3(0)] - 80$$

$$M_{BC} = 0.5EI\theta_B + 0.25EI\theta_C - 80 \tag{3}$$

$$M_{CB} = 2E\left(\frac{I}{8}\right)[2\theta_C + \theta_B - 3(0)] + 80$$

$$M_{CB} = 0.5EI\theta_C + 0.25EI\theta_B + 80 \tag{4}$$

$$M_{CD} = 2E\left(\frac{I}{12}\right)[2\theta_C + 0 - 3(0)] + 0$$

$$M_{CD} = 0.333EI\theta_C \tag{5}$$

$$M_{DC} = 2E\left(\frac{I}{12}\right)[2(0) + \theta_C - 3(0)] + 0$$

$$M_{DC} = 0.1667EI\theta_C \tag{6}$$

Equilibrium Equations. The preceding six equations contain eight unknowns. The remaining two equilibrium equations come from moment equilibrium at joints B and C, Fig. 11–16b. We have

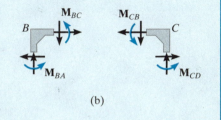

(b)

$$M_{BA} + M_{BC} = 0 \qquad (7)$$

$$M_{CB} + M_{CD} = 0 \qquad (8)$$

To solve these eight equations, substitute Eqs. (2) and (3) into Eq. (7) and substitute Eqs. (4) and (5) into Eq. (8). We get

$$0.833EI\theta_B + 0.25EI\theta_C = 80$$

$$0.833EI\theta_C + 0.25EI\theta_B = -80$$

Solving simultaneously yields

$$\theta_B = -\theta_C = \frac{137.1}{EI}$$

which conforms with the way the frame deflects as shown in Fig. 11–16a. Substituting into Eqs. (1)–(6), we get

$$M_{AB} = 22.9 \text{ kN} \cdot \text{m} \qquad \textit{Ans.}$$

$$M_{BA} = 45.7 \text{ kN} \cdot \text{m} \qquad \textit{Ans.}$$

$$M_{BC} = -45.7 \text{ kN} \cdot \text{m} \qquad \textit{Ans.}$$

$$M_{CB} = 45.7 \text{ kN} \cdot \text{m} \qquad \textit{Ans.}$$

$$M_{CD} = -45.7 \text{ kN} \cdot \text{m} \qquad \textit{Ans.}$$

$$M_{DC} = -22.9 \text{ kN} \cdot \text{m} \qquad \textit{Ans.}$$

Using these results, the reactions at the ends of each member can be determined from the equations of equilibrium, and the moment diagram for the frame can be drawn, Fig. 11–16c.

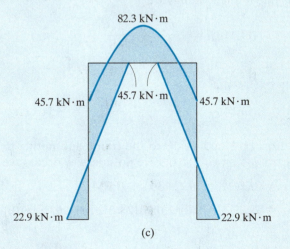

(c)

EXAMPLE 11.6

Determine the internal moments at each joint of the frame shown in Fig. 11–17a. The moment of inertia for each member is given in the figure. Take $E = 29(10^3)$ ksi.

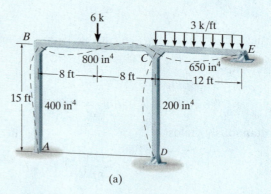

(a)

Fig. 11–17

SOLUTION

Slope-Deflection Equations.   Four spans must be considered in this problem. Equation 11–8 applies to spans AB and BC, and Eq. 11–10 will be applied to CD and CE, because the ends at D and E are pinned. Computing the member stiffnesses, we have

$$k_{AB} = \frac{400}{15(12)^4} = 0.001286 \text{ ft}^3 \qquad k_{CD} = \frac{200}{15(12)^4} = 0.000643 \text{ ft}^3$$

$$k_{BC} = \frac{800}{16(12)^4} = 0.002411 \text{ ft}^3 \qquad k_{CE} = \frac{650}{12(12)^4} = 0.002612 \text{ ft}^3$$

The FEMs due to the loadings are

$$(\text{FEM})_{BC} = -\frac{PL}{8} = -\frac{6(16)}{8} = -12 \text{ k} \cdot \text{ft}$$

$$(\text{FEM})_{CB} = \frac{PL}{8} = \frac{6(16)}{8} = 12 \text{ k} \cdot \text{ft}$$

$$(\text{FEM})_{CE} = -\frac{wL^2}{8} = -\frac{3(12)^2}{8} = -54 \text{ k} \cdot \text{ft}$$

Applying Eqs. 11–8 and 11–10 to the frame and noting that $\theta_A = 0$, $\psi_{AB} = \psi_{BC} = \psi_{CD} = \psi_{CE} = 0$ since no sidesway occurs, we have

$$M_N = 2Ek(2\theta_N + \theta_F - 3\psi) + (\text{FEM})_N$$

$$M_{AB} = 2\left[29(10^3)(12)^2\right](0.001286)[2(0) + \theta_B - 3(0)] + 0$$

$$M_{AB} = 10740.7\theta_B \tag{1}$$

$$M_{BA} = 2\left[29(10^3)(12)^2\right](0.001286)[2\theta_B + 0 - 3(0)] + 0$$

$$M_{BA} = 21\,481.5\theta_B \tag{2}$$

$$M_{BC} = 2\left[29(10^3)(12)^2\right](0.002411)[2\theta_B + \theta_C - 3(0)] - 12$$

$$M_{BC} = 40\,277.8\theta_B + 20\,138.9\theta_C - 12 \tag{3}$$

$$M_{CB} = 2\left[29(10^3)(12)^2\right](0.002411)[2\theta_C + \theta_B - 3(0)] + 12$$

$$M_{CB} = 20\,138.9\theta_B + 40\,277.8\theta_C + 12 \tag{4}$$

$$M_N = 3Ek(\theta_N - \psi) + (\text{FEM})_N$$

$$M_{CD} = 3\left[29(10^3)(12)^2\right](0.000643)[\theta_C - 0] + 0 \tag{5}$$

$$M_{CD} = 8055.6\theta_C$$

$$M_{CE} = 3\left[29(10^3)(12)^2\right](0.002612)[\theta_C - 0] - 54$$

$$M_{CE} = 32\,725.7\theta_C - 54 \tag{6}$$

Equations of Equilibrium. These six equations contain eight unknowns. Two moment equilibrium equations can be written for joints B and C, Fig. 11–17b. We have

$$M_{BA} + M_{BC} = 0 \tag{7}$$

$$M_{CB} + M_{CD} + M_{CE} = 0 \tag{8}$$

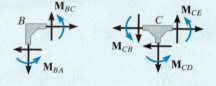

(b)

In order to solve, substitute Eqs. (2) and (3) into Eq. (7), and Eqs. (4)–(6) into Eq. (8). This gives

$$61\,759.3\theta_B + 20\,138.9\theta_C = 12$$

$$20\,138.9\theta_B + 81\,059.0\theta_C = 42$$

Solving these equations simultaneously yields

$$\theta_B = 2.758(10^{-5})\ \text{rad} \qquad \theta_C = 5.113(10^{-4})\ \text{rad}$$

These values, being clockwise, tend to distort the frame as shown in Fig. 11–17a. Substituting these values into Eqs. (1)–(6) and solving, we get

$$M_{AB} = 0.296\ \text{k} \cdot \text{ft} \qquad\qquad Ans.$$

$$M_{BA} = 0.592\ \text{k} \cdot \text{ft} \qquad\qquad Ans.$$

$$M_{BC} = -0.592\ \text{k} \cdot \text{ft} \qquad\qquad Ans.$$

$$M_{CB} = 33.1\ \text{k} \cdot \text{ft} \qquad\qquad Ans.$$

$$M_{CD} = 4.12\ \text{k} \cdot \text{ft} \qquad\qquad Ans.$$

$$M_{CE} = -37.3\ \text{k} \cdot \text{ft} \qquad\qquad Ans.$$

11

11.5   Analysis of Frames: Sidesway

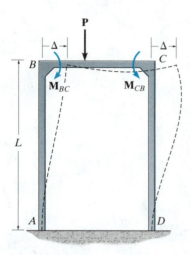

Fig. 11–18

A frame will sidesway, or be displaced to the side, when it or the loading acting on it is nonsymmetric. To illustrate this effect, consider the frame shown in Fig. 11–18. Here the loading **P** causes *unequal* moments M_{BC} and M_{CB} at the joints B and C, respectively. M_{BC} tends to displace joint B to the right, whereas M_{CB} tends to displace joint C to the left. Since M_{BC} is larger than M_{CB}, the net result is a sidesway Δ of both joints B and C to the right, as shown in the figure.* When applying the slope-deflection equation to each column of this frame, we must therefore consider the column rotation ψ (since $\psi = \Delta/L$) as unknown in the equation. As a result an extra equilibrium equation must be included for the solution. In the previous sections it was shown that unknown *angular displacements* θ were related by joint *moment equilibrium equations*. In a similar manner, when unknown joint *linear displacements* Δ (or span rotations ψ) occur, we must write *force equilibrium equations* in order to obtain the complete solution. The unknowns in these equations, however, must only involve the internal *moments* acting at the ends of the columns, since the slope-deflection equations involve these moments. The technique for solving problems for frames with sidesway is best illustrated by examples.

EXAMPLE 11.7

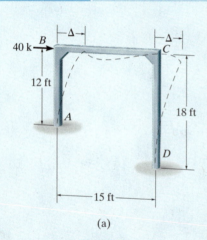

(a)

Fig. 11–19

Determine the moments at each joint of the frame shown in Fig. 11–19a. EI is constant.

SOLUTION

Slope-Deflection Equations.   Since the ends A and D are fixed, Eq. 11–8 applies for all three spans of the frame. Sidesway occurs here since both the applied loading and the geometry of the frame are nonsymmetric. Here the load is applied directly to joint B and therefore no FEMs act at the joints. As shown in Fig. 11–19a, both joints B and C are assumed to be displaced an *equal amount* Δ. Consequently, $\psi_{AB} = \Delta/12$ and $\psi_{DC} = \Delta/18$. Both terms are positive since the cords of members AB and CD "rotate" clockwise. Relating ψ_{AB} to ψ_{DC}, we have $\psi_{AB} = (18/12)\psi_{DC}$. Applying Eq. 11–8 to the frame, we have

$$M_{AB} = 2E\left(\frac{I}{12}\right)\left[2(0) + \theta_B - 3\left(\frac{18}{12}\psi_{DC}\right)\right] + 0 = EI(0.1667\theta_B - 0.75\psi_{DC}) \tag{1}$$

$$M_{BA} = 2E\left(\frac{I}{12}\right)\left[2\theta_B + 0 - 3\left(\frac{18}{12}\psi_{DC}\right)\right] + 0 = EI(0.333\theta_B - 0.75\psi_{DC}) \tag{2}$$

$$M_{BC} = 2E\left(\frac{I}{15}\right)\left[2\theta_B + \theta_C - 3(0)\right] + 0 = EI(0.267\theta_B + 0.133\theta_C) \tag{3}$$

*Recall that the deformation of all three members due to shear and axial force is neglected.

$$M_{CB} = 2E\left(\frac{I}{15}\right)[2\theta_C + \theta_B - 3(0)] + 0 = EI(0.267\theta_C + 0.133\theta_B) \quad (4)$$

$$M_{CD} = 2E\left(\frac{I}{18}\right)[2\theta_C + 0 - 3\psi_{DC}] + 0 = EI(0.222\theta_C - 0.333\psi_{DC}) \quad (5)$$

$$M_{DC} = 2E\left(\frac{I}{18}\right)[2(0) + \theta_C - 3\psi_{DC}] + 0 = EI(0.111\theta_C - 0.333\psi_{DC}) \quad (6)$$

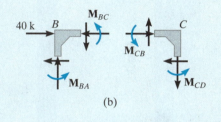

Equations of Equilibrium. The six equations contain nine unknowns. Two moment equilibrium equations for joints B and C, Fig. 11–19b, can be written, namely,

$$M_{BA} + M_{BC} = 0 \quad (7)$$
$$M_{CB} + M_{CD} = 0 \quad (8)$$

(b)

Since a horizontal displacement Δ occurs, we will consider summing forces on the *entire frame* in the x direction. This yields

$$\xrightarrow{+} \Sigma F_x = 0; \qquad 40 - V_A - V_D = 0$$

The horizontal reactions or column shears V_A and V_D can be related to the internal moments by considering the free-body diagram of each column separately, Fig. 11–19c. We have

$$\Sigma M_B = 0; \qquad V_A = -\frac{M_{AB} + M_{BA}}{12}$$

$$\Sigma M_C = 0; \qquad V_D = -\frac{M_{DC} + M_{CD}}{18}$$

Thus,

$$40 + \frac{M_{AB} + M_{BA}}{12} + \frac{M_{DC} + M_{CD}}{18} = 0 \quad (9)$$

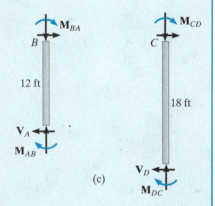

(c)

In order to solve, substitute Eqs. (2) and (3) into Eq. (7), Eqs. (4) and (5) into Eq. (8), and Eqs. (1), (2), (5), (6) into Eq. (9). This yields

$$0.6\theta_B + 0.133\theta_C - 0.75\psi_{DC} = 0$$
$$0.133\theta_B + 0.489\theta_C - 0.333\psi_{DC} = 0$$

$$0.5\theta_B + 0.222\theta_C - 1.944\psi_{DC} = -\frac{480}{EI}$$

Solving simultaneously, we have

$$EI\theta_B = 438.81 \qquad EI\theta_C = 136.18 \qquad EI\psi_{DC} = 375.26$$

Finally, using these results and solving Eqs. (1)–(6) yields

$$M_{AB} = -208 \text{ k} \cdot \text{ft} \qquad\qquad \textit{Ans.}$$
$$M_{BA} = -135 \text{ k} \cdot \text{ft} \qquad\qquad \textit{Ans.}$$
$$M_{BC} = 135 \text{ k} \cdot \text{ft} \qquad\qquad \textit{Ans.}$$
$$M_{CB} = 94.8 \text{ k} \cdot \text{ft} \qquad\qquad \textit{Ans.}$$
$$M_{CD} = -94.8 \text{ k} \cdot \text{ft} \qquad\qquad \textit{Ans.}$$
$$M_{DC} = -110 \text{ k} \cdot \text{ft} \qquad\qquad \textit{Ans.}$$

11

EXAMPLE 11.8

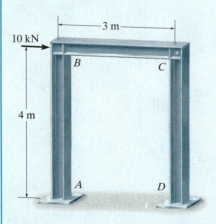

(a)

Fig. 11–20

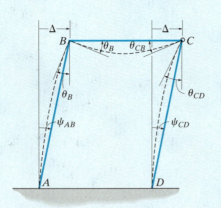

(b)

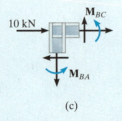

(c)

Determine the moments at each joint of the frame shown in Fig. 11–20a. The supports at A and D are fixed and joint C is assumed pin connected. EI is constant for each member.

SOLUTION

Slope-Deflection Equations.   We will apply Eq. 11–8 to member AB since it is fixed connected at both ends. Equation 11–10 can be applied from B to C and from D to C since the pin at C supports zero moment. As shown by the deflection diagram, Fig. 11–20b, there is an unknown linear displacement Δ of the frame and unknown angular displacement θ_B at joint B.* Due to Δ, the cord members AB and CD rotate clockwise, $\psi = \psi_{AB} = \psi_{DC} = \Delta/4$. Realizing that $\theta_A = \theta_D = 0$ and that there are no FEMs for the members, we have

$$M_N = 2E\left(\frac{I}{L}\right)(2\theta_N + \theta_F - 3\psi) + (FEM)_N$$

$$M_{AB} = 2E\left(\frac{I}{4}\right)[2(0) + \theta_B - 3\psi] + 0 \tag{1}$$

$$M_{BA} = 2E\left(\frac{I}{4}\right)(2\theta_B + 0 - 3\psi) + 0 \tag{2}$$

$$M_N = 3E\left(\frac{I}{L}\right)(\theta_N - \psi) + (FEM)_N$$

$$M_{BC} = 3E\left(\frac{I}{3}\right)(\theta_B - 0) + 0 \tag{3}$$

$$M_{DC} = 3E\left(\frac{I}{4}\right)(0 - \psi) + 0 \tag{4}$$

Equilibrium Equations.   Moment equilibrium of joint B, Fig. 11–20c, requires

$$M_{BA} + M_{BC} = 0 \tag{5}$$

If forces are summed for the *entire frame* in the horizontal direction, we have

$$\overset{+}{\rightarrow}\Sigma F_x = 0; \qquad 10 - V_A - V_D = 0 \tag{6}$$

As shown on the free-body diagram of each column, Fig. 11–20d, we have

$$\Sigma M_B = 0; \qquad V_A = -\frac{M_{AB} + M_{BA}}{4}$$

$$\Sigma M_C = 0; \qquad V_D = -\frac{M_{DC}}{4}$$

*The angular displacements θ_{CB} and θ_{CD} at joint C (pin) are not included in the analysis since Eq. 11–10 is to be used.

11

Thus, from Eq. (6),

$$10 + \frac{M_{AB} + M_{BA}}{4} + \frac{M_{DC}}{4} = 0 \qquad (7)$$

Substituting the slope-deflection equations into Eqs. (5) and (7) and simplifying yields

$$\theta_B = \frac{3}{4}\psi$$

$$10 + \frac{EI}{4}\left(\frac{3}{2}\theta_B - \frac{15}{4}\psi\right) = 0$$

Thus,

$$\theta_B = \frac{240}{21EI} \qquad \psi = \frac{320}{21EI}$$

Substituting these values into Eqs. (1)–(4), we have

$M_{AB} = -17.1 \text{ kN} \cdot \text{m}, \qquad M_{BA} = -11.4 \text{ kN} \cdot \text{m}$ *Ans.*

$M_{BC} = 11.4 \text{ kN} \cdot \text{m}, \qquad M_{DC} = -11.4 \text{ kN} \cdot \text{m}$ *Ans.*

Using these results, the end reactions on each member can be determined from the equations of equilibrium, Fig. 11–20e. The moment diagram for the frame is shown in Fig. 11–20f.

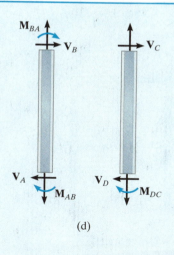

(d)

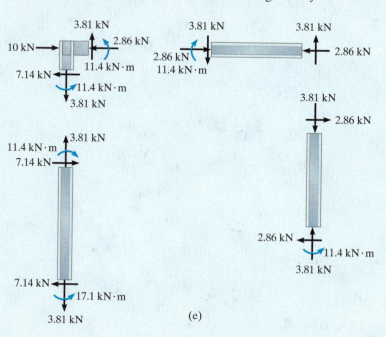

(e)

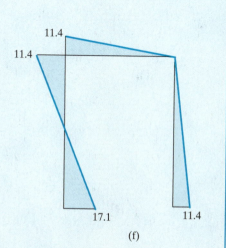

(f)

EXAMPLE | 11.9

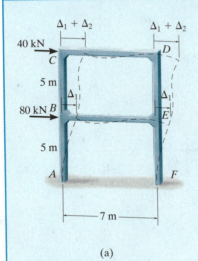

(a)

Fig. 11–21

Explain how the moments in each joint of the two-story frame shown in Fig. 11–21a are determined. EI is constant.

SOLUTION

Slope-Deflection Equation. Since the supports at A and F are fixed, Eq. 11–8 applies for all six spans of the frame. No FEMs have to be calculated, since the applied loading acts at the joints. Here the loading displaces joints B and E an amount Δ_1, and C and D an amount $\Delta_1 + \Delta_2$. As a result, members AB and FE undergo rotations of $\psi_1 = \Delta_1/5$ and BC and ED undergo rotations of $\psi_2 = \Delta_2/5$.

Applying Eq. 11–8 to the frame yields

$$M_{AB} = 2E\left(\frac{I}{5}\right)[2(0) + \theta_B - 3\psi_1] + 0 \tag{1}$$

$$M_{BA} = 2E\left(\frac{I}{5}\right)[2\theta_B + 0 - 3\psi_1] + 0 \tag{2}$$

$$M_{BC} = 2E\left(\frac{I}{5}\right)[2\theta_B + \theta_C - 3\psi_2] + 0 \tag{3}$$

$$M_{CB} = 2E\left(\frac{I}{5}\right)[2\theta_C + \theta_B - 3\psi_2] + 0 \tag{4}$$

$$M_{CD} = 2E\left(\frac{I}{7}\right)[2\theta_C + \theta_D - 3(0)] + 0 \tag{5}$$

$$M_{DC} = 2E\left(\frac{I}{7}\right)[2\theta_D + \theta_C - 3(0)] + 0 \tag{6}$$

$$M_{BE} = 2E\left(\frac{I}{7}\right)[2\theta_B + \theta_E - 3(0)] + 0 \tag{7}$$

$$M_{EB} = 2E\left(\frac{I}{7}\right)[2\theta_E + \theta_B - 3(0)] + 0 \tag{8}$$

$$M_{ED} = 2E\left(\frac{I}{5}\right)[2\theta_E + \theta_D - 3\psi_2] + 0 \tag{9}$$

$$M_{DE} = 2E\left(\frac{I}{5}\right)[2\theta_D + \theta_E - 3\psi_2] + 0 \tag{10}$$

$$M_{FE} = 2E\left(\frac{I}{5}\right)[2(0) + \theta_E - 3\psi_1] + 0 \tag{11}$$

$$M_{EF} = 2E\left(\frac{I}{5}\right)[2\theta_E + 0 - 3\psi_1] + 0 \tag{12}$$

These 12 equations contain 18 unknowns.

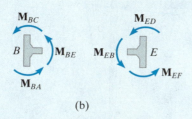

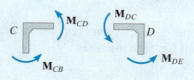

(b)

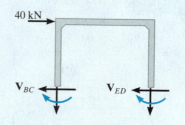

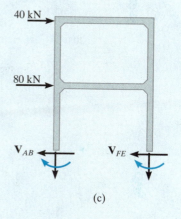

(c)

Equilibrium Equations.   Moment equilibrium of joints B, C, D, and E, Fig. 11–21b, requires

$$M_{BA} + M_{BE} + M_{BC} = 0 \qquad (13)$$

$$M_{CB} + M_{CD} = 0 \qquad (14)$$

$$M_{DC} + M_{DE} = 0 \qquad (15)$$

$$M_{EF} + M_{EB} + M_{ED} = 0 \qquad (16)$$

As in the preceding examples, the shear at the base of all the columns for any story must balance the applied horizontal loads, Fig. 11–21c. This yields

$$\xrightarrow{+} \Sigma F_x = 0; \qquad 40 - V_{BC} - V_{ED} = 0$$

$$40 + \frac{M_{BC} + M_{CB}}{5} + \frac{M_{ED} + M_{DE}}{5} = 0 \qquad (17)$$

$$\xrightarrow{+} \Sigma F_x = 0; \qquad 40 + 80 - V_{AB} - V_{FE} = 0$$

$$120 + \frac{M_{AB} + M_{BA}}{5} + \frac{M_{EF} + M_{FE}}{5} = 0 \qquad (18)$$

Solution requires substituting Eqs. (1)–(12) into Eqs. (13)–(18), which yields six equations having six unknowns, $\psi_1, \psi_2, \theta_B, \theta_C, \theta_D$, and θ_E. These equations can then be solved simultaneously. The results are resubstituted into Eqs. (1)–(12), which yields the moments at the joints.

11

EXAMPLE   11.10

Determine the moments at each joint of the frame shown in Fig. 11–22a. *EI* is constant for each member.

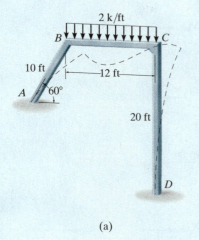

(a)

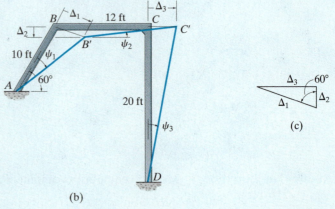

(b)

(c)

Fig. 11–22

SOLUTION

Slope-Deflection Equations.   Equation 11–8 applies to each of the three spans. The FEMs are

$$(\text{FEM})_{BC} = -\frac{wL^2}{12} = -\frac{2(12)^2}{12} = -24 \text{ k} \cdot \text{ft}$$

$$(\text{FEM})_{CB} = \frac{wL^2}{12} = \frac{2(12)^2}{12} = 24 \text{ k} \cdot \text{ft}$$

The sloping member *AB* causes the frame to sidesway to the right as shown in Fig. 11–22a. As a result, joints *B* and *C* are subjected to both rotational *and* linear displacements. The linear displacements are shown in Fig. 11–22b, where *B* moves Δ_1 to *B'* and *C* moves Δ_3 to *C'*. These displacements cause the members' cords to rotate ψ_1, ψ_3 (clockwise) and $-\psi_2$ (counterclockwise) as shown.* Hence,

$$\psi_1 = \frac{\Delta_1}{10} \qquad \psi_2 = -\frac{\Delta_2}{12} \qquad \psi_3 = \frac{\Delta_3}{20}$$

As shown in Fig. 11–22c, the three displacements can be related. For example, $\Delta_2 = 0.5\Delta_1$ and $\Delta_3 = 0.866\Delta_1$. Thus, from the above equations we have

$$\psi_2 = -0.417\psi_1 \qquad \psi_3 = 0.433\psi_1$$

Using these results, the slope-deflection equations for the frame are

---

*Recall that distortions due to axial forces are neglected and the arc displacements *BB'* and *CC'* can be considered as straight lines, since ψ_1 and ψ_3 are actually very small.

$$M_{AB} = 2E\left(\frac{I}{10}\right)[2(0) + \theta_B - 3\psi_1] + 0 \tag{1}$$

$$M_{BA} = 2E\left(\frac{I}{10}\right)[2\theta_B + 0 - 3\psi_1] + 0 \tag{2}$$

$$M_{BC} = 2E\left(\frac{I}{12}\right)[2\theta_B + \theta_C - 3(-0.417\psi_1)] - 24 \tag{3}$$

$$M_{CB} = 2E\left(\frac{I}{12}\right)[2\theta_C + \theta_B - 3(-0.417\psi_1)] + 24 \tag{4}$$

$$M_{CD} = 2E\left(\frac{I}{20}\right)[2\theta_C + 0 - 3(0.433\psi_1)] + 0 \tag{5}$$

$$M_{DC} = 2E\left(\frac{I}{20}\right)[2(0) + \theta_C - 3(0.433\psi_1)] + 0 \tag{6}$$

These six equations contain nine unknowns.

Equations of Equilibrium. Moment equilibrium at joints B and C yields

$$M_{BA} + M_{BC} = 0 \tag{7}$$
$$M_{CD} + M_{CB} = 0 \tag{8}$$

The necessary third equilibrium equation can be obtained by summing moments about point O on the entire frame, Fig. 11–22d. This eliminates the unknown normal forces $\mathbf{N}_A$ and $\mathbf{N}_D$, and therefore

$$\zeta + \Sigma M_O = 0;$$

$$M_{AB} + M_{DC} - \left(\frac{M_{AB} + M_{BA}}{10}\right)(34) - \left(\frac{M_{DC} + M_{CD}}{20}\right)(40.78) - 24(6) = 0$$

$$-2.4M_{AB} - 3.4M_{BA} - 2.04M_{CD} - 1.04M_{DC} - 144 = 0 \tag{9}$$

Substituting Eqs. (2) and (3) into Eq. (7), Eqs. (4) and (5) into Eq. (8), and Eqs. (1), (2), (5), and (6) into Eq. (9) yields

$$0.733\theta_B + 0.167\theta_C - 0.392\psi_1 = \frac{24}{EI}$$

$$0.167\theta_B + 0.533\theta_C + 0.0784\psi_1 = -\frac{24}{EI}$$

$$-1.840\theta_B - 0.512\theta_C + 3.880\psi_1 = \frac{144}{EI}$$

Solving these equations simultaneously yields

$$EI\theta_B = 87.67 \qquad EI\theta_C = -82.3 \qquad EI\psi_1 = 67.83$$

Substituting these values into Eqs. (1)–(6), we have

$M_{AB} = -23.2 \text{ k} \cdot \text{ft} \quad M_{BC} = 5.63 \text{ k} \cdot \text{ft} \quad M_{CD} = -25.3 \text{ k} \cdot \text{ft}$ *Ans.*

$M_{BA} = -5.63 \text{ k} \cdot \text{ft} \quad M_{CB} = 25.3 \text{ k} \cdot \text{ft} \quad M_{DC} = -17.0 \text{ k} \cdot \text{ft}$ *Ans.*

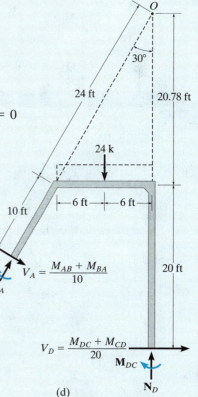

(d)

11

PROBLEMS

Sec. 11.1–11.3

11–1. Determine the moments at A and B, then draw the moment diagram for the beam. EI is constant.

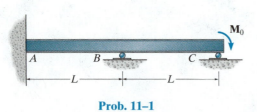

Prob. 11–1

11–2. The continuous beam supports the three concentrated loads. Determine the maximum moment in the beam and then draw the moment diagram. EI is constant.

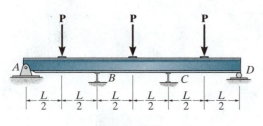

Prob. 11–2

11–3. Determine the moments at B and C of the overhanging beam, then draw the bending moment diagram. EI is constant. Assume the beam is supported by a pin at A and rollers at B and C.

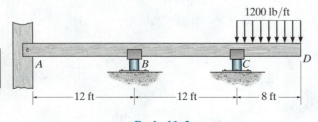

Prob. 11–3

*11–4. Determine the internal moments at the supports A, B, and C, then draw the moment diagram. Assume A is pinned, and B and C are rollers. EI is constant.

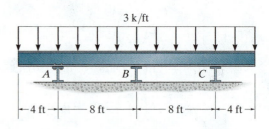

Prob. 11–4

11–5. Determine the moment at B, then draw the moment diagram for the beam. Assume the supports at A and C are fixed. EI is constant.

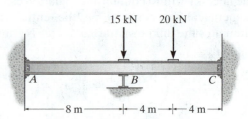

Prob. 11–5

11–6. Determine the reactions at the supports, then draw the moment diagram. Assume A and D are pins and B and C are rollers. The support at B settles 0.03 ft. Take $E = 29(10^3)$ ksi and $I = 4500$ in^2.

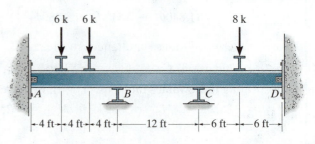

Prob. 11–6

11–7. Determine the internal moments at A and B, then draw the moment diagram. Assume B and C are rollers. EI is constant.

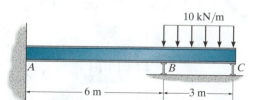

Prob. 11–7

11–10. Determine the moments at A, B, and C. The support at B settles 0.15 ft. $E = 29(10^3)$ ksi and $I = 8000$ in⁴. Assume the supports at B and C are rollers and A is fixed.

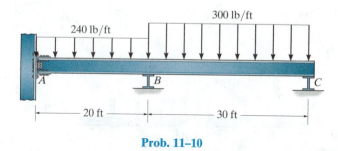

Prob. 11–10

*11–8. Determine the moments at A, B, and C, then draw the moment diagram. EI is constant. Assume the support at B is a roller and A and C are fixed.

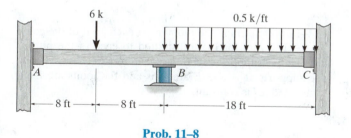

Prob. 11–8

11–11. Determine the moments at A, B, and C, then draw the moment diagram for the beam. Assume the support at A is fixed. B and C are rollers, and D is a pin. EI is constant.

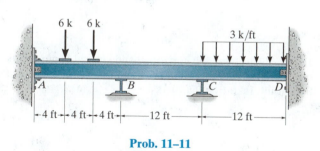

Prob. 11–11

11–9. Determine the reactions at A, B, and C, then draw the moment diagram for the beam. Assume the supports at A and C are pins. EI is constant.

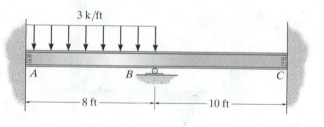

Prob. 11–9

*11–12. Determine the moments at B and C, then draw the moment diagram. Assume A, B, and C are rollers and D is pinned. EI is constant.

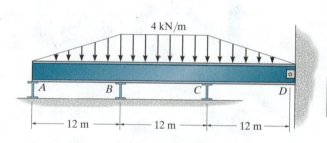

Prob. 11–12

11

Sec. 11.4

11–13. Determine the moments at B, C, and D, then draw the moment diagram for $ABDE$. Assume A is pinned, D is a roller, and C is fixed. EI is constant.

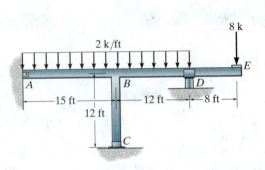

Prob. 11–13

11–14. Determine the horizontal and vertical components of reaction at A and C. Assume A and C are pins and B is a fixed joint. Take $E = 29(10^3)$ ksi.

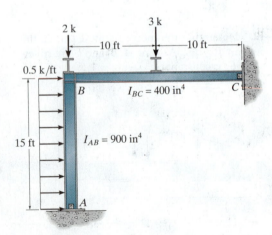

Prob. 11–14

11–15. Determine the moments at each joint and support of the battered-column frame. The joints and supports are fixed connected. EI is constant.

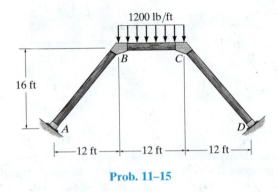

Prob. 11–15

*11–16. Determine the moment at each joint of the gable frame. The roof load is transmitted to each of the purlins over simply supported sections of the roof decking. Assume the supports at A and E are pins and the joints are fixed connected. EI is constant.

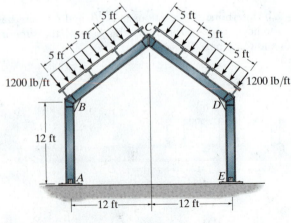

Prob. 11–16

11–17. Determine the moments at the ends of each member of the frame. The supports at A and C and joint B are fixed connected. EI is constant.

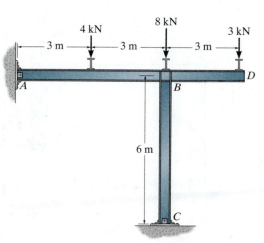

Prob. 11–17

11–18. When the 15 kN/m load is applied to the three-member frame the support at D settles 10 mm. Determine the moment acting at each of the fixed supports A, C, and D. The members are pin connected at B, $E = 200$ GPa, and $I = 800(10^6)$ mm⁴.

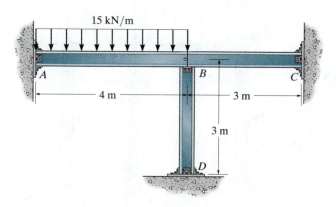

Prob. 11–18

11–19. Determine the moment at B, then draw the moment diagram for each member of the frame. Assume the supports at A and C are pinned and B is a fixed joint. EI is constant.

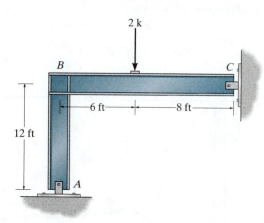

Prob. 11–19

*11–20. The frame at the rear or the truck is made by welding pipe segments together. If the applied load is 1500 lb, determine the moments at the fixed joints B, C, D, and E. Assume the supports at A and F are pinned. EI is constant.

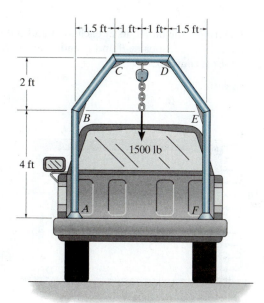

Prob. 11–20

11

Sec. 11.5

11–21. The wood frame is subjected to the load of 6 kN. Determine the moments at the fixed joints A, B, and D. The joint at C is pinned. EI is constant.

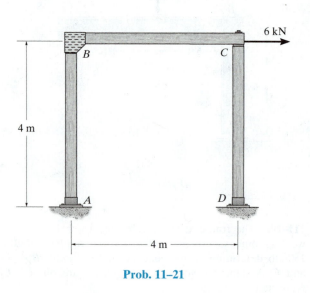

Prob. 11–21

11–22. Determine the moments at each joint and support. There are fixed connections at B and C and fixed supports at A and D. EI is constant.

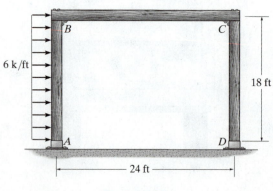

Prob. 11–22

11–23. Determine the moments at each joint and fixed support, then draw the moment diagram. EI is constant.

Prob. 11–23

*11–24. Determine the moments at A, B, C, and D then draw the moment diagram. The members are fixed connected at the supports and joints. EI is constant.

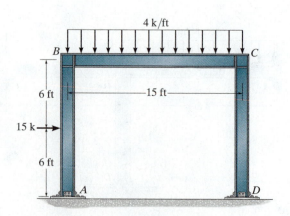

Prob. 11–24

PROJECT PROBLEM

11–1P. The roof is supported by joists that rest on two girders. Each joist can be considered simply supported, and the front girder can be considered attached to the three columns by a pin at *A* and rollers at *B* and *C*. Assume the roof will be made from 3 in.-thick cinder concrete, and each joist has a weight of 550 lb. According to code the roof will be subjected to a snow loading of 25 psf. The joists have a length of 25 ft. Draw the shear and moment diagrams for the girder. Assume the supporting columns are rigid.

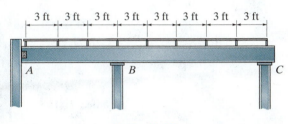

Prob. 11–1P

CHAPTER REVIEW

The unknown displacements of a structure are referred to as the degrees of freedom for the structure. They consist of either joint displacements or rotations.

The slope-deflection equations relate the unknown moments at each joint of a structural member to the unknown rotations that occur there. The following equation is applied twice to each member or span, considering each side as the "near" end and its counterpart as the far end.

$$M_N = 2Ek(2\theta_N + \theta_F - 3\psi) + (FEM)_N$$

For Internal Span or End Span with Far End Fixed

This equation is only applied once, where the "far" end is at the pin or roller support.

$$M_N = 3Ek(\theta_N - \psi) + (FEM)_N$$

Only for End Span with Far End Pinned or Roller Supported

Once the slope-deflection equations are written, they are substituted into the equations of moment equilibrium at each joint and then solved for the unknown displacements. If the structure (frame) has sidesway, then an unknown horizontal displacement at each floor level will occur, and the unknown column shears must be related to the moments at the joints, using both the force and moment equilibrium equations. Once the unknown displacements are obtained, the unknown reactions are found from the load-displacement relations.

11

Chapter 12

© David Grossman/Science Source

Here is an example of a statically indeterminate steel building frame. A portion of the frame can be modeled and then analyzed by the moment distribution method.

Displacement Method of Analysis: Moment Distribution

The moment-distribution method is a displacement method of analysis that is easy to apply once certain elastic constants have been determined. In this chapter we will first state the important definitions and concepts for moment distribution and then apply the method to solve problems involving statically indeterminate beams and frames. Application to multistory frames is discussed in the last part of the chapter.

12.1 General Principles and Definitions

The method of analyzing beams and frames using moment distribution was developed by Hardy Cross, in 1930. At the time this method was first published it attracted immediate attention, and it has been recognized as one of the most notable advances in structural analysis during the twentieth century.

As will be explained in detail later, moment distribution is a method of successive approximations that may be carried out to any desired degree of accuracy. Essentially, the method begins by assuming each joint of a structure is fixed. Then, by unlocking and locking each joint in succession, the internal moments at the joints are "distributed" and balanced until the joints have rotated to their final or nearly final positions. It will be found that this process of calculation is both repetitive and easy to apply. Before explaining the techniques of moment distribution, however, certain definitions and concepts must be presented.

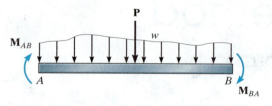

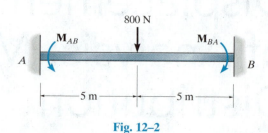

Fig. 12–1

Fig. 12–2

Sign Convention.

We will establish the same sign convention as that established for the slope-deflection equations: *Clockwise moments that act on the member* are considered *positive,* whereas *counterclockwise moments* are *negative,* Fig. 12–1.

Fixed-End Moments (FEMs).

The moments at the "walls" or fixed joints of a loaded member are called *fixed-end moments.* These moments can be determined from the table given on the inside back cover, depending upon the type of loading on the member. For example, the beam loaded as shown in Fig. 12–2 has fixed-end moments of FEM $= PL/8 = 800(10)/8 = 1000 \, \text{N} \cdot \text{m}$. Noting the action of these moments *on the beam* and applying our sign convention, it is seen that $M_{AB} = -1000 \, \text{N} \cdot \text{m}$ and $M_{BA} = +1000 \, \text{N} \cdot \text{m}$.

Member Stiffness Factor.

Consider the beam in Fig. 12–3, which is pinned at one end and fixed at the other. Application of the moment **M** causes the end A to rotate through an angle θ_A. In Chapter 11 we related M to θ_A using the conjugate-beam method. This resulted in Eq. 11–1, that is, $M = (4EI/L)\,\theta_A$. The term in parentheses

$$K = \frac{4EI}{L}$$

Far End Fixed

(12–1)

is referred to as the **stiffness factor** at A and can be defined as the amount of moment M required to rotate the end A of the beam $\theta_A = 1$ rad.

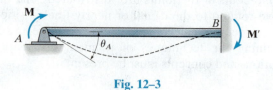

Fig. 12–3

Joint Stiffness Factor.

If several members are fixed connected to a joint and each of their far ends is fixed, then by the principle of superposition, the **total stiffness factor** at the joint is the sum of the member stiffness factors at the joint, that is, $K_T = \Sigma K$. For example, consider the frame joint A in Fig. 12–4a. The numerical value of each member stiffness factor is determined from Eq. 12–1 and listed in the figure. Using these values, the total stiffness factor of joint A is $K_T = \Sigma K = 4000 + 5000 + 1000 = 10\ 000$. This value represents the amount of moment needed to rotate the joint through an angle of 1 rad.

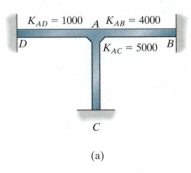

(a)

Distribution Factor (DF).

If a moment **M** is applied to a fixed connected joint, the connecting members will each supply a portion of the resisting moment necessary to satisfy moment equilibrium at the joint. That fraction of the total resisting moment supplied by the member is called the **distribution factor** (DF). To obtain its value, imagine the joint is fixed connected to n members. If an applied moment **M** causes the joint to rotate an amount θ, then each member i rotates by this same amount. If the stiffness factor of the ith member is K_i, then the moment contributed by the member is $M_i = K_i\theta$. Since equilibrium requires $M = M_1 + M_2 + \cdots + M_n = K_1\theta + K_2\theta + \cdots + K_n\theta = \theta\Sigma K_i$ then the distribution factor for the ith member is

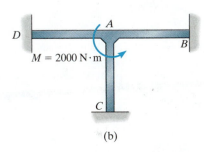

(b)

$$DF_i = \frac{M_i}{M} = \frac{K_i\theta}{\theta\Sigma K_i}$$

Canceling the common term θ, it is seen that the distribution factor for a member is equal to the stiffness factor of the member divided by the total stiffness factor for the joint; that is, in general,

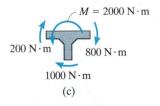

(c)

Fig. 12–4

$$DF = \frac{K}{\Sigma K} \qquad (12\text{–}2)$$

For example, the distribution factors for members AB, AC, and AD at joint A in Fig. 12–4a are

$$DF_{AB} = 4000/10\ 000 = 0.4$$
$$DF_{AC} = 5000/10\ 000 = 0.5$$
$$DF_{AD} = 1000/10\ 000 = 0.1$$

As a result, if $M = 2000\ \text{N} \cdot \text{m}$ acts at joint A, Fig. 12–4b, the equilibrium moments exerted by the members on the joint, Fig. 12–4c, are

$$M_{AB} = 0.4(2000) = 800\ \text{N} \cdot \text{m}$$
$$M_{AC} = 0.5(2000) = 1000\ \text{N} \cdot \text{m}$$
$$M_{AD} = 0.1(2000) = 200\ \text{N} \cdot \text{m}$$

The statically indeterminate loading in bridge girders that are continuous over their piers can be determined using the method of moment distribution.

Member Relative-Stiffness Factor.

Quite often a continuous beam or a frame will be made from the same material so its modulus of elasticity E will be the *same* for all the members. If this is the case, the common factor $4E$ in Eq. 12–1 will *cancel* from the numerator and denominator of Eq. 12–2 when the distribution factor for a joint is determined. Hence, it is *easier* just to determine the member's **relative-stiffness factor**

$$K_R = \frac{I}{L}$$

Far End Fixed

(12–3)

and use this for the computations of the DF.

Carry-Over Factor.

Consider again the beam in Fig. 12–3. It was shown in Chapter 11 that $M_{AB} = (4EI/L)\,\theta_A$ (Eq. 11–1) and $M_{BA} = (2EI/L)\,\theta_A$ (Eq. 11–2). Solving for θ_A and equating these equations we get $M_{BA} = M_{AB}/2$. In other words, the moment $\mathbf{M}$ at the pin induces a moment of $\mathbf{M}' = \frac{1}{2}\mathbf{M}$ at the wall. The **carry-over factor** represents the fraction of $\mathbf{M}$ that is "carried over" from the pin to the wall. Hence, in the case of a beam with *the far end fixed,* the carry-over factor is $+\frac{1}{2}$. The plus sign indicates both moments act in the same direction.

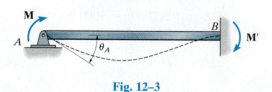

Fig. 12–3

12.2   Moment Distribution for Beams

Moment distribution is based on the principle of successively locking and unlocking the joints of a structure in order to allow the moments at the joints to be distributed and balanced. The best way to explain the method is by examples.

Consider the beam with a constant modulus of elasticity E and having the dimensions and loading shown in Fig. 12–5a. Before we begin, we must first determine the distribution factors at the two ends of each span. Using Eq. 12–1, $K = 4EI/L$, the stiffness factors on either side of B are

$$K_{BA} = \frac{4E(300)}{15} = 4E(20) \text{ in}^4/\text{ft} \qquad K_{BC} = \frac{4E(600)}{20} = 4E(30) \text{ in}^4/\text{ft}$$

Thus, using Eq. 12–2, DF $= K/\Sigma K$, for the ends connected to joint B, we have

$$\text{DF}_{BA} = \frac{4E(20)}{4E(20) + 4E(30)} = 0.4$$

$$\text{DF}_{BC} = \frac{4E(30)}{4E(20) + 4E(30)} = 0.6$$

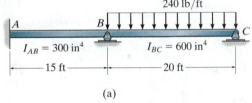

(a)

Fig. 12–5

At the walls, joint A and joint C, the distribution factor depends on the member stiffness factor and the "stiffness factor" of the wall. Since in theory it would take an "infinite" size moment to rotate the wall one radian, the wall stiffness factor is infinite. Thus for joints A and C we have

$$\text{DF}_{AB} = \frac{4E(20)}{\infty + 4E(20)} = 0$$

$$\text{DF}_{CB} = \frac{4E(30)}{\infty + 4E(30)} = 0$$

Note that the above results could also have been obtained if the relative stiffness factor $K_R = I/L$ (Eq. 12–3) had been used for the calculations. Furthermore, as long as a *consistent* set of units is used for the stiffness factor, the DF will always be dimensionless, and at a joint, except where it is located at a fixed wall, the sum of the DFs will always equal 1.

Having computed the DFs, we will now determine the FEMs. Only span BC is loaded, and using the table on the inside back cover for a uniform load, we have

$$(\text{FEM})_{BC} = -\frac{wL^2}{12} = -\frac{240(20)^2}{12} = -8000 \text{ lb} \cdot \text{ft}$$

$$(\text{FEM})_{CB} = \frac{wL^2}{12} = \frac{240(20)^2}{12} = 8000 \text{ lb} \cdot \text{ft}$$

12

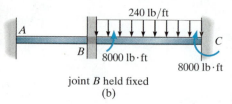

joint B held fixed
(b)

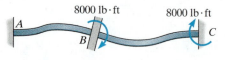

correction moment applied to joint B
(c)

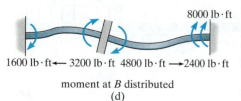

moment at B distributed
(d)

Joint	A	B		C
Member	AB	BA	BC	CB
DF	0	0.4	0.6	0
FEM			−8000	8000
Dist,CO	1600←3200		4800→2400	
ΣM	1600	3200	−3200	10 400

(e)

We begin by assuming joint B is fixed or locked. The fixed-end moment at B then holds span BC in this fixed or locked position as shown in Fig. 12–5b. This, of course, does not represent the actual equilibrium situation at B, since the moments on *each side* of this joint must be equal but opposite. To correct this, we will apply an equal, but opposite moment of 8000 lb · ft to the joint and allow the joint to rotate freely, Fig. 12–5c. As a result, portions of this moment are distributed in spans BC and BA in accordance with the DFs (or stiffness) of these spans at the joint. Specifically, the moment in BA is 0.4(8000) = 3200 lb · ft and the moment in BC is 0.6(8000) = 4800 lb · ft. Finally, due to the released rotation that takes place at B, these moments must be "carried over" since moments are developed at the far ends of the span. Using the carry-over factor of $+\frac{1}{2}$, the results are shown in Fig. 12–5d.

This example indicates the basic steps necessary when distributing moments at a joint: Determine the unbalanced moment acting at the initially "locked" joint, unlock the joint and apply an equal but opposite unbalanced moment to correct the equilibrium, distribute the moment among the connecting spans, and carry the moment in each span over to its other end. The steps are usually presented in tabular form as indicated in Fig. 12–5e. Here the notation Dist, CO indicates a line where moments are distributed, then carried over. In this particular case only *one cycle* of moment distribution is necessary, since the wall supports at A and C "absorb" the moments and no further joints have to be balanced or unlocked to satisfy joint equilibrium. Once distributed in this manner, the moments at each joint are summed, yielding the final results shown on the bottom line of the table in Fig. 12–5e. Notice that joint B is now in equilibrium. Since M_{BC} is negative, this moment is applied to span BC in a counterclockwise sense as shown on free-body diagrams of the beam spans in Fig. 12–5f. With the end moments known, the end shears have been computed from the equations of equilibrium applied to each of these spans.

Consider now the same beam, except the support at C is a rocker, Fig. 12–6a. In this case only *one member* is at joint C, so the distribution factor for member CB at joint C is

$$DF_{CB} = \frac{4E(30)}{4E(30)} = 1$$

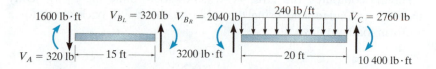

(f)

Fig. 12–5

The other distribution factors and the FEMs are the same as computed previously. They are listed on lines 1 and 2 of the table in Fig. 12–6b. Initially, we will assume joints B and C are locked. We begin by unlocking joint C and placing an equilibrating moment of -8000 lb·ft at the joint. The entire moment is distributed in member CB since $(1)(-8000)$ lb·ft $= -8000$ lb·ft. The arrow on line 3 indicates that $\frac{1}{2}(-8000)$ lb·ft $= -4000$ lb·ft is carried over to joint B since joint C has been allowed to rotate freely. Joint C is now *relocked*. Since the total moment at C is *balanced*, a line is placed under the -8000-lb·ft moment. We will now consider the unbalanced $-12\,000$-lb·ft moment at joint B. Here for equilibrium, a $+12\,000$-lb·ft moment is applied to B and this joint is unlocked such that portions of the moment are distributed into BA and BC, that is, $(0.4)(12\,000) = 4800$ lb·ft and $(0.6)(12\,000) = 7200$ lb·ft as shown on line 4. Also note that $+\frac{1}{2}$ of these moments must be carried over to the fixed wall A and roller C since joint B has rotated. Joint B is now *relocked*. Again joint C is unlocked and the unbalanced moment at the roller is distributed as was done previously. The results are on line 5. Successively locking and unlocking joints B and C will essentially diminish the size of the moment to be balanced until it becomes negligible compared with the original moments, line 14. Each of the steps on lines 3 through 14 should be thoroughly understood. Summing the moments, the final results are shown on line 15, where it is seen that the final moments now satisfy joint equilibrium.

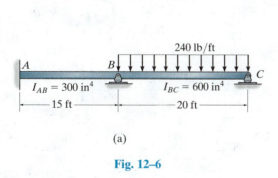

240 lb/ft

A B C

$I_{AB} = 300$ in^4 $I_{BC} = 600$ in^4

15 ft   20 ft

(a)

Fig. 12–6

Joint	A	B		C	
Member	AB	BA	BC	CB	
DF	0	0.4	0.6	1	1
FEM			−8000	8000	2
			−4000	−8000	3
	2400	4800	7200	3600	4
			−1800	−3600	5
	360	720	1080	540	6
			−270	−540	7
	54	108	162	81	8
			−40.5	−81	9
	8.1	16.2	24.3	12.2	10
			−6.1	−12.2	11
	1.2	2.4	3.6	1.8	12
			−0.9	−1.8	13
			0.4	0.5	14
ΣM	2823.3	5647.0	−5647.0	0	15

(b)

Rather than applying the moment distribution process successively to each joint, as illustrated here, it is also possible to apply it to all joints at the *same time*. This scheme is shown in the table in Fig. 12–6c. In this case, we start by fixing all the joints and then balancing and distributing the fixed-end moments at both joints B and C, line 3. Unlocking joints B and C simultaneously (joint A is always fixed), the moments are then carried over to the end of each span, line 4. Again the joints are relocked, and the moments are balanced and distributed, line 5. Unlocking the joints once again allows the moments to be carried over, as shown in line 6. Continuing, we obtain the final results, as before, listed on line 24. By comparison, this method gives a slower convergence to the answer than does the previous method; however, in many cases this method will be more efficient to apply, and for this reason we will use it in the examples that follow. Finally, using the results in either Fig. 12–6b or 12–6c, the free-body diagrams of each beam span are drawn as shown in Fig. 12–6d.

Although several steps were involved in obtaining the final results here, the work required is rather methodical since it requires application of a series of arithmetical steps, rather than solving a set of equations as in the slope deflection method. It should be noted, however, that the

Joint	A	B		C	
Member	AB	BA	BC	CB	
DF	0	0.4	0.6	1	1
FEM			−8000	8000	2
Dist.		3200	4800	−8000	3
CO	1600		−4000	2400	4
Dist.		1600	2400	−2400	5
CO	800		−1200	1200	6
Dist.		480	720	−1200	7
CO	240		−600	360	8
Dist.		240	360	−360	9
CO	120		−180	180	10
Dist.		72	108	−180	11
CO	36		−90	54	12
Dist.		36	54	−54	13
CO	18		−27	27	14
Dist.		10.8	16.2	−27	15
CO	5.4		−13.5	8.1	16
Dist.		5.4	8.1	−8.1	17
CO	2.7		−4.05	4.05	18
Dist.		1.62	2.43	−4.05	19
CO	0.81		−2.02	1.22	20
Dist.		0.80	1.22	−1.22	21
CO	0.40		−0.61	0.61	22
Dist.		0.24	0.37	−0.61	23
ΣM	2823	5647	−5647	0	24

(c)

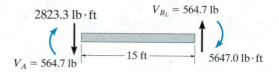

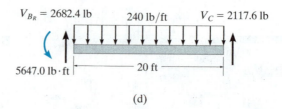

(d)

Fig. 12–6

fundamental process of moment distribution follows the same procedure as any displacement method. There the process is to establish load-displacement relations at each joint and then satisfy joint equilibrium requirements by determining the correct angular displacement for the joint (compatibility). Here, however, the equilibrium and compatibility of rotation at the joint is satisfied *directly*, using a "moment balance" process that incorporates the load-deflection relations (stiffness factors). Further simplification for using moment distribution is possible, and this will be discussed in the next section.

Procedure for Analysis

The following procedure provides a general method for determining the end moments on beam spans using moment distribution.

Distribution Factors and Fixed-End Moments

The joints on the beam should be identified and the stiffness factors for each span at the joints should be calculated. Using these values the distribution factors can be determined from $DF = K/\Sigma K$. Remember that $DF = 0$ for a fixed end and $DF = 1$ for an *end* pin or roller support.

The fixed-end moments for each loaded span are determined using the table given on the inside back cover. Positive FEMs act clockwise on the span and negative FEMs act counterclockwise. For convenience, these values can be recorded in tabular form, similar to that shown in Fig. 12–6c.

Moment Distribution Process

Assume that all joints at which the moments in the connecting spans must be determined are *initially locked*. Then:

1. Determine the moment that is needed to put each joint in equilibrium.

2. Release or "unlock" the joints and distribute the counterbalancing moments into the connecting span at each joint.

3. Carry these moments in each span over to its other end by multiplying each moment by the carry-over factor $+\frac{1}{2}$.

By repeating this cycle of locking and unlocking the joints, it will be found that the moment corrections will diminish since the beam tends to achieve its final deflected shape. When a small enough value for the corrections is obtained, the process of cycling should be stopped with no "carry-over" of the last moments. Each column of FEMs, distributed moments, and carry-over moments should then be added. If this is done correctly, moment equilibrium at the joints will be achieved.

EXAMPLE   12.1

Determine the internal moments at each support of the beam shown in Fig. 12–7a. *EI* is constant.

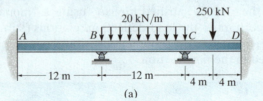

(a)

Fig. 12–7

SOLUTION

The distribution factors at each joint must be computed first.* The stiffness factors for the members are

$$K_{AB} = \frac{4EI}{12} \qquad K_{BC} = \frac{4EI}{12} \qquad K_{CD} = \frac{4EI}{8}$$

Therefore,

$$\mathrm{DF}_{AB} = \mathrm{DF}_{DC} = 0 \qquad \mathrm{DF}_{BA} = \mathrm{DF}_{BC} = \frac{4EI/12}{4EI/12 + 4EI/12} = 0.5$$

$$\mathrm{DF}_{CB} = \frac{4EI/12}{4EI/12 + 4EI/8} = 0.4 \quad \mathrm{DF}_{CD} = \frac{4EI/8}{4EI/12 + 4EI/8} = 0.6$$

The fixed-end moments are

$$(\mathrm{FEM})_{BC} = -\frac{wL^2}{12} = \frac{-20(12)^2}{12} = -240 \text{ kN} \cdot \text{m} \qquad (\mathrm{FEM})_{CB} = \frac{wL^2}{12} = \frac{20(12)^2}{12} = 240 \text{ kN} \cdot \text{m}$$

$$(\mathrm{FEM})_{CD} = -\frac{PL}{8} = \frac{-250(8)}{8} = -250 \text{ kN} \cdot \text{m} \qquad (\mathrm{FEM})_{DC} = \frac{PL}{8} = \frac{250(8)}{8} = 250 \text{ kN} \cdot \text{m}$$

Starting with the FEMs, line 4, Fig. 12–7b, the moments at joints *B* and *C* are distributed *simultaneously*, line 5. These moments are then carried over *simultaneously* to the respective ends of each span, line 6. The resulting moments are again simultaneously distributed and carried over, lines 7 and 8. The process is continued until the resulting moments are diminished an appropriate amount, line 13. The resulting moments are found by summation, line 14.

Placing the moments on each beam span and applying the equations of equilibrium yields the end shears shown in Fig. 12–7c and the bending-moment diagram for the entire beam, Fig. 12–7d.

---

* Here we have used the stiffness factor $4EI/L$; however, the relative stiffness factor I/L could also have been used.

Joint	A	B		C		D	1
Member	AB	BA	BC	CB	CD	DC	2
DF	0	0.5	0.5	0.4	0.6	0	3
FEM			−240	240	−250	250	4
Dist.		120	120	4	6		5
CO	60		2	60		3	6
Dist.		−1	−1	−24	−36		7
CO	−0.5		−12	−0.5		−18	8
Dist.		6	6	0.2	0.3		9
CO	3		0.1	3		0.2	10
Dist.		−0.05	−0.05	−1.2	−1.8		11
CO	−0.02		−0.6	−0.02		−0.9	12
Dist.		0.3	0.3	0.01	0.01		13
ΣM	62.5	125.2	−125.2	281.5	−281.5	234.3	14

(b)

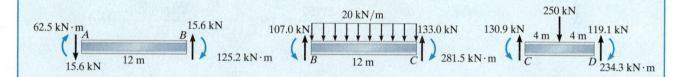

(c)

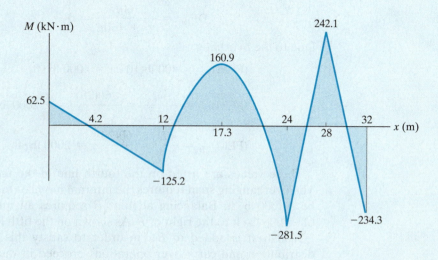

(d)

EXAMPLE 12.2

Determine the internal moment at each support of the beam shown in Fig. 12–8a. The moment of inertia of each span is indicated.

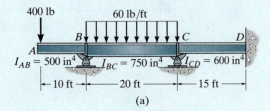

(a)

Fig. 12–8

SOLUTION

In this problem a moment does not get distributed in the overhanging span AB, and so the distribution factor $(DF)_{BA} = 0$. The stiffness of span BC is based on $4EI/L$ since the pin rocker is not at the far end of the beam. The stiffness factors, distribution factors, and fixed-end moments are computed as follows:

$$K_{BC} = \frac{4E(750)}{20} = 150E \qquad K_{CD} = \frac{4E(600)}{15} = 160E$$

$$DF_{BC} = 1 - (DF)_{BA} = 1 - 0 = 1$$

$$DF_{CB} = \frac{150E}{150E + 160E} = 0.484$$

$$DF_{CD} = \frac{160E}{150E + 160E} = 0.516$$

$$DF_{DC} = \frac{160E}{\infty + 160E} = 0$$

Due to the overhang,

$$(FEM)_{BA} = 400\ \text{lb}(10\ \text{ft}) = 4000\ \text{lb} \cdot \text{ft}$$

$$(FEM)_{BC} = -\frac{wL^2}{12} = -\frac{60(20)^2}{12} = -2000\ \text{lb} \cdot \text{ft}$$

$$(FEM)_{CB} = \frac{wL^2}{12} = \frac{60(20)^2}{12} = 2000\ \text{lb} \cdot \text{ft}$$

These values are listed on the fourth line of the table, Fig. 12–8b. The overhanging span requires the internal moment to the left of B to be $+4000\ \text{lb} \cdot \text{ft}$. Balancing at joint B requires an internal moment of $-4000\ \text{lb} \cdot \text{ft}$ to the right of B. As shown on the fifth line of the table $-2000\ \text{lb} \cdot \text{ft}$ is added to BC in order to satisfy this condition. The distribution and carry-over operations proceed in the usual manner as indicated.

Since the internal moments are known, the moment diagram for the beam can be constructed (Fig. 12–8c).

Joint		B	C		D
Member		BC	CB	CD	DC
DF	0	1	0.484	0.516	0
FEM	4000	−2000	2000		
Dist.		−2000	−968	−1032	
CO		−484	−1000		−516
Dist.		484	484	516	
CO		242	242		258
Dist.		−242	−117.1	−124.9	
CO		−58.6	−121		−62.4
Dist.		58.6	58.6	62.4	
CO		29.3	29.3		31.2
Dist.		−29.3	−14.2	−15.1	
CO		−7.1	−14.6		−7.6
Dist.		7.1	7.1	7.6	
CO		3.5	3.5		3.8
Dist.		−3.5	−1.7	−1.8	
CO		−0.8	−1.8		−0.9
Dist.		0.8	0.9	0.9	
CO		0.4	0.4		0.4
Dist.		−0.4	−0.2	−0.2	
CO		−0.1	−0.2		−0.1
Dist.		0.1	0.1	0.1	
ΣM	4000	−4000	587.1	−587.1	−293.6

(b)

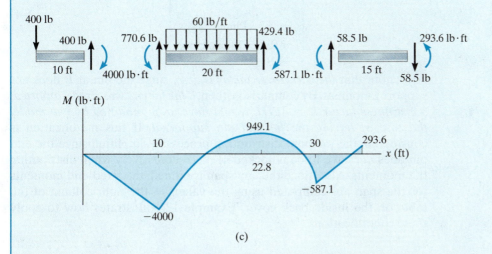

(c)

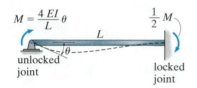

$$M = \frac{4EI}{L}\theta \qquad \frac{1}{2}M$$

unlocked
joint

L

locked
joint

Fig. 12–9

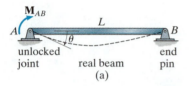

M_{AB}

A L B

θ

unlocked
joint

real beam
(a)

end
pin

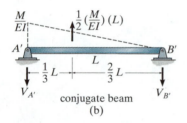

$\frac{M}{EI}$ $\frac{1}{2}\left(\frac{M}{EI}\right)(L)$

A' B'

$\frac{1}{3}L$ | $\frac{2}{3}L$

L

$V_{A'}$ conjugate beam $V_{B'}$
(b)

Fig. 12–10

12.3   Stiffness-Factor Modifications

In the previous examples of moment distribution we have considered each beam span to be constrained by a fixed support (locked joint) at its far end when distributing and carrying over the moments. For this reason we have computed the stiffness factors, distribution factors, and the carry-over factors based on the case shown in Fig. 12–9. Here, of course, the stiffness factor is $K = 4EI/L$ (Eq. 12–1), and the carry-over factor is $+\frac{1}{2}$.

In some cases it is possible to modify the stiffness factor of a particular beam span and thereby simplify the process of moment distribution. Three cases where this frequently occurs in practice will now be considered.

Member Pin Supported at Far End.
Many indeterminate beams have their far end span supported by an end pin (or roller) as in the case of joint B in Fig. 12–10a. Here the applied moment **M** rotates the end A by an amount θ. To determine θ, the shear in the conjugate beam at A' must be determined, Fig. 12–10b. We have

$$\zeta + \Sigma M_{B'} = 0; \quad V_A'(L) - \frac{1}{2}\left(\frac{M}{EI}\right)L\left(\frac{2}{3}L\right) = 0$$

$$V_A' = \theta = \frac{ML}{3EI}$$

or

$$M = \frac{3EI}{L}\theta$$

Thus, the stiffness factor for this beam is

$$K = \frac{3EI}{L}$$

Far End Pinned
or Roller Supported

(12–4)

Also, note that *the carry-over factor is zero*, since the pin at B does not support a moment. By comparison, then, *if the far end was fixed supported, the stiffness factor $K = 4EI/L$ would have to be modified by $\frac{3}{4}$ to model the case of having the far end pin supported.* If this modification is considered, the moment distribution process is simplified since the end pin does *not* have to be unlocked–locked successively when distributing the moments. Also, since the end span is pinned, the fixed-end moments for the span are computed using the values in the right column of the table on the inside back cover. Example 12.4 illustrates how to apply these simplifications.

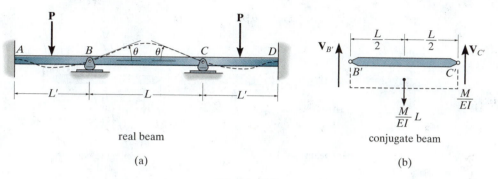

real beam

(a)

conjugate beam

(b)

Fig. 12–11

Symmetric Beam and Loading.

If a beam is symmetric with respect to both its loading and geometry, the bending-moment diagram for the beam will also be symmetric. As a result, a modification of the stiffness factor for the center span can be made, so that moments in the beam only have to be distributed through joints lying on either half of the beam. To develop the appropriate stiffness-factor modification, consider the beam shown in Fig. 12–11a. Due to the symmetry, the internal moments at B and C are equal. Assuming this value to be $\mathbf{M}$, the conjugate beam for span BC is shown in Fig. 12–11b. The slope θ at each end is therefore

$$\zeta + \Sigma M_{C'} = 0; \qquad -V_{B'}(L) + \frac{M}{EI}(L)\left(\frac{L}{2}\right) = 0$$

$$V_{B'} = \theta = \frac{ML}{2EI}$$

or

$$M = \frac{2EI}{L}\theta$$

The stiffness factor for the center span is therefore

$$\boxed{K = \frac{2EI}{L}}$$
$$\text{Symmetric Beam and Loading} \qquad (12\text{--}5)$$

Thus, moments for only half the beam can be distributed provided the stiffness factor for the center span is computed using Eq. 12–5. *By comparison, the center span's stiffness factor will be one half that usually determined using $K = 4EI/L$.*

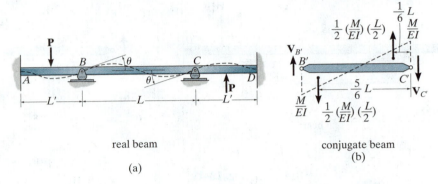

real beam

(a)

conjugate beam

(b)

Fig. 12–12

Symmetric Beam with Antisymmetric Loading. If a symmetric beam is subjected to antisymmetric loading, the resulting moment diagram will be antisymmetric. As in the previous case, we can modify the stiffness factor of the center span so that only one half of the beam has to be considered for the moment-distribution analysis. Consider the beam in Fig. 12–12*a*. The conjugate beam for its center span *BC* is shown in Fig. 12–12*b*. Due to the antisymmetric loading, the internal moment at *B* is equal, but opposite to that at *C*. Assuming this value to be **M**, the slope θ at each end is determined as follows:

$$\zeta + \Sigma M_{C'} = 0; \quad -V_{B'}(L) + \frac{1}{2}\left(\frac{M}{EI}\right)\left(\frac{L}{2}\right)\left(\frac{5L}{6}\right) - \frac{1}{2}\left(\frac{M}{EI}\right)\left(\frac{L}{2}\right)\left(\frac{L}{6}\right) = 0$$

$$V_{B'} = \theta = \frac{ML}{6EI}$$

or

$$M = \frac{6EI}{L}\theta$$

The stiffness factor for the center span is, therefore,

$$K = \frac{6EI}{L}$$

Symmetric Beam with
Antisymmetric Loading

(12–6)

Thus, when the stiffness factor for the beam's center span is computed using Eq. 12–6, the moments in only half the beam have to be distributed. *Here the stiffness factor is one and a half times as large as that determined using* $K = 4EI/L$.

EXAMPLE 12.3

Determine the internal moments at the supports for the beam shown in Fig. 12–13a. *EI* is constant.

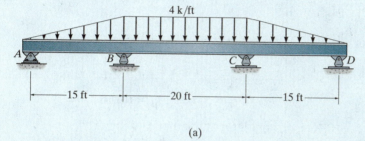

4 k/ft

A B C D

|—15 ft—|—20 ft—|—15 ft—|

(a)

Fig. 12–13

SOLUTION

By inspection, the beam and loading are symmetrical. Thus, we will apply $K = 2EI/L$ to compute the stiffness factor of the center span *BC* and therefore use only the left half of the beam for the analysis. The analysis can be shortened even further by using $K = 3EI/L$ for computing the stiffness factor of segment *AB* since the far end *A* is pinned. Furthermore, the distribution of moment at *A* can be skipped by using the FEM for a triangular loading on a span with one end fixed and the other pinned. Thus,

$$K_{AB} = \frac{3EI}{15} \quad \text{(using Eq. 12–4)}$$

$$K_{BC} = \frac{2EI}{20} \quad \text{(using Eq. 12–5)}$$

$$DF_{AB} = \frac{3EI/15}{3EI/15} = 1$$

$$DF_{BA} = \frac{3EI/15}{3EI/15 + 2EI/20} = 0.667$$

$$DF_{BC} = \frac{2EI/20}{3EI/15 + 2EI/20} = 0.333$$

$$(FEM)_{BA} = \frac{wL^2}{15} = \frac{4(15)^2}{15} = 60 \text{ k} \cdot \text{ft}$$

$$(FEM)_{BC} = -\frac{wL^2}{12} = -\frac{4(20)^2}{12} = -133.3 \text{ k} \cdot \text{ft}$$

Joint	A	B	
Member	AB	BA	BC
DF	1	0.667	0.333
FEM Dist.		60 48.9	−133.3 24.4
ΣM	0	108.9	−108.9

(b)

These data are listed in the table in Fig. 12–13b. Computing the stiffness factors as shown above considerably reduces the analysis, since only joint *B* must be balanced and carry-overs to joints *A* and *C* are not necessary. Obviously, joint *C* is subjected to the same internal moment of 108.9 k·ft.

12

EXAMPLE   12.4

Determine the internal moments at the supports of the beam shown in Fig. 12–14a. The moment of inertia of the two spans is shown in the figure.

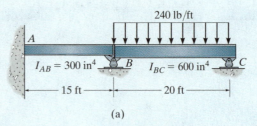

(a)

Fig. 12–14

SOLUTION

Since the beam is roller supported at its far end C, the stiffness of span BC will be computed on the basis of $K = 3EI/L$. We have

$$K_{AB} = \frac{4EI}{L} = \frac{4E(300)}{15} = 80E$$

$$K_{BC} = \frac{3EI}{L} = \frac{3E(600)}{20} = 90E$$

Thus,

$$DF_{AB} = \frac{80E}{\infty + 80E} = 0$$

$$DF_{BA} = \frac{80E}{80E + 90E} = 0.4706$$

$$DF_{BC} = \frac{90E}{80E + 90E} = 0.5294$$

$$DF_{CB} = \frac{90E}{90E} = 1$$

Further simplification of the distribution method for this problem is possible by realizing that a *single* fixed-end moment for the end span BC can be used. Using the right-hand column of the table on the inside back cover for a uniformly loaded span having one side fixed, the other pinned, we have

$$(FEM)_{BC} = -\frac{wL^2}{8} = \frac{-240(20)^2}{8} = -12\,000 \text{ lb} \cdot \text{ft}$$

The foregoing data are entered into the table in Fig. 12–14*b* and the moment distribution is carried out. By comparison with Fig. 12–6*b*, this method considerably simplifies the distribution.

Using the results, the beam's end shears and moment diagrams are shown in Fig. 12–14*c*.

Joint	A	B		C
Member	AB	BA	BC	CB
DF	0	0.4706	0.5294	1
FEM Dist.		5647.2	−12 000 6352.8	
CO	2823.6			
ΣM	2823.6	5647.2	−5647.2	0

(b)

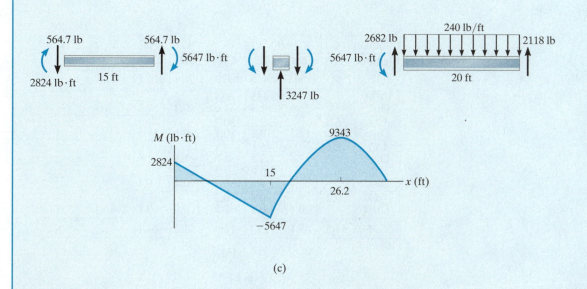

(c)

12.4   Moment Distribution for Frames: No Sidesway

Application of the moment-distribution method for frames having no sidesway follows the same procedure as that given for beams. To minimize the chance for errors, it is suggested that the analysis be arranged in a tabular form, as in the previous examples. Also, the distribution of moments can be shortened if the stiffness factor of a span can be modified as indicated in the previous section.

EXAMPLE | 12.5

Determine the internal moments at the joints of the frame shown in Fig. 12–15a. There is a pin at E and D and a fixed support at A. EI is constant.

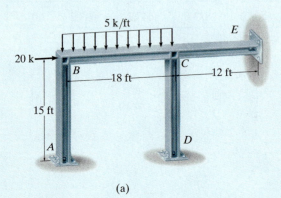

(a)

Joint	A	B		C			D	E
Member	AB	BA	BC	CB	CD	CE	DC	EC
DF	0	0.545	0.455	0.330	0.298	0.372	1	1
FEM			−135	135				
Dist.		73.6	61.4	−44.6	−40.2	−50.2		
CO	36.8		−22.3	30.7				
Dist.		12.2	10.1	−10.1	−9.1	−11.5		
CO	6.1		−5.1	5.1				
Dist.		2.8	2.3	−1.7	−1.5	−1.9		
CO	1.4		−0.8	1.2				
Dist.		0.4	0.4	−0.4	−0.4	−0.4		
CO	0.2		−0.2	0.2				
Dist.		0.1	0.1	−0.1	0.0	−0.1		
ΣM	44.5	89.1	−89.1	115	−51.2	−64.1		

(b)

Fig. 12–15

SOLUTION

By inspection, the pin at E will prevent the frame from sidesway. The stiffness factors of CD and CE can be computed using $K = 3EI/L$ since the far ends are pinned. Also, the 20-k load does not contribute a FEM since it is applied at joint B. Thus,

$$K_{AB} = \frac{4EI}{15} \qquad K_{BC} = \frac{4EI}{18} \qquad K_{CD} = \frac{3EI}{15} \qquad K_{CE} = \frac{3EI}{12}$$

$$DF_{AB} = 0$$

$$DF_{BA} = \frac{4EI/15}{4EI/15 + 4EI/18} = 0.545$$

$$DF_{BC} = 1 - 0.545 = 0.455$$

$$DF_{CB} = \frac{4EI/18}{4EI/18 + 3EI/15 + 3EI/12} = 0.330$$

$$DF_{CD} = \frac{3EI/15}{4EI/18 + 3EI/15 + 3EI/12} = 0.298$$

$$DF_{CE} = 1 - 0.330 - 0.298 = 0.372$$

$$DF_{DC} = 1 \qquad DF_{EC} = 1$$

$$(FEM)_{BC} = \frac{-wL^2}{12} = \frac{-5(18)^2}{12} = -135 \text{ k} \cdot \text{ft}$$

$$(FEM)_{CB} = \frac{wL^2}{12} = \frac{5(18)^2}{12} = 135 \text{ k} \cdot \text{ft}$$

The data are shown in the table in Fig. 12–15b. Here the distribution of moments successively goes to joints B and C. The final moments are shown on the last line.

Using these data, the moment diagram for the frame is constructed in Fig. 12–15c.

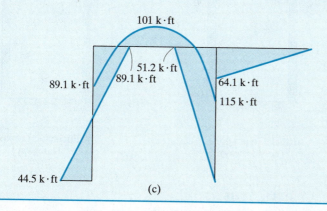

(c)

12.5   Moment Distribution for Frames: Sidesway

It has been shown in Sec. 11.5 that frames that are nonsymmetrical or subjected to nonsymmetrical loadings have a tendency to sidesway. An example of one such case is shown in Fig. 12–16a. Here the applied loading **P** will create unequal moments at joints B and C such that the frame will deflect an amount Δ to the right. To determine this deflection and the internal moments at the joints using moment distribution, we will use the principle of superposition. In this regard, the frame in Fig. 12–16b is first considered held from sidesway by applying an artificial joint support at C. Moment distribution is applied and then by statics the restraining force **R** is determined. The equal, but opposite, restraining force is then applied to the frame, Fig. 12–16c, and the moments in the frame are calculated. One method for doing this last step requires first *assuming* a numerical value for one of the internal moments, say $\mathbf{M}'_{BA}$. Using moment distribution and statics, the deflection Δ' and external force **R**′ corresponding to the assumed value of $\mathbf{M}'_{BA}$ can then be determined. Since linear elastic deformations occur, the force **R**′ develops moments in the frame that are *proportional* to those developed by **R**. For example, if $\mathbf{M}'_{BA}$ and **R**′ are known, the moment at B developed by **R** will be $M_{BA} = M'_{BA}(R/R')$. Addition of the joint moments for both cases, Fig. 12–16b and c, will yield the actual moments in the frame, Fig. 12–16a. Application of this technique is illustrated in Examples 12.6 through 12.8.

Multistory Frames.   Quite often, multistory frameworks may have several *independent* joint displacements, and consequently the moment distribution analysis using the above techniques will involve more computation. Consider, for example, the two-story frame shown in Fig. 12–17a. This structure can have two independent joint displacements, since the sidesway Δ_1 of the first story is independent of any displacement

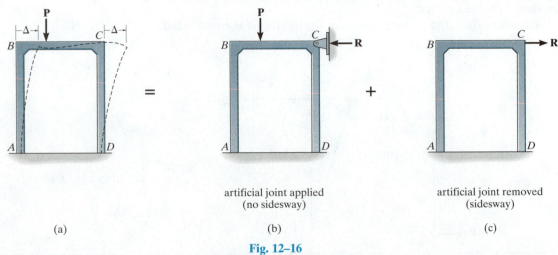

artificial joint applied
(no sidesway)

artificial joint removed
(sidesway)

(a)   (b)   (c)

Fig. 12–16

Δ_2 of the second story. Unfortunately, these displacements are not known initially, so the analysis must proceed on the basis of superposition, in the same manner as discussed previously. In this case, two restraining forces $\mathbf{R_1}$ and $\mathbf{R_2}$ are applied, Fig. 12–17b, and the fixed-end moments are determined and distributed. Using the equations of equilibrium, the numerical values of $\mathbf{R_1}$ and $\mathbf{R_2}$ are then determined. Next, the restraint at the floor of the first story is removed and the floor is given a displacement Δ'. This displacement causes fixed-end moments (FEMs) in the frame, which can be assigned specific numerical values. By distributing these moments and using the equations of equilibrium, the associated numerical values of $\mathbf{R_1'}$ and $\mathbf{R_2'}$ can be determined. In a similar manner, the floor of the second story is then given a displacement Δ'', Fig. 12–17d. Assuming numerical values for the fixed-end moments, the moment distribution and equilibrium analysis will yield specific values of $\mathbf{R_1''}$ and $\mathbf{R_2''}$. Since the last two steps associated with Fig. 12–17c and d depend on *assumed* values of the FEMs, correction factors C' and C'' must be applied to the distributed moments. With reference to the restraining forces in Fig. 12–17c and 12–17d, we require equal but opposite application of $\mathbf{R_1}$ and $\mathbf{R_2}$ to the frame, such that

$$R_2 = -C'R_2' + C''R_2''$$

$$R_1 = +C'R_1' - C''R_1''$$

Simultaneous solution of these equations yields the values of C' and C''. These correction factors are then multiplied by the internal joint moments found from the moment distribution in Fig. 12–17c and 12–17d. The resultant moments are then found by adding these corrected moments to those obtained for the frame in Fig. 12–17b.

Other types of frames having independent joint displacements can be analyzed using this same procedure; however, it must be admitted that the foregoing method does require quite a bit of numerical calculation. Although some techniques have been developed to shorten the calculations, it is best to solve these types of problems on a computer, preferably using a matrix analysis. The techniques for doing this will be discussed in Chapter 16.

This statically indeterminate concrete building frame can be subjected to sidesway due to wind and earthquake loadings.

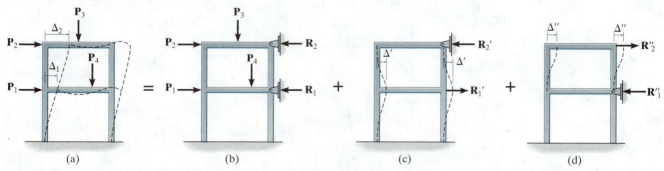

Fig. 12–17

EXAMPLE 12.6

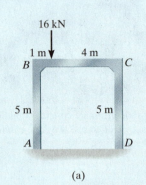

(a)

$\parallel$

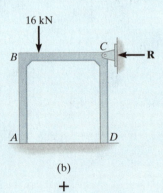

(b)

$+$

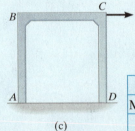

(c)

Fig. 12–18

Determine the moments at each joint of the frame shown in Fig. 12–18a. *EI* is constant.

SOLUTION

First we consider the frame held from sidesway as shown in Fig. 12–18b. We have

$$(\text{FEM})_{BC} = -\frac{16(4)^2(1)}{(5)^2} = -10.24 \text{ kN} \cdot \text{m}$$

$$(\text{FEM})_{CB} = \frac{16(1)^2(4)}{(5)^2} = 2.56 \text{ kN} \cdot \text{m}$$

The stiffness factor of each span is computed on the basis of $4EI/L$ or by using the relative-stiffness factor I/L. The DFs and the moment distribution are shown in the table, Fig. 12–18d. Using these results, the equations of equilibrium are applied to the free-body diagrams of the columns in order to determine $\mathbf{A}_x$ and $\mathbf{D}_x$ Fig. 12–18e. From the free-body diagram of the entire frame (not shown) the joint restraint $\mathbf{R}$ in Fig. 12–18b has a magnitude of

$$\Sigma F_x = 0; \qquad R = 1.73 \text{ kN} - 0.81 \text{ kN} = 0.92 \text{ kN}$$

An equal but opposite value of $R = 0.92$ kN must now be applied to the frame at C and the internal moments computed, Fig. 12–18c. To solve the problem of computing these moments, we will assume a force $\mathbf{R}'$ is applied at C, causing the frame to deflect Δ' as shown in Fig. 12–18f. Here the joints at B and C are *temporarily restrained from rotating,* and as a result the fixed-end moments at the ends of the columns are determined from the formula for deflection found on the inside back cover, that is,

Joint	A	B		C		D
Member	AB	BA	BC	CB	CD	DC
DF	0	0.5	0.5	0.5	0.5	0
FEM			−10.24	2.56		
Dist.		5.12	5.12	−1.28	−1.28	
CO	2.56		−0.64	2.56		−0.64
Dist.		0.32	0.32	−1.28	−1.28	
CO	0.16		−0.64	0.16		−0.64
Dist.		0.32	0.32	−0.08	−0.08	
CO	0.16		−0.04	0.16		−0.04
Dist.		0.02	0.02	−0.08	−0.08	
ΣM	2.88	5.78	−5.78	2.72	−2.72	−1.32

(d)

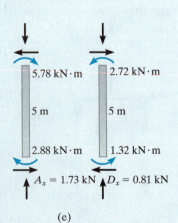

(e)

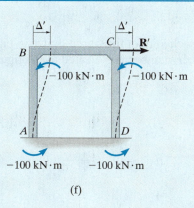

(f)

Joint	A	B		C		D
Member	AB	BA	BC	CB	CD	DC
DF	0	0.5	0.5	0.5	0.5	0
FEM	−100	−100			−100	−100
Dist.		50	50	50	50	
CO	25		25	25		25
Dist.		−12.5	−12.5	−12.5	−12.5	
CO	−6.25		−6.25	−6.25		−6.25
Dist.		3.125	3.125	3.125	3.125	
CO	1.56		1.56	1.56		1.56
Dist.		−0.78	−0.78	−0.78	−0.78	
CO	−0.39		−0.39	−0.39		−0.39
Dist.		0.195	0.195	0.195	0.195	
ΣM	−80.00	−60.00	60.00	60.00	−60.00	−80.00

$$M = \frac{6EI\Delta}{L^2}$$

(g)

Since *both* B and C happen to be displaced the same amount Δ', and AB and DC have the *same* E, I, and L, the FEM in AB will be the *same* as that in DC. As shown in Fig. 12–18f, we will arbitrarily *assume* this fixed-end moment to be

$$(FEM)_{AB} = (FEM)_{BA} = (FEM)_{CD} = (FEM)_{DC} = -100 \text{ kN} \cdot \text{m}$$

A *negative sign* is necessary since the moment must act *counterclockwise* on the column for deflection Δ' to the right. The value of **R′** associated with this −100 kN · m moment can now be determined. The moment distribution of the FEMs is shown in Fig. 12–18g. From equilibrium, the horizontal reactions at A and D are calculated, Fig. 12–18h. Thus, for the entire frame we require

$$\Sigma F_x = 0; \qquad R' = 28 + 28 = 56.0 \text{ kN}$$

Hence, $R' = 56.0$ kN creates the moments tabulated in Fig. 12–18g. Corresponding moments caused by $R = 0.92$ kN can be determined by proportion. Therefore, the resultant moment in the frame, Fig. 12–18a, is equal to the *sum* of those calculated for the frame in Fig. 12–18b plus the proportionate amount of those for the frame in Fig. 12–18c. We have

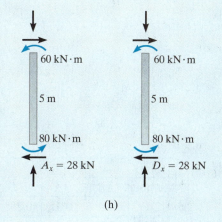

(h)

$$M_{AB} = 2.88 + \tfrac{0.92}{56.0}(-80) = 1.57 \text{ kN} \cdot \text{m} \qquad Ans.$$

$$M_{BA} = 5.78 + \tfrac{0.92}{56.0}(-60) = 4.79 \text{ kN} \cdot \text{m} \qquad Ans.$$

$$M_{BC} = -5.78 + \tfrac{0.92}{56.0}(60) = -4.79 \text{ kN} \cdot \text{m} \qquad Ans.$$

$$M_{CB} = 2.72 + \tfrac{0.92}{56.0}(60) = 3.71 \text{ kN} \cdot \text{m} \qquad Ans.$$

$$M_{CD} = -2.72 + \tfrac{0.92}{56.0}(-60) = -3.71 \text{ kN} \cdot \text{m} \qquad Ans.$$

$$M_{DC} = -1.32 + \tfrac{0.92}{56.0}(-80) = -2.63 \text{ kN} \cdot \text{m} \qquad Ans.$$

12

EXAMPLE | 12.7

Determine the moments at each joint of the frame shown in Fig. 12–19a. The moment of inertia of each member is indicated in the figure.

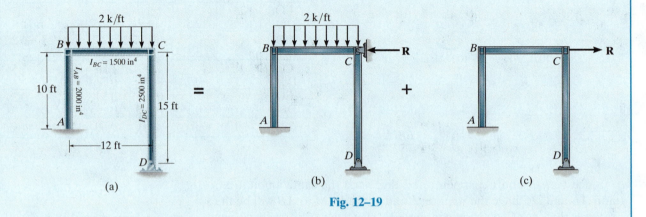

Fig. 12–19

SOLUTION

The frame is first held from sidesway as shown in Fig. 12–19b. The internal moments are computed at the joints as indicated in Fig. 12–19d. Here the stiffness factor of CD was computed using $3EI/L$ since there is a pin at D. Calculation of the horizontal reactions at A and D is shown in Fig. 12–19e. Thus, for the entire frame,

$$\Sigma F_x = 0; \qquad\qquad R = 2.89 - 1.00 = 1.89 \text{ k}$$

Joint	A	B		C		D
Member	AB	BA	BC	CB	CD	DC
DF	0	0.615	0.385	0.5	0.5	1
FEM			−24	24		
Dist.		14.76	9.24	−12	−12	
CO	7.38		−6	4.62		
Dist.		3.69	2.31	−2.31	−2.31	
CO	1.84		−1.16	1.16		
Dist.		0.713	0.447	−0.58	−0.58	
CO	0.357		−0.29	0.224		
Dist.		0.18	0.11	−0.11	−0.11	
ΣM	9.58	19.34	−19.34	15.00	−15.00	0

(d)

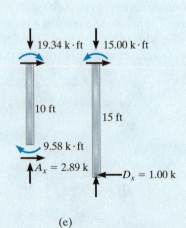

(e)

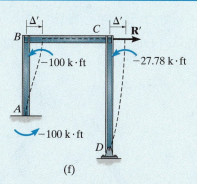

(f)

Joint	A	B		C		D
Member	AB	BA	BC	CB	CD	DC
DF	0	0.615	0.385	0.5	0.5	1
FEM	−100	−100			−27.78	
Dist.		61.5	38.5	13.89	13.89	
CO	30.75			6.94	19.25	
Dist.		−4.27	−2.67	−9.625	−9.625	
CO	−2.14			−4.81	−1.34	
Dist.		2.96	1.85	0.67	0.67	
CO	1.48			0.33	0.92	
Dist.		−0.20	−0.13	−0.46	−0.46	
ΣM	−69.91	−40.01	40.01	23.31	−23.31	0

(g)

The opposite force is now applied to the frame as shown in Fig. 12–19c. As in the previous example, we will consider a force **R′** acting as shown in Fig. 12–19f. As a result, joints B and C are displaced by the same amount Δ′. The fixed-end moments for BA are computed from

$$(\text{FEM})_{AB} = (\text{FEM})_{BA} = -\frac{6EI\Delta}{L^2} = -\frac{6E(2000)\Delta'}{(10)^2}$$

However, from the table on the inside back cover, for CD we have

$$(\text{FEM})_{CD} = -\frac{3EI\Delta}{L^2} = -\frac{3E(2500)\Delta'}{(15)^2}$$

Assuming the FEM for AB is −100 k·ft as shown in Fig. 12–19f, the *corresponding* FEM at C, causing the *same* Δ′, is found by comparison, i.e.,

$$\Delta' = -\frac{(-100)(10)^2}{6E(2000)} = -\frac{(\text{FEM})_{CD}(15)^2}{3E(2500)}$$

$$(\text{FEM})_{CD} = -27.78 \text{ k·ft}$$

Moment distribution for these FEMs is tabulated in Fig. 12–19g. Computation of the horizontal reactions at A and D is shown in Fig. 12–19h. Thus, for the entire frame,

$$\Sigma F_x = 0; \qquad R' = 11.0 + 1.55 = 12.55 \text{ k}$$

The resultant moments in the frame are therefore

$$M_{AB} = 9.58 + \left(\tfrac{1.89}{12.55}\right)(-69.91) = -0.948 \text{ k·ft} \qquad Ans.$$

$$M_{BA} = 19.34 + \left(\tfrac{1.89}{12.55}\right)(-40.01) = 13.3 \text{ k·ft} \qquad Ans.$$

$$M_{BC} = -19.34 + \left(\tfrac{1.89}{12.55}\right)(40.01) = -13.3 \text{ k·ft} \qquad Ans.$$

$$M_{CB} = 15.00 + \left(\tfrac{1.89}{12.55}\right)(23.31) = 18.5 \text{ k·ft} \qquad Ans.$$

$$M_{CD} = -15.00 + \left(\tfrac{1.89}{12.55}\right)(-23.31) = -18.5 \text{ k·ft} \qquad Ans.$$

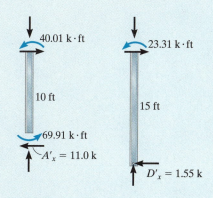

(h)

12

EXAMPLE 12.8

Determine the moments at each joint of the frame shown in Fig. 12–20a. EI is constant.

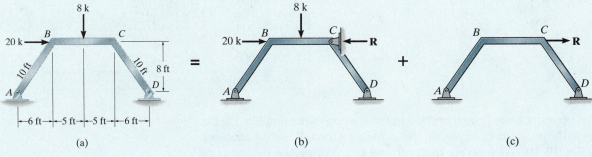

Fig. 12–20

SOLUTION

First sidesway is prevented by the restraining force $\mathbf{R}$, Fig. 12–20b. The FEMs for member BC are

$$(\text{FEM})_{BC} = -\frac{8(10)}{8} = -10 \text{ k} \cdot \text{ft} \qquad (\text{FEM})_{CB} = \frac{8(10)}{8} = 10 \text{ k} \cdot \text{ft}$$

Since spans AB and DC are pinned at their ends, the stiffness factor is computed using $3EI/L$. The moment distribution is shown in Fig. 12–20d.

Using these results, the *horizontal reactions* at A and D must be determined. This is done using an equilibrium analysis of *each member*, Fig. 12–20e. Summing moments about points B and C on each leg, we have

$$\zeta +\Sigma M_B = 0; \qquad -5.97 + A_x(8) - 4(6) = 0 \qquad A_x = 3.75 \text{ k}$$

$$\zeta +\Sigma M_C = 0; \qquad 5.97 - D_x(8) + 4(6) = 0 \qquad D_x = 3.75 \text{ k}$$

Thus, for the entire frame,

$$\Sigma F_x = 0; \qquad R = 3.75 - 3.75 + 20 = 20 \text{ k}$$

Joint	A	B		C		D
Member	AB	BA	BC	CB	CD	DC
DF	1	0.429	0.571	0.571	0.429	1
FEM			−10	10		
Dist.		4.29	5.71	−5.71	−4.29	
CO			−2.86	2.86		
Dist.		1.23	1.63	−1.63	−1.23	
CO			−0.82	0.82		
Dist.		0.35	0.47	−0.47	−0.35	
CO			−0.24	0.24		
Dist.		0.10	0.13	−0.13	−0.10	
ΣM	0	5.97	−5.97	5.97	−5.97	0

(d)

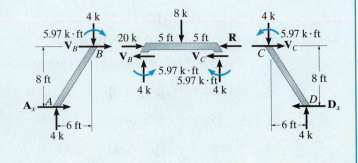

(e)

The opposite force **R** is now applied to the frame as shown in Fig. 12–20c. In order to determine the internal moments developed by **R** we will first consider the force **R'** acting as shown in Fig. 12–20f. Here the dashed lines do not represent the distortion of the frame members; instead, they are constructed as straight lines extended to the final positions B' and C' of points B and C, respectively. Due to the symmetry of the frame, the displacement $BB' = CC' = \Delta'$. Furthermore, these displacements cause BC to rotate. The vertical distance between B' and C' is $1.2\Delta'$, as shown on the displacement diagram, Fig. 12–20g. Since each span undergoes end-point displacements that cause the spans to rotate, fixed-end moments are induced in the spans. These are: $(FEM)_{BA} = (FEM)_{CD} = -3EI\Delta'/(10)^2$, $(FEM)_{BC} = (FEM)_{CB} = 6EI(1.2\Delta')/(10)^2$.

Notice that for BA and CD the moments are *negative* since clockwise rotation of the span causes a *counterclockwise* FEM.

If we arbitrarily assign a value of $(FEM)_{BA} = (FEM)_{CD} = -100\ \text{k}\cdot\text{ft}$, then equating Δ' in the above formulas yields $(FEM)_{BC} = (FEM)_{CB} = 240\ \text{k}\cdot\text{ft}$. These moments are applied to the frame and distributed, Fig. 12–20h. Using these results, the equilibrium analysis is shown in Fig. 12–20i. For each leg, we have

$$\zeta + \Sigma M_B = 0;\qquad -A'_x(8) + 29.36(6) + 146.80 = 0\qquad A'_x = 40.37\ \text{k}$$

$$\zeta + \Sigma M_C = 0;\qquad -D'_x(8) + 29.36(6) + 146.80 = 0\qquad D'_x = 40.37\ \text{k}$$

Thus, for the entire frame,

$$\Sigma F_x = 0;\qquad\qquad R' = 40.37 + 40.37 = 80.74\ \text{k}$$

The resultant moments in the frame are therefore

$$M_{BA} = 5.97 + \left(\tfrac{20}{80.74}\right)(-146.80) = -30.4\ \text{k}\cdot\text{ft}\qquad Ans.$$

$$M_{BC} = -5.97 + \left(\tfrac{20}{80.74}\right)(146.80) = 30.4\ \text{k}\cdot\text{ft}\qquad Ans.$$

$$M_{CB} = 5.97 + \left(\tfrac{20}{80.74}\right)(146.80) = 42.3\ \text{k}\cdot\text{ft}\qquad Ans.$$

$$M_{CD} = -5.97 + \left(\tfrac{20}{80.74}\right)(-146.80) = -42.3\ \text{k}\cdot\text{ft}\qquad Ans.$$

(f)

(g)

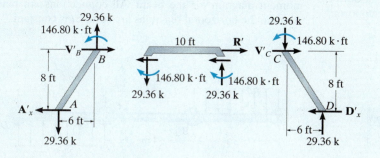

(i)

Joint	A	B		C		D
Member	AB	BA	BC	CB	CD	DC
DF	1	0.429	0.571	0.571	0.429	1
FEM Dist.		-100 -60.06	240 -79.94	240 -79.94	-100 -60.06	
CO Dist.		17.15	-39.97 22.82	-39.97 22.82	17.15	
CO Dist.		-4.89	11.41 -6.52	11.41 -6.52	-4.89	
CO Dist.		1.40	-3.26 1.86	-3.26 1.86	1.40	
CO Dist.		-0.40	0.93 -0.53	0.93 -0.53	-0.40	
ΣM	0	+146.80	146.80	146.80	-146.80	0

(h)

PROBLEMS

Sec. 12.1–12.3

12–1. Determine the moments at A, B, and C, then draw the moment diagram for the beam. The moment of inertia of each span is indicated in the figure. Assume the support at B is a roller and A and C are fixed. $E = 29(10^3)$ ksi.

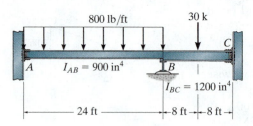

Prob. 12–1

12–2. Determine the moments at A, B, and C, then draw the moment diagram for the beam. Assume the supports at A and C are fixed and B is a roller. EI is constant.

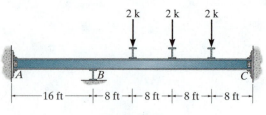

Prob. 12–2

12–3. Determine the moments at A, B, and C, then draw the moment diagram for the beam. Assume the supports at A and C are fixed. EI is constant.

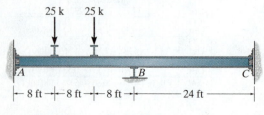

Prob. 12–3

*12–4. Determine the internal moment in the beam at B, then draw the moment diagram. Assume C is a pin. Segment AB has a moment of interia of $I_{AB} = 0.75\, I_{BC}$. E is constant.

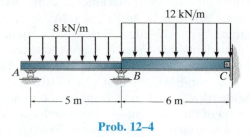

Prob. 12–4

12–5. Determine the moments at A, B, and C, then draw the moment diagram. Assume the support at B is a roller and A and C are fixed. EI is constant.

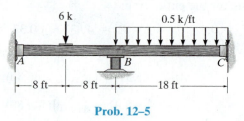

Prob. 12–5

12–6. Determine the moments at B and C, then draw the moment diagram for the beam. All connections are pins. Assume the horizontal reactions are zero. EI is constant.

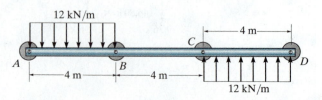

Prob. 12–6

12–7. Determine the moments at the supports, then draw the moment diagram. Assume A and D are fixed. EI is constant.

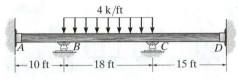

Prob. 12–7

*12–8. The beam is subjected to the loading shown. Determine the reactions at the supports, then draw moment diagram. EI is constant.

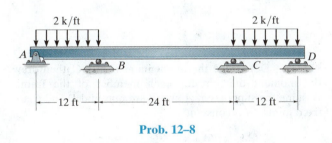

Prob. 12–8

12–9. The bar is pin connected at each indicated point. If the normal force in the bar can be neglected, determine the vertical reaction at each pin. EI is constant.

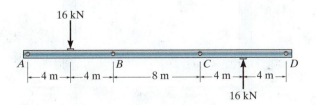

Prob. 12–9

12–10. Determine the moments at A, B, and C, then draw the moment diagram for the girder DE. Assume the support at B is a pin and A and C are rollers. The distributed load rests on simply supported floor boards that transmit the load to the floor beams. EI is constant.

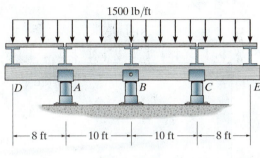

Prob. 12–10

12–11. Determine the moments at A and B, then draw the moment diagram. Assume the support at B is a roller, C is a pin, and A is fixed.

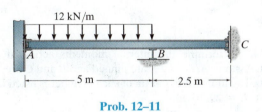

Prob. 12–11

*12–12. Determine the moments at A, B, and C, then draw the moment diagram. EI is constant.

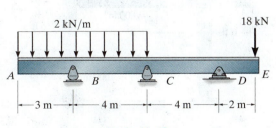

Prob. 12–12

Sec. 12.4

12–13. Determine the moment at B, then draw the moment diagram for each member of the frame. Assume the supports at A and C are pins. EI is constant.

12–15. Determine the reactions at A and D. Assume the supports at A and D are fixed and B and C are fixed connected. EI is constant.

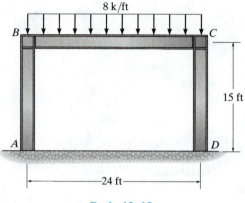

Prob. 12–15

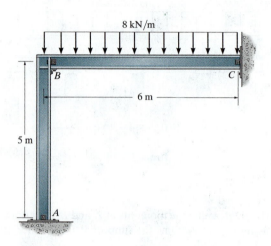

Prob. 12–13

12–14. Determine the internal moments acting at each joint. Assume A, D, and E are pinned and B and C are fixed joints. The moment of inertia of each member is listed in the figure. $E = 29(10^3)$ ksi.

*12–16.** Determine the moments at D and C, then draw the moment diagram for each member of the frame. Assume the supports at A and B are pins and D and C are fixed joints. EI is constant.

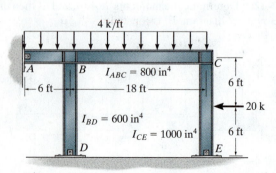

Prob. 12–14

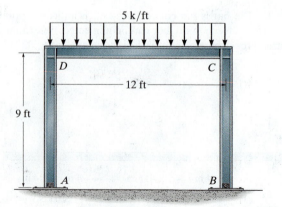

Prob. 12–16

12–17. Determine the moments acting at the ends of each member, then draw the moment diagram. Assume *B* is a fixed joint and *A* and *D* are pin supported and *C* is fixed. $E = 29(10^3)$ ksi. $I_{ABC} = 700$ in^4, and $I_{BD} = 1100$ in^4.

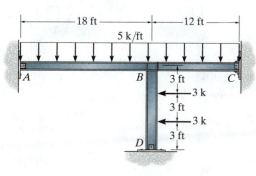

Prob. 12–17

12–19. Determine the moment at *B*, then draw the moment diagram for each member of the frame. Support *A* is pinned. *EI* is constant.

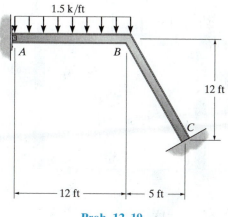

Prob. 12–19

12–18. Determine the moments at each joint of the frame, then draw the moment diagram for member *BCE*. Assume *B*, *C*, and *E* are fixed connected and *A* and *D* are pins $E = 29(10^3)$ ksi.

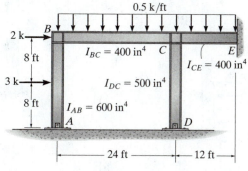

Prob. 12–18

*12–20. Determine the moments at *A*, *C*, and *D*, then draw the moment diagram for each member of the frame. Support *A* and joints *C* and *D* are fixed connected. *EI* is constant.

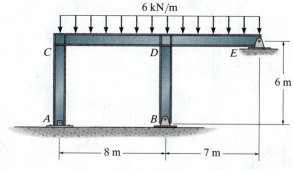

Prob. 12–20

Sec. 12.5

12–21. Determine the moments at B and C and then draw the moment diagram. Assume A and D are pins and B and C are fixed-connected joints. EI is constant.

12–23. Determine the moments at the ends of each member of the frame. The members are fixed connected at the supports and joints. EI is the same for each member.

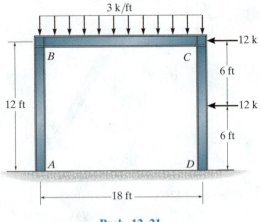

Prob. 12–21

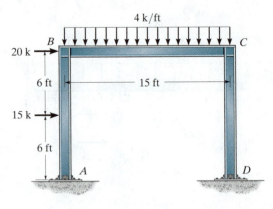

Prob. 12–23

12–22. Determine the moments acting at the ends of each member. Assume the joints are fixed connected and A and D are fixed supports. EI is constant.

*12–24. Determine the moments acting at the ends of each member. Assume the joints are fixed connected and A and B are fixed supports. EI is constant.

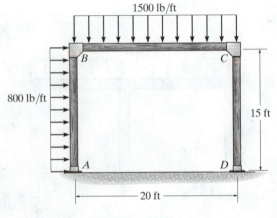

Prob. 12–22

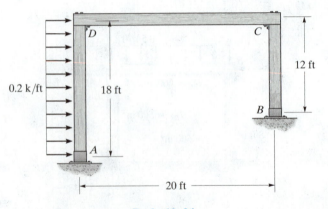

Prob. 12–24

12–25. Determine the horizontal and vertical components of reaction at the pin supports A and D. EI is constant.

12–26. Determine the moments at C and D, then draw the moment diagram for each member of the frame. Assume the supports at A and B are pins. EI is constant.

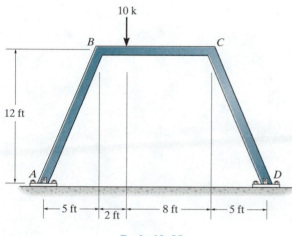

Prob. 12–25

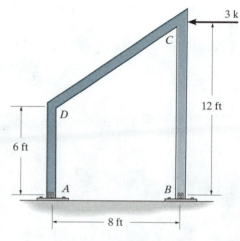

Prob. 12–26

CHAPTER REVIEW

Moment distribution is a method of successive approximations that can be carried out to any desired degree of accuracy. It initially requires locking all the joints of the structure. The equilibrium moment at each joint is then determined, the joints are unlocked and this moment is distributed onto each connecting member, and half its value is carried over to the other side of the span. This cycle of locking and unlocking the joints is repeated until the carry-over moments become acceptably small. The process then stops and the moment at each joint is the sum of the moments from each cycle of locking and unlocking.

The process of moment distribution is conveniently done in tabular form. Before starting, the fixed-end moment for each span must be calculated using the table on the inside back cover of the book. The distribution factors are found by dividing a member's stiffness by the total stiffness of the joint. For members having a far end fixed, use $K = 4EI/L$; for a far-end pinned or roller supported member, $K = 3EI/L$; for a symmetric span and loading, $K = 2EI/L$; and for an antisymmetric loading, $K = 6EI/L$. Remember that the distribution factor for a fixed end is DF $= 0$, and for a pin or roller-supported end, DF $= 1$.

Chapter 13

© Stockphoto Mania/Shutterstock

The use of variable-moment-of-inertia girders has reduced considerably the deadweight loading of each of these spans.

Beams and Frames Having Nonprismatic Members

In this chapter we will apply the slope-deflection and moment-distribution methods to analyze beams and frames composed of nonprismatic members. We will first discuss how the necessary carry-over factors, stiffness factors, and fixed-end moments are obtained. This is followed by a discussion related to using tabular values often published in design literature. Finally, the analysis of statically indeterminate structures using the slope-deflection and moment-distribution methods will be discussed.

13.1   Loading Properties of Nonprismatic Members

Often, to save material, girders used for long spans on bridges and buildings are designed to be nonprismatic, that is, to have a variable moment of inertia. The most common forms of structural members that are nonprismatic have haunches that are either stepped, tapered, or parabolic, Fig. 13–1. Provided we can express the member's moment of inertia as a function of the length coordinate x, then we can use the principle of virtual work or Castigliano's theorem as discussed in Chapter 9 to find its deflection. The equations are

$$\Delta = \int_0^l \frac{Mm}{EI}dx \qquad \text{or} \qquad \Delta = \int_0^l \frac{\partial M}{\partial P}\frac{M}{EI}dx$$

If the member's geometry and loading require evaluation of an integral that cannot be determined in closed form, then Simpson's rule or some other numerical technique will have to be used to carry out the integration.

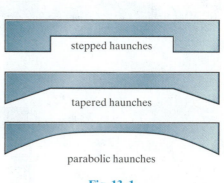

stepped haunches

tapered haunches

parabolic haunches

Fig. 13–1

533

If the slope deflection equations or moment distribution are used to determine the reactions on a nonprismatic member, then we must first calculate the following properties for the member.

Fixed-End Moments (FEM).
The end moment reactions on the member that is assumed fixed supported, Fig. 13–2a.

Stiffness Factor (K).
The magnitude of moment that must be applied to the end of the member such that the end rotates through an angle of $\theta = 1$ rad. Here the moment is applied at the pin support, while the other end is assumed fixed, Fig. 13–2b.

Carry-Over Factor (COF).
Represents the numerical fraction (C) of the moment that is "carried over" from the pin-supported end to the wall, Fig. 13–2c.

Once obtained, the computations for the stiffness and carry-over factors can be checked, in part, by noting an important relationship that exists between them. In this regard, consider the beam in Fig. 13–3 subjected to the loads and deflections shown. Application of the Maxwell-Betti reciprocal theorem requires the work done by the loads in Fig. 13–3a acting through the displacements in Fig. 13–3b be equal to the work of the loads in Fig. 13–3b acting through the displacements in Fig. 13–3a, that is,

$$U_{AB} = U_{BA}$$
$$K_A(0) + C_{AB}K_A(1) = C_{BA}K_B(1) + K_B(0)$$

or

$$C_{AB}K_A = C_{BA}K_B \qquad (13\text{–}1)$$

Hence, once determined, the stiffness and carry-over factors must satisfy Eq. 13–1.

The tapered concrete hammerhead pier is used to support the girders of this highway bridge.

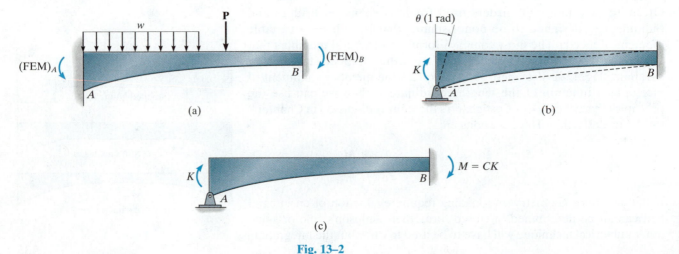

(a)

(b)

(c)

Fig. 13–2

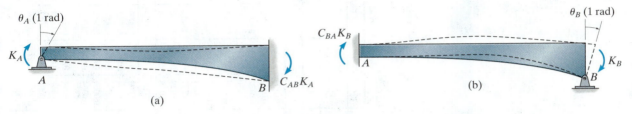

Fig. 13–3

These properties can be obtained using, for example, the conjugate beam method or an energy method. However, considerable labor is often involved in the process. As a result, graphs and tables have been made available to determine this data for common shapes used in structural design. One such source is the *Handbook of Frame Constants,* published by the Portland Cement Association.* A portion of these tables, taken from this publication, is listed here as Tables 13.1 and 13.2. A more complete tabular form of the data is given in the PCA handbook along with the relevant derivations of formulas used.

The nomenclature is defined as follows:

a_A, a_B = ratio of the length of haunch at ends A and B to the length of span.

b = ratio of the distance from the concentrated load to end A to the length of span.

C_{AB}, C_{BA} = carry-over factors of member AB at ends A and B, respectively.

h_A, h_B = depth of member at ends A and B, respectively.

h_C = depth of member at minimum section.

I_C = moment of inertia of section at minimum depth.

k_{AB}, k_{BA} = stiffness factor at ends A and B, respectively.

L = length of member.

M_{AB}, M_{BA} = fixed-end moment at ends A and B, respectively; specified in tables for uniform load w or concentrated force P.

r_A, r_B = ratios for rectangular cross-sectional areas, where $r_A = (h_A - h_C)/h_C, r_B = (h_B - h_C)/h_C.$

As noted, the fixed-end moments and carry-over factors are found from the tables. The absolute stiffness factor can be determined using the tabulated stiffness factors and found from

$$K_A = \frac{k_{AB}EI_C}{L} \qquad K_B = \frac{k_{BA}EI_C}{L} \tag{13–2}$$

Application of the use of the tables will be illustrated in Example 13.1.

Handbook of Frame Constants. Portland Cement Association, Chicago, Illinois.

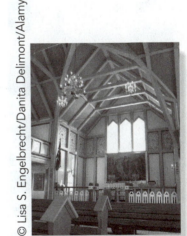

© Lisa S. Engelbrecht/Danita Delimont/Alamy

Timber frames having a variable moment of inertia are often used in the construction of churches.

TABLE 13.1 Tapered Haunches—Constant Width

Note: All carry-over factors are negative and all stiffness factors are positive.

Concentrated Load FEM-Coef. × PL

$a_A = 0.3$

Right Haunch		Carry-over Factors		Stiffness Factors		Unif. Load FEM Coef. × wL^2		0.1		0.3		0.5		0.7		0.9		Haunch Load at Left FEM Coef. × w_AL^2		Right FEM Coef. × w_BL^2	
a_B	r_B	C_{AB}	C_{BA}	k_{AB}	k_{BA}	M_{AB}	M_{BA}	M_{AB}	M_{BA}	M_{AB}	M_{BA}	M_{AB}	M_{BA}	M_{AB}	M_{BA}	M_{AB}	M_{BA}	M_{AB}	M_{BA}	M_{AB}	M_{BA}
								a_B = variable				r_A = 1.0				r_B = variable					
0.2	0.4	0.543	0.765	9.19	6.52	0.1194	0.0791	0.0935	0.0034	0.2185	0.0384	0.1955	0.1147	0.0889	0.1601	0.0096	0.0870	0.0133	0.0008	0.0006	0.0058
	0.6	0.576	0.758	9.53	7.24	0.1152	0.0851	0.0934	0.0038	0.2158	0.0422	0.1883	0.1250	0.0798	0.1729	0.0075	0.0898	0.0133	0.0009	0.0005	0.0060
	1.0	0.622	0.748	10.06	8.37	0.1089	0.0942	0.0931	0.0042	0.2118	0.0480	0.1771	0.1411	0.0668	0.1919	0.0047	0.0935	0.0132	0.0011	0.0004	0.0062
	1.5	0.660	0.740	10.52	9.38	0.1037	0.1018	0.0927	0.0047	0.2085	0.0530	0.1678	0.1550	0.0559	0.2078	0.0028	0.0961	0.0130	0.0012	0.0002	0.0064
	2.0	0.684	0.734	10.83	10.09	0.1002	0.1069	0.0924	0.0050	0.2062	0.0565	0.1614	0.1645	0.0487	0.2185	0.0019	0.0974	0.0129	0.0013	0.0001	0.0065
0.3	0.4	0.579	0.741	9.47	7.40	0.1175	0.0822	0.0934	0.0037	0.2164	0.0419	0.1909	0.1225	0.0856	0.1649	0.0100	0.0861	0.0133	0.0009	0.0022	0.0118
	0.6	0.629	0.726	9.98	8.64	0.1120	0.0902	0.0931	0.0042	0.2126	0.0477	0.1808	0.1379	0.0747	0.1807	0.0080	0.0888	0.0132	0.0010	0.0018	0.0124
	1.0	0.705	0.705	10.85	10.85	0.1034	0.1034	0.0924	0.0052	0.2063	0.0577	0.1640	0.1640	0.0577	0.2063	0.0052	0.0924	0.0131	0.0013	0.0013	0.0131
	1.5	0.771	0.689	11.70	13.10	0.0956	0.1157	0.0917	0.0062	0.2002	0.0675	0.1483	0.1892	0.0428	0.2294	0.0033	0.0953	0.0129	0.0015	0.0008	0.0137
	2.0	0.817	0.678	12.33	14.85	0.0901	0.1246	0.0913	0.0069	0.1957	0.0750	0.1368	0.2080	0.0326	0.2455	0.0022	0.0968	0.0128	0.0017	0.0006	0.0141
								a_B = variable				r_A = 1.5				r_B = variable					
								$a_A = 0.2$													
0.2	0.4	0.569	0.714	7.97	6.35	0.1166	0.0799	0.0966	0.0019	0.2186	0.0377	0.1847	0.1183	0.0821	0.1626	0.0088	0.0873	0.0064	0.0001	0.0006	0.0058
	0.6	0.603	0.707	8.26	7.04	0.1127	0.0858	0.0965	0.0021	0.2163	0.0413	0.1778	0.1288	0.0736	0.1752	0.0068	0.0901	0.0064	0.0001	0.0005	0.0060
	1.0	0.652	0.698	8.70	8.12	0.1069	0.0947	0.0963	0.0023	0.2127	0.0468	0.1675	0.1449	0.0616	0.1940	0.0043	0.0937	0.0064	0.0002	0.0004	0.0062
	1.5	0.691	0.691	9.08	9.08	0.1021	0.1021	0.0962	0.0025	0.2097	0.0515	0.1587	0.1587	0.0515	0.2097	0.0025	0.0962	0.0064	0.0002	0.0002	0.0064
	2.0	0.716	0.686	9.34	9.75	0.0990	0.1071	0.0960	0.0028	0.2077	0.0547	0.1528	0.1681	0.0449	0.2202	0.0017	0.0975	0.0064	0.0002	0.0001	0.0065
0.3	0.4	0.607	0.692	8.21	7.21	0.1148	0.0829	0.0965	0.0021	0.2168	0.0409	0.1801	0.1263	0.0789	0.1674	0.0091	0.0866	0.0064	0.0002	0.0020	0.0118
	0.6	0.699	0.678	8.65	8.40	0.1098	0.0907	0.0964	0.0024	0.2135	0.0464	0.1706	0.1418	0.0688	0.1831	0.0072	0.0892	0.0064	0.0002	0.0017	0.0123
	1.0	0.740	0.660	9.38	10.52	0.1018	0.1037	0.0961	0.0028	0.2078	0.0559	0.1550	0.1678	0.0530	0.2085	0.0047	0.0927	0.0064	0.0002	0.0012	0.0130
	1.5	0.809	0.645	10.09	12.66	0.0947	0.1156	0.0958	0.0033	0.2024	0.0651	0.1403	0.1928	0.0393	0.2311	0.0029	0.0950	0.0063	0.0003	0.0008	0.0137
	2.0	0.857	0.636	10.62	14.32	0.0897	0.1242	0.0955	0.0038	0.1985	0.0720	0.1296	0.2119	0.0299	0.2469	0.0020	0.0968	0.0063	0.0003	0.0005	0.0141

© Straight Haunches-Constant Width, Handbook of Frame Constants. Portland Cement Association, Chicago Illinois, Portland Cement Association.

TABLE 13.2 Parabolic Haunches—Constant Width

Note: All carry-over factors are negative and all stiffness factors are positive.

Right Haunch a_B	r_B	Carry-over Factors C_{AB}	C_{BA}	Stiffness Factors k_{AB}	k_{BA}	Unif. Load FEM Coef. × wL^2 M_{AB}	M_{BA}	Conc. b=0.1 M_{AB}	M_{BA}	b=0.3 M_{AB}	M_{BA}	b=0.5 M_{AB}	M_{BA}	b=0.7 M_{AB}	M_{BA}	b=0.9 M_{AB}	M_{BA}	Haunch Left M_{AB}	M_{BA}	Haunch Right M_{AB}	M_{BA}
								a_B = variable		r_A = 1.0		r_B = variable									
0.2	0.4	0.558	0.627	6.08	5.40	0.1022	0.0841	0.0938	0.0033	0.1891	0.0502	0.1572	0.1261	0.0715	0.1618	0.0073	0.0877	0.0032	0.0001	0.0002	0.0030
	0.6	0.582	0.624	6.21	5.80	0.0995	0.0887	0.0936	0.0036	0.1872	0.0535	0.1527	0.1339	0.0663	0.1708	0.0058	0.0902	0.0032	0.0001	0.0002	0.0031
	1.0	0.619	0.619	6.41	6.41	0.0956	0.0956	0.0935	0.0038	0.1844	0.0584	0.1459	0.1459	0.0584	0.1844	0.0038	0.0935	0.0032	0.0001	0.0001	0.0032
	1.5	0.649	0.614	6.59	6.97	0.0921	0.1015	0.0933	0.0041	0.1819	0.0628	0.1399	0.1563	0.0518	0.1962	0.0025	0.0958	0.0032	0.0001	0.0001	0.0032
	2.0	0.671	0.611	6.71	7.38	0.0899	0.1056	0.0932	0.0044	0.1801	0.0660	0.1358	0.1638	0.0472	0.2042	0.0017	0.0971	0.0032	0.0001	0.0000	0.0033
								a_B = variable		r_A = 1.0		r_B = variable									
0.3	0.4	0.588	0.616	6.22	5.93	0.1002	0.0877	0.0937	0.0035	0.1873	0.0537	0.1532	0.1339	0.0678	0.1686	0.0073	0.0877	0.0032	0.0001	0.0007	0.0063
	0.6	0.625	0.609	6.41	6.58	0.0966	0.0942	0.0935	0.0039	0.1845	0.0587	0.1467	0.1455	0.0609	0.1808	0.0057	0.0902	0.0032	0.0001	0.0005	0.0065
	1.0	0.683	0.598	6.73	7.68	0.0911	0.1042	0.0932	0.0044	0.1801	0.0669	0.1365	0.1643	0.0502	0.2000	0.0037	0.0936	0.0031	0.0001	0.0004	0.0068
	1.5	0.735	0.589	7.02	8.76	0.0862	0.1133	0.0929	0.0050	0.1760	0.0746	0.1272	0.1819	0.0410	0.2170	0.0023	0.0959	0.0031	0.0001	0.0003	0.0070
	2.0	0.772	0.582	7.25	9.61	0.0827	0.1198	0.0927	0.0054	0.1730	0.0805	0.1203	0.1951	0.0345	0.2293	0.0016	0.0972	0.0031	0.0001	0.0002	0.0072
								a_B = variable		r_A = 1.0		r_B = variable									
0.2	0.4	0.488	0.807	9.85	5.97	0.1214	0.0753	0.0929	0.0034	0.2131	0.0371	0.2021	0.1061	0.0979	0.1506	0.0105	0.0863	0.0171	0.0017	0.0003	0.0030
	0.6	0.515	0.803	10.10	6.45	0.1183	0.0795	0.0928	0.0036	0.2110	0.0404	0.1969	0.1136	0.0917	0.1600	0.0083	0.0892	0.0170	0.0018	0.0002	0.0030
	1.0	0.547	0.796	10.51	7.22	0.1138	0.0865	0.0926	0.0040	0.2079	0.0448	0.1890	0.1245	0.0809	0.1740	0.0056	0.0928	0.0168	0.0020	0.0001	0.0031
	1.5	0.571	0.786	10.90	7.90	0.1093	0.0922	0.0923	0.0043	0.2055	0.0485	0.1818	0.1344	0.0719	0.1862	0.0035	0.0951	0.0167	0.0021	0.0001	0.0032
	2.0	0.590	0.784	11.17	8.40	0.1063	0.0961	0.0922	0.0046	0.2041	0.0506	0.1764	0.1417	0.0661	0.1948	0.0025	0.0968	0.0166	0.0022	0.0001	0.0032
								a_B = 0.5		r_A = 1.0		r_B = variable									
0.5	0.4	0.554	0.753	10.42	7.66	0.1170	0.0811	0.0926	0.0040	0.2087	0.0442	0.1924	0.1205	0.0898	0.1595	0.0107	0.0853	0.0169	0.0020	0.0042	0.0145
	0.6	0.606	0.730	10.96	9.12	0.1115	0.0889	0.0922	0.0046	0.2045	0.0506	0.1820	0.1360	0.0791	0.1738	0.0086	0.0878	0.0167	0.0022	0.0036	0.0152
	1.0	0.694	0.694	12.03	12.03	0.1025	0.1025	0.0915	0.0057	0.1970	0.0626	0.1639	0.1639	0.0626	0.1970	0.0057	0.0915	0.0164	0.0028	0.0028	0.0164
	1.5	0.781	0.664	13.12	15.47	0.0937	0.1163	0.0908	0.0070	0.1891	0.0759	0.1456	0.1939	0.0479	0.2187	0.0039	0.0940	0.0160	0.0034	0.0021	0.0174
	2.0	0.850	0.642	14.09	18.64	0.0870	0.1275	0.0901	0.0082	0.1825	0.0877	0.1307	0.2193	0.0376	0.2348	0.0027	0.0957	0.0157	0.0039	0.0016	0.0181

© Parabolic Haunches-Constant Width, Handbook of Frame Constants. Portland Cement Association, Chicago Illinois, Portland Cement Association.

13.2 Moment Distribution for Structures Having Nonprismatic Members

Once the fixed-end moments and stiffness and carry-over factors for the nonprismatic members of a structure have been determined, application of the moment-distribution method follows the same procedure as outlined in Chapter 12. In this regard, recall that the distribution of moments may be shortened if a member stiffness factor is modified to account for conditions of end-span pin support and structure symmetry or antisymmetry. Similar modifications can also be made to nonprismatic members.

Beam Pin Supported at Far End. Consider the beam in Fig. 13–4a, which is pinned at its far end B. The absolute stiffness factor K'_A is the moment applied at A such that it rotates the beam at A, $\theta_A = 1$ rad. It can be determined as follows. First assume that B is temporarily fixed and a moment K_A is applied at A, Fig. 13–4b. The moment induced at B is $C_{AB}K_A$, where C_{AB} is the carry-over factor from A to B. Second, since B is not to be fixed, application of the opposite moment $C_{AB}K_A$ to the beam, Fig. 13–4c, will induce a moment $C_{BA}C_{AB}K_A$ at end A. By superposition, the result of these two applications of moment yields the beam loaded as shown in Fig. 13–4a. Hence it can be seen that the absolute stiffness factor of the beam at A is

$$K'_A = K_A(1 - C_{AB}C_{BA}) \tag{13–3}$$

Here K_A is the absolute stiffness factor of the beam, assuming it to be fixed at the far end B. For example, in the case of a prismatic beam, $K_A = 4EI/L$ and $C_{AB} = C_{BA} = \frac{1}{2}$. Substituting into Eq. 13–3 yields $K'_A = 3EI/L$, the same as Eq. 12–4.

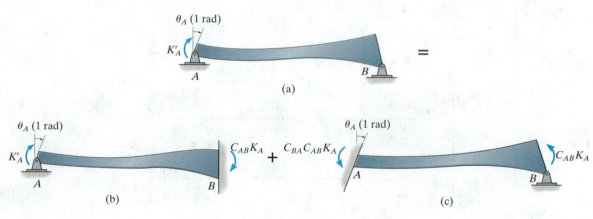

Fig. 13–4

Symmetric Beam and Loading.

Here we must determine the moment K'_A needed to rotate end A, $\theta_A = +1$ rad, while $\theta_B = -1$ rad, Fig. 13–5a. In this case we first assume that end B is fixed and apply the moment K_A at A, Fig. 13–5b. Next we apply a negative moment K_B to end B assuming that end A is fixed. This results in a moment of $C_{BA}K_B$ at end A as shown in Fig. 13–5c. Superposition of these two applications of moment at A yields the results of Fig. 13–5a. We require

$$K'_A = K_A - C_{BA}K_B$$

Using Eq. 13–1 ($C_{BA}K_B = C_{AB}K_A$), we can also write

$$\boxed{K'_A = K_A(1 - C_{AB})} \tag{13–4}$$

In the case of a prismatic beam, $K_A = 4EI/L$ and $C_{AB} = \frac{1}{2}$, so that $K'_A = 2EI/L$, which is the same as Eq. 12–5.

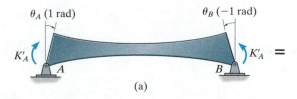

(a)

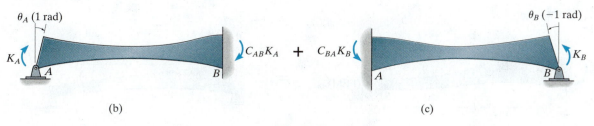

(b)   (c)

Fig. 13–5

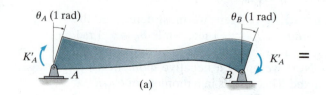

(a)

(b)

(c)

Fig. 13–6

Symmetric Beam with Antisymmetric Loading.

In the case of a symmetric beam with antisymmetric loading, we must determine K'_A such that equal rotations occur at the ends of the beam, Fig. 13–6a. To do this, we first fix end B and apply the moment K_A at A, Fig. 13–6b. Likewise, application of K_B at end B while end A is fixed is shown in Fig. 13–6c. Superposition of both cases yields the results of Fig. 13–6a. Hence,

$$K'_A = K_A + C_{BA}K_B$$

or, using Eq. 13–1 ($C_{BA}K_B = C_{AB}K_A$), we have for the absolute stiffness

$$K'_A = K_A(1 + C_{AB}) \qquad (13\text{–}5)$$

Substituting the data for a prismatic member, $K_A = 4EI/L$ and $C_{AB} = \frac{1}{2}$, yields $K'_A = 6EI/L$, which is the same as Eq. 12–6.

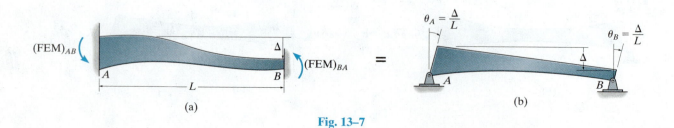

(a)

Fig. 13–7

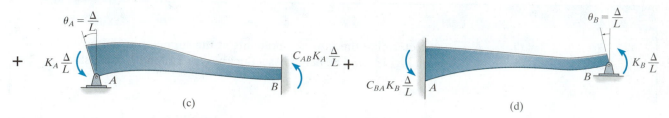

$$\theta_A = \frac{\Delta}{L}$$

$$+ \quad K_A \frac{\Delta}{L} \qquad \qquad C_{AB} K_A \frac{\Delta}{L} + \qquad \qquad C_{BA} K_B \frac{\Delta}{L}$$

$$\theta_B = \frac{\Delta}{L}$$

$$K_B \frac{\Delta}{L}$$

(c)   (d)

Fig. 13–7

Relative Joint Translation of Beam.

Fixed-end moments are developed in a nonprismatic member if it has a relative joint translation Δ between its ends A and B, Fig. 13–7a. In order to determine these moments, we proceed as follows. First consider the ends A and B to be pin connected and allow end B of the beam to be displaced a distance Δ such that the end rotations are $\theta_A = \theta_B = \Delta/L$, Fig. 13–7b. Second, assume that B is fixed and apply a moment of $M_A' = -K_A(\Delta/L)$ to end A such that it rotates the end $\theta_A = -\Delta/L$, Fig. 13–7c. Third, assume that A is fixed and apply a moment $M_B' = -K_B(\Delta/L)$ to end B such that it rotates the end $\theta_B = -\Delta/L$, Fig. 13–7d. Since the total sum of these three operations yields the condition shown in Fig. 13–7a, we have at A

$$(\text{FEM})_{AB} = -K_A \frac{\Delta}{L} - C_{BA} K_B \frac{\Delta}{L}$$

Applying Eq. 13–1 ($C_{BA} K_B = C_{AB} K_A$) yields

$$(\text{FEM})_{AB} = -K_A \frac{\Delta}{L}(1 + C_{AB}) \qquad (13\text{–}6)$$

A similar expression can be written for end B. Recall that for a prismatic member $K_A = 4EI/L$ and $C_{AB} = \frac{1}{2}$. Thus $(\text{FEM})_{AB} = -6EI\Delta/L^2$, which is the same as Eq. 11–5.

If end B is pinned rather than fixed, Fig. 13–8, the fixed-end moment at A can be determined in a manner similar to that described above. The result is

$$(\text{FEM})_{AB}' = -K_A \frac{\Delta}{L}(1 - C_{AB} C_{BA}) \qquad (13\text{–}7)$$

$(\text{FEM})_{AB}'$

Fig. 13–8

Here it is seen that for a prismatic member this equation gives $(\text{FEM})_{AB}' = -3EI\Delta/L^2$, which is the same as that listed on the inside back cover.

The following example illustrates application of the moment-distribution method to structures having nonprismatic members. Once the fixed-end moments and stiffness and carry-over factors have been determined, and the stiffness factor modified according to the equations given above, the procedure for analysis is the same as that discussed in Chapter 12.

13

EXAMPLE 13.1

Determine the internal moments at the supports of the beam shown in Fig. 13–9a. The beam has a thickness of 1 ft and E is constant.

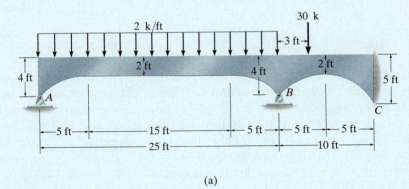

(a)

Fig. 13–9

SOLUTION

Since the haunches are parabolic, we will use Table 13.2 to obtain the moment-distribution properties of the beam.

Span AB

$$a_A = a_B = \frac{5}{25} = 0.2 \qquad r_A = r_B = \frac{4-2}{2} = 1.0$$

Entering Table 13.2 with these ratios, we find

$$C_{AB} = C_{BA} = 0.619$$

$$k_{AB} = k_{BA} = 6.41$$

Using Eqs. 13–2,

$$K_{AB} = K_{BA} = \frac{kEI_C}{L} = \frac{6.41E\left(\frac{1}{12}\right)(1)(2)^3}{25} = 0.171E$$

Since the far end of span BA is pinned, we will modify the stiffness factor of BA using Eq. 13–3. We have

$$K'_{BA} = K_{BA}(1 - C_{AB}C_{BA}) = 0.171E[1 - 0.619(0.619)] = 0.105E$$

Uniform load, Table 13.2,

$$(\text{FEM})_{AB} = -(0.0956)(2)(25)^2 = -119.50 \text{ k} \cdot \text{ft}$$

$$(\text{FEM})_{BA} = 119.50 \text{ k} \cdot \text{ft}$$

Span BC

$$a_B = a_C = \frac{5}{10} = 0.5 \qquad r_B = \frac{4-2}{2} = 1.0$$

$$r_C = \frac{5-2}{2} = 1.5$$

From Table 13.2 we find

$$C_{BC} = 0.781 \qquad C_{CB} = 0.664$$

$$k_{BC} = 13.12 \qquad k_{CB} = 15.47$$

Thus, from Eqs. 13–2,

$$K_{BC} = \frac{kEI_C}{L} = \frac{13.12E\left(\frac{1}{12}\right)(1)(2)^3}{10} = 0.875E$$

$$K_{CB} = \frac{kEI_C}{L} = \frac{15.47E\left(\frac{1}{12}\right)(1)(2)^3}{10} = 1.031E$$

Concentrated load,

$$b = \frac{3}{10} = 0.3$$

$$(FEM)_{BC} = -0.1891(30)(10) = -56.73 \text{ k} \cdot \text{ft}$$

$$(FEM)_{CB} = 0.0759(30)(10) = 22.77 \text{ k} \cdot \text{ft}$$

Using the foregoing values for the stiffness factors, the distribution factors are computed and entered in the table, Fig. 13–9b. The moment distribution follows the same procedure outlined in Chapter 12. The results in k · ft are shown on the last line of the table.

Joint	A	B		C
Member	AB	BA	BC	CB
K	0.171E	0.105E	0.875E	1.031E
DF	1	0.107	0.893	0
COF	0.619	0.619	0.781	0.664
FEM	−119.50	119.50	−56.73	22.77
Dist.	119.50	−6.72	−56.05	
CO		73.97		−43.78
Dist.		−7.91	−66.06	
CO				−51.59
ΣM	0	178.84	−178.84	−72.60

(b)

Fig. 13–9

13.3   Slope-Deflection Equations for Nonprismatic Members

The slope-deflection equations for prismatic members were developed in Chapter 11. In this section we will generalize the form of these equations so that they apply as well to nonprismatic members. To do this, we will use the results of the previous section and proceed to formulate the equations in the same manner discussed in Chapter 11, that is, considering the effects caused by the loads, relative joint displacement, and each joint rotation separately, and then superimposing the results.

Loads. Loads are specified by the fixed-end moments $(\text{FEM})_{AB}$ and $(\text{FEM})_{BA}$ acting at the ends A and B of the span. Positive moments act clockwise.

Relative Joint Translation. When a relative displacement Δ between the joints occurs, the induced moments are determined from Eq. 13–6. At end A this moment is $-\left[K_A \Delta / L \right](1 + C_{AB})$ and at end B it is $-\left[K_B \Delta / L \right](1 + C_{BA})$.

Rotation at A. If end A rotates θ_A, the required moment in the span at A is $K_A \theta_A$. Also, this induces a moment of $C_{AB} K_A \theta_A = C_{BA} K_B \theta_A$ at end B.

Rotation at B. If end B rotates θ_B, a moment of $K_B \theta_B$ must act at end B, and the moment induced at end A is $C_{BA} K_B \theta_B = C_{AB} K_A \theta_B$.

The total end moments caused by these effects yield the generalized slope-deflection equations, which can therefore be written as

$$M_{AB} = K_A \left[\theta_A + C_{AB}\theta_B - \frac{\Delta}{L}(1 + C_{AB}) \right] + (\text{FEM})_{AB}$$

$$M_{BA} = K_B \left[\theta_B + C_{BA}\theta_A - \frac{\Delta}{L}(1 + C_{BA}) \right] + (\text{FEM})_{BA}$$

Since these two equations are similar, we can express them as a single equation. Referring to one end of the span as the near end (N) and the other end as the far end (F), and representing the member rotation as $\psi = \Delta / L$, we have

$$M_N = K_N\big(\theta_N + C_N\theta_F - \psi(1 + C_N)\big) + (\text{FEM})_N \qquad (13\text{–}8)$$

Here

$\qquad M_N =$ internal moment at the near end of the span; this moment is positive clockwise when acting on the span.

$\qquad K_N =$ absolute stiffness of the near end determined from tables or by calculation.

θ_N, θ_F = near- and far-end slopes of the span at the supports; the angles are measured in *radians* and are *positive clockwise*.

ψ = span cord rotation due to a linear displacement, $\psi = \Delta/L$; this angle is measured in *radians* and is *positive clockwise*.

$(\text{FEM})_N$ = fixed-end moment at the near-end support; the moment is *positive* clockwise when acting on the span and is obtained from tables or by calculations.

Application of the equation follows the same procedure outlined in Chapter 11 and therefore will not be discussed here. In particular, note that Eq. 13–8 reduces to Eq. 11–8 when applied to members that are prismatic.

13

Light-weight metal buildings are often designed using frame members having variable moments of inertia.

A continuous, reinforced-concrete highway bridge.

PROBLEMS

Sec. 13.1–13.3

13–1. Determine the moments at A, B, and C by the moment-distribution method. Assume the supports at A and C are fixed and a roller support at B is on a rigid base. The girder has a thickness of 4 ft. Use Table 13.1. E is constant. The haunches are tapered.

13–2. Solve Prob. 13–1 using the slope-deflection equations.

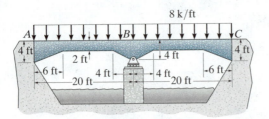

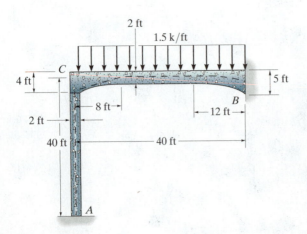

Probs. 13–1/2

13–3. Apply the moment-distribution method to determine the moment at each joint of the parabolic haunched frame. Supports A and B are fixed. Use Table 13.2. The members are each 1 ft thick. E is constant.

*13–4. Solve Prob. 13–3 using the slope-deflection equations.

Probs. 13–3/4

13–5. Use the moment-distribution method to determine the moment at each joint of the bridge frame. E is constant. Use Table 13.1 to compute the necessary frame constants. The members are each 1 ft thick.

13–6. Solve Prob. 13–5 using the slope-deflection equations.

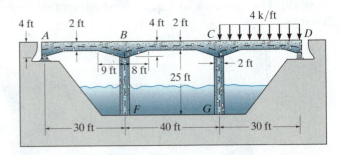

Probs. 13–5/6

13–7. Apply the moment-distribution method to determine the moment at each joint of the symmetric parabolic haunched frame. Supports A and D are fixed. Use Table 13.2. The members are each 1 ft thick. E is constant.

*13–8. Solve Prob. 13–7 using the slope-deflection equations.

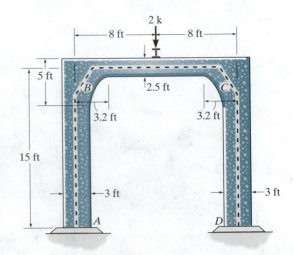

Probs. 13–7/8

13–9. Use the moment-distribution method to determine the moment at each joint of the frame. Assume that E is constant and the members have a thickness of 1 ft. The supports at A and C are pinned and the joints at B and D are fixed connected. Use Table 13.1.

13–10. Solve Prob. 13–9 using the slope-deflection equations.

13–11. Use the moment-distribution method to determine the moment at each joint of the symmetric bridge frame. Supports F and E are fixed and B and C are fixed connected. The haunches are tapered so use Table 13.2. Assume E is constant and the members are each 1 ft thick.

*13–12. Solve Prob. 13–11 using the slope-deflection equations.

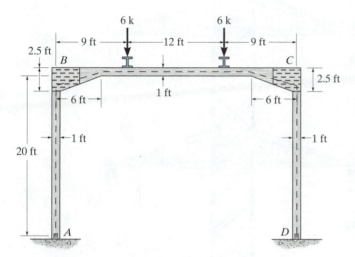

Probs. 13–9/10

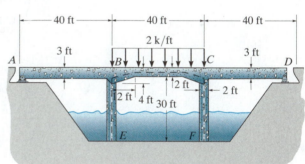

Probs. 13–11/12

CHAPTER REVIEW

Non-prismatic members having a variable moment of inertia are often used on long-span bridges and building frames to save material.

A structural analysis using non-prismatic members can be performed using either the slope-deflection equations or moment distribution. If this is done, it then becomes necessary to obtain the fixed-end moments, stiffness factors, and carry-over factors for the member. One way to obtain these values is to use the conjugate beam method, although the work is somewhat tedious. It is also possible to obtain these values from tabulated data, such as published by the Portland Cement Association.

If the moment distribution method is used, then the process can be simplified if the stiffness of some of the members is modified.

Chapter 14

© P.G. Bowater/Science Source

The space-truss analysis of electrical transmission towers can be performed using the stiffness method.

Truss Analysis Using the Stiffness Method

In this chapter we will explain the basic fundamentals of using the stiffness method for analyzing structures. It will be shown that this method, although tedious to do by hand, is quite suited for use on a computer. Examples of specific applications to planar trusses will be given. The method will then be expanded to include space-truss analysis. Beams and framed structures will be discussed in the next chapters.

14.1   Fundamentals of the Stiffness Method

There are essentially two ways in which structures can be analyzed using matrix methods. The stiffness method, to be used in this and the following chapters, is a displacement method of analysis. A force method, called the flexibility method, as outlined in Sec. 9.1, can also be used to analyze structures; however, this method will not be presented in this text. There are several reasons for this. Most important, the stiffness method can be used to analyze both statically determinate and indeterminate structures, whereas the flexibility method requires a different procedure for each of these two cases. Also, the stiffness method yields the displacements and forces directly, whereas with the flexibility method the displacements are not obtained directly. Furthermore, it is generally much easier to formulate the necessary matrices for the computer operations using the stiffness method; and once this is done, the computer calculations can be performed efficiently.

Application of the stiffness method requires subdividing the structure into a series of discrete **finite elements** and identifying their end points as *nodes*. For truss analysis, the finite elements are represented by each of the members that compose the truss, and the nodes represent the joints. The force-displacement properties of each element are determined and then related to one another using the force equilibrium equations written at the nodes. These relationships, for the entire structure, are then grouped together into what is called the *structure stiffness matrix* **K**. Once it is established, the unknown displacements of the nodes can then be determined for any given loading on the structure. When these displacements are known, the external and internal forces in the structure can be calculated using the force-displacement relations for each member.

Before developing a formal procedure for applying the stiffness method, it is first necessary to establish some preliminary definitions and concepts.

Member and Node Identification.

One of the first steps when applying the stiffness method is to identify the elements or members of the structure and their nodes. We will specify each member by a number enclosed within a square, and use a number enclosed within a circle to identify the nodes. Also, the "near" and "far" ends of the member must be identified. This will be done using an arrow written along the member, with the head of the arrow directed toward the far end. Examples of member, node, and "direction" identification for a truss are shown in Fig. 14–1a. These assignments have all been done *arbitrarily*.*

Global and Member Coordinates.

Since loads and displacements are vector quantities, it is necessary to establish a coordinate system in order to specify their correct sense of direction. Here we will use two different types of coordinate systems. A single *global* or *structure coordinate system, x, y*, will be used to specify the sense of each of the *external* force and displacement components at the nodes, Fig. 14–1a. A *local* or *member coordinate system* will be used for each member to specify the sense of direction of its displacements and *internal* loadings. This system will be identified using x', y' axes with the origin at the "near" node and the x' axis extending toward the "far" node. An example for truss member 4 is shown in Fig. 14–1b.

---

*For large trusses, matrix manipulations using **K** are actually more efficient using selective numbering of the members in a wave pattern, that is, starting from top to bottom, then bottom to top, etc.

Kinematic Indeterminacy. As discussed in Sec. 11.1, the unconstrained degrees of freedom for the truss represent the primary unknowns of any displacement method, and therefore these must be identified. As a general rule there are two degrees of freedom, or two possible displacements, for each joint (node). For application, each degree of freedom will be specified on the truss using a code number, shown at the joint or node, and referenced to its positive global coordinate direction using an associated arrow. For example, the truss in Fig. 14–1a has eight degrees of freedom, which have been identified by the "code numbers" 1 through 8 as shown. The truss is kinematically indeterminate to the fifth degree because of these eight possible displacements: 1 through 5 represent unknown or **unconstrained degrees of freedom**, and 6 through 8 represent **constrained degrees of freedom**. Due to the constraints, the displacements here are zero. For later application, *the lowest code numbers will always be used to identify the unknown displacements (unconstrained degrees of freedom) and the highest code numbers will be used to identify the known displacements (constrained degrees of freedom).* The reason for choosing this method of identification has to do with the convenience of later partitioning the structure stiffness matrix, so that the unknown displacements can be found in the most direct manner.

Once the truss is labeled and the code numbers are specified, the structure stiffness matrix **K** can then be determined. To do this we must first establish a *member stiffness matrix* **k′** for each member of the truss. This matrix is used to express the member's load-displacement relations in terms of the *local coordinates*. Since all the members of the truss are not in the same direction, we must develop a means for transforming these quantities from each member's local x′, y′ coordinate system to the structure's global x, y coordinate system. This can be done using *force and displacement transformation matrices*. Once established, the elements of the member stiffness matrix are transformed from local to global coordinates and then assembled to create the structure stiffness matrix. Using **K**, as stated previously, we can determine the node displacements first, followed by the support reactions and the member forces. We will now elaborate on the development of this method.

© Bethlehem Steel Corporation

The structural framework of this aircraft hangar is constructed entirely of trusses, in order to significantly reduce the weight of the structure. (Courtesy of Bethlehem Steel Corporation).

14

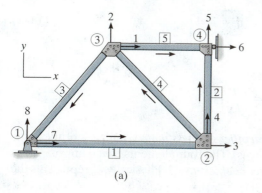

(a)

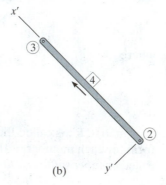

(b)

Fig. 14–1

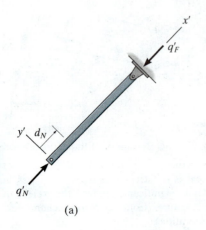

(a)

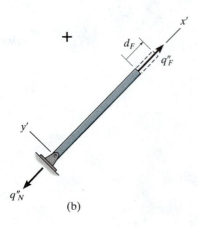

+

(b)

||

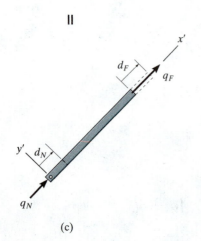

(c)

Fig. 14–2

14.2   Member Stiffness Matrix

In this section we will establish the stiffness matrix for a single truss member using local x', y' coordinates, oriented as shown in Fig. 14–2. The terms in this matrix will represent the load-displacement relations for the member.

A truss member can only be displaced along its axis (x' axis) since the loads are applied along this axis. Two independent displacements are therefore possible. When a positive displacement d_N is imposed on the near end of the member while the far end is held pinned, Fig. 14–2a, the forces developed at the ends of the members are

$$q'_N = \frac{AE}{L} d_N \quad q'_F = -\frac{AE}{L} d_N$$

Note that q'_F is negative since for equilibrium it acts in the negative x' direction. Likewise, a positive displacement d_F at the far end, keeping the near end pinned, Fig. 14–2b, results in member forces of

$$q''_N = -\frac{AE}{L} d_F \quad q''_F = \frac{AE}{L} d_F$$

By superposition, Fig. 14–2c, the resultant forces caused by both displacements are

$$q_N = \frac{AE}{L} d_N - \frac{AE}{L} d_F \tag{14–1}$$

$$q_F = -\frac{AE}{L} d_N + \frac{AE}{L} d_F \tag{14–2}$$

These load-displacement equations may be written in matrix form* as

$$\begin{bmatrix} q_N \\ q_F \end{bmatrix} = \frac{AE}{L} \begin{bmatrix} 1 & -1 \\ -1 & 1 \end{bmatrix} \begin{bmatrix} d_N \\ d_F \end{bmatrix}$$

or

$$\mathbf{q = k'd} \tag{14–3}$$

where

$$\mathbf{k'} = \frac{AE}{L} \begin{bmatrix} 1 & -1 \\ -1 & 1 \end{bmatrix} \tag{14–4}$$

This matrix, $\mathbf{k'}$, is called the *member stiffness matrix*, and it is of the same form for each member of the truss. The four elements that comprise it are called *member stiffness influence coefficients*, k'_{ij}. Physically, k'_{ij} represents the

*A review of matrix algebra is given in Appendix A.

force at joint i when a *unit displacement* is imposed at joint j. For example, if $i = j = 1$, then k'_{11} is the force at the near joint when the far joint is held fixed, and the near joint undergoes a displacement of $d_N = 1$, i.e.,

$$q_N = k'_{11} = \frac{AE}{L}$$

Likewise, the force at the far joint is determined from $i = 2$, $j = 1$, so that

$$q_F = k'_{21} = -\frac{AE}{L}$$

These two terms represent the first column of the member stiffness matrix. In the same manner, the second column of this matrix represents the forces in the member only when the far end of the member undergoes a unit displacement.

14.3 Displacement and Force Transformation Matrices

Since a truss is composed of many members (elements), we will now develop a method for transforming the member forces **q** and displacements **d** defined in local coordinates to global coordinates. For the sake of convention, we will consider the global coordinates positive x to the right and positive y upward. The smallest angles between the *positive x, y* global axes and the *positive x'* local axis will be defined as θ_x and θ_y as shown in Fig. 14–3. The cosines of these angles will be used in the matrix analysis that follows. These will be identified as $\lambda_x = \cos \theta_x$, $\lambda_y = \cos \theta_y$. Numerical values for λ_x and λ_y can easily be generated by a computer once the x, y coordinates of the near end N and far end F of the member have been specified. For example, consider member NF of the truss shown in Fig. 14–4. Here the coordinates of N and F are (x_N, y_N) and (x_F, y_F), respectively.* Thus,

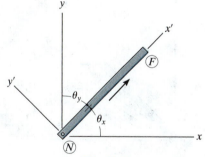

Fig. 14–3

$$\lambda_x = \cos \theta_x = \frac{x_F - x_N}{L} = \frac{x_F - x_N}{\sqrt{(x_F - x_N)^2 + (y_F - y_N)^2}} \tag{14–5}$$

$$\lambda_y = \cos \theta_y = \frac{y_F - y_N}{L} = \frac{y_F - y_N}{\sqrt{(x_F - x_N)^2 + (y_F - y_N)^2}} \tag{14–6}$$

The algebraic signs in these "generalized" equations will automatically account for members that are oriented in other quadrants of the x–y plane.

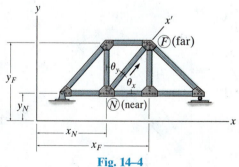

Fig. 14–4

*The origin can be located at any convenient point. Usually, however, it is located where the x, y coordinates of all the nodes will be *positive*, as shown in Fig. 14–4.

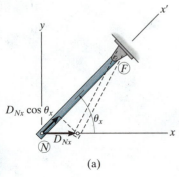

(a)

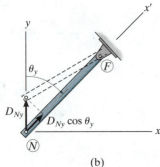

(b)

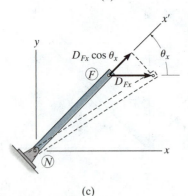

(c)

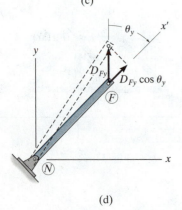

(d)

Fig. 14–5

14

Displacement Transformation Matrix.   In global coordinates each end of the member can have two degrees of freedom or independent displacements; namely, joint N has D_{Nx} and D_{Ny}, Figs. 14–5a and 14–5b, and joint F has D_{Fx} and D_{Fy}, Figs. 14–5c and 14–5d. We will now consider each of these displacements separately, in order to determine its component displacement along the member. When the far end is held pinned and the near end is given a global displacement D_{Nx}, Fig. 14–5a, the corresponding displacement (deformation) along the member is $D_{Nx} \cos \theta_x$.* Likewise, a displacement D_{Ny} will cause the member to be displaced $D_{Ny} \cos \theta_y$ along the x' axis, Fig. 14–5b. The effect of *both* global displacements causes the member to be displaced

$$d_N = D_{Nx} \cos \theta_x + D_{Ny} \cos \theta_y$$

In a similar manner, positive displacements D_{Fx} and D_{Fy} successively applied at the far end F, while the near end is held pinned, Figs. 14–5c and 14–5d, will cause the member to be displaced

$$d_F = D_{Fx} \cos \theta_x + D_{Fy} \cos \theta_y$$

Letting $\lambda_x = \cos \theta_x$ and $\lambda_y = \cos \theta_y$ represent the *direction cosines* for the member, we have

$$d_N = D_{Nx} \lambda_x + D_{Ny} \lambda_y$$

$$d_F = D_{Fx} \lambda_x + D_{Fy} \lambda_y$$

which can be written in matrix form as

$$\begin{bmatrix} d_N \\ d_F \end{bmatrix} = \begin{bmatrix} \lambda_x & \lambda_y & 0 & 0 \\ 0 & 0 & \lambda_x & \lambda_y \end{bmatrix} \begin{bmatrix} D_{Nx} \\ D_{Ny} \\ D_{Fx} \\ D_{Fy} \end{bmatrix} \qquad (14\text{–}7)$$

or

$$\boxed{\mathbf{d} = \mathbf{TD}} \qquad (14\text{–}8)$$

where

$$\mathbf{T} = \begin{bmatrix} \lambda_x & \lambda_y & 0 & 0 \\ 0 & 0 & \lambda_x & \lambda_y \end{bmatrix} \qquad (14\text{–}9)$$

From the above derivation, $\mathbf{T}$ transforms the four global x, y displacements $\mathbf{D}$ into the two local x' displacements $\mathbf{d}$. Hence, $\mathbf{T}$ is referred to as the *displacement transformation matrix*.

*The change in θ_x or θ_y will be neglected, since it is very small.

Force Transformation Matrix.   Consider now application of the force q_N to the near end of the member, the far end held pinned, Fig. 14–6a. Here the global force components of q_N at N are

$$Q_{Nx} = q_N \cos \theta_x \qquad Q_{Ny} = q_N \cos \theta_y$$

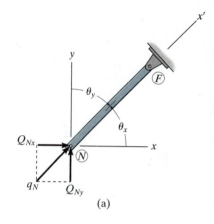

(a)

Likewise, if q_F is applied to the bar, Fig. 14–6b, the global force components at F are

$$Q_{Fx} = q_F \cos \theta_x \qquad Q_{Fy} = q_F \cos \theta_y$$

Using the direction cosines $\lambda_x = \cos \theta_x$, $\lambda_y = \cos \theta_y$, these equations become

$$Q_{Nx} = q_N \lambda_x \qquad Q_{Ny} = q_N \lambda_y$$

$$Q_{Fx} = q_F \lambda_x \qquad Q_{Fy} = q_F \lambda_y$$

which can be written in matrix form as

$$\begin{bmatrix} Q_{Nx} \\ Q_{Ny} \\ Q_{Fx} \\ Q_{Fy} \end{bmatrix} = \begin{bmatrix} \lambda_x & 0 \\ \lambda_y & 0 \\ 0 & \lambda_x \\ 0 & \lambda_y \end{bmatrix} \begin{bmatrix} q_N \\ q_F \end{bmatrix} \qquad (14\text{–}10)$$

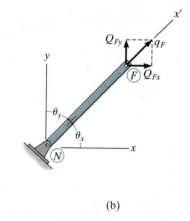

(b)

Fig. 14–6

or

$$\boxed{\mathbf{Q} = \mathbf{T}^T \mathbf{q}} \qquad (14\text{–}11)$$

where

$$\mathbf{T}^T = \begin{bmatrix} \lambda_x & 0 \\ \lambda_y & 0 \\ 0 & \lambda_x \\ 0 & \lambda_y \end{bmatrix} \qquad (14\text{–}12)$$

In this case $\mathbf{T}^T$ transforms the two local (x') forces $\mathbf{q}$ acting at the ends of the member into the four global (x, y) force components $\mathbf{Q}$. By comparison, this *force transformation matrix* is the transpose of the displacement transformation matrix, Eq. 14–9.

14.4   Member Global Stiffness Matrix

We will now combine the results of the preceding sections and determine the stiffness matrix for a member which relates the member's global force components $\mathbf{Q}$ to its global displacements $\mathbf{D}$. If we substitute Eq. 14–8 ($\mathbf{d} = \mathbf{TD}$) into Eq. 14–3 ($\mathbf{q} = \mathbf{k'd}$), we can determine the member's forces $\mathbf{q}$ in terms of the global displacements $\mathbf{D}$ at its end points, namely,

$$\mathbf{q} = \mathbf{k'TD} \tag{14–13}$$

Substituting this equation into Eq. 14–11, $\mathbf{Q} = \mathbf{T}^T\mathbf{q}$, yields the final result,

$$\mathbf{Q} = \mathbf{T}^T\mathbf{k'TD}$$

or

$$\mathbf{Q} = \mathbf{kD} \tag{14–14}$$

where

$$\mathbf{k} = \mathbf{T}^T\mathbf{k'T} \tag{14–15}$$

The matrix $\mathbf{k}$ is the **member stiffness matrix** in global coordinates. Since $\mathbf{T}^T$, $\mathbf{T}$, and $\mathbf{k'}$ are known, we have

$$\mathbf{k} = \begin{bmatrix} \lambda_x & 0 \\ \lambda_y & 0 \\ 0 & \lambda_x \\ 0 & \lambda_y \end{bmatrix} \frac{AE}{L} \begin{bmatrix} 1 & -1 \\ -1 & 1 \end{bmatrix} \begin{bmatrix} \lambda_x & \lambda_y & 0 & 0 \\ 0 & 0 & \lambda_x & \lambda_y \end{bmatrix}$$

Performing the matrix operations yields

$$\mathbf{k} = \frac{AE}{L} \begin{array}{c} \begin{array}{cccc} N_x & \quad N_y & \quad F_x & \quad F_y \end{array} \\ \begin{bmatrix} \lambda_x^2 & \lambda_x\lambda_y & -\lambda_x^2 & -\lambda_x\lambda_y \\ \lambda_x\lambda_y & \lambda_y^2 & -\lambda_x\lambda_y & -\lambda_y^2 \\ -\lambda_x^2 & -\lambda_x\lambda_y & \lambda_x^2 & \lambda_x\lambda_y \\ -\lambda_x\lambda_y & -\lambda_y^2 & \lambda_x\lambda_y & \lambda_y^2 \end{bmatrix} \begin{array}{c} N_x \\ N_y \\ F_x \\ F_y \end{array} \end{array} \tag{14–16}$$

The *location* of each element in this 4×4 symmetric matrix is referenced with each global degree of freedom associated with the near end N, followed by the far end F. This is indicated by the code number notation along the rows and columns, that is, N_x, N_y, F_x, F_y. Here **k** represents the force-displacement relations for the member when the components of force and displacement at the ends of the member are in the global or x, y directions. Each of the terms in the matrix is therefore a *stiffness influence coefficient* $\mathbf{k}_{ij}$, which denotes the x or y force component at i needed to cause an associated *unit x or y* displacement component at j. As a result, each identified column of the matrix represents the four force components developed at the ends of the member when the identified end undergoes a unit displacement related to its matrix column. For example, a unit displacement $D_{Nx} = 1$ will create the four force components on the member shown in the first column of the matrix.

14.5   Truss Stiffness Matrix

Once all the member stiffness matrices are formed in global coordinates, it becomes necessary to assemble them in the proper order so that the stiffness matrix **K** for the entire truss can be found. This process of combining the member matrices depends on careful identification of the elements in each member matrix. As discussed in the previous section, this is done by designating the rows and columns of the matrix by the *four* code numbers N_x, N_y, F_x, F_y used to identify the two global degrees of freedom that can occur at each end of the member (see Eq. 14–16). The structure stiffness matrix will then have an order that will be equal to the highest code number assigned to the truss, since this represents the total number of degrees of freedom for the structure. When the **k** matrices are assembled, each element in **k** will then be placed in its *same* row and column designation in the structure stiffness matrix **K**. In particular, when two or more members are *connected* to the *same* joint or node, then some of the elements of each member's **k** matrix will be assigned to the same position in the **K** matrix. When this occurs, the elements assigned to the common location must be added together algebraically. The reason for this becomes clear if one realizes that each element of the **k** matrix represents the resistance of the member to an applied force at its end. In this way, adding these resistances in the x or y direction when forming the **K** matrix determines the *total resistance* of each joint to a unit displacement in the x or y direction.

This method of assembling the member matrices to form the structure stiffness matrix will now be demonstrated by two numerical examples. Although this process is somewhat tedious when done by hand, it is rather easy to program on a computer.

14

EXAMPLE | **14.1**

Determine the structure stiffness matrix for the two-member truss shown in Fig. 14–7a. AE is constant.

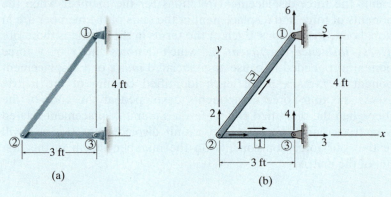

(a)   (b)

Fig. 14–7

SOLUTION

By inspection, ② will have two unknown displacement components, whereas joints ① and ③ are constrained from displacement. Consequently, the displacement components at joint ② are code numbered first, followed by those at joints ③ and ①, Fig. 14–7b. The origin of the global coordinate system can be located at any point. For convenience, we will choose joint ② as shown. The members are identified arbitrarily and arrows are written along the two members to identify the near and far ends of each member. The direction cosines and the stiffness matrix for each member can now be determined.

Member 1. Since ② is the near end and ③ is the far end, then by Eqs. 14–5 and 14–6, we have

$$\lambda_x = \frac{3 - 0}{3} = 1 \qquad \lambda_y = \frac{0 - 0}{3} = 0$$

Using Eq. 14–16, dividing each element by $L = 3$ ft, we have

$$
\mathbf{k_1} = AE
\begin{array}{c}
\begin{array}{cccc} 1 & 2 & 3 & 4 \end{array} \\
\begin{bmatrix}
0.333 & 0 & -0.333 & 0 \\
0 & 0 & 0 & 0 \\
-0.333 & 0 & 0.333 & 0 \\
0 & 0 & 0 & 0
\end{bmatrix}
\begin{array}{c} 1 \\ 2 \\ 3 \\ 4 \end{array}
\end{array}
$$

The calculations can be checked in part by noting that $\mathbf{k_1}$ is *symmetric*. Note that the rows and columns in $\mathbf{k_1}$ are identified by the x, y degrees of freedom at the near end, followed by the far end, that is, 1, 2, 3, 4, respectively, for member 1, Fig. 14–7b. This is done in order to identify the elements for later assembly into the $\mathbf{K}$ matrix.

Member 2.   Since ② is the near end and ① is the far end, we have

$$\lambda_x = \frac{3-0}{5} = 0.6 \qquad \lambda_y = \frac{4-0}{5} = 0.8$$

Thus Eq. 14–16 with $L = 5$ ft becomes

$$
k_2 = AE
\begin{array}{c}

\begin{array}{cccc}
1 & 2 & 5 & 6
\end{array}\\
\left[
\begin{array}{cccc}
0.072 & 0.096 & -0.072 & -0.096\\
0.096 & 0.128 & -0.096 & -0.128\\
-0.072 & -0.096 & 0.072 & 0.096\\
-0.096 & -0.128 & 0.096 & 0.128
\end{array}
\right]
\begin{array}{c}
1\\2\\5\\6
\end{array}
\end{array}
$$

Here the rows and columns are identified as 1, 2, 5, 6, since these numbers represent, respectively, the x, y degrees of freedom at the near and far ends of member 2.

Structure Stiffness Matrix.   This matrix has an order of 6×6 since there are six designated degrees of freedom for the truss, Fig. 14–7b. Corresponding elements of the above two matrices are added algebraically to form the structure stiffness matrix. Perhaps the assembly process is easier to see if the missing numerical columns and rows in $\mathbf{k}_1$ and $\mathbf{k}_2$ are expanded with zeros to form two 6×6 matrices. Then

$$\mathbf{K} = \mathbf{k}_1 + \mathbf{k}_2$$

$$
\mathbf{K} = AE
\begin{array}{c}
\begin{array}{cccccc}
1 & 2 & 3 & 4 & 5 & 6
\end{array}\\
\left[
\begin{array}{cccccc}
0.333 & 0 & -0.333 & 0 & 0 & 0\\
0 & 0 & 0 & 0 & 0 & 0\\
-0.333 & 0 & 0.333 & 0 & 0 & 0\\
0 & 0 & 0 & 0 & 0 & 0\\
0 & 0 & 0 & 0 & 0 & 0\\
0 & 0 & 0 & 0 & 0 & 0
\end{array}
\right]
\begin{array}{c}
1\\2\\3\\4\\5\\6
\end{array}
\end{array}
+ AE
\begin{array}{c}
\begin{array}{cccccc}
1 & 2 & 3 & 4 & 5 & 6
\end{array}\\
\left[
\begin{array}{cccccc}
0.072 & 0.096 & 0 & 0 & -0.072 & -0.096\\
0.096 & 0.128 & 0 & 0 & -0.096 & -0.128\\
0 & 0 & 0 & 0 & 0 & 0\\
0 & 0 & 0 & 0 & 0 & 0\\
-0.072 & -0.096 & 0 & 0 & 0.072 & 0.096\\
-0.096 & -0.128 & 0 & 0 & 0.096 & 0.128
\end{array}
\right]
\begin{array}{c}
1\\2\\3\\4\\5\\6
\end{array}
\end{array}
$$

$$
\mathbf{K} = AE
\left[
\begin{array}{cccccc}
0.405 & 0.096 & -0.333 & 0 & -0.072 & -0.096\\
0.096 & 0.128 & 0 & 0 & -0.096 & -0.128\\
-0.333 & 0 & 0.333 & 0 & 0 & 0\\
0 & 0 & 0 & 0 & 0 & 0\\
-0.072 & -0.096 & 0 & 0 & 0.072 & 0.096\\
-0.096 & -0.128 & 0 & 0 & 0.096 & 0.128
\end{array}
\right]
$$

If a computer is used for this operation, generally one starts with $\mathbf{K}$ having all zero elements; then as the member global stiffness matrices are generated, they are placed directly into their respective element positions in the $\mathbf{K}$ matrix, rather than developing the member stiffness matrices, storing them, then assembling them.

14

EXAMPLE | 14.2

Determine the structure stiffness matrix for the truss shown in Fig. 14–8a. *AE* is constant.

SOLUTION

Although the truss is statically indeterminate to the first degree, this will present no difficulty for obtaining the structure stiffness matrix. Each joint and member are arbitrarily identified numerically, and the near and far ends are indicated by the arrows along the members. As shown in Fig. 14–8b, the *unconstrained displacements* are *code numbered first*. There are eight degrees of freedom for the truss, and so **K** will be an 8 × 8 matrix. In order to keep all the joint coordinates positive, the origin of the global coordinates is chosen at ①. Equations 14–5, 14–6, and 14–16 will now be applied to each member.

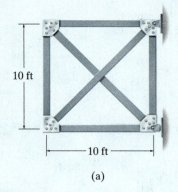

10 ft

10 ft

(a)

Member 1. Here $L = 10$ ft, so that

$$\lambda_x = \frac{10 - 0}{10} = 1 \qquad \lambda_y = \frac{0 - 0}{10} = 0$$

$$\mathbf{k_1} = AE \begin{bmatrix} 0.1 & 0 & -0.1 & 0 \\ 0 & 0 & 0 & 0 \\ -0.1 & 0 & 0.1 & 0 \\ 0 & 0 & 0 & 0 \end{bmatrix} \begin{matrix} 1 \\ 2 \\ 6 \\ 5 \end{matrix}$$

with column headers $\begin{matrix} 1 & 2 & 6 & 5 \end{matrix}$

Member 2. Here $L = 10\sqrt{2}$ ft, so that

$$\lambda_x = \frac{10 - 0}{10\sqrt{2}} = 0.707 \qquad \lambda_y = \frac{10 - 0}{10\sqrt{2}} = 0.707$$

$$\mathbf{k_2} = AE \begin{bmatrix} 0.035 & 0.035 & -0.035 & -0.035 \\ 0.035 & 0.035 & -0.035 & -0.035 \\ -0.035 & -0.035 & 0.035 & 0.035 \\ -0.035 & -0.035 & 0.035 & 0.035 \end{bmatrix} \begin{matrix} 1 \\ 2 \\ 7 \\ 8 \end{matrix}$$

with column headers $\begin{matrix} 1 & 2 & 7 & 8 \end{matrix}$

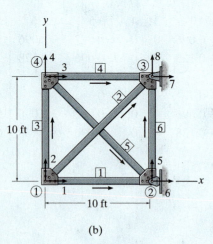

(b)

Fig. 14–8

Member 3. Here $L = 10$ ft, so that

$$\lambda_x = \frac{0 - 0}{10} = 0 \qquad \lambda_y = \frac{10 - 0}{10} = 1$$

$$\mathbf{k_3} = AE \begin{bmatrix} 0 & 0 & 0 & 0 \\ 0 & 0.1 & 0 & -0.1 \\ 0 & 0 & 0 & 0 \\ 0 & -0.1 & 0 & 0.1 \end{bmatrix} \begin{matrix} 1 \\ 2 \\ 3 \\ 4 \end{matrix}$$

with column headers $\begin{matrix} 1 & 2 & 3 & 4 \end{matrix}$

Member 4. Here $L = 10$ ft, so that

$$\lambda_x = \frac{10 - 0}{10} = 1 \qquad \lambda_y = \frac{10 - 10}{10} = 0$$

$$\mathbf{k}_4 = AE \begin{array}{c} \\ \end{array} \begin{array}{cccc} 3 & 4 & 7 & 8 \\ \left[\begin{array}{cccc} 0.1 & 0 & -0.1 & 0 \\ 0 & 0 & 0 & 0 \\ -0.1 & 0 & 0.1 & 0 \\ 0 & 0 & 0 & 0 \end{array} \right] & \begin{array}{c} 3 \\ 4 \\ 7 \\ 8 \end{array} \end{array}$$

Member 5. Here $L = 10\sqrt{2}$ ft, so that

$$\lambda_x = \frac{10 - 0}{10\sqrt{2}} = 0.707 \qquad \lambda_y = \frac{0 - 10}{10\sqrt{2}} = -0.707$$

$$\mathbf{k}_5 = AE \begin{array}{cccc} 3 & 4 & 6 & 5 \\ \left[\begin{array}{cccc} 0.035 & -0.035 & -0.035 & 0.035 \\ -0.035 & 0.035 & 0.035 & -0.035 \\ -0.035 & 0.035 & 0.035 & -0.035 \\ 0.035 & -0.035 & -0.035 & 0.035 \end{array} \right] & \begin{array}{c} 3 \\ 4 \\ 6 \\ 5 \end{array} \end{array}$$

Member 6. Here $L = 10$ ft, so that

$$\lambda_x = \frac{10 - 10}{10} = 0 \qquad \lambda_y = \frac{10 - 0}{10} = 1$$

$$\mathbf{k}_6 = AE \begin{array}{cccc} 6 & 5 & 7 & 8 \\ \left[\begin{array}{cccc} 0 & 0 & 0 & 0 \\ 0 & 0.1 & 0 & -0.1 \\ 0 & 0 & 0 & 0 \\ 0 & -0.1 & 0 & 0.1 \end{array} \right] & \begin{array}{c} 6 \\ 5 \\ 7 \\ 8 \end{array} \end{array}$$

Structure Stiffness Matrix. The foregoing six matrices can now be assembled into the 8×8 $\mathbf{K}$ matrix by algebraically adding their corresponding elements. For example, since $(k_{11})_1 = AE(0.1)$, $(k_{11})_2 = AE(0.035)$, $(k_{11})_3 = (k_{11})_4 = (k_{11})_5 = (k_{11})_6 = 0$, then, $K_{11} = AE(0.1 + 0.035) = AE(0.135)$, and so on. The final result is thus,

$$\mathbf{K} = AE \begin{array}{cccccccc} 1 & 2 & 3 & 4 & 5 & 6 & 7 & 8 \\ \left[\begin{array}{cccccccc} 0.135 & 0.035 & 0 & 0 & 0 & -0.1 & -0.035 & -0.035 \\ 0.035 & 0.135 & 0 & -0.1 & 0 & 0 & -0.035 & -0.035 \\ 0 & 0 & 0.135 & -0.035 & 0.035 & -0.035 & -0.1 & 0 \\ 0 & -0.1 & -0.035 & 0.135 & -0.035 & 0.035 & 0 & 0 \\ 0 & 0 & 0.035 & -0.035 & 0.135 & -0.035 & 0 & -0.1 \\ -0.1 & 0 & -0.035 & 0.035 & -0.035 & 0.135 & 0 & 0 \\ -0.035 & -0.035 & -0.1 & 0 & 0 & 0 & 0.135 & 0.035 \\ -0.035 & -0.035 & 0 & 0 & -0.1 & 0 & 0.035 & 0.135 \end{array} \right] & \begin{array}{c} 1 \\ 2 \\ 3 \\ 4 \\ 5 \\ 6 \\ 7 \\ 8 \end{array} \end{array}$$

Ans.

14.6   Application of the Stiffness Method for Truss Analysis

Once the structure stiffness matrix is formed, the global force components **Q** acting on the truss can then be related to its global displacements **D** using

$$\mathbf{Q} = \mathbf{KD} \qquad (14\text{–}17)$$

This equation is referred to as the ***structure stiffness equation***. Since we have always assigned the lowest code numbers to identify the uncon-strained degrees of freedom, this will allow us now to partition this equation in the following form*:

$$\left[\begin{array}{c} \mathbf{Q}_k \\ \hline \mathbf{Q}_u \end{array}\right] = \left[\begin{array}{c:c} \mathbf{K}_{11} & \mathbf{K}_{12} \\ \hdashline \mathbf{K}_{21} & \mathbf{K}_{22} \end{array}\right] \left[\begin{array}{c} \mathbf{D}_u \\ \hline \mathbf{D}_k \end{array}\right] \qquad (14\text{–}18)$$

Here

$\mathbf{Q}_k, \mathbf{D}_k = $ *known* external loads and displacements; the loads here exist on the truss as part of the problem, and the displacements are generally specified as zero due to support constraints such as pins or rollers.

$\mathbf{Q}_u, \mathbf{D}_u = $ *unknown* loads and displacements; the loads here represent the unknown support reactions, and the displacements are at joints where motion is unconstrained in a particular direction.

$\mathbf{K} = $ *structure* stiffness matrix, which is partitioned to be compatible with the partitioning of **Q** and **D**.

Expanding Eq. 14–18 yields

$$\mathbf{Q}_k = \mathbf{K}_{11}\mathbf{D}_u + \mathbf{K}_{12}\mathbf{D}_k \qquad (14\text{–}19)$$
$$\mathbf{Q}_u = \mathbf{K}_{21}\mathbf{D}_u + \mathbf{K}_{22}\mathbf{D}_k \qquad (14\text{–}20)$$

Most often $\mathbf{D}_k = \mathbf{0}$ since the supports are not displaced. Provided this is the case, Eq. 14–19 becomes

$$\mathbf{Q}_k = \mathbf{K}_{11}\mathbf{D}_u$$

Since the elements in the partitioned matrix $\mathbf{K}_{11}$ represent the *total resistance* at a truss joint to a unit displacement in either the *x* or *y* direction, then the above equation symbolizes the collection of all the *force equilibrium equations* applied to the joints where the external loads are zero or have a known value ($\mathbf{Q}_k$). Solving for $\mathbf{D}_u$, we have

$$\mathbf{D}_u = [\mathbf{K}_{11}]^{-1}\mathbf{Q}_k \qquad (14\text{–}21)$$

From this equation we can obtain a direct solution for all the unknown joint displacements; then using Eq. 14–20 with $\mathbf{D}_k = \mathbf{0}$ yields

$$\mathbf{Q}_u = \mathbf{K}_{21}\mathbf{D}_u \qquad (14\text{–}22)$$

from which the unknown support reactions can be determined. The member forces can be determined using Eq. 14–13, namely

$$\mathbf{q} = \mathbf{k}'\mathbf{TD}$$

*This partitioning scheme will become obvious in the numerical examples that follow.

Expanding this equation yields

$$\begin{bmatrix} q_N \\ q_F \end{bmatrix} = \frac{AE}{L} \begin{bmatrix} 1 & -1 \\ -1 & 1 \end{bmatrix} \begin{bmatrix} \lambda_x & \lambda_y & 0 & 0 \\ 0 & 0 & \lambda_x & \lambda_y \end{bmatrix} \begin{bmatrix} D_{Nx} \\ D_{Ny} \\ D_{Fx} \\ D_{Fy} \end{bmatrix}$$

Since $q_N = -q_F$ for equilibrium, only one of the forces has to be found. Here we will determine q_F, the one that exerts tension in the member, Fig. 14–2c.

$$q_F = \frac{AE}{L} [-\lambda_x \quad -\lambda_y \quad \lambda_x \quad \lambda_y] \begin{bmatrix} D_{Nx} \\ D_{Ny} \\ D_{Fx} \\ D_{Fy} \end{bmatrix} \qquad (14\text{–}23)$$

In particular, if the computed result using this equation is negative, the member is then in compression.

Procedure for Analysis

The following method provides a means for determining the unknown displacements and support reactions for a truss using the stiffness method.

Notation

- Establish the x, y global coordinate system. The origin is usually located at the joint for which the coordinates for all the other joints are positive.

- Identify each joint and member numerically, and arbitrarily specify the near and far ends of each member symbolically by directing an arrow along the member with the head directed towards the far end.

- Specify the two code numbers at each joint, using the *lowest numbers* to identify *unconstrained degrees of freedom*, followed by the *highest numbers* to identify the *constrained degrees of freedom*.

- From the problem, establish $\mathbf{D}_k$ and $\mathbf{Q}_k$.

Structure Stiffness Matrix

- For each member determine λ_x and λ_y and the member stiffness matrix using Eq. 14–16.

- Assemble these matrices to form the stiffness matrix for the entire truss as explained in Sec. 14.5. As a partial check of the calculations, the member and structure stiffness matrices should be *symmetric*.

Displacements and Loads

- Partition the structure stiffness matrix as indicated by Eq. 14–18.

- Determine the unknown joint displacements $\mathbf{D}_u$ using Eq. 14–21, the support reactions $\mathbf{Q}_u$ using Eq. 14–22, and each member force $\mathbf{q}_F$ using Eq. 14–23.

EXAMPLE   14.3

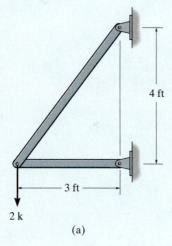

4 ft

3 ft

2 k

(a)

Fig. 14–9

Determine the force in each member of the two-member truss shown in Fig. 14–9a. AE is constant.

SOLUTION

Notation. The origin of x, y and the numbering of the joints and members are shown in Fig. 14–9b. Also, the near and far ends of each member are identified by arrows, and code numbers are used at each joint. By inspection it is seen that the known external displacements are $D_3 = D_4 = D_5 = D_6 = 0$. Also, the known external loads are $Q_1 = 0$, $Q_2 = -2$ k. Hence,

$$\mathbf{D}_k = \begin{bmatrix} 0 \\ 0 \\ 0 \\ 0 \end{bmatrix} \begin{matrix} 3 \\ 4 \\ 5 \\ 6 \end{matrix} \qquad \mathbf{Q}_k = \begin{bmatrix} 0 \\ -2 \end{bmatrix} \begin{matrix} 1 \\ 2 \end{matrix}$$

Structure Stiffness Matrix. Using the same notation as used here, this matrix has been developed in Example 14.1.

Displacements and Loads. Writing Eq. 14–17, $\mathbf{Q} = \mathbf{KD}$, for the truss we have

$$\begin{bmatrix} 0 \\ -2 \\ Q_3 \\ Q_4 \\ Q_5 \\ Q_6 \end{bmatrix} = AE \begin{bmatrix} 0.405 & 0.096 & \vdots & -0.333 & 0 & -0.072 & -0.096 \\ 0.096 & 0.128 & \vdots & 0 & 0 & -0.096 & -0.128 \\ \hdotsfor{7} \\ -0.333 & 0 & \vdots & 0.333 & 0 & 0 & 0 \\ 0 & 0 & \vdots & 0 & 0 & 0 & 0 \\ -0.072 & -0.096 & \vdots & 0 & 0 & 0.072 & 0.096 \\ -0.096 & -0.128 & \vdots & 0 & 0 & 0.096 & 0.128 \end{bmatrix} \begin{bmatrix} D_1 \\ D_2 \\ 0 \\ 0 \\ 0 \\ 0 \end{bmatrix} \quad (1)$$

From this equation we can now identify $\mathbf{K}_{11}$ and thereby determine $\mathbf{D}_u$. It is seen that the matrix multiplication, like Eq. 14–19, yields

$$\begin{bmatrix} 0 \\ -2 \end{bmatrix} = AE \begin{bmatrix} 0.405 & 0.096 \\ 0.096 & 0.128 \end{bmatrix} \begin{bmatrix} D_1 \\ D_2 \end{bmatrix} + \begin{bmatrix} 0 \\ 0 \end{bmatrix}$$

Here it is easy to solve by a direct expansion,

$$0 = AE(0.405D_1 + 0.096D_2)$$

$$-2 = AE(0.096D_1 + 0.128D_2)$$

Physically these equations represent $\Sigma F_x = 0$ and $\Sigma F_y = 0$ applied to joint ②. Solving, we get

$$D_1 = \frac{4.505}{AE} \qquad D_2 = \frac{-19.003}{AE}$$

By inspection of Fig. 14–9b, one would indeed expect a rightward and downward displacement to occur at joint ② as indicated by the positive and negative signs of these answers.

Using these results, the support reactions are now obtained from Eq. (1), written in the form of Eq. 14–20 (or Eq. 14–22) as

$$
\begin{bmatrix} Q_3 \\ Q_4 \\ Q_5 \\ Q_6 \end{bmatrix} = AE \begin{bmatrix} -0.333 & 0 \\ 0 & 0 \\ -0.072 & -0.096 \\ -0.096 & -0.128 \end{bmatrix} \frac{1}{AE} \begin{bmatrix} 4.505 \\ -19.003 \end{bmatrix} + \begin{bmatrix} 0 \\ 0 \\ 0 \\ 0 \end{bmatrix}
$$

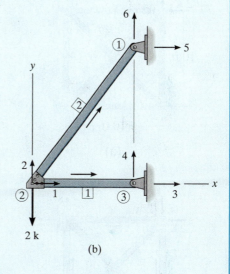

Expanding and solving for the reactions,

$Q_3 = -0.333(4.505) = -1.5$ k

$Q_4 = 0$

$Q_5 = -0.072(4.505) - 0.096(-19.003) = 1.5$ k

$Q_6 = -0.096(4.505) - 0.128(-19.003) = 2.0$ k

The force in each member is found from Eq. 14–23. Using the data for λ_x and λ_y in Example 14.1, we have

Member 1: $\lambda_x = 1, \lambda_y = 0, L = 3$ ft

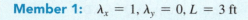

$$
q_1 = \frac{AE}{3} \begin{matrix} 1 & 2 & 3 & 4 \end{matrix} \begin{bmatrix} -1 & 0 & 1 & 0 \end{bmatrix} \frac{1}{AE} \begin{bmatrix} 4.505 \\ -19.003 \\ 0 \\ 0 \end{bmatrix} \begin{matrix} 1 \\ 2 \\ 3 \\ 4 \end{matrix}
$$

$$
= \frac{1}{3} [-4.505] = -1.5 \text{ k} \qquad\qquad Ans.
$$

Member 2: $\lambda_x = 0.6, \lambda_y = 0.8, L = 5$ ft

$$
q_2 = \frac{AE}{5} \begin{matrix} 1 & 2 & 5 & 6 \end{matrix} \begin{bmatrix} -0.6 & -0.8 & 0.6 & 0.8 \end{bmatrix} \frac{1}{AE} \begin{bmatrix} 4.505 \\ -19.003 \\ 0 \\ 0 \end{bmatrix} \begin{matrix} 1 \\ 2 \\ 5 \\ 6 \end{matrix}
$$

$$
= \frac{1}{5} [-0.6(4.505) - 0.8(-19.003)] = 2.5 \text{ k} \qquad\qquad Ans.
$$

These answers can of course be verified by equilibrium, applied at joint ②.

EXAMPLE   14.4

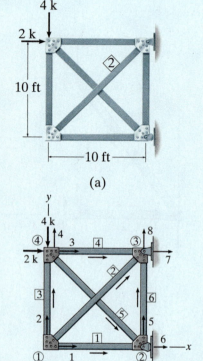

4 k

2 k

10 ft

10 ft

(a)

②

Determine the support reactions and the force in member 2 of the truss shown in Fig. 14–10a. AE is constant.

SOLUTION

Notation. The joints and members are numbered and the origin of the x, y axes is established at ①, Fig. 14–10b. Also, arrows are used to reference the near and far ends of each member. Using the code numbers, where the lowest numbers denote the unconstrained degrees of freedom, Fig. 14–10b, we have

$$
\mathbf{D}_k = \begin{bmatrix} 0 \\ 0 \\ 0 \\ 0 \end{bmatrix} \begin{matrix} 6 \\ 7 \\ 8 \end{matrix} \qquad
\mathbf{Q}_k = \begin{bmatrix} 0 \\ 0 \\ 2 \\ -4 \\ 0 \end{bmatrix} \begin{matrix} 1 \\ 2 \\ 3 \\ 4 \\ 5 \end{matrix}
$$

Structure Stiffness Matrix. This matrix has been determined in Example 14.2 using the same notation as in Fig. 14–10b.

Displacements and Loads. For this problem $\mathbf{Q} = \mathbf{KD}$ is

(b)

Fig. 14–10

$$
\begin{bmatrix} 0 \\ 0 \\ 2 \\ -4 \\ 0 \\ Q_6 \\ Q_7 \\ Q_8 \end{bmatrix}
= AE
\begin{bmatrix}
0.135 & 0.035 & 0 & 0 & 0 & -0.1 & -0.035 & -0.035 \\
0.035 & 0.135 & 0 & -0.1 & 0 & 0 & -0.035 & -0.035 \\
0 & 0 & 0.135 & -0.035 & 0.035 & -0.035 & -0.1 & 0 \\
0 & -0.1 & -0.035 & 0.135 & -0.035 & 0.035 & 0 & 0 \\
0 & 0 & 0.035 & -0.035 & 0.135 & -0.035 & 0 & -0.1 \\
-0.1 & 0 & -0.035 & 0.035 & -0.035 & 0.135 & 0 & 0 \\
-0.035 & -0.035 & -0.1 & 0 & 0 & 0 & 0.135 & 0.035 \\
-0.035 & -0.035 & 0 & 0 & -0.1 & 0 & 0.035 & 0.135
\end{bmatrix}
\begin{bmatrix} D_1 \\ D_2 \\ D_3 \\ D_4 \\ D_5 \\ 0 \\ 0 \\ 0 \end{bmatrix}
\qquad (1)
$$

Multiplying so as to formulate the unknown displacement equation 14–18, we get

$$
\begin{bmatrix} 0 \\ 0 \\ 2 \\ -4 \\ 0 \end{bmatrix}
= AE
\begin{bmatrix}
0.135 & 0.035 & 0 & 0 & 0 \\
0.035 & 0.135 & 0 & -0.1 & 0 \\
0 & 0 & 0.135 & -0.035 & 0.035 \\
0 & -0.1 & -0.035 & 0.135 & -0.035 \\
0 & 0 & 0.035 & -0.035 & 0.135
\end{bmatrix}
\begin{bmatrix} D_1 \\ D_2 \\ D_3 \\ D_4 \\ D_5 \end{bmatrix}
+
\begin{bmatrix} 0 \\ 0 \\ 0 \\ 0 \\ 0 \end{bmatrix}
$$

Expanding and solving the equations for the displacements yields

$$\begin{bmatrix} D_1 \\ D_2 \\ D_3 \\ D_4 \\ D_5 \end{bmatrix} = \frac{1}{AE} \begin{bmatrix} 17.94 \\ -69.20 \\ -2.06 \\ -87.14 \\ -22.06 \end{bmatrix}$$

Developing Eq. 14–20 from Eq. (1) using the calculated results, we have

$$\begin{bmatrix} Q_6 \\ Q_7 \\ Q_8 \end{bmatrix} = AE \begin{bmatrix} -0.1 & 0 & -0.035 & 0.035 & -0.035 \\ -0.035 & -0.035 & -0.1 & 0 & 0 \\ -0.035 & -0.035 & 0 & 0 & -0.1 \end{bmatrix} \frac{1}{AE} \begin{bmatrix} 17.94 \\ -69.20 \\ -2.06 \\ -87.14 \\ -22.06 \end{bmatrix} + \begin{bmatrix} 0 \\ 0 \\ 0 \end{bmatrix}$$

Expanding and computing the support reactions yields

$$Q_6 = -4.0 \text{ k} \qquad\qquad Ans.$$

$$Q_7 = 2.0 \text{ k} \qquad\qquad Ans.$$

$$Q_8 = 4.0 \text{ k} \qquad\qquad Ans.$$

The negative sign for Q_6 indicates that the rocker support reaction acts in the negative x direction. The force in member 2 is found from Eq. 14–23, where from Example 14.2, $\lambda_x = 0.707$, $\lambda_y = 0.707$, $L = 10\sqrt{2}$ ft. Thus,

$$q_2 = \frac{AE}{10\sqrt{2}} [-0.707 \quad -0.707 \quad 0.707 \quad 0.707] \frac{1}{AE} \begin{bmatrix} 17.94 \\ -69.20 \\ 0 \\ 0 \end{bmatrix}$$

$$= 2.56 \text{ k} \qquad\qquad Ans.$$

EXAMPLE | 14.5

Determine the force in member 2 of the assembly in Fig. 14–11a if the support at joint ① settles *downward* 25 mm. Take $AE = 8(10^3)$ kN.

SOLUTION

Notation. For convenience the origin of the global coordinates in Fig. 14–11b is established at joint ③, and as usual the lowest code numbers are used to reference the unconstrained degrees of freedom. Thus,

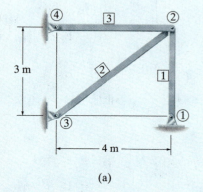

(a)

$$D_k = \begin{bmatrix} 0 \\ -0.025 \\ 0 \\ 0 \\ 0 \\ 0 \end{bmatrix} \begin{matrix} 3 \\ 4 \\ 5 \\ 6 \\ 7 \\ 8 \end{matrix} \qquad Q_k = \begin{bmatrix} 0 \\ 0 \end{bmatrix} \begin{matrix} 1 \\ 2 \end{matrix}$$

Structure Stiffness Matrix.   Using Eq. 14–16, we have

Member 1: $\lambda_x = 0, \lambda_y = 1, L = 3$ m, so that

$$k_1 = AE \begin{bmatrix} \overset{3}{0} & \overset{4}{0} & \overset{1}{0} & \overset{2}{0} \\ 0 & 0.333 & 0 & -0.333 \\ 0 & 0 & 0 & 0 \\ 0 & -0.333 & 0 & 0.333 \end{bmatrix} \begin{matrix} 3 \\ 4 \\ 1 \\ 2 \end{matrix}$$

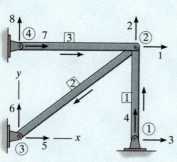

(b)

Fig. 14–11

Member 2: $\lambda_x = -0.8, \lambda_y = -0.6, L = 5$ m, so that

$$k_2 = AE \begin{bmatrix} \overset{1}{0.128} & \overset{2}{0.096} & \overset{5}{-0.128} & \overset{6}{-0.096} \\ 0.096 & 0.072 & -0.096 & -0.072 \\ -0.128 & -0.096 & 0.128 & 0.096 \\ -0.096 & -0.072 & 0.096 & 0.072 \end{bmatrix} \begin{matrix} 1 \\ 2 \\ 5 \\ 6 \end{matrix}$$

Member 3: $\lambda_x = 1, \lambda_y = 0, L = 4$ m, so that

$$k_3 = AE \begin{bmatrix} \overset{7}{0.25} & \overset{8}{0} & \overset{1}{-0.25} & \overset{2}{0} \\ 0 & 0 & 0 & 0 \\ -0.25 & 0 & 0.25 & 0 \\ 0 & 0 & 0 & 0 \end{bmatrix} \begin{matrix} 7 \\ 8 \\ 1 \\ 2 \end{matrix}$$

By assembling these matrices, the structure stiffness matrix becomes

$$K = AE \begin{bmatrix}
\overset{1}{0.378} & \overset{2}{0.096} & \overset{3}{0} & \overset{4}{0} & \overset{5}{-0.128} & \overset{6}{-0.096} & \overset{7}{-0.25} & \overset{8}{0} \\
0.096 & 0.405 & 0 & -0.333 & -0.096 & -0.072 & 0 & 0 \\
0 & 0 & 0 & 0 & 0 & 0 & 0 & 0 \\
0 & -0.333 & 0 & 0.333 & 0 & 0 & 0 & 0 \\
-0.128 & -0.096 & 0 & 0 & 0.128 & 0.096 & 0 & 0 \\
-0.096 & -0.072 & 0 & 0 & 0.096 & 0.072 & 0 & 0 \\
-0.25 & 0 & 0 & 0 & 0 & 0 & 0.25 & 0 \\
0 & 0 & 0 & 0 & 0 & 0 & 0 & 0
\end{bmatrix} \begin{matrix} 1 \\ 2 \\ 3 \\ 4 \\ 5 \\ 6 \\ 7 \\ 8 \end{matrix}$$

Displacements and Loads. Here $\mathbf{Q} = \mathbf{KD}$ yields

$$
\begin{bmatrix} 0 \\ 0 \\ Q_3 \\ Q_4 \\ Q_5 \\ Q_6 \\ Q_7 \\ Q_8 \end{bmatrix} = AE \begin{bmatrix} 0.378 & 0.096 & 0 & 0 & -0.128 & -0.096 & -0.25 & 0 \\ 0.096 & 0.405 & 0 & -0.333 & -0.096 & -0.072 & 0 & 0 \\ 0 & 0 & 0 & 0 & 0 & 0 & 0 & 0 \\ 0 & -0.333 & 0 & 0.333 & 0 & 0 & 0 & 0 \\ -0.128 & -0.096 & 0 & 0 & 0.128 & 0.096 & 0 & 0 \\ -0.096 & -0.072 & 0 & 0 & 0.096 & 0.072 & 0 & 0 \\ -0.25 & 0 & 0 & 0 & 0 & 0 & 0.25 & 0 \\ 0 & 0 & 0 & 0 & 0 & 0 & 0 & 0 \end{bmatrix} \begin{bmatrix} D_1 \\ D_2 \\ 0 \\ -0.025 \\ 0 \\ 0 \\ 0 \\ 0 \end{bmatrix}
$$

Developing the solution for the displacements, Eq. 14–19, we have

$$
\begin{bmatrix} 0 \\ 0 \end{bmatrix} = AE \begin{bmatrix} 0.378 & 0.096 \\ 0.096 & 0.405 \end{bmatrix} \begin{bmatrix} D_1 \\ D_2 \end{bmatrix} + AE \begin{bmatrix} 0 & 0 & -0.128 & -0.096 & -0.25 & 0 \\ 0 & -0.333 & -0.096 & -0.072 & 0 & 0 \end{bmatrix} \begin{bmatrix} 0 \\ -0.025 \\ 0 \\ 0 \\ 0 \\ 0 \end{bmatrix}
$$

which yields

$$0 = AE\big[(0.378D_1 + 0.096D_2) + 0\big]$$

$$0 = AE\big[(0.096D_1 + 0.405D_2) + 0.00833\big]$$

Solving these equations simultaneously gives

$$D_1 = 0.00556 \text{ m}$$

$$D_2 = -0.021875 \text{ m}$$

Although the support reactions do not have to be calculated, if needed they can be found from the expansion defined by Eq. 14–20. Using Eq. 14–23 to determine the force in member 2 yields

Member 2: $\lambda_x = -0.8$, $\lambda_y = -0.6$, $L = 5$ m, $AE = 8(10^3)$ kN, so that

$$
q_2 = \frac{8(10^3)}{5} [0.8 \quad 0.6 \quad -0.8 \quad -0.6] \begin{bmatrix} 0.00556 \\ -0.021875 \\ 0 \\ 0 \end{bmatrix}
$$

$$= \frac{8(10^3)}{5}(0.00444 - 0.0131) = -13.9 \text{ kN} \qquad \textit{Ans.}$$

Using the same procedure, show that the force in member 1 is $q_1 = 8.34$ kN and in member 3, $q_3 = 11.1$ kN. The results are shown on the free-body diagram of joint ②, Fig. 14–11c, which can be checked to be in equilibrium.

(c)

14

14.7 Nodal Coordinates

On occasion a truss can be supported by a roller placed on an *incline,* and when this occurs the constraint of zero deflection at the support (node) *cannot* be directly defined using a single horizontal and vertical global coordinate system. For example, consider the truss in Fig. 14–12a. The condition of zero displacement at node ① is defined only along the y'' axis, and because the roller can displace along the x'' axis this node will have displacement *components* along *both* global coordinate axes, x, y. For this reason we cannot include the zero displacement condition at this node when writing the global stiffness equation for the truss using x, y axes without making some modifications to the matrix analysis procedure.

To solve this problem, so that it can easily be incorporated into a computer analysis, we will use a set of **nodal coordinates** x'', y'' located at the inclined support. These axes are oriented such that the reactions and support displacements are along each of the coordinate axes, Fig. 14–12a. In order to determine the global stiffness equation for the truss, it then becomes necessary to develop force and displacement transformation matrices for each of the connecting members at this support so that the results can be summed within the same global x, y coordinate system. To show how this is done, consider truss member 1 in Fig. 14–12b, having a global coordinate system x, y at the near node Ⓝ, and a nodal coordinate system x'', y'' at the far node Ⓕ. When displacements **D** occur so that they have components along each of these axes as shown in Fig. 14–12c, the displacements **d** in the x' direction along the ends of the member become

$$d_N = D_{Nx} \cos \theta_x + D_{Ny} \cos \theta_y$$

$$d_F = D_{Fx''} \cos \theta_{x''} + D_{Fy''} \cos \theta_{y''}$$

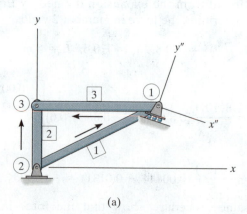

(a)

Fig. 14–12

These equations can be written in matrix form as

$$
\begin{bmatrix} d_N \\ d_F \end{bmatrix} = \begin{bmatrix} \lambda_x & \lambda_y & 0 & 0 \\ 0 & 0 & \lambda_{x''} & \lambda_{y''} \end{bmatrix} \begin{bmatrix} D_{Nx} \\ D_{Ny} \\ D_{Fx''} \\ D_{Fy''} \end{bmatrix}
$$

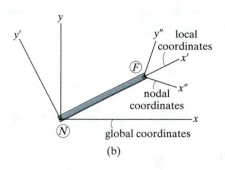

(b)

Likewise, forces **q** at the near and far ends of the member, Fig. 14–12d, have components **Q** along the global axes of

$$
Q_{Nx} = q_N \cos \theta_x \qquad Q_{Ny} = q_N \cos \theta_y
$$

$$
Q_{Fx''} = q_F \cos \theta_{x''} \qquad Q_{Fy''} = q_F \cos \theta_{y''}
$$

which can be expressed as

$$
\begin{bmatrix} Q_{Nx} \\ Q_{Ny} \\ Q_{Fx''} \\ Q_{Fy''} \end{bmatrix} = \begin{bmatrix} \lambda_x & 0 \\ \lambda_y & 0 \\ 0 & \lambda_{x''} \\ 0 & \lambda_{y''} \end{bmatrix} \begin{bmatrix} q_N \\ q_F \end{bmatrix}
$$

The displacement and force transformation matrices in the above equations are used to develop the member stiffness matrix for this situation. Applying Eq. 14–15, we have

$$
\mathbf{k} = \mathbf{T}^T \mathbf{k}' \mathbf{T}
$$

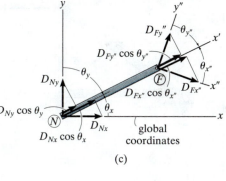

(c)

$$
\mathbf{k} = \begin{bmatrix} \lambda_x & 0 \\ \lambda_y & 0 \\ 0 & \lambda_{x''} \\ 0 & \lambda_{y''} \end{bmatrix} \frac{AE}{L} \begin{bmatrix} 1 & -1 \\ -1 & 1 \end{bmatrix} \begin{bmatrix} \lambda_x & \lambda_y & 0 & 0 \\ 0 & 0 & \lambda_{x''} & \lambda_{y''} \end{bmatrix}
$$

Performing the matrix operations yields,

$$
\mathbf{k} = \frac{AE}{L} \begin{bmatrix} \lambda_x^2 & \lambda_x\lambda_y & -\lambda_x\lambda_{x''} & -\lambda_x\lambda_{y''} \\ \lambda_x\lambda_y & \lambda_y^2 & -\lambda_y\lambda_{x''} & -\lambda_y\lambda_{y''} \\ -\lambda_x\lambda_{x''} & -\lambda_y\lambda_{x''} & \lambda_{x''}^2 & \lambda_{x''}\lambda_{y''} \\ -\lambda_x\lambda_{y''} & -\lambda_y\lambda_{y''} & \lambda_{x''}\lambda_{y''} & \lambda_{y''}^2 \end{bmatrix} \qquad (14\text{–}24)
$$

This stiffness matrix is then used for each member that is connected to an inclined roller support, and the process of assembling the matrices to form the structure stiffness matrix follows the standard procedure. The following example problem illustrates its application.

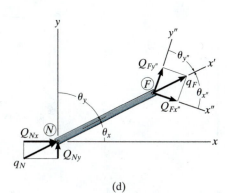

(d)

EXAMPLE 14.6

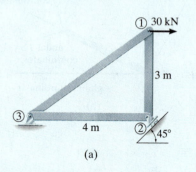

(a)

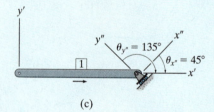

(b)

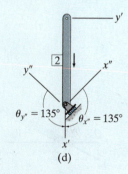

(c)

(d)

Fig. 14–13

Determine the support reactions for the truss shown in Fig. 14–13a.

SOLUTION

Notation. Since the roller support at ② is on an incline, we must use nodal coordinates at this node. The joints and members are numbered and the global x, y axes are established at node ③, Fig. 14–13b. Notice that the code numbers 3 and 4 are along the x'', y'' axes in order to use the condition that $D_4 = 0$.

Member Stiffness Matrices. The stiffness matrices for members 1 and 2 must be developed using Eq. 14–24 since these members have code numbers in the direction of global and nodal axes. The stiffness matrix for member 3 is determined in the usual manner.

Member 1. Fig. 14–13c, $\lambda_x = 1$, $\lambda_y = 0$, $\lambda_{x''} = 0.707$, $\lambda_{y''} = -0.707$

$$
\mathbf{k}_1 = AE
\begin{array}{cccc}
\quad 5 \quad & \quad 6 \quad & \quad 3 \quad & \quad 4 \quad \\
\end{array}
\begin{bmatrix}
0.25 & 0 & -0.17675 & 0.17675 \\
0 & 0 & 0 & 0 \\
-0.17675 & 0 & 0.125 & -0.125 \\
0.17675 & 0 & -0.125 & 0.125
\end{bmatrix}
\begin{array}{c}
5 \\ 6 \\ 3 \\ 4
\end{array}
$$

Member 2. Fig. 14–13d, $\lambda_x = 0$, $\lambda_y = -1$, $\lambda_{x''} = -0.707$, $\lambda_{y''} = -0.707$

$$
\mathbf{k}_2 = AE
\begin{array}{cccc}
\quad 1 \quad & \quad 2 \quad & \quad 3 \quad & \quad 4 \quad \\
\end{array}
\begin{bmatrix}
0 & 0 & 0 & 0 \\
0 & 0.3333 & -0.2357 & -0.2357 \\
0 & -0.2357 & 0.1667 & 0.1667 \\
0 & -0.2357 & 0.1667 & 0.1667
\end{bmatrix}
\begin{array}{c}
1 \\ 2 \\ 3 \\ 4
\end{array}
$$

Member 3. $\lambda_x = 0.8$, $\lambda_y = 0.6$

$$
\mathbf{k}_3 = AE
\begin{array}{cccc}
\quad 5 \quad & \quad 6 \quad & \quad 1 \quad & \quad 2 \quad \\
\end{array}
\begin{bmatrix}
0.128 & 0.096 & -0.128 & -0.096 \\
0.096 & 0.072 & -0.096 & -0.072 \\
-0.128 & -0.096 & 0.128 & 0.096 \\
-0.096 & -0.072 & 0.096 & 0.072
\end{bmatrix}
\begin{array}{c}
5 \\ 6 \\ 1 \\ 2
\end{array}
$$

Structure Stiffness Matrix.   Assembling these matrices to determine the structure stiffness matrix, we have

$$
\begin{bmatrix} 30 \\ 0 \\ 0 \\ \hdashline Q_4 \\ Q_5 \\ Q_6 \end{bmatrix} = AE \begin{bmatrix} 0.128 & 0.096 & 0 & 0 & -0.128 & -0.096 \\ 0.096 & 0.4053 & -0.2357 & -0.2357 & -0.096 & -0.072 \\ 0 & -0.2357 & 0.2917 & 0.0417 & -0.17675 & 0 \\ \hdashline 0 & -0.2357 & 0.0417 & 0.2917 & 0.17675 & 0 \\ -0.128 & -0.096 & -0.17675 & 0.17675 & 0.378 & 0.096 \\ -0.096 & -0.072 & 0 & 0 & 0.096 & 0.072 \end{bmatrix} \begin{bmatrix} D_1 \\ D_2 \\ D_3 \\ \hdashline 0 \\ 0 \\ 0 \end{bmatrix} \quad (1)
$$

Carrying out the matrix multiplication of the upper partitioned matrices, the three unknown displacements **D** are determined from solving the resulting simultaneous equations, i.e.,

$$
D_1 = \frac{352.5}{AE}
$$

$$
D_2 = \frac{-157.5}{AE}
$$

$$
D_3 = \frac{-127.3}{AE}
$$

The unknown reactions **Q** are obtained from the multiplication of the lower partitioned matrices in Eq. (1). Using the computed displacements, we have,

$$
Q_4 = 0(352.5) - 0.2357(-157.5) + 0.0417(-127.3)
$$
$$
= 31.8 \text{ kN} \qquad\qquad\qquad\qquad\qquad\qquad Ans.
$$

$$
Q_5 = -0.128(352.5) - 0.096(-157.5) - 0.17675(-127.3)
$$
$$
= -7.5 \text{ kN} \qquad\qquad\qquad\qquad\qquad\qquad Ans.
$$

$$
Q_6 = -0.096(352.5) - 0.072(-157.5) + 0(-127.3)
$$
$$
= -22.5 \text{ kN} \qquad\qquad\qquad\qquad\qquad\qquad Ans.
$$

14.8   Trusses Having Thermal Changes and Fabrication Errors

If some of the members of the truss are subjected to an increase or decrease in length due to thermal changes or fabrication errors, then it is necessary to use the method of superposition to obtain the solution. This requires three steps. First, the fixed-end forces necessary to *prevent* movement of the nodes as caused by temperature or fabrication are calculated. Second, the equal but opposite forces are placed on the truss at the nodes and the displacements of the nodes are calculated using the matrix analysis. Finally, the actual forces in the members and the reactions on the truss are determined by superposing these two results. This procedure, of course, is only necessary if the truss is statically indeterminate. If the truss is statically determinate, the displacements at the nodes can be found by this method; however, the temperature changes and fabrication errors will not affect the reactions and the member forces since the truss is free to adjust to these changes of length.

Thermal Effects.   If a truss member of length L is subjected to a temperature increase ΔT, the member will undergo an increase in length of $\Delta L = \alpha \Delta T L$, where α is the coefficient of thermal expansion. A compressive force q_0 applied to the member will cause a decrease in the member's length of $\Delta L' = q_0 L / AE$. If we equate these two displacements, then $q_0 = AE\alpha\Delta T$. This force will hold the nodes of the member fixed as shown in Fig. 14–14, and so we have

$$(q_N)_0 = AE\alpha\Delta T$$

$$(q_F)_0 = -AE\alpha\Delta T$$

Realize that if a temperature decrease occurs, then ΔT becomes negative and these forces reverse direction to hold the member in equilibrium.

We can transform these two forces into global coordinates using Eq. 14–10, which yields

$$\begin{bmatrix} (Q_{Nx})_0 \\ (Q_{Ny})_0 \\ (Q_{Fx})_0 \\ (Q_{Fy})_0 \end{bmatrix} = \begin{bmatrix} \lambda_x & 0 \\ \lambda_y & 0 \\ 0 & \lambda_x \\ 0 & \lambda_y \end{bmatrix} AE\alpha\Delta T \begin{bmatrix} 1 \\ -1 \end{bmatrix} = AE\alpha\Delta T \begin{bmatrix} \lambda_x \\ \lambda_y \\ -\lambda_x \\ -\lambda_y \end{bmatrix} \quad (14\text{–}25)$$

Fabrication Errors.   If a truss member is made too long by an amount ΔL before it is fitted into a truss, then the force q_0 needed to keep the member at its design length L is $q_0 = AE\Delta L / L$, and so for the member in Fig. 14–14, we have

$$(q_N)_0 = \frac{AE\Delta L}{L}$$

$$(q_F)_0 = -\frac{AE\Delta L}{L}$$

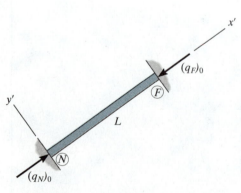

Fig. 14–14

If the member is originally too short, then ΔL becomes negative and these forces will reverse.

In global coordinates, these forces are

$$\begin{bmatrix} (Q_{Nx})_0 \\ (Q_{Ny})_0 \\ (Q_{Fx})_0 \\ (Q_{Fy})_0 \end{bmatrix} = \frac{AE\Delta L}{L} \begin{bmatrix} \lambda_x \\ \lambda_y \\ -\lambda_x \\ -\lambda_y \end{bmatrix} \qquad (14\text{--}26)$$

Matrix Analysis.   In the general case, with the truss subjected to applied forces, temperature changes, and fabrication errors, the initial force-displacement relationship for the truss then becomes

$$\mathbf{Q} = \mathbf{KD} + \mathbf{Q}_0 \qquad (14\text{--}27)$$

Here $\mathbf{Q}_0$ is a column matrix for the entire truss of the initial fixed-end forces caused by the temperature changes and fabrication errors of the members defined in Eqs. 14–25 and 14–26. We can partition this equation in the following form

$$\begin{bmatrix} \mathbf{Q}_k \\ \cdots \\ \mathbf{Q}_u \end{bmatrix} = \begin{bmatrix} \mathbf{K}_{11} & \vdots & \mathbf{K}_{12} \\ \cdots & & \cdots \\ \mathbf{K}_{21} & \vdots & \mathbf{K}_{22} \end{bmatrix} \begin{bmatrix} \mathbf{D}_u \\ \cdots \\ \mathbf{D}_k \end{bmatrix} + \begin{bmatrix} (\mathbf{Q}_k)_0 \\ \cdots \\ (\mathbf{Q}_u)_0 \end{bmatrix}$$

Carrying out the multiplication, we obtain

$$\mathbf{Q}_k = \mathbf{K}_{11}\mathbf{D}_u + \mathbf{K}_{12}\mathbf{D}_k + (\mathbf{Q}_k)_0 \qquad (14\text{--}28)$$

$$\mathbf{Q}_u = \mathbf{K}_{21}\mathbf{D}_u + \mathbf{K}_{22}\mathbf{D}_k + (\mathbf{Q}_u)_0 \qquad (14\text{--}29)$$

According to the superposition procedure described above, the unknown displacements $\mathbf{D}_u$ are determined from the first equation by subtracting $\mathbf{K}_{12}\mathbf{D}_k$ and $(\mathbf{Q}_k)_0$ from both sides and then solving for $\mathbf{D}_u$. This yields

$$\mathbf{D}_u = \mathbf{K}_{11}^{-1}(\mathbf{Q}_k - \mathbf{K}_{12}\mathbf{D}_k - (\mathbf{Q}_k)_0)$$

Once these nodal displacements are obtained, the member forces are then determined by superposition, i.e.,

$$\mathbf{q} = \mathbf{k}'\mathbf{TD} + \mathbf{q}_0$$

If this equation is expanded to determine the force at the far end of the member, we obtain

$$q_F = \frac{AE}{L}[-\lambda_x \quad -\lambda_y \quad \lambda_x \quad \lambda_y] \begin{bmatrix} D_{Nx} \\ D_{Ny} \\ D_{Fx} \\ D_{Fy} \end{bmatrix} + (q_F)_0 \qquad (14\text{--}30)$$

This result is similar to Eq. 14–23, except here we have the additional term $(q_F)_0$ which represents the initial fixed-end member force due to temperature changes and/or fabrication error as defined previously. Realize that if the computed result from this equation is negative, the member will be in compression.

The following two examples illustrate application of this procedure.

EXAMPLE 14.7

14

Determine the force in members 1 and 2 of the pin-connected assembly of Fig. 14–15 if member 2 was made 0.01 m too short before it was fitted into place. Take $AE = 8(10^3)$ kN.

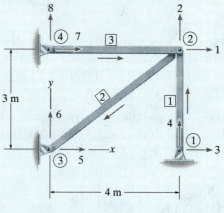

Fig. 14–15

SOLUTION

Since the member is short, then $\Delta L = -0.01$ m, and therefore applying Eq. 14–26 to member 2, with $\lambda_x = -0.8$, $\lambda_y = -0.6$, we have

$$\begin{bmatrix} (Q_1)_0 \\ (Q_2)_0 \\ (Q_5)_0 \\ (Q_6)_0 \end{bmatrix} = \frac{AE(-0.01)}{5} \begin{bmatrix} -0.8 \\ -0.6 \\ 0.8 \\ 0.6 \end{bmatrix} = AE \begin{bmatrix} 0.0016 \\ 0.0012 \\ -0.0016 \\ -0.0012 \end{bmatrix} \begin{matrix} 1 \\ 2 \\ 5 \\ 6 \end{matrix}$$

The structure stiffness matrix for this assembly has been established in Example 14.5. Applying Eq. 14–27, we have

$$\begin{bmatrix} 0 \\ 0 \\ \hdashline Q_3 \\ Q_4 \\ Q_5 \\ Q_6 \\ Q_7 \\ Q_8 \end{bmatrix} = AE \begin{bmatrix} 0.378 & 0.096 & \vdots & 0 & 0 & -0.128 & -0.096 & -0.25 & 0 \\ 0.096 & 0.405 & \vdots & 0 & -0.333 & -0.096 & -0.072 & 0 & 0 \\ \hdashline 0 & 0 & \vdots & 0 & 0 & 0 & 0 & 0 & 0 \\ 0 & -0.333 & \vdots & 0 & 0.333 & 0 & 0 & 0 & 0 \\ -0.128 & -0.096 & \vdots & 0 & 0 & 0.128 & 0.096 & 0 & 0 \\ -0.096 & -0.072 & \vdots & 0 & 0 & 0.096 & 0.072 & 0 & 0 \\ -0.25 & 0 & \vdots & 0 & 0 & 0 & 0 & 0.25 & 0 \\ 0 & 0 & \vdots & 0 & 0 & 0 & 0 & 0 & 0 \end{bmatrix} \begin{bmatrix} D_1 \\ D_2 \\ \hdashline 0 \\ 0 \\ 0 \\ 0 \\ 0 \\ 0 \end{bmatrix} + AE \begin{bmatrix} 0.0016 \\ 0.0012 \\ \hdashline 0 \\ 0 \\ -0.0016 \\ -0.0012 \\ 0 \\ 0 \end{bmatrix} \quad (1)$$

Partitioning the matrices as shown and carrying out the multiplication
to obtain the equations for the unknown displacements yields

$$
\begin{bmatrix} 0 \\ 0 \end{bmatrix} = AE \begin{bmatrix} 0.378 & 0.096 \\ 0.096 & 0.405 \end{bmatrix} \begin{bmatrix} D_1 \\ D_2 \end{bmatrix} + AE \begin{bmatrix} 0 & 0 & -0.128 & -0.096 & -0.25 & 0 \\ 0 & -0.333 & -0.096 & -0.072 & 0 & 0 \end{bmatrix} \begin{bmatrix} 0 \\ 0 \\ 0 \\ 0 \\ 0 \\ 0 \end{bmatrix} + AE \begin{bmatrix} 0.0016 \\ 0.0012 \end{bmatrix}
$$

which gives

$$0 = AE\,[0.378D_1 + 0.096D_2] + AE[0] + AE[0.0016]$$

$$0 = AE\,[0.096D_1 + 0.405D_2] + AE[0] + AE[0.0012]$$

Solving these equations simultaneously,

$$D_1 = -0.003704 \text{ m}$$

$$D_2 = -0.002084 \text{ m}$$

Although not needed, the reactions **Q** can be found from the expansion
of Eq. (1) following the format of Eq. 14–29.
   In order to determine the force in members 1 and 2 we must apply
Eq. 14–30, in which case we have

Member 1. $\lambda_x = 0$, $\lambda_y = 1$, $L = 3$ m, $AE = 8(10^3)$ kN, so that

$$
q_1 = \frac{8(10^3)}{3} \begin{bmatrix} 0 & -1 & 0 & 1 \end{bmatrix} \begin{bmatrix} 0 \\ 0 \\ -0.003704 \\ -0.002084 \end{bmatrix} + [0]
$$

$$q_1 = -5.56 \text{ kN} \hspace{4cm} \textit{Ans.}$$

Member 2. $\lambda_x = -0.8$, $\lambda_y = -0.6$, $L = 5$ m, $AE = 8(10^3)$ kN, so

$$
q_2 = \frac{8(10^3)}{5} \begin{bmatrix} 0.8 & 0.6 & -0.8 & -0.6 \end{bmatrix} \begin{bmatrix} -0.003704 \\ -0.002084 \\ 0 \\ 0 \end{bmatrix} - \frac{8(10^3)\,(-0.01)}{5}
$$

$$q_2 = 9.26 \text{ kN} \hspace{4cm} \textit{Ans.}$$

EXAMPLE   14.8

Member 2 of the truss shown in Fig. 14–16 is subjected to an increase in temperature of 150°F. Determine the force developed in member 2. Take $\alpha = 6.5(10^{-6})/°F$, $E = 29(10^6)$ lb/in². Each member has a cross-sectional area of $A = 0.75$ in².

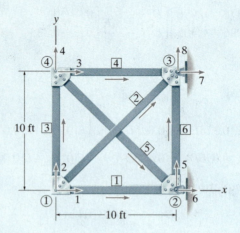

Fig. 14–16

SOLUTION

Since there is a temperature increase, $\Delta T = +150°F$. Applying Eq. 14–25 to member 2, where $\lambda_x = 0.7071$, $\lambda_y = 0.7071$, we have

$$
\begin{bmatrix} (Q_1)_0 \\ (Q_2)_0 \\ (Q_7)_0 \\ (Q_8)_0 \end{bmatrix} = AE(6.5)\,(10^{-6})\,(150) \begin{bmatrix} 0.7071 \\ 0.7071 \\ -0.7071 \\ -0.7071 \end{bmatrix} = AE \begin{bmatrix} 0.000689325 \\ 0.000689325 \\ -0.000689325 \\ -0.000689325 \end{bmatrix} \begin{matrix} 1 \\ 2 \\ 7 \\ 8 \end{matrix}
$$

The stiffness matrix for this truss has been developed in Example 14.2.

$$
\begin{bmatrix} 0 \\ 0 \\ 0 \\ 0 \\ 0 \\ \hdashline Q_6 \\ Q_7 \\ Q_8 \end{bmatrix} = AE \begin{bmatrix} 0.135 & 0.035 & 0 & 0 & 0 & -0.1 & -0.035 & -0.035 \\ 0.035 & 0.135 & 0 & -0.1 & 0 & 0 & -0.035 & -0.035 \\ 0 & 0 & 0.135 & -0.035 & 0.035 & -0.035 & -0.1 & 0 \\ 0 & -0.1 & -0.035 & 0.135 & -0.035 & 0.035 & 0 & 0 \\ 0 & 0 & 0.035 & -0.035 & 0.135 & -0.035 & 0 & -0.1 \\ \hdashline -0.1 & 0 & -0.035 & 0.035 & -0.035 & 0.135 & 0 & 0 \\ -0.035 & -0.035 & -0.1 & 0 & 0 & 0 & 0.135 & 0.035 \\ -0.035 & -0.035 & 0 & 0 & -0.1 & 0 & 0.035 & 0.135 \end{bmatrix} \begin{bmatrix} D_1 \\ D_2 \\ D_3 \\ D_4 \\ D_5 \\ 0 \\ 0 \\ 0 \end{bmatrix} + AE \begin{bmatrix} 0.000689325 \\ 0.000689325 \\ 0 \\ 0 \\ 0 \\ 0 \\ -0.000689325 \\ -0.000689325 \end{bmatrix} \begin{matrix} 1 \\ 2 \\ 3 \\ 4 \\ 5 \\ 6 \\ 7 \\ 8 \end{matrix} \quad (1)
$$

Expanding to determine the equations of the unknown displacements, and solving these equations simultaneously, yields

$$D_1 = -0.002027 \text{ ft}$$

$$D_2 = -0.01187 \text{ ft}$$

$$D_3 = -0.002027 \text{ ft}$$

$$D_4 = -0.009848 \text{ ft}$$

$$D_5 = -0.002027 \text{ ft}$$

Using Eq. 14–30 to determine the force in member 2, we have

$$q_2 = \frac{0.75\left[29(10^6)\right]}{10\sqrt{2}}\begin{bmatrix}-0.707 & -0.707 & 0.707 & 0.707\end{bmatrix}\begin{bmatrix}-0.002027 \\ -0.01187 \\ 0 \\ 0\end{bmatrix} - 0.75\left[29(10^6)\right]\left[6.5(10^{-6})\right](150)$$

$$= -6093 \text{ lb} = -6.09 \text{ k} \qquad \textit{Ans.}$$

Note that the temperature increase of member 2 will not cause any reactions on the truss since externally the truss is statically determinate. To show this, consider the matrix expansion of Eq. (1) for determining the reactions. Using the results for the displacements, we have

$$Q_6 = AE[-0.1(-0.002027) + 0 - 0.035(-0.002027)$$

$$+ 0.035(-0.009848) - 0.035(-0.002027)] + AE[0] = 0$$

$$Q_7 = AE[-0.035(-0.002027) - 0.035(-0.01187)$$

$$- 0.1(-0.002027) + 0 + 0] + AE[-0.000689325] = 0$$

$$Q_8 = AE[-0.035(-0.002027) - 0.035(-0.01187) + 0$$

$$+ 0 - 0.1(-0.002027)] + AE[-0.000689325] = 0$$

14.9   Space-Truss Analysis

The analysis of both statically determinate and indeterminate space trusses can be performed by using the same procedure discussed previously. To account for the three-dimensional aspects of the problem, however, additional elements must be included in the transformation matrix **T**. In this regard, consider the truss member shown in Fig. 14–17. The stiffness matrix for the member defined in terms of the local coordinate x' is given by Eq. 14–4. Furthermore, by inspection of Fig. 14–17, the direction cosines between the global and local coordinates can be found using equations analogous to Eqs. 14–5 and 14–6, that is,

$$\lambda_x = \cos \theta_x = \frac{x_F - x_N}{L}$$

$$= \frac{x_F - x_N}{\sqrt{(x_F - x_N)^2 + (y_F - y_N)^2 + (z_F - z_N)^2}} \tag{14–31}$$

$$\lambda_y = \cos \theta_y = \frac{y_F - y_N}{L}$$

$$= \frac{y_F - y_N}{\sqrt{(x_F - x_N)^2 + (y_F - y_N)^2 + (z_F - z_N)^2}} \tag{14–32}$$

$$\lambda_z = \cos \theta_z = \frac{z_F - z_N}{L}$$

$$= \frac{z_F - z_N}{\sqrt{(x_F - x_N)^2 + (y_F - y_N)^2 + (z_F - z_N)^2}} \tag{14–33}$$

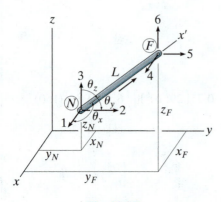

Fig. 14–17

As a result of the third dimension, the transformation matrix, Eq. 14–9, becomes

$$\mathbf{T} = \begin{bmatrix} \lambda_x & \lambda_y & \lambda_z & 0 & 0 & 0 \\ 0 & 0 & 0 & \lambda_x & \lambda_y & \lambda_z \end{bmatrix}$$

Substituting this and Eq. 14–4 into Eq. 14–15, $\mathbf{k} = \mathbf{T}^T \mathbf{k}' \mathbf{T}$, yields

$$\mathbf{k} = \begin{bmatrix} \lambda_x & 0 \\ \lambda_y & 0 \\ \lambda_z & 0 \\ 0 & \lambda_x \\ 0 & \lambda_y \\ 0 & \lambda_z \end{bmatrix} \frac{AE}{L} \begin{bmatrix} 1 & -1 \\ -1 & 1 \end{bmatrix} \begin{bmatrix} \lambda_x & \lambda_y & \lambda_z & 0 & 0 & 0 \\ 0 & 0 & 0 & \lambda_x & \lambda_y & \lambda_z \end{bmatrix}$$

Carrying out the matrix multiplication yields the *symmetric* matrix

$$\mathbf{k} = \frac{AE}{L} \begin{bmatrix} \lambda_x^2 & \lambda_x\lambda_y & \lambda_x\lambda_z & -\lambda_x^2 & -\lambda_x\lambda_y & -\lambda_x\lambda_z \\ \lambda_y\lambda_x & \lambda_y^2 & \lambda_y\lambda_z & -\lambda_y\lambda_x & -\lambda_y^2 & -\lambda_y\lambda_z \\ \lambda_z\lambda_x & \lambda_z\lambda_y & \lambda_z^2 & -\lambda_z\lambda_x & -\lambda_z\lambda_y & -\lambda_z^2 \\ -\lambda_x^2 & -\lambda_x\lambda_y & -\lambda_x\lambda_z & \lambda_x^2 & \lambda_x\lambda_y & \lambda_x\lambda_z \\ -\lambda_y\lambda_x & -\lambda_y^2 & -\lambda_y\lambda_z & \lambda_y\lambda_x & \lambda_y^2 & \lambda_y\lambda_z \\ -\lambda_z\lambda_x & -\lambda_z\lambda_y & -\lambda_z^2 & \lambda_z\lambda_x & \lambda_z\lambda_y & \lambda_z^2 \end{bmatrix} \begin{matrix} N_x \\ N_y \\ N_z \\ F_x \\ F_y \\ F_z \end{matrix} \quad (14\text{–}34)$$

with column headings $N_x \quad N_y \quad N_z \quad F_x \quad F_y \quad F_z$.

This equation represents the **member stiffness matrix** expressed in *global coordinates*. The code numbers along the rows and columns reference the x, y, z directions at the near end, N_x, N_y, N_z, followed by those at the far end, F_x, F_y, F_z.

For computer programming, it is generally more efficient to use Eq. 14–34 than to carry out the matrix multiplication $\mathbf{T}^T\mathbf{k}'\mathbf{T}$ for each member. Computer storage space is saved if the "structure" stiffness matrix $\mathbf{K}$ is first initialized with all zero elements; then as the elements of each member stiffness matrix are generated, they are placed directly into their respective positions in $\mathbf{K}$. After the structure stiffness matrix has been developed, the same procedure outlined in Sec. 14.6 can be followed to determine the joint displacements, support reactions, and internal member forces.

The roof of this building is supported by a series of space trusses.

PROBLEMS

14–1. Determine the stiffness matrix **K** for the truss. AE is constant.

14–2. Determine the force in each member of the truss in Prob. 14–1. AE is constant.

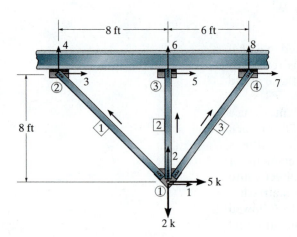

Probs. 14–1/2

14–3. Determine the stiffness matrix **K** for the truss. Take $A = 0.0015 \ m^2$ and $E = 200$ GPa for each member.

*14–4. Determine the vertical deflection at joint ② and the force in member 4 of the truss in Prob. 14–3. Take $A = 0.0015 \ m^2$ and $E = 200$ GPa for each member.

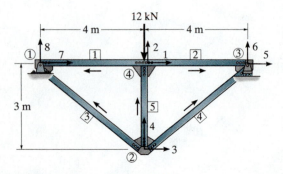

Probs. 14–3/4

14–5. Determine the stiffness matrix **K** for the truss. AE is constant.

14–6. Determine the horizontal displacement of joint ③ and the force in member 1. AE is constant.

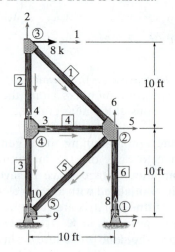

Probs. 14–5/6

14–7. Determine the stiffness matrix **K** for the truss. AE is constant.

*14–8. Determine the force in members 1 and 5 of the truss in Prob. 14–7. AE is constant.

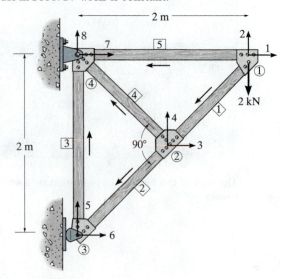

Probs. 14–7/8

14–9. Determine the stiffness matrix **K** for the truss. Take $A = 0.0015\ m^2$ and $E = 200\ GPa$ for each member.

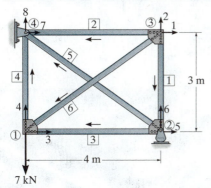

Prob. 14–9

14–11. Determine the stiffness matrix **K** for the truss. AE is constant.

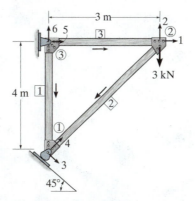

Prob. 14–11

14–10. Determine the force in member 6 of the truss in Prob. 14–9. Take $A = 0.0015\ m^2$ and $E = 200\ GPa$ for each member.

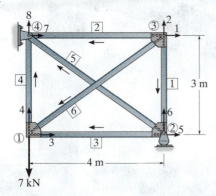

Prob. 14–10

*14–12. Determine the vertical displacement of joint ② and the support reactions. AE is constant.

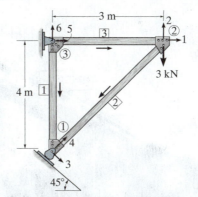

Prob. 14–12

CHAPTER REVIEW

The stiffness method is the preferred method for analyzing structures using a computer. It first requires identifying the number of structural elements and their nodes. The global coordinates for the entire structure are then established, and each member's local coordinate system is located so that its origin is at a selected near end, such that the positive x' axis extends towards the far end.

Formulation of the method first requires that each member stiffness matrix **k′** be constructed. It relates the loads at the ends of the member, **q**, to their displacements, **d**, where $\mathbf{q} = \mathbf{k'd}$. Then, using the transformation matrix **T**, the local displacements **d** are related to the global displacements **D**, where $\mathbf{d} = \mathbf{TD}$. Also, the local forces **q** are transformed into the global forces **Q** using the transformation matrix **T**, i.e., $\mathbf{Q} = \mathbf{T}^T\mathbf{q}$. When these matrices are combined, one obtains the member's stiffness matrix **K** in global coordinates, $\mathbf{k} = \mathbf{T}^T\mathbf{k'T}$. Assembling all the member stiffness matrices yields the stiffness matrix **K** for the entire structure.

The displacements and loads on the structure are then obtained by partitioning $\mathbf{Q} = \mathbf{KD}$, such that the unknown displacements are obtained from $\mathbf{D}_u = [\mathbf{K}_{11}]^{-1}\mathbf{Q}_k$, provided the supports do not displace. Finally, the support reactions are obtained from $\mathbf{Q}_u = \mathbf{K}_{21}\mathbf{D}_u$, and each member force is found from $\mathbf{q} = \mathbf{k'TD}$.

Chapter 15

© Bright/Fotolia

The statically indeterminate loading in bridge girders that are continuous over their piers can be determined using the stiffness method.

Beam Analysis Using the Stiffness Method

The concepts presented in the previous chapter will be extended here and applied to the analysis of beams. It will be shown that once the member stiffness matrix and the transformation matrix have been developed, the procedure for application is exactly the same as that for trusses. Special consideration will be given to cases of differential settlement and temperature.

15.1 Preliminary Remarks

Before we show how the stiffness method applies to beams, we will first discuss some preliminary concepts and definitions related to these members.

Member and Node Identification. In order to apply the stiffness method to beams, we must first determine how to subdivide the beam into its component finite elements. In general, each element must be free from load and have a prismatic cross section. For this reason the nodes of each element are located at a support or at points where members are connected together, where an external force is applied, where the cross-sectional area suddenly changes, or where the vertical or rotational displacement at a point is to be determined. For example, consider the beam in Fig. 15–1a. Using the same scheme as that for trusses, four nodes are specified numerically within a circle, and the three elements are identified numerically within a square. Also, notice that the "near" and "far" ends of each element are identified by the arrows written alongside each element.

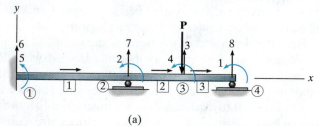

(a)

Fig. 15–1

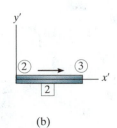

(b)

Fig. 15–1 (cont.)

15

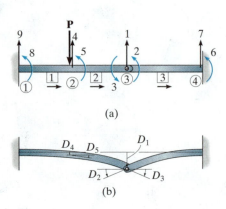

(a)

(b)

Fig. 15–2

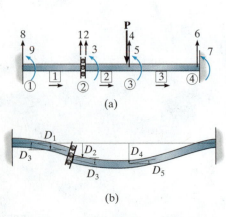

(a)

(b)

Fig. 15–3

Global and Member Coordinates. The global coordinate system will be identified using x, y, z axes that generally have their origin at a node and are positioned so that the nodes at other points on the beam all have positive coordinates, Fig. 15–1a. The local or member x', y', z' coordinates have their origin at the "near" end of each element, and the positive x' axis is directed towards the "far" end. Figure 15–1b shows these coordinates for element 2. In both cases we have used a right-handed coordinate system, so that if the fingers of the right hand are curled from the x (x') axis towards the y (y') axis, the thumb points in the positive direction of the z (z') axis, which is directed out of the page. Notice that for each beam element the x and x' axes will be collinear and the global and member coordinates will all be parallel. Therefore, unlike the case for trusses, here we will not need to develop transformation matrices between these coordinate systems.

Kinematic Indeterminacy. Once the elements and nodes have been identified, and the global coordinate system has been established, the degrees of freedom for the beam and its kinematic determinacy can be determined. If we consider the effects of both bending and shear, then *each node* on a beam can have two degrees of freedom, namely, a vertical displacement and a rotation. As in the case of trusses, these linear and rotational displacements will be identified by code numbers. The lowest code numbers will be used to identify the unknown displacements (unconstrained degrees of freedom), and the highest numbers are used to identify the known displacements (constrained degrees of freedom). Recall that the reason for choosing this method of identification has to do with the convenience of later partitioning the structure stiffness matrix, so that the unknown displacements can be found in the most direct manner.

To show an example of code-number labeling, consider again the continuous beam in Fig. 15–1a. Here the beam is kinematically indeterminate to the fourth degree. There are eight degrees of freedom, for which code numbers 1 through 4 represent the unknown displacements, and numbers 5 through 8 represent the known displacements, which in this case are all zero. As another example, the beam in Fig. 15–2a can be subdivided into three elements and four nodes. In particular, notice that the internal hinge at node 3 deflects the same for both elements 2 and 3; however, the rotation at the end of each element is different. For this reason three code numbers are used to show these deflections. Here there are nine degrees of freedom, five of which are unknown, as shown in Fig. 15–2b, and four known; again they are all zero. Finally, consider the slider mechanism used on the beam in Fig. 15–3a. Here the deflection of the beam is shown in Fig. 15–3b, and so there are five unknown deflection components labeled with the lowest code numbers. The beam is kinematically indeterminate to the fifth degree.

Development of the stiffness method for beams follows a similar procedure as that used for trusses. First we must establish the stiffness matrix for each element, and then these matrices are combined to form the beam or structure stiffness matrix. Using the structure matrix equation, we can then proceed to determine the unknown displacements at the nodes and from this determine the reactions on the beam and the internal shear and moment at the nodes.

15.2   Beam-Member Stiffness Matrix

In this section we will develop the stiffness matrix for a beam element or member having a constant cross-sectional area and referenced from the local x', y', z' coordinate system, Fig. 15–4. The origin of the coordinates is placed at the "near" end N, and the positive x' axis extends toward the "far" end F. There are two reactions at each end of the element, consisting of shear forces $q_{Ny'}$ and $q_{Fy'}$ and bending moments $q_{Nz'}$ and $q_{Fz'}$. These loadings all act in the positive coordinate directions. In particular, the moments $q_{Nz'}$ and $q_{Fz'}$ are positive *counterclockwise*, since by the right-hand rule the moment vectors are then directed along the positive z' axis, which is out of the page.

   Linear and angular displacements associated with these loadings also follow this same positive sign convention. We will now impose each of these displacements separately and then determine the loadings acting on the member caused by each displacement.

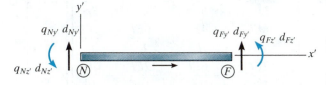

positive sign convention

Fig. 15–4

Displacements.   When a positive displacement $d_{Ny'}$ is imposed while other possible displacements are prevented, the resulting shear forces and bending moments that are created are shown in Fig. 15–5a. In particular, the moment has been developed in Sec. 11.2 as Eq. 11–5. Likewise, when $d_{Fy'}$ is imposed, the required shear forces and bending moments are given in Fig. 15–5b.

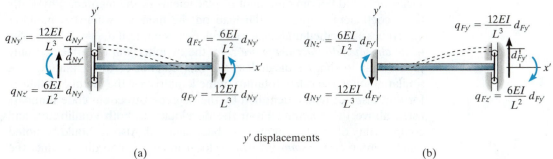

y' displacements

(a)   (b)

Fig. 15–5

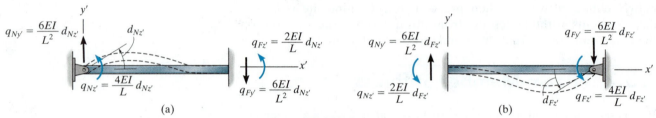

Fig. 15–6

z Rotations. If a positive rotation $d_{Nz'}$ is imposed while all other possible displacements are prevented, the required shear forces and moments necessary for the deformation are shown in Fig. 15–6a. In particular, the moment results have been developed in Sec. 11.2 as Eqs. 11–1 and 11–2. Likewise, when $d_{Fz'}$ is imposed, the resultant loadings are shown in Fig. 15–6b.

By superposition, if the above results in Figs. 15–5 and 15–6 are added, the resulting four load-displacement relations for the member can be expressed in matrix form as

$$
\begin{bmatrix} q_{Ny'} \\ q_{Nz'} \\ q_{Fy'} \\ q_{Fz'} \end{bmatrix}
=
\begin{matrix}
N_{y'} \quad\; N_{z'} \quad\;\; F_{y'} \quad\;\; F_{z'} \\
\begin{bmatrix}
\dfrac{12EI}{L^3} & \dfrac{6EI}{L^2} & -\dfrac{12EI}{L^3} & \dfrac{6EI}{L^2} \\[2mm]
\dfrac{6EI}{L^2} & \dfrac{4EI}{L} & -\dfrac{6EI}{L^2} & \dfrac{2EI}{L} \\[2mm]
-\dfrac{12EI}{L^3} & -\dfrac{6EI}{L^2} & \dfrac{12EI}{L^3} & -\dfrac{6EI}{L^2} \\[2mm]
\dfrac{6EI}{L^2} & \dfrac{2EI}{L} & -\dfrac{6EI}{L^2} & \dfrac{4EI}{L}
\end{bmatrix}
\end{matrix}
\begin{bmatrix} d_{Ny'} \\ d_{Nz'} \\ d_{Fy'} \\ d_{Fz'} \end{bmatrix}
\tag{15–1}
$$

These equations can also be written in abbreviated form as

$$\mathbf{q} = \mathbf{kd} \tag{15–2}$$

The symmetric matrix $\mathbf{k}$ in Eq. 15–1 is referred to as the *member stiffness matrix.* The 16 influence coefficients $\mathbf{k}_{ij}$ that comprise it account for the shear-force and bending-moment displacements of the member. Physically these coefficients represent the load on the member when the member undergoes a specified unit displacement. For example, if $d_{Ny'} = 1$, Fig. 15–5a, *while all other displacements are zero,* the member will be subjected only to the four loadings indicated in the first column of the $\mathbf{k}$ matrix. In a similar manner, the other columns of the $\mathbf{k}$ matrix are the member loadings for unit displacements identified by the degree-of-freedom code numbers listed above the columns. From the development, both equilibrium and compatibility of displacements have been satisfied. Also, it should be noted that this matrix is the *same* in both the local and global coordinates since the x', y', z' axes are parallel to x, y, z and, therefore, transformation matrices are not needed between the coordinates.

15.3   Beam-Structure Stiffness Matrix

Once all the member stiffness matrices have been found, we must assemble them into the structure stiffness matrix **K**. This process depends on first knowing the *location* of each element in the member stiffness matrix. Here the rows and columns of each **k** matrix (Eq. 15–1) are identified by the two code numbers at the near end of the member $(N_{y'}, N_{z'})$ followed by those at the far end $(F_{y'}, F_{z'})$. Therefore, when assembling the matrices, each element must be placed in the same location of the **K** matrix. In this way, **K** will have an order that will be equal to the highest code number assigned to the beam, since this represents the total number of degrees of freedom. Also, where several members are connected to a node, their member stiffness influence coefficients will have the same position in the **K** matrix and therefore must be algebraically added together to determine the nodal stiffness influence coefficient for the structure. This is necessary since each coefficient represents the nodal resistance of the structure in a particular direction (y' or z') when a unit displacement (y' or z') occurs either at the same or at another node. For example, K_{23} represents the load in the direction and at the location of code number "2" when a unit displacement occurs in the direction and at the location of code number "3."

15.4   Application of the Stiffness Method for Beam Analysis

After the structure stiffness matrix is determined, the loads at the nodes of the beam can be related to the displacements using the structure stiffness equation

$$\mathbf{Q} = \mathbf{KD}$$

Here **Q** and **D** are column matrices that represent both the known and unknown loads and displacements. Partitioning the stiffness matrix into the known and unknown elements of load and displacement, we have

$$\begin{bmatrix} \mathbf{Q}_k \\ \mathbf{Q}_u \end{bmatrix} = \begin{bmatrix} \mathbf{K}_{11} & \mathbf{K}_{12} \\ \mathbf{K}_{21} & \mathbf{K}_{22} \end{bmatrix} \begin{bmatrix} \mathbf{D}_u \\ \mathbf{D}_k \end{bmatrix}$$

which when expanded yields the two equations

$$\mathbf{Q}_k = \mathbf{K}_{11}\mathbf{D}_u + \mathbf{K}_{12}\mathbf{D}_k \tag{15–3}$$

$$\mathbf{Q}_u = \mathbf{K}_{21}\mathbf{D}_u + \mathbf{K}_{22}\mathbf{D}_k \tag{15–4}$$

The unknown displacements $\mathbf{D}_u$ are determined from the first of these equations. Using these values, the support reactions $\mathbf{Q}_u$ are computed for the second equation.

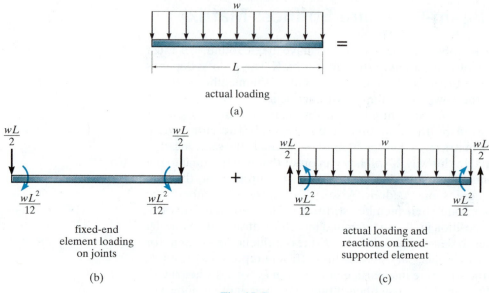

Fig. 15–7

Intermediate Loadings.

For application, it is important that the elements of the beam be free of loading along its length. This is necessary since the stiffness matrix for each element was developed for loadings applied only at its ends. (See Fig. 15–4.) Oftentimes, however, beams will support a distributed loading, and this condition will require modification in order to perform the matrix analysis.

To handle this case, we will use the principle of superposition in a manner similar to that used for trusses discussed in Sec. 14.8. To show its application, consider the beam element of length L in Fig. 15–7a, which is subjected to the uniform distributed load w. First we will apply fixed-end moments and reactions to the element, which will be used in the stiffness method, Fig. 15–7b. We will refer to these loadings as a column matrix $-\mathbf{q}_0$. Then the distributed loading and its reactions $\mathbf{q}_0$ are applied, Fig. 15–7c. The actual loading within the beam is determined by adding these two results. The fixed-end reactions for other cases of loading are given on the inside back cover. In addition to solving problems involving lateral loadings such as this, we can also use this method to solve problems involving temperature changes or fabrication errors.

Member Forces.

The shear and moment at the ends of each beam element can be determined using Eq. 15–2 and adding on any fixed-end reactions $\mathbf{q}_0$ if the element is subjected to an intermediate loading. We have

$$\mathbf{q} = \mathbf{k}\mathbf{d} + \mathbf{q}_0 \qquad (15\text{–}5)$$

If the results are negative, it indicates that the loading acts in the opposite direction to that shown in Fig. 15–4.

Procedure for Analysis

The following method provides a means of determining the displacements, support reactions, and internal loadings for the members or finite elements of a statically determinate or statically indeterminate beam.

Notation

- Divide the beam into finite elements and arbitrarily identify each element and its nodes. Use a number written in a circle for a node and a number written in a square for a member. Usually an element extends between points of support, points of concentrated loads, and joints, or to points where internal loadings or displacements are to be determined. Also, E and I for the elements must be constants.

- Specify the near and far ends of each element symbolically by directing an arrow along the element, with the head directed toward the far end.

- At each nodal point specify numerically the y and z code numbers. In all cases use the *lowest code numbers* to identify all the unconstrained degrees of freedom, followed by the remaining or highest numbers to identify the degrees of freedom that are constrained.

- From the problem, establish the known displacements $\mathbf{D}_k$ and known external loads $\mathbf{Q}_k$ Include any *reversed* fixed-end loadings if an element supports an intermediate load.

Structure Stiffness Matrix

- Apply Eq. 15–1 to determine the stiffness matrix for each element expressed in global coordinates.

- After each member stiffness matrix is determined, and the rows and columns are identified with the appropriate code numbers, assemble the matrices to determine the structure stiffness matrix $\mathbf{K}$. As a partial check, the member *and* structure stiffness matrices should all be *symmetric*.

Displacements and Loads

- Partition the structure stiffness equation and carry out the matrix multiplication in order to determine the unknown displacements $\mathbf{D}_u$ and support reactions $\mathbf{Q}_u$.

- The internal shear and moment $\mathbf{q}$ at the ends of each beam element can be determined from Eq. 15–5, accounting for the additional fixed-end loadings.

15

EXAMPLE 15.1

Determine the reactions at the supports of the beam shown in Fig. 15–8a. EI is constant.

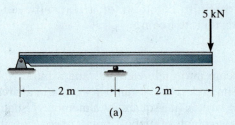

(a)

Fig. 15–8

SOLUTION

Notation. The beam has two elements and three nodes, which are identified in Fig. 15–8b. The code numbers 1 through 6 are indicated such that the *lowest numbers 1–4 identify the unconstrained degrees of freedom.*

The known load and displacement matrices are

$$
\mathbf{Q}_k = \begin{bmatrix} 0 \\ -5 \\ 0 \\ 0 \end{bmatrix} \begin{matrix} 1 \\ 2 \\ 3 \\ 4 \end{matrix} \qquad \mathbf{D}_k = \begin{bmatrix} 0 \\ 0 \end{bmatrix} \begin{matrix} 5 \\ 6 \end{matrix}
$$

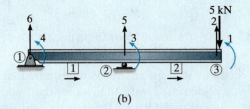

(b)

Member Stiffness Matrices. Each of the two member stiffness matrices is determined from Eq. 15–1. Note carefully how the code numbers for each column and row are established.

$$
\mathbf{k}_1 = EI \begin{matrix} 6 & 4 & 5 & 3 \\ \begin{bmatrix} 1.5 & 1.5 & -1.5 & 1.5 \\ 1.5 & 2 & -1.5 & 1 \\ -1.5 & -1.5 & 1.5 & -1.5 \\ 1.5 & 1 & -1.5 & 2 \end{bmatrix} \begin{matrix} 6 \\ 4 \\ 5 \\ 3 \end{matrix} \end{matrix}
\qquad
\mathbf{k}_2 = EI \begin{matrix} 5 & 3 & 2 & 1 \\ \begin{bmatrix} 1.5 & 1.5 & -1.5 & 1.5 \\ 1.5 & 2 & -1.5 & 1 \\ -1.5 & -1.5 & 1.5 & -1.5 \\ 1.5 & 1 & -1.5 & 2 \end{bmatrix} \begin{matrix} 5 \\ 3 \\ 2 \\ 1 \end{matrix} \end{matrix}
$$

Displacements and Loads.   We can now assemble these elements into the structure stiffness matrix. For example, element $\mathbf{K}_{11} = 0 + 2 = 2$, $\mathbf{K}_{55} = 1.5 + 1.5 = 3$, etc. Thus,

$$\mathbf{Q} = \mathbf{KD}$$

$$
\begin{bmatrix} 0 \\ -5 \\ 0 \\ 0 \\ \hline Q_5 \\ Q_6 \end{bmatrix}
= EI
\begin{array}{cccccc}
1 & 2 & 3 & 4 & 5 & 6
\end{array}
\begin{bmatrix}
2 & -1.5 & 1 & 0 & \vdots & 1.5 & 0 \\
-1.5 & 1.5 & -1.5 & 0 & \vdots & -1.5 & 0 \\
1 & -1.5 & 4 & 1 & \vdots & 0 & 1.5 \\
0 & 0 & 1 & 2 & \vdots & -1.5 & 1.5 \\
\hline
1.5 & -1.5 & 0 & -1.5 & \vdots & 3 & -1.5 \\
0 & 0 & 1.5 & 1.5 & \vdots & -1.5 & 1.5
\end{bmatrix}
\begin{bmatrix} D_1 \\ D_2 \\ D_3 \\ D_4 \\ \hline 0 \\ 0 \end{bmatrix}
$$

The matrices are partitioned as shown. Carrying out the multiplication for the first four rows, we have

$$0 = 2D_1 - 1.5D_2 + D_3 + 0$$

$$-\frac{5}{EI} = -1.5D_1 + 1.5D_2 - 1.5D_3 + 0$$

$$0 = D_1 - 1.5D_2 + 4D_3 + D_4$$

$$0 = 0 + 0 + D_3 + 2D_4$$

Solving,

$$D_1 = -\frac{16.67}{EI}$$

$$D_2 = -\frac{26.67}{EI}$$

$$D_3 = -\frac{6.67}{EI}$$

$$D_4 = \frac{3.33}{EI}$$

Using these results, and multiplying the last two rows, gives

$$Q_5 = 1.5EI\left(-\frac{16.67}{EI}\right) - 1.5EI\left(-\frac{26.67}{EI}\right) + 0 - 1.5EI\left(\frac{3.33}{EI}\right)$$

$$= 10 \text{ kN} \qquad\qquad\qquad\qquad\qquad\qquad\qquad\qquad Ans.$$

$$Q_6 = 0 + 0 + 1.5EI\left(-\frac{6.67}{EI}\right) + 1.5EI\left(\frac{3.33}{EI}\right)$$

$$= -5 \text{ kN} \qquad\qquad\qquad\qquad\qquad\qquad\qquad\qquad Ans.$$

15

EXAMPLE 15.2

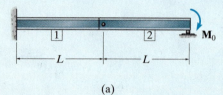

(a)

Fig. 15-9

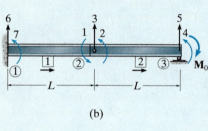

(b)

Determine the internal shear and moment in member 1 of the compound beam shown in Fig. 15–9a. EI is constant.

SOLUTION

Notation. When the beam deflects, the internal pin will allow a single deflection, however, the slope of each connected member will be different. Also, a slope at the roller will occur. These four unknown degrees of freedom are labeled with the code numbers 1, 2, 3, and 4, Fig. 15–9b.

$$Q_k = \begin{bmatrix} 0 \\ 0 \\ 0 \\ -M_0 \end{bmatrix} \begin{matrix} 1 \\ 2 \\ 3 \\ 4 \end{matrix} \qquad D_k = \begin{bmatrix} 0 \\ 0 \\ 0 \end{bmatrix} \begin{matrix} 5 \\ 6 \\ 7 \end{matrix}$$

Member Stiffness Matrices. Applying Eq. 15–1 to each member, in accordance with the code numbers shown in Fig. 15–9b, we have

$$k_1 = EI \begin{matrix} & 6 & 7 & 3 & 1 \\ & \begin{bmatrix} \dfrac{12}{L^3} & \dfrac{6}{L^2} & -\dfrac{12}{L^3} & \dfrac{6}{L^2} \\[2mm] \dfrac{6}{L^2} & \dfrac{4}{L} & -\dfrac{6}{L^2} & \dfrac{2}{L} \\[2mm] -\dfrac{12}{L^3} & -\dfrac{6}{L^2} & \dfrac{12}{L^3} & -\dfrac{6}{L^2} \\[2mm] \dfrac{6}{L^2} & \dfrac{2}{L} & -\dfrac{6}{L^2} & \dfrac{4}{L} \end{bmatrix} & \begin{matrix} 6 \\ 7 \\ 3 \\ 1 \end{matrix} \end{matrix}$$

$$k_2 = EI \begin{matrix} & 3 & 2 & 5 & 4 \\ & \begin{bmatrix} \dfrac{12}{L^3} & \dfrac{6}{L^2} & -\dfrac{12}{L^3} & \dfrac{6}{L^2} \\[2mm] \dfrac{6}{L^2} & \dfrac{4}{L} & -\dfrac{6}{L^2} & \dfrac{2}{L} \\[2mm] -\dfrac{12}{L^3} & -\dfrac{6}{L^2} & \dfrac{12}{L^3} & -\dfrac{6}{L^2} \\[2mm] \dfrac{6}{L^2} & \dfrac{2}{L} & -\dfrac{6}{L^2} & \dfrac{4}{L} \end{bmatrix} & \begin{matrix} 3 \\ 2 \\ 5 \\ 4 \end{matrix} \end{matrix}$$

Displacements and Loads. The structure stiffness matrix is formed by assembling the elements of the member stiffness matrices. Applying the structure matrix equation, we have

$$Q = KD$$

$$\begin{bmatrix} 0 \\ 0 \\ 0 \\ -M_0 \\ \hline Q_5 \\ Q_6 \\ Q_7 \end{bmatrix} = EI \begin{bmatrix} \dfrac{4}{L} & 0 & -\dfrac{6}{L^2} & 0 & 0 & \dfrac{6}{L^2} & \dfrac{2}{L} \\[2mm] 0 & \dfrac{4}{L} & \dfrac{6}{L^2} & \dfrac{2}{L} & -\dfrac{6}{L^2} & 0 & 0 \\[2mm] -\dfrac{6}{L^2} & \dfrac{6}{L^2} & \dfrac{24}{L^3} & \dfrac{6}{L^2} & -\dfrac{12}{L^3} & \dfrac{12}{L^3} & \dfrac{6}{L^2} \\[2mm] 0 & \dfrac{2}{L} & \dfrac{6}{L^2} & \dfrac{4}{L} & -\dfrac{6}{L^2} & 0 & 0 \\[2mm] \hline 0 & -\dfrac{6}{L^2} & -\dfrac{12}{L^3} & -\dfrac{6}{L^2} & \dfrac{12}{L^3} & 0 & 0 \\[2mm] \dfrac{6}{L^2} & 0 & -\dfrac{12}{L^3} & 0 & 0 & \dfrac{12}{L^3} & \dfrac{6}{L^2} \\[2mm] \dfrac{2}{L} & 0 & -\dfrac{6}{L^2} & 0 & 0 & \dfrac{6}{L^2} & \dfrac{4}{L} \end{bmatrix} \begin{bmatrix} D_1 \\ D_2 \\ D_3 \\ D_4 \\ \hline 0 \\ 0 \\ 0 \end{bmatrix}$$

Multiplying the first four rows to determine the displacement yields

$$0 = \frac{4}{L}D_1 - \frac{6}{L^2}D_3$$

$$0 = \frac{4}{L}D_2 + \frac{6}{L^2}D_3 + \frac{2}{L}D_4$$

$$0 = -\frac{6}{L^2}D_1 + \frac{6}{L^2}D_2 + \frac{24}{L^3}D_3 + \frac{6}{L^2}D_4$$

$$-M_0 = \frac{2}{L}D_2 + \frac{6}{L^2}D_3 + \frac{4}{L}D_4$$

So that

$$D_1 = \frac{M_0 L}{2EI}$$

$$D_2 = -\frac{M_0 L}{6EI}$$

$$D_3 = \frac{M_0 L^2}{3EI}$$

$$D_4 = -\frac{2M_0 L}{3EI}$$

Using these results, the reaction Q_5 is obtained from the multiplication of the fifth row.

$$Q_5 = -\frac{6EI}{L^2}\left(-\frac{M_0 L}{6EI}\right) - \frac{12EI}{L^3}\left(\frac{M_0 L^2}{3EI}\right) - \frac{6EI}{L^2}\left(-\frac{2M_0 L}{3EI}\right)$$

$$Q_5 = \frac{M_0}{L} \qquad\qquad\qquad \textit{Ans.}$$

This result can be easily checked by statics applied to member 2.

EXAMPLE   15.3

The beam in Fig. 15–10a is subjected to the two couple moments. If the center support ② settles 1.5 mm, determine the reactions at the supports. Assume the roller supports at ① and ③ can pull down or push up on the beam. Take $E = 200$ GPa and $I = 22(10^{-6})$ m^4.

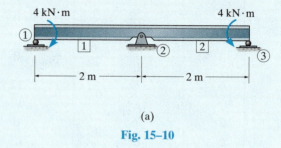

(a)

Fig. 15–10

SOLUTION

Notation.   The beam has two elements and three unknown degrees of freedom. These are labeled with the lowest code numbers, Fig. 15–10b. Here the known load and displacement matrices are

$$\mathbf{Q}_k = \begin{bmatrix} 4 \\ 0 \\ -4 \end{bmatrix} \begin{matrix} 1 \\ 2 \\ 3 \end{matrix} \qquad \mathbf{D}_k = \begin{bmatrix} 0 \\ -0.0015 \\ 0 \end{bmatrix} \begin{matrix} 4 \\ 5 \\ 6 \end{matrix}$$

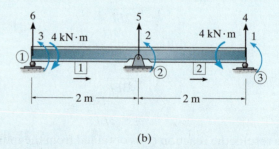

(b)

Member Stiffness Matrices.   The member stiffness matrices are determined using Eq. 15–1 in accordance with the code numbers and member directions shown in Fig. 15–10b. We have,

$$
\mathbf{k_1} = EI
\begin{array}{c}
\begin{array}{cccc} 6 & 3 & 5 & 2 \end{array} \\
\left[
\begin{array}{cccc}
1.5 & 1.5 & -1.5 & 1.5 \\
1.5 & 2 & -1.5 & 1 \\
-1.5 & -1.5 & 1.5 & -1.5 \\
1.5 & 1 & -1.5 & 2
\end{array}
\right]
\begin{array}{c} 6 \\ 3 \\ 5 \\ 2 \end{array}
\end{array}
$$

$$
\mathbf{k_2} = EI
\begin{array}{c}
\begin{array}{cccc} 5 & 2 & 4 & 1 \end{array} \\
\left[
\begin{array}{cccc}
1.5 & 1.5 & -1.5 & 1.5 \\
1.5 & 2 & -1.5 & 1 \\
-1.5 & -1.5 & 1.5 & -1.5 \\
1.5 & 1 & -1.5 & 2
\end{array}
\right]
\begin{array}{c} 5 \\ 2 \\ 4 \\ 1 \end{array}
\end{array}
$$

Displacements and Loads. Assembling the structure stiffness matrix and writing the stiffness equation for the structure, yields

$$
\begin{bmatrix}
4 \\
0 \\
-4 \\
\hdashline
Q_4 \\
Q_5 \\
Q_6
\end{bmatrix}
= EI
\begin{array}{c}
\begin{array}{cccccc} 1 & 2 & 3 & 4 & 5 & 6 \end{array} \\
\left[
\begin{array}{ccc:ccc}
2 & 1 & 0 & -1.5 & 1.5 & 0 \\
1 & 4 & 1 & -1.5 & 0 & 1.5 \\
0 & 1 & 2 & 0 & -1.5 & 1.5 \\
\hdashline
-1.5 & -1.5 & 0 & 1.5 & -1.5 & 0 \\
1.5 & 0 & -1.5 & -1.5 & 3 & -1.5 \\
0 & 1.5 & 1.5 & 0 & -1.5 & 1.5
\end{array}
\right]
\end{array}
\begin{bmatrix}
D_1 \\
D_2 \\
D_3 \\
\hdashline
0 \\
-0.0015 \\
0
\end{bmatrix}
$$

Solving for the unknown displacements,

$$
\frac{4}{EI} = 2D_1 + D_2 + 0D_3 - 1.5(0) + 1.5(-0.0015) + 0
$$

$$
0 = 1D_1 + 4D_2 + 1D_3 - 1.5(0) + 0 + 0
$$

$$
\frac{-4}{EI} = 0D_1 + 1D_2 + 2D_3 + 0 - 1.5(-0.0015) + 0
$$

Substituting $EI = 200(10^6)(22)(10^{-6})$, and solving,

$$
D_1 = 0.001580 \text{ rad}, \quad D_2 = 0, \quad D_3 = -0.001580 \text{ rad}
$$

Using these results, the support reactions are therefore

$Q_4 = 200(10^6)22(10^{-6})[-1.5(0.001580) - 1.5(0) + 0 + 1.5(0) - 1.5(-0.0015) + 0] = -0.525 \text{ kN}$ *Ans.*

$Q_5 = 200(10^6)22(10^{-6})[1.5(0.001580) + 0 - 1.5(-0.001580) - 1.5(0) + 3(-0.0015) - 1.5(0)] = 1.05 \text{ kN}$ *Ans.*

$Q_6 = 200(10^6)22(10^{-6})[0 + 1.5(0) + 1.5(-0.001580) + 0 - 1.5(-0.0015) + 1.5(0)] = -0.525 \text{ kN}$ *Ans.*

EXAMPLE   15.4

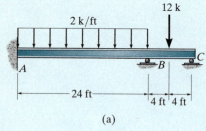

(a)

Determine the moment developed at support A of the beam shown in Fig. 15–11a. Assume the roller supports can pull down or push up on the beam. Take $E = 29(10^3)$ ksi, $I = 510$ in^4.

SOLUTION

Notation.   Here the beam has two unconstrained degrees of freedom, identified by the code numbers 1 and 2.

The matrix analysis requires that the external loading be applied at the nodes, and therefore the distributed and concentrated loads are replaced by their equivalent fixed-end moments, which are determined from the table on the inside back cover. (See Example 11.2.) Note that no external loads are placed at ① and no external vertical forces are placed at ② since the reactions at code numbers 3, 4 and 5 *are to be unknowns* in the load matrix. Using superposition, the results of the matrix analysis for the loading in Fig. 15–11b will later be modified by the loads shown in Fig. 15–11c. From Fig. 15–11b, the known displacement and load matrices are

$$\mathbf{D}_k = \begin{bmatrix} 0 \\ 0 \\ 0 \end{bmatrix} \begin{matrix} 4 \\ 5 \\ 6 \end{matrix} \qquad \mathbf{Q}_k = \begin{bmatrix} 144 \\ 1008 \end{bmatrix} \begin{matrix} 1 \\ 2 \end{matrix}$$

Member Stiffness Matrices.   Each of the two member stiffness matrices is determined from Eq. 15–1.

Member 1:

$$\frac{12EI}{L^3} = \frac{12(29)(10^3)(510)}{[24(12)]^3} = 7.430$$

$$\frac{6EI}{L^2} = \frac{6(29)(10^3)(510)}{[24(12)]^2} = 1069.9$$

$$\frac{4EI}{L} = \frac{4(29)(10^3)(510)}{24(12)} = 205\,417$$

$$\frac{2EI}{L} = \frac{2(29)(10^3)(510)}{24(12)} = 102\,708$$

$$\mathbf{k}_1 = \begin{matrix} & 4 & 3 & 5 & 2 & \\ & 7.430 & 1069.9 & -7.430 & 1069.9 & 4 \\ & 1069.9 & 205\,417 & -1069.9 & 102\,708 & 3 \\ & -7.430 & -1069.9 & 7.430 & -1069.9 & 5 \\ & 1069.9 & 102\,708 & -1069.9 & 205\,417 & 2 \end{matrix}$$

Member 2:

$$\frac{12EI}{L^3} = \frac{12(29)(10^3)(510)}{[8(12)]^3} = 200.602$$

$$\frac{6EI}{L^2} = \frac{6(29)(10^3)(510)}{[8(12)]^2} = 9628.91$$

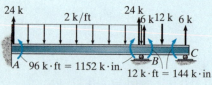

4   5   6
2
3 ⟨   ⟧ 1
①   1   ② 2   ③
96 k·ft − 12 k·ft = 1008 k·in.   12 k·ft = 144 k·in.

beam to be analyzed by stiffness method
(b)

24 k   24 k
2 k/ft   6 k 12 k 6 k
C
A 96 k·ft = 1152 k·in.   B
12 k·ft = 144 k·in.

beam subjected to actual load and
fixed-supported reactions
(c)

Fig. 15–11

$$\frac{4EI}{L} = \frac{4(29)(10^3)(510)}{8(12)} = 616\,250$$

$$\frac{2EI}{L} = \frac{2(29)(10^3)(510)}{8(12)} = 308\,125$$

$$\mathbf{k}_2 = \begin{array}{c} \\ \\ \begin{bmatrix} 200.602 & 9628.91 & -200.602 & 9628.91 \\ 9628.91 & 616\,250 & -9628.91 & 308\,125 \\ -200.602 & -9628.91 & 200.602 & -9628.91 \\ 9628.91 & 308\,125 & -9628.91 & 616\,250 \end{bmatrix} \end{array} \begin{array}{c} 5 \\ 2 \\ 6 \\ 1 \end{array}$$

$$\begin{array}{cccc} 5 & 2 & 6 & 1 \end{array}$$

Displacements and Loads.   We require

$$\mathbf{Q} = \mathbf{KD}$$

$$\begin{array}{ccccccc} & 1 & 2 & 3 & 4 & 5 & 6 \end{array}$$

$$\begin{bmatrix} 144 \\ \hline 1008 \\ \hline Q_3 \\ Q_4 \\ Q_5 \\ Q_6 \end{bmatrix} = \begin{bmatrix} 616\,250 & 308\,125 & 0 & 0 & 9628.91 & -9628.91 \\ 308\,125 & 821\,667 & 102\,708 & 1069.9 & 8559.01 & -9628.91 \\ \hline 0 & 102\,708 & 205\,417 & 1069.9 & -1069.9 & 0 \\ 0 & 1069.9 & 1069.9 & 7.430 & -7.430 & 0 \\ 9628.91 & 8559.01 & -1069.9 & -7.430 & 208.03 & -200.602 \\ -9628.91 & -9628.91 & 0 & 0 & -200.602 & 200.602 \end{bmatrix} \begin{bmatrix} D_1 \\ D_2 \\ \hline 0 \\ 0 \\ 0 \\ 0 \end{bmatrix}$$

Solving in the usual manner,

$$144 = 616\,250D_1 + 308\,125D_2$$

$$1008 = 308\,125D_1 + 821\,667D_2$$

$$D_1 = -0.4673(10^{-3})\text{ in.}$$

$$D_2 = 1.40203(10^{-3})\text{ in.}$$

Thus,

$$Q_3 = 0 + 102\,708(1.40203)(10^{-3}) = 144\text{ k}\cdot\text{in.} = 12\text{ k}\cdot\text{ft}$$

The actual moment at A must include the fixed-supported *reaction* of $+96$ k $\cdot$ ft shown in Fig. 15–11c, along with the calculated result for Q_3. Thus,

$$M_{AB} = 12\text{ k}\cdot\text{ft} + 96\text{ k}\cdot\text{ft} = 108\text{ k}\cdot\text{ft} \circlearrowright \qquad \textit{Ans.}$$

This result compares with that determined in Example 11.2.

   Although not required here, we can determine the internal moment and shear at B by considering, for example, member 1, node 2, Fig. 15–11b. The result requires expanding

$$\mathbf{q}_1 = \mathbf{k}_1\mathbf{d} + (\mathbf{q}_0)_1$$

$$\begin{array}{cccc} 4 & 3 & 5 & 2 \end{array}$$

$$\begin{bmatrix} q_4 \\ q_3 \\ q_5 \\ q_2 \end{bmatrix} = \begin{bmatrix} 7.430 & 1069.9 & -7.430 & 1069.9 \\ 1069.9 & 205\,417 & -1069.9 & 102\,708 \\ -7.430 & -1069.9 & 7.430 & -1069.9 \\ 1069.9 & 102\,708 & -1069.9 & 205\,417 \end{bmatrix} \begin{bmatrix} 0 \\ 0 \\ 0 \\ 1.40203 \end{bmatrix} (10^{-3}) + \begin{bmatrix} 24 \\ 1152 \\ 24 \\ -1152 \end{bmatrix}$$

EXAMPLE   15.5

Determine the deflection at ① and the reactions on the beam shown in Fig. 15–12a. *EI* is constant.

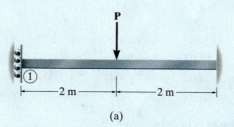

(a)

Fig. 15–12

15

SOLUTION

Notation.   The beam is divided into two elements and the nodes and members are identified along with the directions from the near to far ends, Fig. 15–12b. The unknown deflections are shown in Fig. 15–12c. In particular, notice that a rotational displacement D_4 does not occur because of the roller constraint.

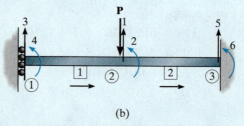

(b)

Member Stiffness Matrices.   Since *EI* is constant and the members are of equal length, the member stiffness matrices are identical. Using the code numbers to identify each column and row in accordance with Eq. 15–1 and Fig. 15–12b, we have

$$
\mathbf{k}_1 = EI
\begin{array}{c}
\begin{array}{cccc} 3 & 4 & 1 & 2 \end{array} \\
\begin{bmatrix}
1.5 & 1.5 & -1.5 & 1.5 \\
1.5 & 2 & -1.5 & 1 \\
-1.5 & -1.5 & 1.5 & -1.5 \\
1.5 & 1 & -1.5 & 2
\end{bmatrix}
\end{array}
\begin{array}{c} 3 \\ 4 \\ 1 \\ 2 \end{array}
$$

$$
\mathbf{k}_2 = EI
\begin{array}{c}
\begin{array}{cccc} 1 & 2 & 5 & 6 \end{array} \\
\begin{bmatrix}
1.5 & 1.5 & -1.5 & 1.5 \\
1.5 & 2 & -1.5 & 1 \\
-1.5 & -1.5 & 1.5 & -1.5 \\
1.5 & 1 & -1.5 & 2
\end{bmatrix}
\end{array}
\begin{array}{c} 1 \\ 2 \\ 5 \\ 6 \end{array}
$$

Displacements and Loads. Assembling the member stiffness matrices into the structure stiffness matrix, and applying the structure stiffness matrix equation, we have

$$\mathbf{Q} = \mathbf{KD}$$

$$
\begin{bmatrix} -P \\ 0 \\ 0 \\ \hline Q_4 \\ Q_5 \\ Q_6 \end{bmatrix} = EI
\begin{bmatrix}
\overset{1}{3} & \overset{2}{0} & \overset{3}{-1.5} & \overset{4}{-1.5} & \overset{5}{-1.5} & \overset{6}{1.5} \\
0 & 4 & 1.5 & 1 & -1.5 & 1 \\
-1.5 & 1.5 & 1.5 & 1.5 & 0 & 0 \\
\hline
-1.5 & 1 & 1.5 & 2 & 0 & 0 \\
-1.5 & -1.5 & 0 & 0 & 1.5 & -1.5 \\
1.5 & 1 & 0 & 0 & -1.5 & 2
\end{bmatrix}
\begin{bmatrix} D_1 \\ D_2 \\ D_3 \\ \hline 0 \\ 0 \\ 0 \end{bmatrix}
$$

Solving for the displacements yields

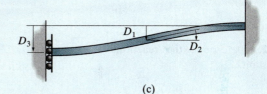

(c)

$$-\frac{P}{EI} = 3D_1 + 0D_2 - 1.5D_3$$

$$0 = 0D_1 + 4D_2 + 1.5D_3$$

$$0 = -1.5D_1 + 1.5D_2 + 1.5D_3$$

$$D_1 = -\frac{1.667P}{EI}$$

$$D_2 = \frac{P}{EI}$$

$$D_3 = -\frac{2.667P}{EI} \qquad\qquad Ans.$$

Note that the signs of the results match the directions of the deflections shown in Fig. 15–12c. Using these results, the reactions therefore are

$$Q_4 = -1.5EI\left(-\frac{1.667P}{EI}\right) + 1EI\left(\frac{P}{EI}\right) + 1.5EI\left(-\frac{2.667P}{EI}\right)$$

$$= -0.5P \qquad\qquad Ans.$$

$$Q_5 = -1.5EI\left(-\frac{1.667P}{EI}\right) - 1.5EI\left(\frac{P}{EI}\right) + 0\left(-\frac{2.667P}{EI}\right)$$

$$= P \qquad\qquad Ans.$$

$$Q_6 = 1.5EI\left(-\frac{1.667P}{EI}\right) + 1EI\left(\frac{P}{EI}\right) + 0\left(-\frac{2.667P}{EI}\right)$$

$$= -1.5P \qquad\qquad Ans.$$

15

PROBLEMS

15–1. Determine the internal moment in the beam at ① and ②. Assume ② is a roller and ③ is a pin. EI is constant.

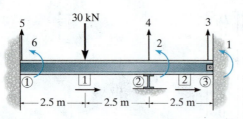

Prob. 15–1

15–2. Determine the reactions at the supports. EI is constant.

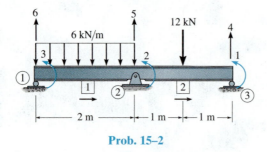

Prob. 15–2

15–3. Determine the reactions at the supports ①, ②, and ③. Assume ① is pinned, ② and ③ are rollers. EI is constant.

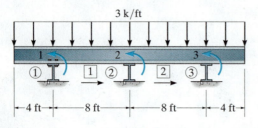

Prob. 15–3

*15–4. Determine the reactions at the supports ①, ②, and ③. Assume ① is pinned ② and ③ are rollers. EI is constant.

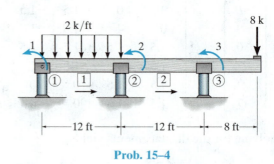

Prob. 15–4

15–5. Determine the moments at ② and ③. Assume ② and ③ are rollers and ① and ④ are pins. EI is constant.

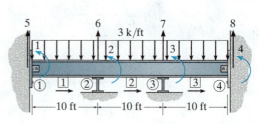

Prob. 15–5

15–6. Determine the internal moment in the beam at ① and ②. EI is constant. Assume ② and ③ are rollers.

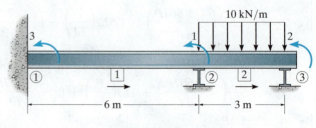

Prob. 15–6

15–7. Determine the moments at the supports ① and ③. *EI* is constant. Assume joint ② is a roller.

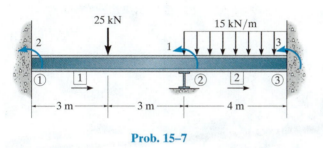

Prob. 15–7

15–9. Determine the reactions at the supports. There is a slider at ①. *EI* is constant.

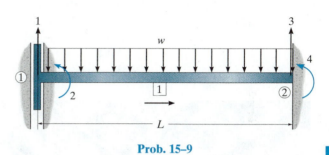

Prob. 15–9

15

15–10. Determine the moments at ① and ③. Assume ② is a roller and ① and ③ are fixed. Also, here *EI* is constant.

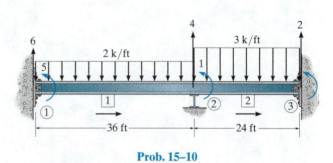

Prob. 15–10

***15–8.** Determine the moments at the supports. Assume ② is a roller. *EI* is constant.

15–11. Determine the moments at ① and ③ if the support ② settles 0.1 ft. Assume ② is a roller and ① and ③ are fixed. $EI = 9500 \text{ k} \cdot \text{ft}^2$.

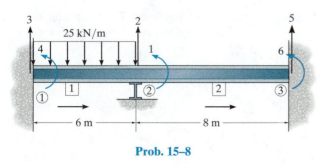

Prob. 15–8

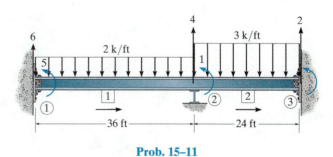

Prob. 15–11

Chapter **16**

© Joe Gough/Fotolia

The frame of this building is statically indeterminate. The force analysis can be done using the stiffness method.

Plane Frame Analysis Using the Stiffness Method

The concepts presented in the previous chapters on trusses and beams will be extended in this chapter and applied to the analysis of frames. It will be shown that the procedure for the solution is similar to that for beams, but will require the use of transformation matrices since frame members are oriented in different directions.

16.1 Frame-Member Stiffness Matrix

In this section we will develop the stiffness matrix for a prismatic frame member referenced from the local x', y', z' coordinate system, Fig. 16–1. Here the member is subjected to axial loads $q_{Nx'}$, $q_{Fx'}$, shear loads $q_{Ny'}$, $q_{Fy'}$, and bending moments $q_{Nz'}$, $q_{Fz'}$ at its near and far ends, respectively. These loadings all act in the positive coordinate directions along with their associated displacements. As in the case of beams, the moments $q_{Nz'}$ and $q_{Fz'}$ are positive counterclockwise, since by the right-hand rule the moment vectors are then directed along the positive z' axis, which is out of the page.

We have considered each of the load-displacement relationships caused by these loadings in the previous chapters. The axial load was discussed in reference to Fig. 14–2, the shear load in reference to Fig. 15–5, and the bending moment in reference to Fig. 15–6. By superposition, if these

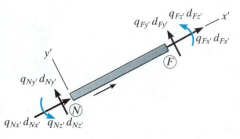

Fig. 16–1

16

This pedestrian bridge takes the form of a "Vendreel truss." Strictly not a truss since the diagonals are absent, it forms a statically indeterminate box framework, which can be analyzed using the stiffness method.

results are added, the resulting six load-displacement relations for the member can be expressed in matrix form as

$$
\begin{bmatrix} q_{Nx'} \\ q_{Ny'} \\ q_{Nz'} \\ q_{Fx'} \\ q_{Fy'} \\ q_{Fz'} \end{bmatrix}
=
\begin{array}{c}
\begin{matrix} N_{x'} & N_{y'} & N_{z'} & F_{x'} & F_{y'} & F_{z'} \end{matrix} \\
\begin{bmatrix}
\dfrac{AE}{L} & 0 & 0 & -\dfrac{AE}{L} & 0 & 0 \\[2mm]
0 & \dfrac{12EI}{L^3} & \dfrac{6EI}{L^2} & 0 & -\dfrac{12EI}{L^3} & \dfrac{6EI}{L^2} \\[2mm]
0 & \dfrac{6EI}{L^2} & \dfrac{4EI}{L} & 0 & -\dfrac{6EI}{L^2} & \dfrac{2EI}{L} \\[2mm]
-\dfrac{AE}{L} & 0 & 0 & \dfrac{AE}{L} & 0 & 0 \\[2mm]
0 & -\dfrac{12EI}{L^3} & -\dfrac{6EI}{L^2} & 0 & \dfrac{12EI}{L^3} & -\dfrac{6EI}{L^2} \\[2mm]
0 & \dfrac{6EI}{L^2} & \dfrac{2EI}{L} & 0 & -\dfrac{6EI}{L^2} & \dfrac{4EI}{L}
\end{bmatrix}
\end{array}
\begin{bmatrix} d_{Nx'} \\ d_{Ny'} \\ d_{Nz'} \\ d_{Fx'} \\ d_{Fy'} \\ d_{Fz'} \end{bmatrix}
\qquad (16\text{–}1)
$$

or in abbreviated form as

$$\mathbf{q} = \mathbf{k'd} \qquad (16\text{–}2)$$

The member stiffness matrix $\mathbf{k'}$ consists of thirty-six influence coefficients that physically represent the load on the member when the member undergoes a specified unit displacement. Specifically, each column in the matrix represents the member loadings for unit displacements identified by the degree-of-freedom coding listed above the columns. From the assembly, both equilibrium and compatibility of displacements have been satisfied.

16.2 Displacement and Force Transformation Matrices

As in the case for trusses, we must be able to transform the internal member loads **q** and deformations **d** from local x', y', z' coordinates to global x, y, z coordinates. For this reason transformation matrices are needed.

Displacement Transformation Matrix. Consider the frame member shown in Fig. 16–2a. Here it is seen that a global coordinate displacement D_{Nx} creates local coordinate displacements

$$d_{Nx'} = D_{Nx} \cos \theta_x \qquad d_{Ny'} = -D_{Nx} \cos \theta_y$$

Likewise, a global coordinate displacement D_{Ny}, Fig. 16–2b, creates local coordinate displacements of

$$d_{Nx'} = D_{Ny} \cos \theta_y \qquad d_{Ny'} = D_{Ny} \cos \theta_x$$

Finally, since the z' and z axes are coincident, that is, directed out of the page, a rotation D_{Nz} about z causes a corresponding rotation $d_{Nz'}$ about z'. Thus,

$$d_{Nz'} = D_{Nz}$$

In a similar manner, if global displacements D_{Fx} in the x direction, D_{Fy} in the y direction, and a rotation D_{Fz} are imposed on the far end of the member, the resulting transformation equations are, respectively,

$$d_{Fx'} = D_{Fx} \cos \theta_x \qquad d_{Fy'} = -D_{Fx} \cos \theta_y$$
$$d_{Fx'} = D_{Fy} \cos \theta_y \qquad d_{Fy'} = D_{Fy} \cos \theta_x$$
$$d_{Fz'} = D_{Fz}$$

Letting $\lambda_x = \cos \theta_x$, $\lambda_y = \cos \theta_y$ represent the direction cosines of the member, we can write the superposition of displacements in matrix form as

$$
\begin{bmatrix} d_{Nx'} \\ d_{Ny'} \\ d_{Nz'} \\ d_{Fx'} \\ d_{Fy'} \\ d_{Fz'} \end{bmatrix} =
\begin{bmatrix}
\lambda_x & \lambda_y & 0 & 0 & 0 & 0 \\
-\lambda_y & \lambda_x & 0 & 0 & 0 & 0 \\
0 & 0 & 1 & 0 & 0 & 0 \\
0 & 0 & 0 & \lambda_x & \lambda_y & 0 \\
0 & 0 & 0 & -\lambda_y & \lambda_x & 0 \\
0 & 0 & 0 & 0 & 0 & 1
\end{bmatrix}
\begin{bmatrix} D_{Nx} \\ D_{Ny} \\ D_{Nz} \\ D_{Fx} \\ D_{Fy} \\ D_{Fz} \end{bmatrix}
\qquad (16\text{–}3)
$$

or

$$\mathbf{d} = \mathbf{T D} \qquad (16\text{–}4)$$

By inspection, **T** transforms the six global x, y, z displacements **D** into the six local x', y', z' displacements **d**. Hence **T** is referred to as the *displacement transformation matrix*.

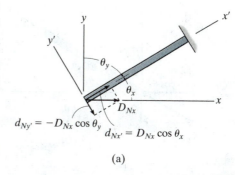

$$d_{Ny'} = -D_{Nx} \cos \theta_y$$
$$d_{Nx'} = D_{Nx} \cos \theta_x$$

(a)

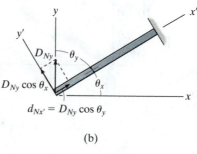

$$d_{Ny'} = D_{Ny} \cos \theta_x$$
$$d_{Nx'} = D_{Ny} \cos \theta_y$$

(b)

Fig. 16–2

16

Force Transformation Matrix. If we now apply each component of load to the near end of the member, we can determine how to transform the load components from local to global coordinates. Applying $q_{Nx'}$, Fig. 16–3a, it can be seen that

$$Q_{Nx} = q_{Nx'} \cos \theta_x \qquad Q_{Ny} = q_{Nx'} \cos \theta_y$$

If $q_{Ny'}$ is applied, Fig. 16–3b, then its components are

$$Q_{Nx} = -q_{Ny'} \cos \theta_y \qquad Q_{Ny} = q_{Ny'} \cos \theta_x$$

Finally, since $q_{Nz'}$ is collinear with Q_{Nz}, we have

$$Q_{Nz} = q_{Nz'}$$

In a similar manner, end loads of $q_{Fx'}$, $q_{Fy'}$, $q_{Fz'}$ will yield the following respective components:

$$Q_{Fx} = q_{Fx'} \cos \theta_x \qquad Q_{Fy} = q_{Fx'} \cos \theta_y$$

$$Q_{Fx} = -q_{Fy'} \cos \theta_y \qquad Q_{Fy} = q_{Fy'} \cos \theta_x$$

$$Q_{Fz} = q_{Fz'}$$

These equations, assembled in matrix form with $\lambda_x = \cos \theta_x$, $\lambda_y = \cos \theta_y$, yield

$$
\begin{bmatrix} Q_{Nx} \\ Q_{Ny} \\ Q_{Nz} \\ Q_{Fx} \\ Q_{Fy} \\ Q_{Fz} \end{bmatrix}
=
\begin{bmatrix}
\lambda_x & -\lambda_y & 0 & 0 & 0 & 0 \\
\lambda_y & \lambda_x & 0 & 0 & 0 & 0 \\
0 & 0 & 1 & 0 & 0 & 0 \\
0 & 0 & 0 & \lambda_x & -\lambda_y & 0 \\
0 & 0 & 0 & \lambda_y & \lambda_x & 0 \\
0 & 0 & 0 & 0 & 0 & 1
\end{bmatrix}
\begin{bmatrix} q_{Nx'} \\ q_{Ny'} \\ q_{Nz'} \\ q_{Fx'} \\ q_{Fy'} \\ q_{Fz'} \end{bmatrix}
\tag{16–5}
$$

or

$$\mathbf{Q} = \mathbf{T}^T \mathbf{q} \tag{16–6}$$

Here, as stated, $\mathbf{T}^T$ transforms the six member loads expressed in local coordinates into the six loadings expressed in global coordinates.

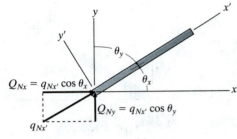

$Q_{Nx} = q_{Nx'} \cos \theta_x$

$Q_{Ny} = q_{Nx'} \cos \theta_y$

$q_{Nx'}$

(a)

$Q_{Nx} = -q_{Ny'} \cos \theta_y \quad Q_{Ny} = q_{Ny'} \cos \theta_x$

$q_{Ny'}$

(b)

Fig. 16–3

16.3 Frame-Member Global Stiffness Matrix

The results of the previous section will now be combined in order to determine the stiffness matrix for a member that relates the global loadings $\mathbf{Q}$ to the global displacements $\mathbf{D}$. To do this, substitute Eq. 16–4 ($\mathbf{d} = \mathbf{TD}$) into Eq. 16–2 ($\mathbf{q} = \mathbf{k'd}$). We have

$$\mathbf{q} = \mathbf{k'TD} \tag{16–7}$$

Here the member forces $\mathbf{q}$ are related to the global displacements $\mathbf{D}$. Substituting this result into Eq. 16–6 ($\mathbf{Q} = \mathbf{T}^T\mathbf{q}$) yields the final result,

$$\mathbf{Q} = \mathbf{T}^T\mathbf{k'TD} \tag{16–8}$$

or

$$\mathbf{Q} = \mathbf{kD}$$

where

$$\mathbf{k} = \mathbf{T}^T\mathbf{k'T} \tag{16–9}$$

Here $\mathbf{k}$ represents the global stiffness matrix for the member. We can obtain its value in generalized form using Eqs. 16–5, 16–1, and 16–3 and performing the matrix operations. This yields the final result,

$$
\mathbf{k} =
\begin{array}{cccccc}
N_x & N_y & N_z & F_x & F_y & F_z
\end{array}
$$

$$
\mathbf{k} =
\left[
\begin{array}{cccccc}
\left(\dfrac{AE}{L}\lambda_x^2 + \dfrac{12EI}{L^3}\lambda_y^2\right) & \left(\dfrac{AE}{L} - \dfrac{12EI}{L^3}\right)\lambda_x\lambda_y & -\dfrac{6EI}{L^2}\lambda_y & -\left(\dfrac{AE}{L}\lambda_x^2 + \dfrac{12EI}{L^3}\lambda_y^2\right) & -\left(\dfrac{AE}{L} - \dfrac{12EI}{L^3}\right)\lambda_x\lambda_y & -\dfrac{6EI}{L^2}\lambda_y \\[2mm]
\left(\dfrac{AE}{L} - \dfrac{12EI}{L^3}\right)\lambda_x\lambda_y & \left(\dfrac{AE}{L}\lambda_y^2 + \dfrac{12EI}{L^3}\lambda_x^2\right) & \dfrac{6EI}{L^2}\lambda_x & -\left(\dfrac{AE}{L} - \dfrac{12EI}{L^3}\right)\lambda_x\lambda_y & -\left(\dfrac{AE}{L}\lambda_y^2 + \dfrac{12EI}{L^3}\lambda_x^2\right) & \dfrac{6EI}{L^2}\lambda_x \\[2mm]
-\dfrac{6EI}{L^2}\lambda_y & \dfrac{6EI}{L^2}\lambda_x & \dfrac{4EI}{L} & \dfrac{6EI}{L^2}\lambda_y & -\dfrac{6EI}{L^2}\lambda_x & \dfrac{2EI}{L} \\[2mm]
-\left(\dfrac{AE}{L}\lambda_x^2 + \dfrac{12EI}{L^3}\lambda_y^2\right) & -\left(\dfrac{AE}{L} - \dfrac{12EI}{L^3}\right)\lambda_x\lambda_y & \dfrac{6EI}{L^2}\lambda_y & \left(\dfrac{AE}{L}\lambda_x^2 + \dfrac{12EI}{L^3}\lambda_y^2\right) & \left(\dfrac{AE}{L} - \dfrac{12EI}{L^3}\right)\lambda_x\lambda_y & \dfrac{6EI}{L^2}\lambda_y \\[2mm]
-\left(\dfrac{AE}{L} - \dfrac{12EI}{L^3}\right)\lambda_x\lambda_y & -\left(\dfrac{AE}{L}\lambda_y^2 + \dfrac{12EI}{L^3}\lambda_x^2\right) & -\dfrac{6EI}{L^2}\lambda_x & \left(\dfrac{AE}{L} - \dfrac{12EI}{L^3}\right)\lambda_x\lambda_y & \left(\dfrac{AE}{L}\lambda_y^2 + \dfrac{12EI}{L^3}\lambda_x^2\right) & -\dfrac{6EI}{L^2}\lambda_x \\[2mm]
-\dfrac{6EI}{L^2}\lambda_y & \dfrac{6EI}{L^2}\lambda_x & \dfrac{2EI}{L} & \dfrac{6EI}{L^2}\lambda_y & -\dfrac{6EI}{L^2}\lambda_x & \dfrac{4EI}{L}
\end{array}
\right]
\begin{array}{c}
N_x \\[2mm] N_y \\[2mm] N_z \\[2mm] F_x \\[2mm] F_y \\[2mm] F_z
\end{array}
$$

$$\tag{16–10}$$

Note that this 6×6 matrix is *symmetric*. Furthermore, the location of each element is associated with the coding at the near end, N_x, N_y, N_z, followed by that of the far end, F_x, F_y, F_z, which is listed at the top of the columns and along the rows. Like the $\mathbf{k'}$ matrix, each column of the $\mathbf{k}$ matrix represents the coordinate loads on the member at the nodes that are necessary to resist a unit displacement in the direction defined by the coding of the column. For example, the first column of $\mathbf{k}$ represents the global coordinate loadings at the near and far ends caused by a *unit displacement* at the near end in the x direction, that is, D_{Nx}.

16.4   Application of the Stiffness Method for Frame Analysis

Once the member stiffness matrices are established, they may be assembled into the structure stiffness matrix in the usual manner. By writing the structure matrix equation, the displacements at the unconstrained nodes can be determined, followed by the reactions and internal loadings at the nodes. Lateral loads acting on a member, fabrication errors, temperature changes, inclined supports, and internal supports are handled in the same manner as was outlined for trusses and beams.

Procedure for Analysis

The following method provides a means of finding the displacements, support reactions, and internal loadings for members of statically determinate and indeterminate frames.

Notation

- Divide the structure into finite elements and arbitrarily identify each element and its nodes. Elements usually extend between points of support, points of concentrated loads, corners or joints, or to points where internal loadings or displacements are to be determined.

- Establish the *x, y, z,* global coordinate system, usually for convenience with the origin located at a nodal point on one of the elements and the axes located such that all the nodes have positive coordinates.

- At each nodal point of the frame, specify numerically the three *x, y, z* coding components. In all cases use the *lowest code numbers* to identify all the *unconstrained degrees of freedom,* followed by the remaining or *highest code numbers* to identify the *constrained degrees of freedom.*

- From the problem, establish the known displacements $\mathbf{D}_k$ and known external loads $\mathbf{Q}_k$. When establishing $\mathbf{Q}_k$ be sure to include any *reversed* fixed-end loadings if an element supports an intermediate load.

Structure Stiffness Matrix

- Apply Eq. 16–10 to determine the stiffness matrix for each element expressed in global coordinates. In particular, the direction cosines λ_x and λ_y are determined from the *x, y* coordinates of at the ends of the element, Eqs. 14–5 and 14–6.

- After each member stiffness matrix is written, and the six rows and columns are identified with the near and far code numbers, merge the matrices to form the structure stiffness matrix **K**. As a partial check, the element and structure stiffness matrices should all be *symmetric.*

Displacements and Loads

- Partition the stiffness matrix as indicated by Eq. 14–18. Expansion then leads to

$$\mathbf{Q}_k = \mathbf{K}_{11}\mathbf{D}_u + \mathbf{K}_{12}\mathbf{D}_k$$
$$\mathbf{Q}_u = \mathbf{K}_{21}\mathbf{D}_u + \mathbf{K}_{22}\mathbf{D}_k$$

The unknown displacements $\mathbf{D}_u$ are determined from the first of these equations. Using these values, the support reactions $\mathbf{Q}_u$ are computed from the second equation. Finally, the internal loadings $\mathbf{q}$ at the ends of the members can be computed from Eq. 16–7, namely

$$\mathbf{q} = \mathbf{k}'\mathbf{T}\mathbf{D}$$

If the results of any of the unknowns are calculated as negative quantities, it indicates they act in the negative coordinate directions.

EXAMPLE 16.1

Determine the loadings at the joints of the two-member frame shown in Fig. 16–4a. Take $I = 500$ in^4, $A = 10$ in^2, and $E = 29(10^3)$ ksi for both members.

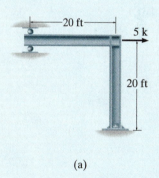

(a)

SOLUTION

Notation. By inspection, the frame has two elements and three nodes, which are identified as shown in Fig. 16–4b. The origin of the global coordinate system is located at ①. The code numbers at the nodes are specified with the *unconstrained degrees of freedom numbered first*. From the constraints at ① and ③, and the applied loading, we have

$$\mathbf{D}_k = \begin{bmatrix} 0 \\ 0 \\ 0 \\ 0 \end{bmatrix} \begin{matrix} 6 \\ 7 \\ 8 \\ 9 \end{matrix} \qquad \mathbf{Q}_k = \begin{bmatrix} 5 \\ 0 \\ 0 \\ 0 \\ 0 \end{bmatrix} \begin{matrix} 1 \\ 2 \\ 3 \\ 4 \\ 5 \end{matrix}$$

Structure Stiffness Matrix. The following terms are common to both element stiffness matrices:

$$\frac{AE}{L} = \frac{10\left[29(10^3)\right]}{20(12)} = 1208.3 \text{ k/in.}$$

$$\frac{12EI}{L^3} = \frac{12\left[29(10^3)(500)\right]}{\left[20(12)\right]^3} = 12.6 \text{ k/in.}$$

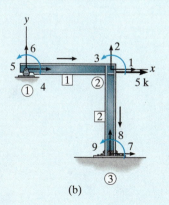

(b)

Fig. 16–4

$$\frac{6EI}{L^2} = \frac{6\left[29(10^3)(500)\right]}{\left[20(12)\right]^2} = 1510.4 \text{ k}$$

$$\frac{4EI}{L} = \frac{4\left[29(10^3)(500)\right]}{20(12)} = 241.7(10^3) \text{ k} \cdot \text{in.}$$

$$\frac{2EI}{L} = \frac{2\left[29(10^3)(500)\right]}{20(12)} = 120.83(10^3) \text{ k} \cdot \text{in.}$$

Member 1:

$$\lambda_x = \frac{20 - 0}{20} = 1 \qquad \lambda_y = \frac{0 - 0}{20} = 0$$

Substituting the data into Eq. 16–10, we have

$$
\mathbf{k_1} =
\begin{array}{cccccc}
4 & 6 & 5 & 1 & 2 & 3
\end{array}
$$

$$
\mathbf{k_1} =
\left[
\begin{array}{cccccc}
1208.3 & 0 & 0 & -1208.3 & 0 & 0 \\
0 & 12.6 & 1510.4 & 0 & -12.6 & 1510.4 \\
0 & 1510.4 & 241.7(10^3) & 0 & -1510.4 & 120.83(10^3) \\
-1208.3 & 0 & 0 & 1208.3 & 0 & 0 \\
0 & -12.6 & -1510.4 & 0 & 12.6 & -1510.4 \\
0 & 1510.4 & 120.83(10^3) & 0 & -1510.4 & 241.7(10^3)
\end{array}
\right]
\begin{array}{c}
4 \\ 6 \\ 5 \\ 1 \\ 2 \\ 3
\end{array}
$$

The rows and columns of this 6 × 6 matrix are identified by the three x, y, z code numbers, first at the near end and followed by the far end, that is, 4, 6, 5, 1, 2, 3, respectively, Fig. 16–4b. This is done for later assembly of the elements.

Member 2:

$$\lambda_x = \frac{20 - 20}{20} = 0 \qquad \lambda_y = \frac{-20 - 0}{20} = -1$$

Substituting the data into Eq. 16–10 yields

$$
\begin{array}{cccccc}
1 & 2 & 3 & 7 & 8 & 9
\end{array}
$$

$$
\mathbf{k_2} =
\left[
\begin{array}{cccccc}
12.6 & 0 & 1510.4 & -12.6 & 0 & 1510.4 \\
0 & 1208.3 & 0 & 0 & -1208.3 & 0 \\
1510.4 & 0 & 241.7(10^3) & -1510.4 & 0 & 120.83(10^3) \\
-12.6 & 0 & -1510.4 & 12.6 & 0 & -1510.4 \\
0 & -1208.3 & 0 & 0 & 1208.3 & 0 \\
1510.4 & 0 & 120.83(10^3) & -1510.4 & 0 & 241.7(10^3)
\end{array}
\right]
\begin{array}{c}
1 \\ 2 \\ 3 \\ 7 \\ 8 \\ 9
\end{array}
$$

As usual, column and row identification is referenced by the three code numbers in x, y, z sequence for the near and far ends, respectively, that is, 1, 2, 3, then 7, 8, 9, Fig. 16–4b.

The structure stiffness matrix is determined by assembling $\mathbf{k}_1$ and $\mathbf{k}_2$. The result, shown partitioned, as $\mathbf{Q} = \mathbf{KD}$, is

$$
\begin{bmatrix} 5 \\ 0 \\ 0 \\ 0 \\ 0 \\ \cdots \\ Q_6 \\ Q_7 \\ Q_8 \\ Q_9 \end{bmatrix}
=
\begin{bmatrix}
\overset{1}{1220.9} & \overset{2}{0} & \overset{3}{1510.4} & \overset{4}{-1208.3} & \overset{5}{0} & \vdots & \overset{6}{0} & \overset{7}{-12.6} & \overset{8}{0} & \overset{9}{1510.4} \\
0 & 1220.9 & -1510.4 & 0 & -1510.4 & \vdots & -12.6 & 0 & -1208.3 & 0 \\
1510.4 & -1510.4 & 483.3(10^3) & 0 & 120.83(10^3) & \vdots & 1510.4 & -1510.4 & 0 & 120.83(10^3) \\
-1208.3 & 0 & 0 & 1208.3 & 0 & \vdots & 0 & 0 & 0 & 0 \\
0 & -1510.4 & 120.83(10^3) & 0 & 241.7(10^3) & \vdots & 1510.4 & 0 & 0 & 0 \\
\cdots & & & & & & & & & \\
0 & -12.6 & 1510.4 & 0 & 1510.4 & \vdots & 12.6 & 0 & 0 & 0 \\
-12.6 & 0 & -1510.4 & 0 & 0 & \vdots & 0 & 12.6 & 0 & -1510.4 \\
0 & -1208.3 & 0 & 0 & 0 & \vdots & 0 & 0 & 1208.3 & 0 \\
1510.4 & 0 & 120.83(10^3) & 0 & 0 & \vdots & 0 & -1510.4 & 0 & 241.7(10^3)
\end{bmatrix}
\begin{bmatrix} D_1 \\ D_2 \\ D_3 \\ D_4 \\ D_5 \\ \cdots \\ 0 \\ 0 \\ 0 \\ 0 \end{bmatrix} \quad (1)
$$

Displacements and Loads. Expanding to determine the displacements yields

$$
\begin{bmatrix} 5 \\ 0 \\ 0 \\ 0 \\ 0 \end{bmatrix}
=
\begin{bmatrix}
1220.9 & 0 & 1510.4 & -1208.3 & 0 \\
0 & 1220.9 & -1510.4 & 0 & -1510.4 \\
1510.4 & -1510.4 & 483.3(10^3) & 0 & 120.83(10^3) \\
-1208.3 & 0 & 0 & 1208.3 & 0 \\
0 & -1510.4 & 120.83(10^3) & 0 & 241.7(10^3)
\end{bmatrix}
\begin{bmatrix} D_1 \\ D_2 \\ D_3 \\ D_4 \\ D_5 \end{bmatrix}
+
\begin{bmatrix} 0 \\ 0 \\ 0 \\ 0 \\ 0 \end{bmatrix}
$$

Solving, we obtain

$$
\begin{bmatrix} D_1 \\ D_2 \\ D_3 \\ D_4 \\ D_5 \end{bmatrix}
=
\begin{bmatrix}
0.696 \text{ in.} \\
-1.55(10^{-3}) \text{ in.} \\
-2.488(10^{-3}) \text{ rad} \\
0.696 \text{ in.} \\
1.234(10^{-3}) \text{ rad}
\end{bmatrix}
$$

Using these results, the support reactions are determined from Eq. (1) as follows:

$$
\begin{bmatrix} Q_6 \\ Q_7 \\ Q_8 \\ Q_9 \end{bmatrix}
=
\begin{bmatrix}
\overset{1}{0} & \overset{2}{-12.6} & \overset{3}{1510.4} & \overset{4}{0} & \overset{5}{1510.4} \\
-12.6 & 0 & -1510.4 & 0 & 0 \\
0 & -1208.3 & 0 & 0 & 0 \\
1510.4 & 0 & 120.83(10^3) & 0 & 0
\end{bmatrix}
\begin{bmatrix}
0.696 \\
-1.55(10^{-3}) \\
-2.488(10^{-3}) \\
0.696 \\
1.234(10^{-3})
\end{bmatrix}
+
\begin{bmatrix} 0 \\ 0 \\ 0 \\ 0 \end{bmatrix}
=
\begin{bmatrix}
-1.87 \text{ k} \\
-5.00 \text{ k} \\
1.87 \text{ k} \\
750 \text{ k} \cdot \text{in.}
\end{bmatrix} \quad Ans.
$$

16

The internal loadings at node ② can be determined by applying Eq. 16–7 to member 1. Here $\mathbf{k_1'}$ is defined by Eq. 16–1 and $\mathbf{T_1}$ by Eq. 16–3. Thus,

$$
\mathbf{q_1} = \mathbf{k_1}\mathbf{T_1}\mathbf{D} =
\begin{bmatrix}
1208.3 & 0 & 0 & -1208.3 & 0 & 0 \\
0 & 12.6 & 1510.4 & 0 & -12.6 & 1510.4 \\
0 & 1510.4 & 241.7(10^3) & 0 & -1510.4 & 120.83(10^3) \\
-1208.3 & 0 & 0 & 1208.3 & 0 & 0 \\
0 & -12.6 & -1510.4 & 0 & 12.6 & -1510.4 \\
0 & 1510.4 & 120.83(10^3) & 0 & -1510.4 & 241.7(10^3)
\end{bmatrix}
\begin{matrix}4\\6\\5\\1\\2\\3\end{matrix}
\begin{bmatrix}
1 & 0 & 0 & 0 & 0 & 0 \\
0 & 1 & 0 & 0 & 0 & 0 \\
0 & 0 & 1 & 0 & 0 & 0 \\
0 & 0 & 0 & 1 & 0 & 0 \\
0 & 0 & 0 & 0 & 1 & 0 \\
0 & 0 & 0 & 0 & 0 & 1
\end{bmatrix}
\begin{bmatrix}
0.696 \\
0 \\
1.234(10^{-3}) \\
0.696 \\
-1.55(10^{-3}) \\
-2.488(10^{-3})
\end{bmatrix}
\begin{matrix}4\\6\\5\\1\\2\\3\end{matrix}
$$

(column code numbers: 4 6 5 1 2 3)

Note the appropriate arrangement of the elements in the matrices as indicated by the code numbers alongside the columns and rows. Solving yields

$$
\begin{bmatrix}
q_4 \\
q_6 \\
q_5 \\
q_1 \\
q_2 \\
q_3
\end{bmatrix}
=
\begin{bmatrix}
0 \\
-1.87 \text{ k} \\
0 \\
0 \\
1.87 \text{ k} \\
-450 \text{ k} \cdot \text{in.}
\end{bmatrix}
$$

Ans.

The above results are shown in Fig. 16–4c. The directions of these vectors are in accordance with the positive directions defined in Fig. 16–1. Furthermore, the origin of the local x', y', z' axes is at the near end of the member. In a similar manner, the free-body diagram of member 2 is shown in Fig. 16–4d.

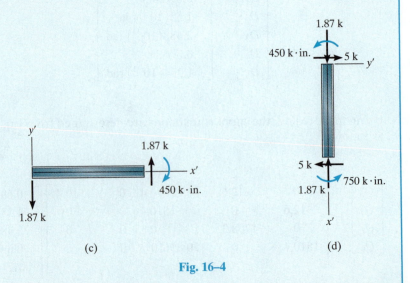

(c)

(d)

Fig. 16–4

EXAMPLE 16.2

Determine the loadings at the ends of each member of the frame shown in Fig. 16–5a. Take $I = 600$ in⁴, $A = 12$ in², and $E = 29(10^3)$ ksi for each member.

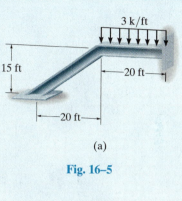

(a)

Fig. 16–5

SOLUTION

Notation. To perform a matrix analysis, the distributed loading acting on the horizontal member will be replaced by equivalent end moments and shears computed from statics and the table listed on the inside back cover. (Note that no external force of 30 k or moment of 1200 k · in. is placed at ③ since the reactions at code numbers 8 and 9 *are to be unknowns* in the load matrix.) Then using superposition, the results obtained for the frame in Fig. 16–5b will be modified for this member by the loads shown in Fig. 16–5c.

As shown in Fig. 16–5b, the nodes and members are numbered and the origin of the global coordinate system is placed at node ①. As usual, the code numbers are specified with numbers assigned first to the unconstrained degrees of freedom. Thus,

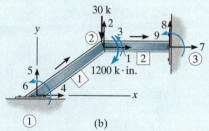

(b)

$$\mathbf{D}_k = \begin{bmatrix} 0 \\ 0 \\ 0 \\ 0 \\ 0 \\ 0 \end{bmatrix} \begin{matrix} 4 \\ 5 \\ 6 \\ 7 \\ 8 \\ 9 \end{matrix} \qquad \mathbf{Q}_k = \begin{bmatrix} 0 \\ -30 \\ -1200 \end{bmatrix} \begin{matrix} 1 \\ 2 \\ 3 \end{matrix}$$

$+$

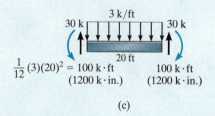

$$\frac{1}{12}(3)(20)^2 = 100 \text{ k·ft}$$
(1200 k·in.)   (1200 k·in.)

(c)

Structure Stiffness Matrix

Member 1:

$$\frac{AE}{L} = \frac{12[29(10^3)]}{25(12)} = 1160 \text{ k/in.}$$

$$\frac{12EI}{L^3} = \frac{12[29(10^3)]600}{[25(12)]^3} = 7.73 \text{ k/in.}$$

$$\frac{6EI}{L^2} = \frac{6[29(10^3)]600}{[25(12)]^2} = 1160 \text{ k}$$

$$\frac{4EI}{L} = \frac{4[29(10^3)]600}{25(12)} = 232(10^3) \text{ k·in.}$$

$$\frac{2EI}{L} = \frac{2[29(10^3)]600}{25(12)} = 116(10^3) \text{ k·in.}$$

$$\lambda_x = \frac{20 - 0}{25} = 0.8 \qquad \lambda_y = \frac{15 - 0}{25} = 0.6$$

Applying Eq. 16–10, we have

$$
\mathbf{k}_1 = \begin{array}{cccccc}
4 & 5 & 6 & 1 & 2 & 3 \\
\end{array}
$$

$$
\mathbf{k}_1 = \left[
\begin{array}{cccccc}
745.18 & 553.09 & -696 & -745.18 & -553.09 & -696 \\
553.09 & 422.55 & 928 & -553.09 & -422.55 & 928 \\
-696 & 928 & 232(10^3) & 696 & -928 & 116(10^3) \\
-745.18 & -553.09 & 696 & 745.18 & 553.09 & 696 \\
-553.09 & -422.55 & -928 & 553.09 & 422.55 & -928 \\
-696 & 928 & 116(10^3) & 696 & -928 & 232(10^3)
\end{array}
\right]
\begin{array}{c}
4 \\ 5 \\ 6 \\ 1 \\ 2 \\ 3
\end{array}
$$

Member 2:

$$
\frac{AE}{L} = \frac{12\left[29(10^3)\right]}{20(12)} = 1450 \text{ k/in.}
$$

$$
\frac{12EI}{L^3} = \frac{12\left[29(10^3)\right]600}{\left[20(12)\right]^3} = 15.10 \text{ k/in.}
$$

$$
\frac{6EI}{L^2} = \frac{6\left[29(10^3)\right]600}{\left[20(12)\right]^2} = 1812.50 \text{ k}
$$

$$
\frac{4EI}{L} = \frac{4\left[29(10^3)\right]600}{20(12)} = 2.90(10^5) \text{ k·in.}
$$

$$
\frac{2EI}{L} = \frac{2\left[29(10^3)\right]600}{\left[20(12)\right]} = 1.45(10^5) \text{ k·in.}
$$

$$
\lambda_x = \frac{40-20}{20} = 1 \qquad \lambda_y = \frac{15-15}{20} = 0
$$

Thus, Eq. 16–10 becomes

$$
\mathbf{k}_2 = \begin{array}{cccccc}
1 & 2 & 3 & 7 & 8 & 9 \\
\end{array}
$$

$$
\mathbf{k}_2 = \left[
\begin{array}{cccccc}
1450 & 0 & 0 & -1450 & 0 & 0 \\
0 & 15.10 & 1812.50 & 0 & -15.10 & 1812.50 \\
0 & 1812.50 & 290(10^3) & 0 & -1812.50 & 145(10^3) \\
-1450 & 0 & 0 & 1450 & 0 & 0 \\
0 & -15.10 & -1812.50 & 0 & 15.10 & -1812.50 \\
0 & 1812.50 & 145(10^3) & 0 & -1812.50 & 290(10^3)
\end{array}
\right]
\begin{array}{c}
1 \\ 2 \\ 3 \\ 7 \\ 8 \\ 9
\end{array}
$$

The structure stiffness matrix, included in $\mathbf{Q} = \mathbf{KD}$, becomes

$$
\begin{bmatrix} 0 \\ -30 \\ -1200 \\ \hline Q_4 \\ Q_5 \\ Q_6 \\ Q_7 \\ Q_8 \\ Q_9 \end{bmatrix} =
\begin{bmatrix}
2195.18 & 553.09 & 696 & -745.18 & -553.09 & 696 & -1450 & 0 & 0 \\
553.09 & 437.65 & 884.5 & -553.09 & -422.55 & -928 & 0 & -15.10 & 1812.50 \\
696 & 884.5 & 522(10^3) & -696 & 928 & 116(10^3) & 0 & -1812.50 & 145(10^3) \\
-745.18 & -553.09 & -696 & 745.18 & 553.09 & -696 & 0 & 0 & 0 \\
-553.09 & -422.55 & 928 & 553.09 & 422.55 & 928 & 0 & 0 & 0 \\
696 & -928 & 116(10^3) & -696 & 928 & 232(10^3) & 0 & 0 & 0 \\
-1450 & 0 & 0 & 0 & 0 & 0 & 1450 & 0 & 0 \\
0 & -15.10 & -1812.50 & 0 & 0 & 0 & 0 & 15.10 & -1812.50 \\
0 & 1812.50 & 145(10^3) & 0 & 0 & 0 & 0 & -1812.50 & 290(10^3)
\end{bmatrix}
\begin{bmatrix} D_1 \\ D_2 \\ D_3 \\ 0 \\ 0 \\ 0 \\ 0 \\ 0 \\ 0 \end{bmatrix} \quad (1)
$$

(Column headers: 1, 2, 3, 4, 5, 6, 7, 8, 9)

Displacements and Loads. Expanding to determine the displacements, and solving, we have

$$
\begin{bmatrix} 0 \\ -30 \\ -1200 \end{bmatrix} =
\begin{bmatrix}
2195.18 & 553.09 & 696 \\
553.09 & 437.65 & 884.5 \\
696 & 884.5 & 522(10^3)
\end{bmatrix}
\begin{bmatrix} D_1 \\ D_2 \\ D_3 \end{bmatrix} +
\begin{bmatrix} 0 \\ 0 \\ 0 \end{bmatrix}
$$

$$
\begin{bmatrix} D_1 \\ D_2 \\ D_3 \end{bmatrix} =
\begin{bmatrix} 0.0247 \text{ in.} \\ -0.0954 \text{ in.} \\ -0.00217 \text{ rad} \end{bmatrix}
$$

Using these results, the support reactions are determined from Eq. (1) as follows:

$$
\begin{bmatrix} Q_4 \\ Q_5 \\ Q_6 \\ Q_7 \\ Q_8 \\ Q_9 \end{bmatrix} =
\begin{bmatrix}
-745.18 & -553.09 & -696 \\
-553.09 & -422.55 & 928 \\
696 & -928 & 116(10^3) \\
-1450 & 0 & 0 \\
0 & -15.10 & -1812.50 \\
0 & 1812.50 & 145(10^3)
\end{bmatrix}
\begin{bmatrix} 0.0247 \\ -0.0954 \\ -0.00217 \end{bmatrix} +
\begin{bmatrix} 0 \\ 0 \\ 0 \\ 0 \\ 0 \\ 0 \end{bmatrix} =
\begin{bmatrix}
35.85 \text{ k} \\
24.63 \text{ k} \\
-145.99 \text{ k} \cdot \text{in.} \\
-35.85 \text{ k} \\
5.37 \text{ k} \\
-487.60 \text{ k} \cdot \text{in.}
\end{bmatrix}
$$

16

The internal loadings can be determined from Eq. 16–7 applied to members 1 and 2. In the case of member 1, $\mathbf{q} = \mathbf{k}_1'\mathbf{T}_1\mathbf{D}$, where $\mathbf{k}_1'$ is determined from Eq. 16–1, and $\mathbf{T}_1$ from Eq. 16–3. Thus,

$$\begin{bmatrix} q_4 \\ q_5 \\ q_6 \\ q_1 \\ q_2 \\ q_3 \end{bmatrix} = \begin{array}{c} \begin{matrix} 4 & 5 & 6 & 1 & 2 & 3 \end{matrix} \\ \begin{bmatrix} 1160 & 0 & 0 & -1160 & 0 & 0 \\ 0 & 7.73 & 1160 & 0 & -7.73 & 1160 \\ 0 & 1160 & 232(10^3) & 0 & -1160 & 116(10^3) \\ -1160 & 0 & 0 & 1160 & 0 & 0 \\ 0 & -7.73 & -1160 & 0 & 7.73 & -1160 \\ 0 & 1160 & 116(10^3) & 0 & -1160 & 232(10^3) \end{bmatrix} \end{array} \begin{bmatrix} 0.8 & 0.6 & 0 & 0 & 0 & 0 \\ -0.6 & 0.8 & 0 & 0 & 0 & 0 \\ 0 & 0 & 1 & 0 & 0 & 0 \\ 0 & 0 & 0 & 0.8 & 0.6 & 0 \\ 0 & 0 & 0 & -0.6 & 0.8 & 0 \\ 0 & 0 & 0 & 0 & 0 & 1 \end{bmatrix} \begin{bmatrix} 0 \\ 0 \\ 0 \\ 0.0247 \\ -0.0954 \\ -0.00217 \end{bmatrix} \begin{matrix} 4 \\ 5 \\ 6 \\ 1 \\ 2 \\ 3 \end{matrix}$$

Here the code numbers indicate the rows and columns for the near and far ends of the member, respectively, that is, 4, 5, 6, then 1, 2, 3, Fig. 16–5*b*. Thus,

$$\begin{bmatrix} q_4 \\ q_5 \\ q_6 \\ q_1 \\ q_2 \\ q_3 \end{bmatrix} = \begin{bmatrix} 43.5\text{ k} \\ -1.81\text{ k} \\ -146\text{ k}\cdot\text{in.} \\ -43.5\text{ k} \\ 1.81\text{ k} \\ -398\text{ k}\cdot\text{in.} \end{bmatrix} \qquad \textit{Ans.}$$

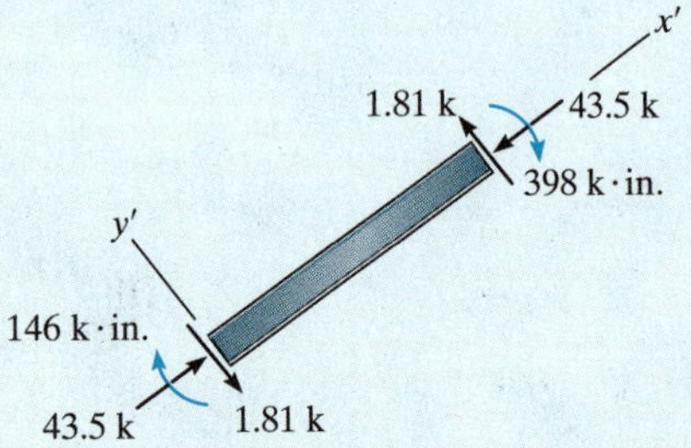

(d)

These results are shown in Fig. 16–5*d*.

A similar analysis is performed for member 2. The results are shown at the left in Fig. 16–5*e*. For this member we must superimpose the loadings of Fig. 16–5*c*, so that the final results for member 2 are shown to the right.

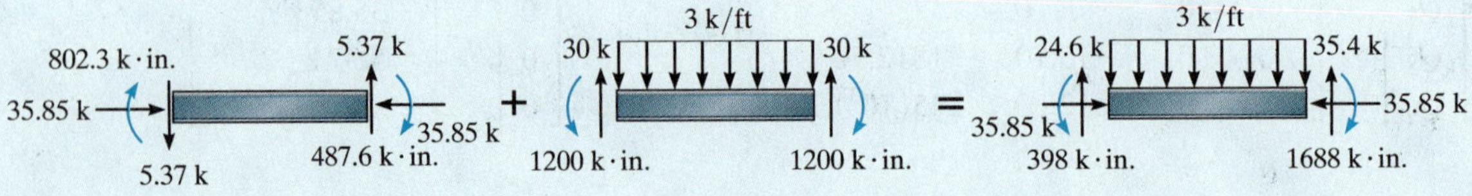

(e)

Fig. 16–5

PROBLEMS

16–1. Determine the structure stiffness matrix **K** for each member of the frame. Assume ③ is pinned and ① is fixed. Take $E = 200$ GPa, $I = 300(10^6)$ mm⁴, $A = 21(10^3)$ mm² for each member.

16–2. Determine the support reactions at ① and ③. Assume ③ is pinned and ① is fixed. Take $E = 200$ GPa, $I = 300(10^6)$ mm⁴, $A = 21(10^3)$ mm² for each member.

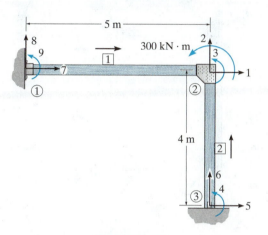

Probs. 16–1/2

16–3. Determine the structure stiffness matrix **K** for each member of the frame. Take $E = 29(10^3)$ ksi, $I = 450$ in⁴, $A = 8$ in² for each member. All joints are fixed connected.

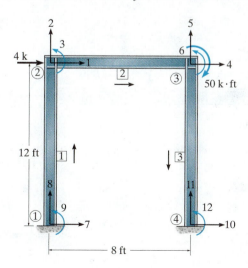

Prob. 16–3

*16–4. Determine the horizontal displacement of joint ②. Also compute the support reactions. Take $E = 29(10^3)$ ksi, $I = 450$ in⁴, $A = 8$ in² for each member. All joints are fixed connected.

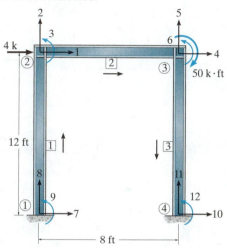

Prob. 16–4

16–5. Determine the structure stiffness matrix **K** for each member of the frame. Take $E = 29(10^3)$ ksi, $I = 700$ in⁴, $A = 30$ in² for each member.

16–6. Determine the internal loadings at the ends of each member. Take $E = 29(10^3)$ ksi, $I = 700$ in⁴, $A = 30$ in² for each member.

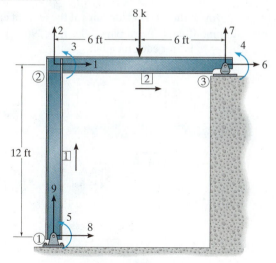

Probs. 16–5/6

16

16–7. Determine the structure stiffness matrix **K** for the frame. Take $E = 29(10^3)$ ksi, $I = 650$ in^4, $A = 20$ in^2 for each member.

*16–8. Determine the components of displacement at ①. Take $E = 29(10^3)$ ksi, $I = 650$ in^4, $A = 20$ in^2 for each member.

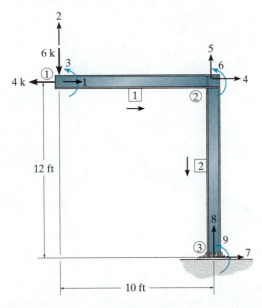

Probs. 16–7/8

16–9. Determine the structure stiffness matrix **K** for the frame. Assume ① and ③ are pins. Take $E = 29(10^3)$ ksi, $I = 600$ in^4, $A = 10$ in^2 for each member.

16–10. Determine the internal loadings at the ends of each member. Assume ① and ③ are pins. Take $E = 29(10^3)$ ksi, $I = 600$ in^4, $A = 10$ in^2 for each member.

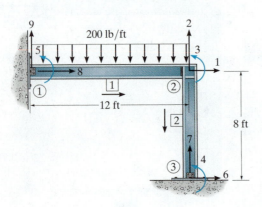

Probs. 16–9/10

16–11. Determine the structure stiffness matrix **K**, for each member of the frame. Take $E = 29(10^3)$ ksi, $I = 700$ in^4, $A = 30$ in^2 for each member. Joint ① is pin connected.

*16–12. Determine the support reactions at ① and ③. Take $E = 29(10^3)$ ksi, $I = 700$ in^4, $A = 30$ in^2 for each member. Joint ① is pin connected.

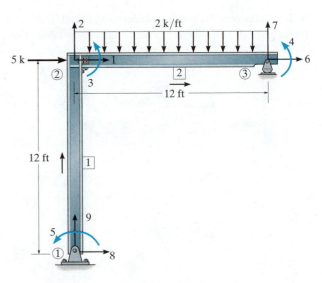

Probs. 16–11/12

16–13. Determine the structure stiffness matrix **K** for the frame. Take $E = 29(10^3)$ ksi, $I = 600$ in^4, $A = 10$ in^2 for each member. Assume joints ① and ③ are pinned; joint ② is fixed.

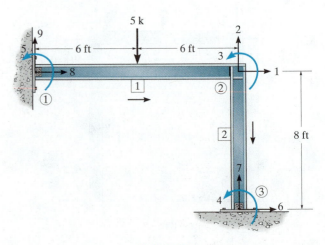

Prob. 16–13

16–14.   Determine the rotation at ① and ③ and the support reactions. Take $E = 29(10^3)$ ksi, $I = 600$ in^4, $A = 10$ in^2 for each member. Assume joints ① and ③ are pinned; joint ② is fixed.

16–16.   Determine the structure stiffness matrix **K** for the two-member frame. Take $E = 200$ GPa, $I = 350(10^6)$ mm^4, $A = 20(10^3)$ mm^2 for each member. Joints 1 and 3 are pinned and joint 2 is fixed.

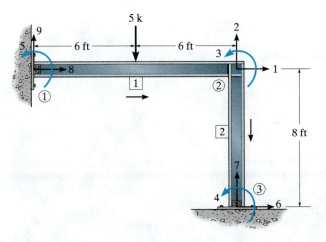

Prob. 16–14

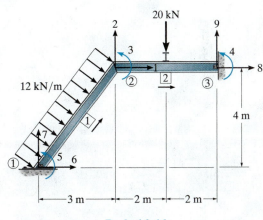

Prob. 16–16

16–15.   Determine the reactions at the supports ① and ④. Joints ① and ④ are pin connected and ② and ③ are fixed connected. Take $E = 29(10^3)$ ksi, $I = 700$ in^4, $A = 15$ in^2 for each member.

16–17.   Determine the support reactions at ① and ③ in. Take $E = 200$ GPa, $I = 350(10^6)$ mm^4, $A = 20(10^3)$ mm^2 for each member. Joints ① and ③ are pinned and joint ② is fixed.

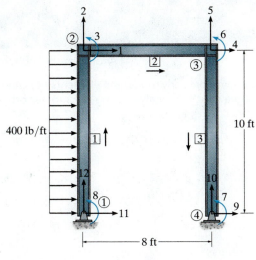

Prob. 16–15

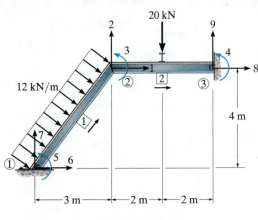

Prob. 16–17

16

Chapter 17

© Andrea Izzotti/Shutterstock

It is important to know how to properly model the elements of a structure before doing a structural analysis using a computer program.

Structural Modeling and Computer Analysis

Throughout the text we have presented some ideas as to how to model a structure so that a reliable structural analysis can be performed. In this chapter we will extend these ideas and show how it applies when the analysis is done on a computer. This process will include a description of how geometric, load, and material data are assembled and then used as input for one of many different structural analysis computer programs currently available.

17.1   General Structural Modeling

In a general sense, a structural analysis is performed in order to determine the internal loadings within a structure, and to find the deflection of its components, when the structure is subjected to a variety of loadings. To do this it is first necessary to select a form for the structure, such as a truss or frame, and then develop a model of this form that can be used for the analysis. The model must account for the geometry of each of the members, the types of connections, the loadings, and the material properties. The modeling process must be such that reasonable yet conservative results are obtained. This is especially true for structures that are occupied by large groups of people, such as assembly halls, schools, and hospitals. Keep in mind that a computer analysis may be accurate when calculating a numerical answer, but the final results can lead to disastrous consequences if the prepared model provides the wrong computer input.

As noted in the previous chapters, there are many different types of structural forms that may be used to support a loading. Choosing the correct form to fulfill a specific function is both a science and an art. Oftentimes, models for several different forms have to be considered, and each analyzed to find the one that is economically feasible, and at the same time provides both safety and reliability. A typical example would be choosing a steel frame structure having light gage metal walls and roof, versus one built from masonry and wood. The behavior of each of these structures is different under load, and the model for each also depends upon the way it is constructed. The selection of some structural forms has a limited use. For example, truss and girder bridges are often used for short spans, and suspension or cable-stayed bridges for longer spans.

Not only is it necessary to select a particular model for the form of a structure, but the elements that make it up may not have a unique pattern. For example, if a truss bridge is selected, then the form of the truss, such as a Pratt or Warren truss, must also be determined, as discussed in Sec. 3.1. Floor systems in buildings also vary in their details, as noted in Sec. 2.1, and models for each must be clearly specified. Proper selection for complex projects comes from experience, and normally requires a team effort, working in close contact with the architects who have conceived the project.

To ensure public safety, the building criteria for the design of some structures requires the structure to remain stable after some of its primary supporting members are *removed*. This requirement follows in the aftermath of the tragic 1995 bombing of the Murrah Federal Building in Oklahoma City. Investigators concluded that the majority of deaths were the result not of the blast, but of the progressive collapse of portions of all the floors in the front of the structure. As a result, the design of many federal buildings, and some high-rise commercial buildings, now require the structure to remain in a stable position when possible loss of its primary members occurs. A complete structural analysis will therefore requires a careful investigation of the load paths for several different cases of structural support, and a model of each case.

In the following sections we will review this modeling process as it applies to basic structural elements, various supports and connections, loadings, and materials. Once the model is constructed, and a structural analysis performed, the computed results should be checked to be sure they parallel our intuition about the structural behavior. If this does not occur, then we may have to improve the modeling process, or justify the calculations based on professional judgment.

17.2 Modeling a Structure and its Members

The various types of structural members have been described in Sec. 1.2. Here we will present a summary description of these members, and illustrate how each can be modeled.

Tie Rods. Sometimes called *bracing struts*, these members are intended to only support a tensile force. They have many applications in structures, and an example, along with its support connection, is shown in the photos in Fig. 17–1a. Because they are slender, the supports for these members are always *assumed* to be pin connections. Consequently, the model of this element is shown in Fig. 17–1b.

(a)

(b)

Fig. 17–1

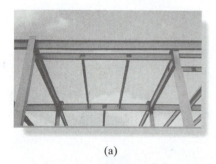

(a)

(b)

Fig. 17–2

Beams and Girders.   *Beams* are normally prismatic members that support loadings applied perpendicular to their length. A *girder* provides support for beams that are connected to it, as in the case of a building girder that supports a series of floor beams, Fig. 17–2*a*. One sometimes has to be careful about selecting the proper support for these members. If the support is a simple bolted connection, as in Fig. 17–2*b*, it should be modeled as a pin. This is because codes generally restrict the elastic deflection of a beam, and so the support rotation will generally be *very small*. Choosing pin supports will lead to a more conservative approach to the design of the member. For example, consider the moment diagrams for the simply-supported, partially fixed, and fixed-supported beams that carry the same uniform distributed loading, Fig. 17–3. The internal moment is *largest* in the simply-supported case, and so this beam must have a higher strength and stiffness to resist the loading compared to the other two cases.

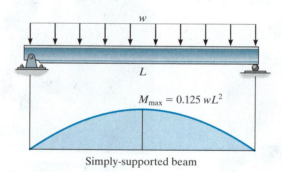

$$M_{max} = 0.125 \, wL^2$$

Simply-supported beam

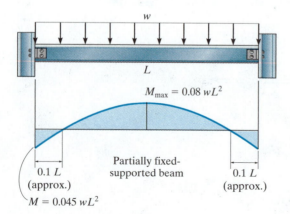

$$M_{max} = 0.08 \, wL^2$$

Partially fixed-supported beam

0.1 *L* (approx.)

0.1 *L* (approx.)

$$M = 0.045 \, wL^2$$

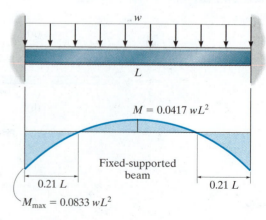

$$M = 0.0417 \, wL^2$$

Fixed-supported beam

0.21 *L*

0.21 *L*

$$M_{max} = 0.0833 \, wL^2$$

Fig. 17–3

Beams can also have cross sections that are tapered or haunched, Fig. 17–4a, or they can be built up by adding plates to their top and bottom, Fig. 17–4b. Apart from its supports, a model of the haunched beam to be used for a computer analysis can be represented by a series of fixed-connected prismatic segments, where "nodes" are placed at the joints of each segment. Using good judgment, the number of segments selected for this division should be reasonable. It would be conservative to select the *smallest* end of each tapered size to represent the thickness of each prismatic segment, as shown in Fig. 17–4a. Treatment of beams in this manner also applies to cases where an unusual distributed load is applied to the beam. Computer software usually accommodates uniform, triangular, and trapezoidal loadings. If a unique loading is not incorporated into the computer program, then it can be approximated by a series of segmented uniform distributed loadings, acting on joined segments of the beam. A conservative approach would be to select the highest intensity of the distributed loading within each segment, as shown in Fig. 17–5.

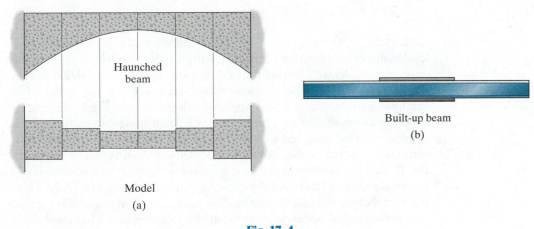

Haunched
beam

Built-up beam

(b)

Model

(a)

Fig. 17–4

Actual loading

Model of loading

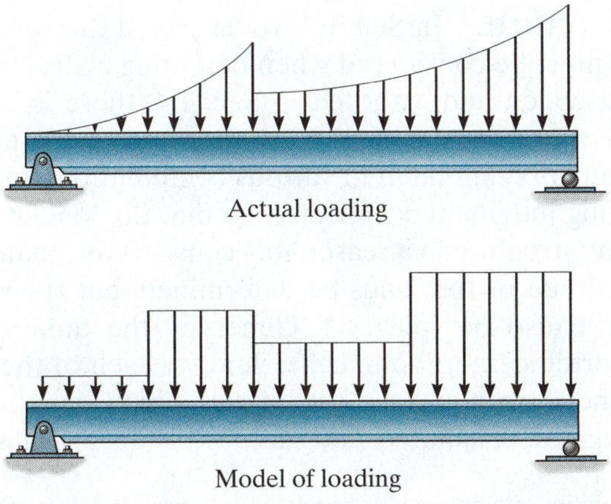

Fig. 17–5

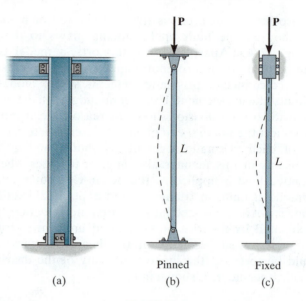

	Pinned	Fixed
(a)	(b)	(c)

Fig. 17–6

Beam column

(a)

(b)

Fig. 17–7

Columns.
As discussed in Sec. 1.2, a column will carry a compressive load in direct bearing, that is, the load will pass through the centroid of the cross section. As with beams, the end supports should be modeled so that the results will provide a conservative approach to the design. For example, if a concentric load is applied to the column in Fig. 17–6a, and its supports are modeled as a pin, Fig. 17–6b, then the cross section of the column must be designed to have a greater stiffness to prevent buckling, compared to the same column modeled as having fixed supports, Fig. 17–6c.* As another example, if a *corbel* is attached to a column it will carry an eccentric load, thereby creating a ***beam column***, Fig. 17–7a. Here the model of the column should be dimensioned so that the load is applied at a conservative distance away from the bending axis of the column's cross section, thus ensuring further safety against buckling, Fig. 17–7b.

General Structure.
In Sec. 1.3 we discussed the various dead and live loads that must be considered when designing a structure. These are all specified in codes, and as noted in Sec. 1.4, there is a trend to use probability theory to account for the uncertainty of the loads by using load factors and applying them in various combinations. Once obtained, good engineering judgment is expected, so that the loading is applied to the model of the structure in a reasonably conservative manner. Not only must the magnitude of the loads be determined, but their locations on the model must also be specified. Generally the dimensions for the model are reported centerline to centerline for each of the members. In the case of uncertainty, always use *larger dimensions*, so that larger internal loadings are calculated, and thereby produce a safe design.

*The sudden instability of a column, or buckling, is discussed in *Mechanics of Materials*, 9/e, R.C. Hibbeler, Pearson Education.

By establishing the load path one can make reasonable assumptions for load transference from one member to another. See Sec. 2.2. For example, a roof that is supported by trusses is often attached to purlins, but sometimes it is directly applied to the entire top cord of the truss, Fig. 17–8a. When this occurs, the resultant of the distribution of load between the joints can generally be divided equally and be considered as a point loading on each of the joints. Fig. 17–8b.

As a second example, consider the effect of the wind on the front of a metal building, modeled as shown in Fig. 17–9a. Cross bracing, using tie rods between each bay, prevents the building from *racking* or leaning, as shown by the dashed lines. In the actual building, the purlins would also restrain the building since they are attached to each of the frames, and the roof would also be firmly attached to the purlins. In Sec. 7.2, using approximate analysis, we have shown how to reduce a statically indeterminate cross-braced truss system to one that is statically determinate in order to simplify the analysis. For the case here, the tie rods can be conservatively designed by neglecting the support provided by the roof assembly, and assuming *only the rods in tension* provide the necessary resistance against collapse. In other words, the rods are assumed not to support a significant compressive force since their cross section is small, and so they can easily buckle. Only the four rods shown on the model in Fig. 17–9b are assumed to resist the wind loading. The other four rods provide support if the wind loading acts on the opposite side.

Actual structure

(a)

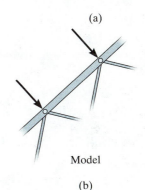

Model

(b)

Fig. 17–8

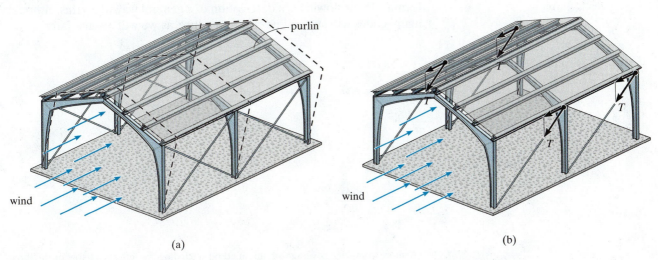

(a)

(b)

Fig. 17–9

Consideration should also be given to the material strength and stiffness of the structure. Strength properties include choosing allowable stresses that ensure the elastic limit or the ultimate stress is not exceeded, and possibly a stress limit that prevents fatigue or fracture in the case of cyclic loadings or temperature variations.

The material properties that are relevant for an elastic analysis include the modulus of elasticity, the shear modulus, Poisson's ratio, the coefficient of thermal expansion, and the specific weight or the density of the material. In particular, the stiffness of the structure is dependent upon its modulus of elasticity E. For steel, this property remains fairly constant from one specimen of steel to another, unless the material undergoes drastic changes in temperature. Care, however, must be given to the selection of E for concrete and wood, because of the variability that can occur within these materials. As time passes, all these material properties can be affected by *atmospheric corrosion*, as in the case of steel and concrete, and decay, in case of wood. As mentioned in Sec. 1.4, a resistance factor is often used to account for the variability of a material as it relates to its strength and stiffness properties.

17.3   General Application of a Structural Analysis Computer Program

Once the model of the structure is established and the load and material properties are specified, then all this data should be tabulated for use in an available computer program. The most popular structural analysis programs currently available, such as STAAD, RISA, and SAP, are all based on the stiffness method of matrix analysis, described in Chapters 13 through 15.* Although each of these programs has a slightly different interface, they all require the engineer to input the data using a specified format. The following is a description of a general way to do this, although many codes will streamline this procedure, as we will discuss later.

---

*A more complete coverage of this method, including the effects of torsion in three-dimensional frames, is given in books on matrix analysis.

Preliminary Steps.   Before using any program it is first necessary to numerically identify the members and joints, called *nodes*, of the structure and establish both global and local coordinate systems in order to specify the structure's geometry and loading. To do this, you may want to make a sketch of the structure and specify each member with a number enclosed within a square, and use a number enclosed within a circle to identify the nodes. In some programs, the "near" and "far" ends of the member must be identified. This is done using an arrow written along the member, with the head of the arrow directed toward the far end. Member, node, and "direction" identification for a plane truss, beam, and plane frame are shown in Figs. 17–10, 17–11, and 17–12. In Fig. 17–10 node ② is at the "near end" of member 4 and node ③ is at its "far end." These assignments can all be done arbitrarily. Notice, however, that the nodes on the truss are always at the joints, since this is where the loads are applied and the displacements and member forces are to be determined. For beams and frames, the nodes are at the supports, at a corner or joint, at an internal pin, or at a point where the linear or rotational displacement is to be determined, Figs. 17–11 and 17–12.

Since loads and displacements are vector quantities, it is necessary to establish a coordinate system in order to specify their correct sense of direction. Here we must use two types of coordinate systems.

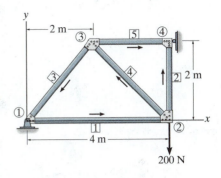

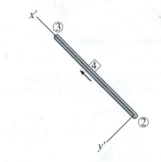

Fig. 17–10

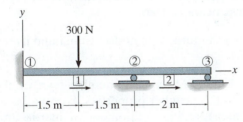

Fig. 17–11

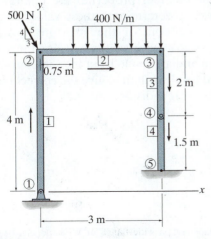

Fig. 17–12

17

Global Coordinates. A single **global** or **structure coordinate system**, using right-handed x, y, z axes, is used to specify the location of each node relative to the origin, and to identify the sense of each of the external load and displacement components at the nodes. It is convenient to locate the origin at a node so that all the other nodes have positive coordinates. See each figure.

Local Coordinates. A **local** or **member coordinate system** is used to specify the location and direction of external loadings acting on beam and frame members and for any structure, to provide a means of interpreting the computed results of internal loadings acting at the nodes of each member. This system can be identified using right-handed x', y', z' axes with the origin at the "near" node and the x' axis extending along the member toward the "far" node. Examples for truss member 4 and frame member 3 are shown in Figs. 17–10 and 17–12, respectively.

Program Operation.
When any program is executed, a menu should appear which allows various selections for inputting the data and getting the results. The following explains the items used for input data. For any problem, be sure to use a consistent set of units for numerical quantities.

General Structure Information. This item should generally be selected first in order to assign a problem title and identify the type of structure to be analyzed—truss, beam, or frame.

Node Data. Enter, in turn, each node number and its far and near end global coordinates.

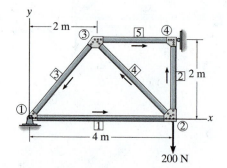

Fig. 17–10 (repeated)

Member Data. Enter, in turn, each member number, the near and far node numbers, and the member properties, E (modulus of elasticity), A (cross-sectional area), and/or I (moment of inertia and/or the polar moment of inertia or other suitable torsional constant required for three-dimensional frames*). If these member properties are unknown then provided the structure is statically determinate, these values can be set equal to one. This can also be done if the structure is statically indeterminate, provided there is no support settlement, and the members all have the same cross section and are made from the same material. In both these cases the computed results will then give the correct reactions and internal forces, but not the correct displacements.

---

*Quite often a pre-set structural shape, e.g., a wide-flange or W shape, can be selected when the program has a database of its geometric properties.

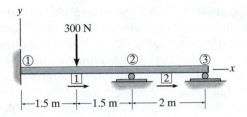

Fig. 17–11 (repeated)

If an internal hinge or pin connects two members of a beam or frame, then the release of moment must be specified at that node. For example, member 3 of the frame in Fig. 17–12 has a pin at the far node, ④. In a like manner, this pin can also be identified at the near node of member 4.

Support Data. Enter, in turn, each node located at a support, and specify the called for global coordinate directions in which restraint occurs. For example, since node ⑤ of the frame in Fig. 17–12 is a fixed support, a zero is entered for the x, y, and z (rotational) directions; however, if this support settles downward 0.003 m, then the value entered for y would be -0.003.

Load Data. Loads are specified either at nodes, or on members. Enter the algebraic values of *nodal loadings* relative to the *global coordinates*. For example, for the truss in Fig. 17–10 the loading at node ② is in the y direction and has a value of -200. For beam and frame *members* the loadings and their location are generally referenced using the *local coordinates*. For example, the distributed loading on member 2 of the frame in Fig. 17–12 is specified with an intensity of -400 N/m located 0.75 m from the near node ② and -400 N/m located 3 m from this node.

Results. Once all the data is entered, then the problem can be solved. One obtains the external reactions on the structure and the displacements and internal loadings at each node, along with a graphic of the deflected structure. As a partial check of the results a statics check is often given at each of the nodes. *It is very important that you never fully trust the results obtained. Instead, it would be wise to perform an intuitive structural analysis using one of the many classical methods discussed in the text to further check the output. After all, the structural engineer must take full responsibility for both the modeling and computing of final results.*

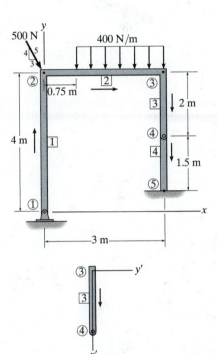

Fig. 17–12 (repeated)

Automatic Assembly.   All the above steps of structural layout, establishing the global and local coordinate systems, and specifying the load data, are sometimes *automatically* incorporated within the program. For example, one can construct a *scaled drawing* of the structure on the monitor, either by specifying the end point coordinates of each member, or by mouse clicking the global coordinates of the member's end points. Once that is done, the program will automatically establish the near and far sides of each member, along with its local coordinates. Another approach, if allowed, would be to use a drafting program, such as AutoCAD, to build the structure, and then input this graphic into the structural analysis program.

The load data can also be entered onto this graphical interface. This way the operator can visually check to be sure the load is applied to the structure in the correct direction. Specifications for the load can either be in local or global coordinates. For example, if the structure is a gabled frame, a distributed wind loading would be applied normal to the inclined roof member, and so *local coordinates* would be selected when this loading in entered, Fig. 17–13*a*. Since gravity loads, such as dead load, floor and roof live loads, and snow load, all act vertically downward, they would be entered by selecting horizontal and vertical global coordinates, Fig. 17–13*b*.

Many programs used for structural analysis and design also have a load combination feature. The engineer simply specifics the type of loading, such a dead loads, wind load, snow load, etc., and then the program will calculate each of these loadings according to the equations and requirements of the code, such as ASCE 7. Finally, the combinations of the loadings, such as those described in Sec. 1.4, are then applied to the structure and used for the analysis.

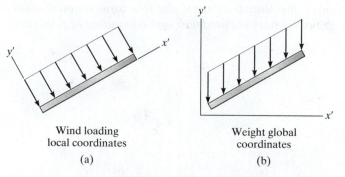

Wind loading
local coordinates

(a)

Weight global
coordinates

(b)

Fig. 17–13

COMPUTER PROBLEMS

*C17–1. Use a computer program and determine the reactions on the truss and the force in each member. *AE* is constant.

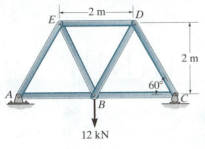

12 kN

C17–1

*C17–2. Use a computer problem and determine the reactions on the truss and the force in each member. *AE* is constant.

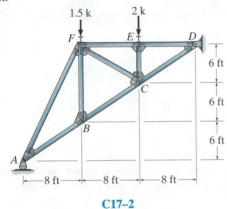

C17–2

*C17–3. Use a computer program to determine the force in member 14 of the truss. *AE* is constant.

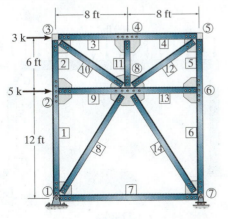

C17–3

*C17–4. Use a computer program to determine the reactions the beam. Assume *A* is fixed. *EI* is constant.

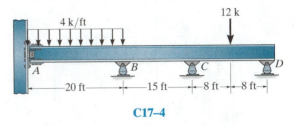

C17–4

*C17–5. Use a computer program to determine the reactions on the beam. Assume *A* and *D* are pins and *B* and *C* are rollers. *EI* is constant.

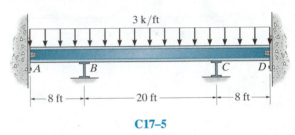

C17–5

*C17–6. Use a computer program to determine the reactions on the beam. Assume *A* is fixed. *EI* is constant.

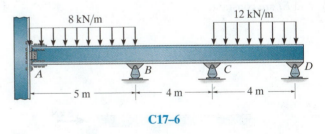

C17–6

*C17–7. Use a computer program to determine the reactions on the beam. Assume *A* and *D* are pins and *B* and *C* are rollers. *EI* is constant.

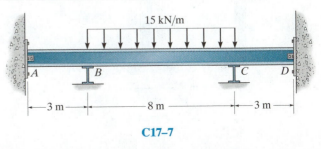

C17–7

*C17–8. Use a computer program to determine the reactions on the frame. *AE* and *EI* are constant.

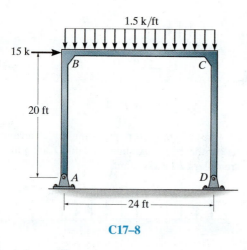

C17–8

*C17–10. Use a computer program to determine the reactions on the frame. Assume *A*, *B*, and *C* are pins. *AE* and *EI* are constant.

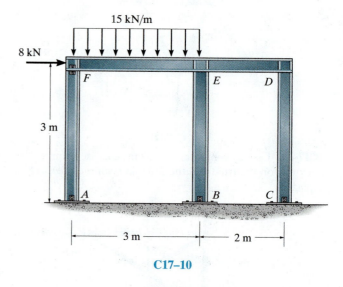

C17–10

*C17–9. Use a computer program to determine the reactions on the frame. Assume *A*, *B*, *D* and *F* are pins. *AE* and *EI* are constant.

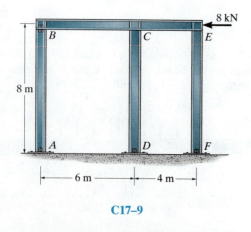

C17–9

*C17–11. Use a computer program to determine the reactions on the frame. *AE* and *EI* are constant.

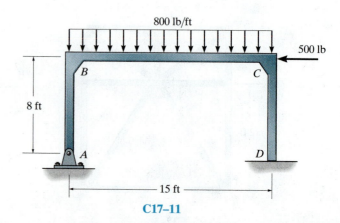

C17–11

PROJECT PROBLEMS

P17–1.   The photo shows a structural assembly designed to support medical equipment in an operating room of a hospital. It is anticipated that this equipment weighs 1 k. Each of the side beams has a cross-sectional area of 3.50 in^2 and moment of inertia of 22.0 in^4. The ends of these beams are welded to the girders as shown. The assembly is bolted to the side beams. Model one of the side beams, and justify any assumptions you have made. Perform a computer analysis and use the results to draw the moment diagram of one of the side beams. As a partial check of the results, determine the maximum moment in this beam using a classical method such as moment distribution, or use the deflection tables. Neglect the weight of the members, and take $E = 29(10^3)$ ksi.

17

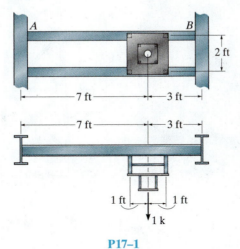

P17–1

P17–2.   The concrete bridge pier, shown in the photo, supports a portion of a highway bridge. The centerline dimensions and the anticipated loading on the cap or top beam are shown in the figure. This beam has the cross section shown. The columns each have a diameter of 400 mm and are fixed-connected to the beam. Establish a structural model of the pier and justify any assumptions you have made. Perform a computer analysis to determine the maximum moment in the beam. As a partial check of the results, determine this moment using a classical method such as moment distribution or the deflection tables. Neglect the weight of the members and the effect of the steel reinforcement within the concrete. Take $E = 29.0$ GPa.

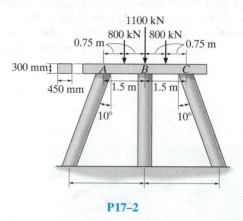

P17–2

P17–3.   The steel-trussed bent shown in the photo is used to support a portion of the pedestrian bridge. It is constructed using two wide-flange columns, each having a cross-sectional area of 4.44 in^2 and a moment of inertia of 48.0 in^4. A similar member is used at the top to support the bridge loading, estimated to be 8 k as shown in the figure. The ends of this member are welded to the columns, and the bottom of the columns are welded to base plates which in turn are bolted into the concrete. Each truss member has a cross-sectional area of 2.63 in^2. and is bolted at its ends to gusset plates. These plates are welded to the web of each column. Establish a structural model of the bent and justify any assumptions you have made. Using this model, determine the forces in the truss members and find the largest axial force in the columns using a computer program for the structural analysis. Neglect the weight of the members and use the centerline dimensions shown in the figure. Compare your results with those obtained using the method of joints to calculate the force in some of the members. Take $E = 29(10^3)$ ksi.

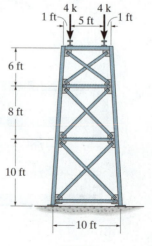

P17–3

P17–4. The load capacity of the historical Pratt truss shown in the photo is to be investigated when the load on the bottom girder AB is 600 lb/ft. A drawing of the bridge shows the centerline dimensions of the members. All the members are bolted to gusset plates. The vertical and half-diagonal members each have a cross sectional area of 3.60 in², and each of the main diagonal members and the top cord have a cross-sectional area of 5.80 in². The side girder has a cross sectional area of 13.2 in² and a moment of inertia of 350 in⁴. Establish a structural model of the bridge truss and justify any assumptions you have made. Perform a computer analysis to determine the force in each member of the truss. From the results, draw the moment diagram for the side girder. Check your results by using the method of sections to determine the force in some of the truss members. Neglect the weight of the members and take $E = 29(10^3)$ ksi.

15 ft
15 ft A B

15 ft 15 ft 15 ft 15 ft 15 ft 15 ft

P17–4

P17–5. The pavilion shown in the photo consists of an open roof supported by five trussed frames. Each of these frames consists of members CD and DE that have a cross-sectional area of $8.08(10^{-3})$ m² and a moment of inertia of $45.8(10^{-6})$ m⁴. All the other members have a cross-sectional area of $5.35(10^{-3})$ m² and moment of inertia of $4.46(10^{-6})$ m⁴. The joints are fully welded, and the columns rest on base plates that are bolted onto a concrete foundation. The wind load acting on one side of the roof and distributed to the frame is 1.20 kN/m. The centerline geometry of the structure is shown in the figure. Establish a structural model of the frame and justify any assumptions you have made. Perform a computer analysis to determine the force at each joint. Use the results and draw the moment diagram for the horizontal member AB. As a partial check of the results, calculate the tensile force in this member assuming the frame system is pin connected and acts as a truss. Neglect the weight of the members and take $E = 200$ GPa.

17

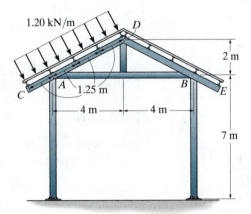

1.20 kN/m D

2 m

A 1.25 m B
C E

4 m   4 m

7 m

P17–5

Matrix Algebra for Structural Analysis

A.1 Basic Definitions and Types of Matrices

With the accessibility of desk top computers, application of matrix algebra for the analysis of structures has become widespread. Matrix algebra provides an appropriate tool for this analysis, since it is relatively easy to formulate the solution in a concise form and then perform the actual matrix manipulations using a computer. For this reason it is important that the structural engineer be familiar with the fundamental operations of this type of mathematics.

Matrix. A *matrix* is a rectangular arrangement of numbers having m rows and n columns. The numbers, which are called *elements*, are assembled within brackets. For example, the $\mathbf{A}$ matrix is written as:

$$\mathbf{A} = \begin{bmatrix} a_{11} & a_{12} & \cdots & a_{1n} \\ a_{21} & a_{22} & \cdots & a_{2n} \\ & & \vdots & \\ a_{m1} & a_{m2} & \cdots & a_{mn} \end{bmatrix}$$

Such a matrix is said to have an *order* of $m \times n$ (m by n). Notice that the first subscript for an element denotes its row position and the second subscript denotes its column position. In general, then, a_{ij} is the element located in the ith row and jth column.

Row Matrix. If the matrix consists only of elements in a single row, it is called a *row matrix*. For example, a $1 \times n$ row matrix is written as

$$\mathbf{A} = [a_1 \quad a_2 \quad \cdots \quad a_n]$$

Here only a single subscript is used to denote an element, since the row subscript is always understood to be equal to 1, that is, $a_1 = a_{11}, a_2 = a_{12}$, and so on.

Column Matrix.   A matrix with elements stacked in a single column is called a ***column matrix***. The $m \times 1$ column matrix is

$$\mathbf{A} = \begin{bmatrix} a_1 \\ a_2 \\ \vdots \\ a_m \end{bmatrix}$$

Here the subscript notation symbolizes $a_1 = a_{11}$, $a_2 = a_{21}$, and so on.

Square Matrix.   When the number of rows in a matrix equals the number of columns, the matrix is referred to as a ***square matrix***. An $n \times n$ square matrix would be

$$\mathbf{A} = \begin{bmatrix} a_{11} & a_{12} & \cdots & a_{1n} \\ a_{21} & a_{22} & \cdots & a_{2n} \\ & & \vdots & \\ a_{n1} & a_{n2} & \cdots & a_{nn} \end{bmatrix}$$

A

Diagonal Matrix.   When all the elements of a square matrix are zero except along the main diagonal, running down from left to right, the matrix is called a ***diagonal matrix***. For example,

$$\mathbf{A} = \begin{bmatrix} a_{11} & 0 & 0 \\ 0 & a_{22} & 0 \\ 0 & 0 & a_{33} \end{bmatrix}$$

Unit or Identity Matrix.   The ***unit*** or ***identity matrix*** is a diagonal matrix with all the diagonal elements equal to unity. For example,

$$\mathbf{I} = \begin{bmatrix} 1 & 0 & 0 \\ 0 & 1 & 0 \\ 0 & 0 & 1 \end{bmatrix}$$

Symmetric Matrix.   A ***square matrix*** is symmetric provided $a_{ij} = a_{ji}$. For example,

$$\mathbf{A} = \begin{bmatrix} 3 & 5 & 2 \\ 5 & -1 & 4 \\ 2 & 4 & 8 \end{bmatrix}$$

A.2   Matrix Operations

Equality of Matrices.   Matrices $\mathbf{A}$ and $\mathbf{B}$ are said to be equal if they are of the same order and each of their corresponding elements are equal, that is, $a_{ij} = b_{ij}$. For example, if

$$\mathbf{A} = \begin{bmatrix} 2 & 6 \\ 4 & -3 \end{bmatrix} \quad \mathbf{B} = \begin{bmatrix} 2 & 6 \\ 4 & -3 \end{bmatrix}$$

then $\mathbf{A} = \mathbf{B}$.

Addition and Subtraction of Matrices.   Two matrices can be added together or subtracted from one another if they are of the same order. The result is obtained by adding or subtracting the corresponding elements. For example, if

$$\mathbf{A} = \begin{bmatrix} 6 & 7 \\ 2 & -1 \end{bmatrix} \quad \mathbf{B} = \begin{bmatrix} -5 & 8 \\ 1 & 4 \end{bmatrix}$$

then

$$\mathbf{A} + \mathbf{B} = \begin{bmatrix} 1 & 15 \\ 3 & 3 \end{bmatrix} \quad \mathbf{A} - \mathbf{B} = \begin{bmatrix} 11 & -1 \\ 1 & -5 \end{bmatrix}$$

Multiplication by a Scalar.   When a matrix is multiplied by a scalar, each element of the matrix is multiplied by the scalar. For example, if

$$\mathbf{A} = \begin{bmatrix} 4 & 1 \\ 6 & -2 \end{bmatrix} \quad k = -6$$

then

$$k\mathbf{A} = \begin{bmatrix} -24 & -6 \\ -36 & 12 \end{bmatrix}$$

Matrix Multiplication.   Two matrices $\mathbf{A}$ and $\mathbf{B}$ can be multiplied together only if they are **conformable**. This condition is satisfied if the number of *columns* in $\mathbf{A}$ *equals* the number of *rows* in $\mathbf{B}$. For example, if

$$\mathbf{A} = \begin{bmatrix} a_{11} & a_{12} \\ a_{21} & a_{22} \end{bmatrix} \quad \mathbf{B} = \begin{bmatrix} b_{11} & b_{12} & b_{13} \\ b_{21} & b_{22} & b_{23} \end{bmatrix} \tag{A–1}$$

then $\mathbf{AB}$ can be determined since $\mathbf{A}$ has two columns and $\mathbf{B}$ has two rows. Notice, however, that $\mathbf{BA}$ is not possible. Why?

If matrix **A** having an order of $(m \times n)$ is multiplied by matrix **B** having an order of $(n \times q)$ it will yield a matrix **C** having an order of $(m \times q)$, that is,

$$\begin{matrix} \mathbf{A} & \mathbf{B} & = & \mathbf{C} \\ (m \times n)(n \times q) & & & (m \times q) \end{matrix}$$

The elements of matrix **C** are found using the elements a_{ij} of **A** and b_{ij} of **B** as follows:

$$c_{ij} = \sum_{k=1}^{n} a_{ik} b_{kj} \tag{A–2}$$

The methodology of this formula can be explained by a few simple examples. Consider

$$\mathbf{A} = \begin{bmatrix} 2 & 4 & 3 \\ -1 & 6 & 1 \end{bmatrix} \qquad \mathbf{B} = \begin{bmatrix} 2 \\ 6 \\ 7 \end{bmatrix}$$

By inspection, the product $\mathbf{C} = \mathbf{AB}$ is possible since the matrices are conformable, that is, **A** has three columns and **B** has three rows. By Eq. A–2, the multiplication will yield matrix **C** having two rows and one column. The results are obtained as follows:

c_{11}: Multiply the elements in the first row of **A** by corresponding elements in the column of **B** and add the results; that is,

$$c_{11} = c_1 = 2(2) + 4(6) + 3(7) = 49$$

c_{21}: Multiply the elements in the second row of **A** by corresponding elements in the column of **B** and add the results; that is,

$$c_{21} = c_2 = -1(2) + 6(6) + 1(7) = 41$$

Thus

$$\mathbf{C} = \begin{bmatrix} 49 \\ 41 \end{bmatrix}$$

As a second example, consider

$$\mathbf{A} = \begin{bmatrix} 5 & 3 \\ 4 & 1 \\ -2 & 8 \end{bmatrix} \qquad \mathbf{B} = \begin{bmatrix} 2 & 7 \\ -3 & 4 \end{bmatrix}$$

Here again the product $\mathbf{C} = \mathbf{AB}$ can be found since $\mathbf{A}$ has two columns and $\mathbf{B}$ has two rows. The resulting matrix $\mathbf{C}$ will have three rows and two columns. The elements are obtained as follows:

$c_{11} = 5(2) + 3(-3) = 1$ (first row of $\mathbf{A}$ times first column of $\mathbf{B}$)

$c_{12} = 5(7) + 3(4) = 47$ (first row of $\mathbf{A}$ times second column of $\mathbf{B}$)

$c_{21} = 4(2) + 1(-3) = 5$ (second row of $\mathbf{A}$ times first column of $\mathbf{B}$)

$c_{22} = 4(7) + 1(4) = 32$ (second row of $\mathbf{A}$ times second column of $\mathbf{B}$)

$c_{31} = -2(2) + 8(-3) = -28$ (third row of $\mathbf{A}$ times first column of $\mathbf{B}$)

$c_{32} = -2(7) + 8(4) = 18$ (third row of $\mathbf{A}$ times second column of $\mathbf{B}$)

The scheme for multiplication follows application of Eq. A–2. Thus,

$$\mathbf{C} = \begin{bmatrix} 1 & 47 \\ 5 & 32 \\ -28 & 18 \end{bmatrix}$$

Notice also that $\mathbf{BA}$ does not exist, since written in this manner the matrices are nonconformable.

The following rules apply to matrix multiplication.

1. In general the product of two matrices is not commutative:

$$\mathbf{AB} \neq \mathbf{BA} \qquad \text{(A–3)}$$

2. The distributive law is valid:

$$\mathbf{A}(\mathbf{B} + \mathbf{C}) = \mathbf{AB} + \mathbf{AC} \qquad \text{(A–4)}$$

3. The associative law is valid:

$$\mathbf{A}(\mathbf{BC}) = (\mathbf{AB})\mathbf{C} \qquad \text{(A–5)}$$

Transposed Matrix.   A matrix may be transposed by interchanging its rows and columns. For example, if

$$\mathbf{A} = \begin{bmatrix} a_{11} & a_{12} & a_{13} \\ a_{21} & a_{22} & a_{23} \\ a_{31} & a_{32} & a_{33} \end{bmatrix} \qquad \mathbf{B} = \begin{bmatrix} b_1 & b_2 & b_3 \end{bmatrix}$$

Then

$$\mathbf{A}^T = \begin{bmatrix} a_{11} & a_{21} & a_{31} \\ a_{12} & a_{22} & a_{32} \\ a_{13} & a_{23} & a_{33} \end{bmatrix} \qquad \mathbf{B}^T = \begin{bmatrix} b_1 \\ b_2 \\ b_3 \end{bmatrix}$$

Notice that $\mathbf{AB}$ is nonconformable and so the product does not exist. ($\mathbf{A}$ has three columns and $\mathbf{B}$ has one row.) Alternatively, multiplication $\mathbf{AB}^T$ is possible since here the matrices are conformable ($\mathbf{A}$ has three columns and $\mathbf{B}^T$ has three rows). The following properties for transposed matrices hold:

$$(\mathbf{A} + \mathbf{B})^T = \mathbf{A}^T + \mathbf{B}^T \tag{A-6}$$

$$(k\mathbf{A})^T = k\mathbf{A}^T \tag{A-7}$$

$$(\mathbf{AB})^T = \mathbf{B}^T\mathbf{A}^T \tag{A-8}$$

This last identity will be illustrated by example. If

$$\mathbf{A} = \begin{bmatrix} 6 & 2 \\ 1 & -3 \end{bmatrix} \qquad \mathbf{B} = \begin{bmatrix} 4 & 3 \\ 2 & 5 \end{bmatrix}$$

Then, by Eq. A–8,

$$\left(\begin{bmatrix} 6 & 2 \\ 1 & -3 \end{bmatrix} \begin{bmatrix} 4 & 3 \\ 2 & 5 \end{bmatrix} \right)^T = \begin{bmatrix} 4 & 2 \\ 3 & 5 \end{bmatrix} \begin{bmatrix} 6 & 1 \\ 2 & -3 \end{bmatrix}$$

$$\left(\begin{bmatrix} 28 & 28 \\ -2 & -12 \end{bmatrix} \right)^T = \begin{bmatrix} 28 & -2 \\ 28 & -12 \end{bmatrix}$$

$$\begin{bmatrix} 28 & -2 \\ 28 & -12 \end{bmatrix} = \begin{bmatrix} 28 & -2 \\ 28 & -12 \end{bmatrix}$$

Matrix Partitioning. A matrix can be subdivided into submatrices by partitioning. For example,

$$\mathbf{A} = \begin{bmatrix} a_{11} & a_{12} & a_{13} & a_{14} \\ a_{21} & a_{22} & a_{23} & a_{24} \\ a_{31} & a_{32} & a_{33} & a_{34} \end{bmatrix} = \begin{bmatrix} \mathbf{A}_{11} & \mathbf{A}_{12} \\ \mathbf{A}_{21} & \mathbf{A}_{22} \end{bmatrix}$$

Here the submatrices are

$$\mathbf{A}_{11} = [a_{11}] \qquad \mathbf{A}_{12} = [a_{12} \quad a_{13} \quad a_{14}]$$

$$\mathbf{A}_{21} = \begin{bmatrix} a_{21} \\ a_{31} \end{bmatrix} \qquad \mathbf{A}_{22} = \begin{bmatrix} a_{22} & a_{23} & a_{24} \\ a_{32} & a_{33} & a_{34} \end{bmatrix}$$

The rules of matrix algebra apply to partitioned matrices provided the partitioning is conformable. For example, corresponding submatrices of **A** and **B** can be added or subtracted provided they have an equal number of rows and columns. Likewise, matrix multiplication is possible provided the respective number of columns and rows of both **A** and **B** and their submatrices are equal. For instance, if

$$\mathbf{A} = \begin{bmatrix} 4 & 1 & -1 \\ -2 & 0 & -5 \\ 6 & 3 & 8 \end{bmatrix} \qquad \mathbf{B} = \begin{bmatrix} 2 & -1 \\ 0 & 8 \\ 7 & 4 \end{bmatrix}$$

then the product **AB** exists, since the number of columns of **A** equals the number of rows of **B** (three). Likewise, the partitioned matrices are conformable for multiplication since **A** is partitioned into two columns and **B** is partitioned into two rows, that is,

$$\mathbf{AB} = \begin{bmatrix} \mathbf{A}_{11} & \mathbf{A}_{12} \\ \mathbf{A}_{21} & \mathbf{A}_{22} \end{bmatrix} \begin{bmatrix} \mathbf{B}_{11} \\ \mathbf{B}_{21} \end{bmatrix} = \begin{bmatrix} \mathbf{A}_{11}\mathbf{B}_{11} + \mathbf{A}_{12}\mathbf{B}_{21} \\ \mathbf{A}_{21}\mathbf{B}_{11} + \mathbf{A}_{22}\mathbf{B}_{21} \end{bmatrix}$$

Multiplication of the submatrices yields

$$\mathbf{A}_{11}\mathbf{B}_{11} = \begin{bmatrix} 4 & 1 \\ -2 & 0 \end{bmatrix} \begin{bmatrix} 2 & -1 \\ 0 & 8 \end{bmatrix} = \begin{bmatrix} 8 & 4 \\ -4 & 2 \end{bmatrix}$$

$$\mathbf{A}_{12}\mathbf{B}_{21} = \begin{bmatrix} -1 \\ -5 \end{bmatrix} [7 \quad 4] = \begin{bmatrix} -7 & -4 \\ -35 & -20 \end{bmatrix}$$

$$\mathbf{A}_{21}\mathbf{B}_{11} = [6 \quad 3] \begin{bmatrix} 2 & -1 \\ 0 & 8 \end{bmatrix} = [12 \quad 18]$$

$$\mathbf{A}_{22}\mathbf{B}_{21} = [8][7 \quad 4] = [56 \quad 32]$$

$$\mathbf{AB} = \begin{bmatrix} \begin{bmatrix} 8 & 4 \\ -4 & 2 \end{bmatrix} + \begin{bmatrix} -7 & -4 \\ -35 & -20 \end{bmatrix} \\ [12 \quad 18] + [56 \quad 32] \end{bmatrix} = \begin{bmatrix} 1 & 0 \\ -39 & -18 \\ 68 & 50 \end{bmatrix}$$

A.3   Determinants

In the next section we will discuss how to invert a matrix. Since this operation requires an evaluation of the determinant of the matrix, we will now discuss some of the basic properties of determinants.

A *determinant* is a square array of numbers enclosed within vertical bars. For example, an nth-order determinant, having n rows and n columns, is

$$|A| = \begin{vmatrix} a_{11} & a_{12} & \cdots & a_{1n} \\ a_{21} & a_{22} & \cdots & a_{2n} \\ & & \vdots & \\ a_{n1} & a_{n2} & \cdots & a_{nn} \end{vmatrix} \qquad (A\text{–}9)$$

Evaluation of this determinant leads to a single numerical value which can be determined using **Laplace's expansion**. This method makes use of the determinant's minors and cofactors. Specifically, each element a_{ij} of a determinant of nth order has a **minor** M_{ij} which is a determinant of order $n - 1$. This determinant (minor) remains when the ith row and jth column in which the a_{ij} element is contained is canceled out. If the minor is multiplied by $(-1)^{i+j}$ it is called the cofactor of a_{ij} and is denoted as

$$C_{ij} = (-1)^{i+j} M_{ij} \qquad \text{(A-10)}$$

For example, consider the third-order determinant

$$\begin{vmatrix} a_{11} & a_{12} & a_{13} \\ a_{21} & a_{22} & a_{23} \\ a_{31} & a_{32} & a_{33} \end{vmatrix}$$

The cofactors for the elements in the first row are

$$C_{11} = (-1)^{1+1} \begin{vmatrix} a_{22} & a_{23} \\ a_{32} & a_{33} \end{vmatrix} = \begin{vmatrix} a_{22} & a_{23} \\ a_{32} & a_{33} \end{vmatrix}$$

$$C_{12} = (-1)^{1+2} \begin{vmatrix} a_{21} & a_{23} \\ a_{31} & a_{33} \end{vmatrix} = -\begin{vmatrix} a_{21} & a_{23} \\ a_{31} & a_{33} \end{vmatrix}$$

$$C_{13} = (-1)^{1+3} \begin{vmatrix} a_{21} & a_{22} \\ a_{31} & a_{32} \end{vmatrix} = \begin{vmatrix} a_{21} & a_{22} \\ a_{31} & a_{32} \end{vmatrix}$$

Laplace's expansion for a determinant of order n, Eq. A–9, states that the numerical value represented by the determinant is equal to the sum of the products of the elements of any row or column and their respective cofactors, i.e.,

$$D = a_{i1}C_{i1} + a_{i2}C_{i2} + \cdots + a_{in}C_{in} \qquad (i = 1, 2, \ldots, \text{or } n)$$

or $\qquad\qquad\qquad\qquad\qquad\qquad\qquad\qquad\qquad\qquad$ (A–11)

$$D = a_{1j}C_{1j} + a_{2j}C_{2j} + \cdots + a_{nj}C_{nj} \qquad (j = 1, 2, \ldots, \text{or } n)$$

For application, it is seen that due to the cofactors the number D is defined in terms of n determinants (cofactors) of order $n - 1$ each. These determinants can each be reevaluated using the same formula, whereby one must then evaluate $(n - 1)$ determinants of order $(n - 2)$, and so on. The process of evaluation continues until the remaining determinants to be evaluated reduce to the second order, whereby the cofactors of the elements are single elements of D. Consider, for example, the following second-order determinant

$$D = \begin{vmatrix} 3 & 5 \\ -1 & 2 \end{vmatrix}$$

We can evaluate D along the top row of elements, which yields

$$D = 3(-1)^{1+1}(2) + 5(-1)^{1+2}(-1) = 11$$

Or, for example, using the second column of elements, we have

$$D = 5(-1)^{1+2}(-1) + 2(-1)^{2+2}(3) = 11$$

Rather than using Eqs. A–11, it is perhaps easier to realize that the evaluation of a second-order determinant can be performed by multiplying the elements of the diagonal, from top left down to right, and subtract from this the product of the elements from top right down to left, i.e., follow the arrow,

$$D = \begin{vmatrix} 3 & 5 \\ -1 & 2 \end{vmatrix} = 3(2) - 5(-1) = 11$$

Consider next the third-order determinant

$$|D| = \begin{vmatrix} 1 & 3 & -1 \\ 4 & 2 & 6 \\ -1 & 0 & 2 \end{vmatrix}$$

Using Eq. A–11, we can evaluate $|D|$ using the elements either along the top row or the first column, that is

$$D = (1)(-1)^{1+1}\begin{vmatrix} 2 & 6 \\ 0 & 2 \end{vmatrix} + (3)(-1)^{1+2}\begin{vmatrix} 4 & 6 \\ -1 & 2 \end{vmatrix} + (-1)(-1)^{1+3}\begin{vmatrix} 4 & 2 \\ -1 & 0 \end{vmatrix}$$

$$= 1(4 - 0) - 3(8 + 6) - 1(0 + 2) = -40$$

$$D = 1(-1)^{1+1}\begin{vmatrix} 2 & 6 \\ 0 & 2 \end{vmatrix} + 4(-1)^{2+1}\begin{vmatrix} 3 & -1 \\ 0 & 2 \end{vmatrix} + (-1)(-1)^{3+1}\begin{vmatrix} 3 & -1 \\ 2 & 6 \end{vmatrix}$$

$$= 1(4 - 0) - 4(6 - 0) - 1(18 + 2) = -40$$

As an exercise try to evaluate $|D|$ using the elements along the second row.

A.4   Inverse of a Matrix

Consider the following set of three linear equations:

$$a_{11}x_1 + a_{12}x_2 + a_{13}x_3 = c_1$$

$$a_{21}x_1 + a_{22}x_2 + a_{23}x_3 = c_2$$

$$a_{31}x_1 + a_{32}x_2 + a_{33}x_3 = c_3$$

which can be written in matrix form as

$$\begin{bmatrix} a_{11} & a_{12} & a_{13} \\ a_{21} & a_{22} & a_{23} \\ a_{31} & a_{32} & a_{33} \end{bmatrix}\begin{bmatrix} x_1 \\ x_2 \\ x_3 \end{bmatrix} = \begin{bmatrix} c_1 \\ c_2 \\ c_3 \end{bmatrix} \qquad (A-12)$$

$$\mathbf{Ax} = \mathbf{C} \qquad (A-13)$$

One would think that a solution for x could be determined by dividing $\mathbf{C}$ by $\mathbf{A}$; however, division is not possible in matrix algebra. Instead, one multiplies by the inverse of the matrix. The **inverse of the matrix** $\mathbf{A}$ is another matrix of the same order and symbolically written as $\mathbf{A}^{-1}$. It has the following property,

$$\mathbf{AA}^{-1} = \mathbf{A}^{-1}\mathbf{A} = \mathbf{I}$$

where $\mathbf{I}$ is an identity matrix. Multiplying both sides of Eq. A–13 by $\mathbf{A}^{-1}$, we obtain

$$\mathbf{A}^{-1}\mathbf{Ax} = \mathbf{A}^{-1}\mathbf{C}$$

Since $\mathbf{A}^{-1}\mathbf{Ax} = \mathbf{Ix} = \mathbf{x}$, we have

$$\mathbf{x} = \mathbf{A}^{-1}\mathbf{C} \tag{A–14}$$

Provided $\mathbf{A}^{-1}$ can be obtained, a solution for $\mathbf{x}$ is possible.

For hand calculation the method used to formulate $\mathbf{A}^{-1}$ can be developed using Cramer's rule. The development will not be given here; instead, only the results are given.* In this regard, the elements in the matrices of Eq. A–14 can be written as

$$\mathbf{x} = \mathbf{A}^{-1}\mathbf{C}$$

$$\begin{bmatrix} x_1 \\ x_2 \\ x_3 \end{bmatrix} = \frac{1}{|A|} \begin{bmatrix} C_{11} & C_{21} & C_{31} \\ C_{12} & C_{22} & C_{32} \\ C_{13} & C_{23} & C_{33} \end{bmatrix} \begin{bmatrix} c_1 \\ c_2 \\ c_3 \end{bmatrix} \tag{A–15}$$

Here $|A|$ is an evaluation of the determinant of the coefficient matrix $\mathbf{A}$, which is determined using the Laplace expansion discussed in Sec. A.3. The square matrix containing the cofactors C_{ij} is called the **adjoint matrix**. By comparison it can be seen that the inverse matrix $\mathbf{A}^{-1}$ is obtained from $\mathbf{A}$ by first replacing each element a_{ij} by its cofactor C_{ij}, then transposing the resulting matrix, yielding the adjoint matrix, and finally multiplying the adjoint matrix by $1/|A|$.

To illustrate how to obtain $\mathbf{A}^{-1}$ numerically, we will consider the solution of the following set of linear equations:

$$\begin{aligned} x_1 - x_2 + x_3 &= -1 \\ -x_1 + x_2 + x_3 &= -1 \\ x_1 + 2x_2 - 2x_3 &= 5 \end{aligned} \tag{A–16}$$

Here

$$\mathbf{A} = \begin{bmatrix} 1 & -1 & 1 \\ -1 & 1 & 1 \\ 1 & 2 & -2 \end{bmatrix}$$

*See Kreyszig, E., *Advanced Engineering Mathematics*, John Wiley & Sons, Inc., New York.

The cofactor matrix for **A** is

$$
\mathbf{C} = \begin{bmatrix}
\begin{vmatrix} 1 & 1 \\ 2 & -2 \end{vmatrix} & -\begin{vmatrix} -1 & 1 \\ 1 & -2 \end{vmatrix} & \begin{vmatrix} -1 & 1 \\ 1 & 2 \end{vmatrix} \\[2mm]
-\begin{vmatrix} -1 & 1 \\ 2 & -2 \end{vmatrix} & \begin{vmatrix} 1 & 1 \\ 1 & -2 \end{vmatrix} & -\begin{vmatrix} 1 & -1 \\ 1 & 2 \end{vmatrix} \\[2mm]
\begin{vmatrix} -1 & 1 \\ 1 & 1 \end{vmatrix} & -\begin{vmatrix} 1 & 1 \\ -1 & 1 \end{vmatrix} & \begin{vmatrix} 1 & -1 \\ -1 & 1 \end{vmatrix}
\end{bmatrix}
$$

Evaluating the determinants and taking the transpose, the adjoint matrix is

$$
\mathbf{C}^T = \begin{bmatrix} -4 & 0 & -2 \\ -1 & -3 & -2 \\ -3 & -3 & 0 \end{bmatrix}
$$

Since

$$
A = \begin{vmatrix} 1 & -1 & 1 \\ -1 & 1 & 1 \\ 1 & 2 & -2 \end{vmatrix} = -6
$$

The inverse of **A** is, therefore,

$$
\mathbf{A}^{-1} = -\frac{1}{6} \begin{bmatrix} -4 & 0 & -2 \\ -1 & -3 & -2 \\ -3 & -3 & 0 \end{bmatrix}
$$

Solution of Eqs. A–16 yields

$$
\begin{bmatrix} x_1 \\ x_2 \\ x_3 \end{bmatrix} = -\frac{1}{6} \begin{bmatrix} -4 & 0 & -2 \\ -1 & -3 & -2 \\ -3 & -3 & 0 \end{bmatrix} \begin{bmatrix} -1 \\ -1 \\ 5 \end{bmatrix}
$$

$$
x_1 = -\tfrac{1}{6}\big[(-4)(-1) + 0(-1) + (-2)(5)\big] = 1
$$

$$
x_2 = -\tfrac{1}{6}\big[(-1)(-1) + (-3)(-1) + (-2)(5)\big] = 1
$$

$$
x_3 = -\tfrac{1}{6}\big[(-3)(-1) + (-3)(-1) + (0)(5)\big] = -1
$$

Obviously, the numerical calculations are quite expanded for larger sets of equations. For this reason, computers are used in structural analysis to determine the inverse of matrices.

A.5   The Gauss Method for Solving Simultaneous Equations

When many simultaneous linear equations have to be solved, the Gauss elimination method may be used because of its numerical efficiency. Application of this method requires solving one of a set of n equations for an unknown, say x_1, in terms of all the other unknowns, $x_2, x_3, \ldots, x_n$. Substituting this so-called *pivotal equation* into the remaining equations leaves a set of $n - 1$ equations with $n - 1$ unknowns. Repeating the process by solving one of these equations for x_2 in terms of the $n - 2$ remaining unknowns $x_3, x_4, \ldots, x_n$ forms the second pivotal equation. This equation is then substituted into the other equations, leaving a set of $n - 3$ equations with $n - 3$ unknowns. The process is repeated until one is left with a pivotal equation having one unknown, which is then solved. The other unknowns are then determined by successive back substitution into the other pivotal equations. To improve the accuracy of solution, when developing each pivotal equation one should always select the equation of the set having the *largest* numerical coefficient for the unknown one is trying to eliminate. The process will now be illustrated by an example.

Solve the following set of equations using Gauss elimination:

$$-2x_1 + 8x_2 + 2x_3 = 2 \qquad \text{(A–17)}$$

$$2x_1 - x_2 + x_3 = 2 \qquad \text{(A–18)}$$

$$4x_1 - 5x_2 + 3x_3 = 4 \qquad \text{(A–19)}$$

We will begin by eliminating x_1. The largest coefficient of x_1 is in Eq. A–19; hence, we will take it to be the pivotal equation. Solving for x_1, we have

$$x_1 = 1 + 1.25x_2 - 0.75x_3 \qquad \text{(A–20)}$$

Substituting into Eqs. A–17 and A–18 and simplifying yields

$$2.75x_2 + 1.75x_3 = 2 \qquad \text{(A–21)}$$

$$1.5x_2 - 0.5x_3 = 0 \qquad \text{(A–22)}$$

Next we eliminate x_2. Choosing Eq. A–21 for the pivotal equation since the coefficient of x_2 is largest here, we have

$$x_2 = 0.727 - 0.636x_3 \qquad \text{(A–23)}$$

Substituting this equation into Eq. A–22 and simplifying yields the final pivotal equation, which can be solved for x_3. This yields $x_3 = 0.75$. Substituting this value into the pivotal Eq. A–23 gives $x_2 = 0.25$. Finally, from pivotal Eq. A–20 we get $x_1 = 0.75$.

A

PROBLEMS

A–1. If $\mathbf{A} = \begin{bmatrix} 4 & 2 & -3 \\ 6 & 1 & 5 \end{bmatrix}$ and $\mathbf{B} = \begin{bmatrix} 4 & -1 & 0 \\ 2 & 0 & 8 \end{bmatrix}$, determine $\mathbf{A} + \mathbf{B}$ and $\mathbf{A} - 2\mathbf{B}$.

A–2. If $\mathbf{A} = \begin{bmatrix} 3 & 5 \\ -2 & 7 \end{bmatrix}$, determine $\mathbf{A} + \mathbf{A}^T$.

A–3. If $\mathbf{A} = [6 \quad 1 \quad 3]$ and $\mathbf{B} = [1 \quad 6 \quad 3]$ show that $(\mathbf{A} + \mathbf{B})^T = \mathbf{A}^T + \mathbf{B}^T$.

A–4. If $\mathbf{A} = \begin{bmatrix} 1 \\ 0 \\ 5 \end{bmatrix}$ and $\mathbf{B} = [2 \quad -1 \quad 3]$, determine $\mathbf{AB}$.

A–5. If $\mathbf{A} = \begin{bmatrix} 6 & 2 & 2 \\ -5 & 1 & 1 \\ 0 & 3 & 1 \end{bmatrix}$ and $\mathbf{B} = \begin{bmatrix} -1 & 3 & 1 \\ 2 & -5 & 1 \\ 0 & 7 & 5 \end{bmatrix}$, determine $\mathbf{AB}$.

A–6. Determine $\mathbf{BA}$ for the matrices of Prob. A–5.

A–7. If $\mathbf{A} = \begin{bmatrix} 5 & 7 \\ -2 & 1 \end{bmatrix}$ and $\mathbf{B} = \begin{bmatrix} 6 \\ 7 \end{bmatrix}$, determine $\mathbf{AB}$.

A–8. If $\mathbf{A} = \begin{bmatrix} 1 & 8 & 4 \\ 1 & 2 & 3 \end{bmatrix}$ and $\mathbf{B} = \begin{bmatrix} 3 \\ 2 \\ -6 \end{bmatrix}$, determine $\mathbf{AB}$.

A–9. If $\mathbf{A} = \begin{bmatrix} 2 & 7 & 3 \\ -2 & 1 & 0 \end{bmatrix}$ and $\mathbf{B} = \begin{bmatrix} 6 \\ 9 \\ -1 \end{bmatrix}$, determine $\mathbf{AB}$.

A–10. If $\mathbf{A} = \begin{bmatrix} 6 & 4 & 2 \\ 2 & 1 & 1 \\ 0 & -3 & 1 \end{bmatrix}$ and $\mathbf{B} = \begin{bmatrix} -1 & 3 & -2 \\ 2 & 4 & 1 \\ 0 & 7 & 5 \end{bmatrix}$, determine $\mathbf{AB}$.

A

A–11. If $A = \begin{bmatrix} 2 & 5 \\ 1 & 3 \end{bmatrix}$, determine AA^T.

A–12. Show that the distributive law is valid, i.e., $A(B + C) = AB + AC$, if

$$A = \begin{bmatrix} 2 & 1 & 6 \\ 4 & 5 & 3 \end{bmatrix}, B = \begin{bmatrix} 3 \\ 1 \\ -6 \end{bmatrix}, C = \begin{bmatrix} 5 \\ -1 \\ 2 \end{bmatrix}.$$

A–13. Show that the associative law is valid, i.e., $A(BC) = (AB)C$, if

$$A = \begin{bmatrix} 2 & 1 & 6 \\ 4 & 5 & 3 \end{bmatrix}, B = \begin{bmatrix} 3 \\ 1 \\ -6 \end{bmatrix}, C = [5 \quad -1 \quad 2].$$

A–14. Evaluate the determinants $\begin{vmatrix} 2 & 5 \\ 7 & 1 \end{vmatrix}$ and $\begin{vmatrix} 1 & 3 & 5 \\ 2 & 7 & 1 \\ 3 & 8 & 6 \end{vmatrix}$.

A–15. If $A = \begin{bmatrix} 5 & 1 \\ 3 & -2 \end{bmatrix}$, determine A^{-1}.

A–16. If $A = \begin{bmatrix} 0 & 1 & 3 \\ 2 & 5 & 0 \\ 1 & -1 & 2 \end{bmatrix}$, determine A^{-1}.

A–17. Solve the equations $-x_1 + 4x_2 + x_3 = 1$, $2x_1 - x_2 + x_3 = 2$, and $4x_1 - 5x_2 + 3x_3 = 4$ using the matrix equation $X = A^{-1}C$.

A–18. Solve the equations in Prob. A–17 using the Gauss elimination method.

A–19. Solve the equations $x_1 - x_2 + x_3 = -1$, $-x_1 + x_2 + x_3 = -1$, and $x_1 + 2x_2 - 2x_3 = 5$ using the matrix equation $X = A^{-1}B$.

A–20. Solve the equations in Prob. A–19 using the Gauss elimination method.

Fundamental Solutions

Chapter 2

F2–1. $\zeta + \Sigma M_A = 0;\quad 60 - F_{BC}\left(\frac{3}{5}\right)(4) = 0\quad F_{BC} = 25.0$ kN

$\zeta + \Sigma M_B = 0;\quad 60 - A_y(4) = 0$

$A_y = 15.0$ kN *Ans.*

$\xrightarrow{+} \Sigma F_x = 0;\quad A_x - 25.0\left(\frac{4}{5}\right) = 0$

$A_x = 20.0$ kN *Ans.*

$B_x = C_x = 25.0\left(\frac{4}{5}\right) = 20.0$ kN

$B_y = C_y = 25.0\left(\frac{3}{5}\right) = 15.0$ kN *Ans.*

F2–2. $\zeta + \Sigma M_A = 0;\quad F_{BC}\sin 45°(4) - 10(4)(2) = 0$

$F_{BC} = \dfrac{20}{\sin 45°}$ kN

$\zeta + \Sigma M_B = 0;\quad 10(4)(2) - A_y(4) = 0$

$A_y = 20.0$ kN *Ans.*

$\xrightarrow{+} \Sigma F_x = 0;\quad A_x - \left(\dfrac{20}{\sin 45°}\right)(\cos 45°) = 0$

$A_x = 20.0$ kN *Ans.*

$B_x = C_x = \left(\dfrac{20}{\sin 45°}\right)(\cos 45°) = 20.0$ kN *Ans.*

$B_y = C_y = \left(\dfrac{20}{\sin 45°}\right)(\sin 45°) = 20.0$ kN *Ans.*

F2–3. $\zeta + \Sigma M_A = 0;\quad F_{BC}\sin 60°(4) - 10(2)(1) = 0$

$F_{BC} = \dfrac{5}{\sin 60°}$ kN

$\zeta + \Sigma M_B = 0;\quad 10(2)(3) - A_y(4) = 0$

$A_y = 15.0$ kN *Ans.*

$\xrightarrow{+} \Sigma F_x = 0;\quad \left(\dfrac{5}{\sin 60°}\right)(\cos 60°) - A_x = 0$

$A_x = 2.89$ kN *Ans.*

$B_x = C_x = \left(\dfrac{5}{\sin 60°}\right)(\cos 60°) = 2.89$ kN *Ans.*

$B_y = C_y = \left(\dfrac{5}{\sin 60°}\right)(\sin 60°) = 5.00$ kN *Ans.*

F2–4. Member AC

$\zeta + \Sigma M_C = 0;\quad 10(3) - N_A(4) = 0\quad N_A = 7.50$ kN *Ans.*

$\zeta + \Sigma M_A = 0;\quad C_y(4) - 10(1) = 0\quad C_y = 2.50$ kN

Member BC

$\xrightarrow{+} \Sigma F_x = 0;\quad B_x = 0$ *Ans.*

$+\uparrow \Sigma F_y = 0;\quad B_y - 2.50 - 8(2) = 0\quad B_y = 18.5$ kN *Ans.*

$\zeta + \Sigma M_B = 0;\quad 2.50(2) + 8(2)(1) - M_B = 0$

$M_B = 21.0$ kN · m *Ans.*

F2–5. $\zeta + \Sigma M_A = 0;\quad F_{BC}\left(\frac{3}{5}\right)(4) + F_{BC}\left(\frac{4}{5}\right)(3) - 300(2) = 0$

$F_{BC} = 125$ lb

$\xrightarrow{+} \Sigma F_x = 0;\quad A_x - 125\left(\frac{4}{5}\right) = 0\quad A_x = 100$ lb *Ans.*

$+\uparrow \Sigma F_y = 0;\quad A_y + 125\left(\frac{3}{5}\right) - 300 = 0$

$A_y = 225$ lb *Ans.*

$B_x = C_x = 125\left(\frac{4}{5}\right) = 100$ lb *Ans.*

$B_y = C_y = 125\left(\frac{3}{5}\right) = 75.0$ lb *Ans.*

F2–6. $\zeta + \Sigma M_C = 0;\quad 6(2) + 2(2) - N_A(4) = 0$

$N_A = 4.00$ kN *Ans.*

$\xrightarrow{+} \Sigma F_x = 0;\quad C_x - 2 = 0\quad C_x = 2.00$ kN *Ans.*

$+\uparrow \Sigma F_y = 0;\quad C_y + 4.00 - 6 = 0\quad C_y = 2.00$ kN *Ans.*

F2–7.

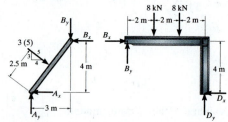

Member AB

$\zeta + \Sigma M_A = 0;\quad B_x(4) - B_y(3) - 3(5)(2.5) = 0$

Member BCD

$\zeta + \Sigma M_D = 0;\quad 8(2) + 8(4) - B_x(4) - B_y(6) = 0$

$B_x = 10.25$ kN $\quad B_y = 1.167$ kN $= 1.17$ kN *Ans.*

Member AB

$\xrightarrow{+} \Sigma F_x = 0;\quad -A_x + 3(5)\left(\frac{4}{5}\right) - 10.25 = 0$

$A_x = 1.75$ kN *Ans.*

$+\uparrow \Sigma F_y = 0;\quad A_y - (3)(5)\left(\frac{3}{5}\right) - 1.167 = 0$

$A_y = 10.167$ kN $= 10.2$ kN *Ans.*

Member BCD

$\xrightarrow{+} \Sigma F_x = 0;\quad 10.25 - D_x = 0\quad D_x = 10.25$ kN *Ans.*

$+\uparrow \Sigma F_y = 0;\quad D_y + 1.167 - 8 - 8 = 0$

$D_y = 14.833$ kN $= 14.8$ kN *Ans.*

F2–8.

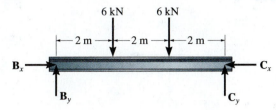

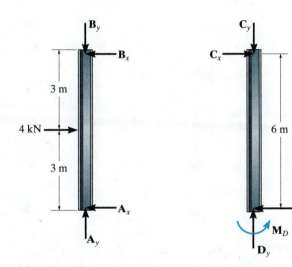

F2–9.

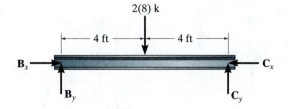

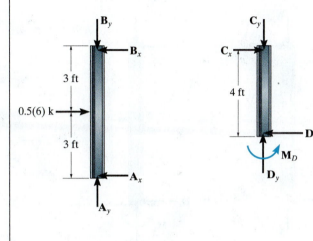

Member AB

$\zeta + \Sigma M_A = 0$; $B_x(6) - 4(3) = 0$ $B_x = 2.00$ kN   *Ans.*

$\zeta + \Sigma M_B = 0$; $4(3) - A_x(6) = 0$ $A_x = 2.00$ kN   *Ans.*

Member BC

$\xrightarrow{+} \Sigma F_x = 0$; $2.00 - C_x = 0$ $C_x = 2.00$ kN   *Ans.*

$\zeta + \Sigma M_C = 0$; $6(2) + 6(4) - B_y(6) = 0$ $B_y = 6.00$ kN *Ans.*

$\zeta + \Sigma M_B = 0$; $C_y(6) - 6(2) - 6(4) = 0$ $C_y = 6.00$ kN *Ans.*

Member AB

$+ \uparrow \Sigma F_y = 0$; $A_y - 6.00 = 0$ $A_y = 6.00$ kN   *Ans.*

Member CD

$\xrightarrow{+} \Sigma F_x = 0$; $2.00 - D_x = 0$ $D_x = 2.00$ kN   *Ans.*

$+ \uparrow \Sigma F_y = 0$; $D_y - 6.00 = 0$ $D_y = 6.00$ kN   *Ans.*

$\zeta + \Sigma M_D = 0$; $M_D - 2.00(6) = 0$ $M_D = 12.0$ kN · m *Ans.*

Member AB

$\zeta + \Sigma M_A = 0$; $B_x(6) - 0.5(6)(3) = 0$ $B_x = 1.50$ k   *Ans.*

$\zeta + \Sigma M_B = 0$; $0.5(6)(3) - A_x(6) = 0$ $A_x = 1.50$ k   *Ans.*

Member BC

$\zeta + \Sigma M_C = 0$; $2(8)(4) - B_y(8) = 0$ $B_y = 8.00$ k   *Ans.*

$\zeta + \Sigma M_B = 0$; $C_y(8) - 2(8)(4) = 0$ $C_y = 8.00$ k   *Ans.*

$\xrightarrow{+} \Sigma F_x = 0$; $1.50 - C_x = 0$ $C_x = 1.50$ k   *Ans.*

Member AB

$+ \uparrow \Sigma F_y = 0$; $A_y - 8.00 = 0$ $A_y = 8.00$ k   *Ans.*

Member CD

$\xrightarrow{+} \Sigma F_x = 0$; $1.50 - D_x = 0$ $D_x = 1.50$ k   *Ans.*

$+ \uparrow \Sigma F_y = 0$; $D_y - 8.00 = 0$ $D_y = 8.00$ k   *Ans.*

$\zeta + \Sigma M_D = 0$; $M_D - 1.50(4) = 0$ $M_D = 6.00$ k · ft *Ans.*

F2–10.

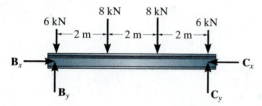

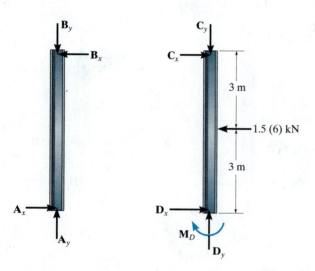

Member BC

$\zeta + \Sigma M_B = 0;$ $C_y(6) - 8(2) - 8(4) - 6(6) = 0$

$C_y = 14.0$ kN *Ans.*

$\zeta + \Sigma M_C = 0;$ $8(2) + 8(4) + 6(6) - B_y(6) = 0$

$B_y = 14.0$ kN *Ans.*

Member AB

$\zeta + \Sigma M_A = 0;$ $B_x = 0$ *Ans.*

$\overset{+}{\rightarrow} \Sigma F_x = 0;$ $A_x = 0$ *Ans.*

$+\uparrow \Sigma F_y = 0;$ $A_y - 14.0 = 0$ $A_y = 14.0$ kN *Ans.*

Member BC

$\overset{+}{\rightarrow} \Sigma F_x = 0;$ $C_x = 0$ *Ans.*

Member CD

$\overset{+}{\rightarrow} \Sigma F_x = 0;$ $D_x - 1.5(6) = 0$ $D_x = 9.00$ kN *Ans.*

$+\uparrow \Sigma F_y = 0;$ $D_y - 14.0 = 0$ $D_y = 14.0$ kN *Ans.*

$\zeta + \Sigma M_D = 0;$ $1.5(6)(3) - M_D = 0$ $M_D = 27.0$ kN · m *Ans.*

Chapter 3

F3–1. Joint *C*

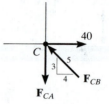

$\overset{+}{\rightarrow} \Sigma F_x = 0;$ $40 - F_{CB}\left(\frac{4}{5}\right) = 0$ $F_{CB} = 50.0$ kN (C) *Ans.*

$+\uparrow \Sigma F_y = 0;$ $50.0\left(\frac{3}{5}\right) - F_{CA} = 0$ $F_{CA} = 30.0$ kN (T) *Ans.*

Joint *B*

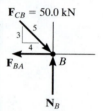

$\overset{+}{\rightarrow} \Sigma F_x = 0;$ $50.0\left(\frac{4}{5}\right) - F_{BA} = 0$ $F_{BA} = 40.0$ kN (T) *Ans.*

$+\uparrow \Sigma F_y = 0;$ $N_B - 50.0\left(\frac{3}{5}\right) = 0$ $N_B = 30.0$ kN

F3–2. Joint *B*

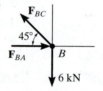

$+\uparrow \Sigma F_y = 0;$ $F_{BC} \sin 45° - 6 = 0$

$F_{BC} = 8.485$ kN (T) = 8.49 kN (T) *Ans.*

$\overset{+}{\rightarrow} \Sigma F_x = 0;$ $F_{BA} - 8.485 \cos 45° = 0$

$F_{BA} = 6.00$ kN (C) *Ans.*

Joint *C*

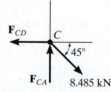

$\overset{+}{\rightarrow} \Sigma F_x = 0;$ $8.485 \cos 45° - F_{CD} = 0$

$F_{CD} = 6.00$ kN (T) *Ans.*

$+\uparrow \Sigma F_y = 0;$ $F_{CA} - 8.485 \sin 45° = 0$

$F_{CA} = 6.00$ kN (C) *Ans.*

F3–3. Joint C

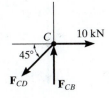

$\xrightarrow{+} \Sigma F_x = 0;$ $10 - F_{CD} \cos 45° = 0$

$F_{CD} = 14.14 \text{ kN (T)} = 14.1 \text{ kN (T)}$ *Ans.*

$+\uparrow \Sigma F_y = 0;$ $F_{CB} - 14.14 \sin 45° = 0$

$F_{CB} = 10.0 \text{ kN (C)}$ *Ans.*

Joint D

$+\nearrow \Sigma F_{x'} = 0;$ $14.14 - F_{DA} = 0$

$F_{DA} = 14.14 \text{ kN (T)} = 14.1 \text{ kN (T)}$ *Ans.*

$\nwarrow + \Sigma F_{y'} = 0;$ $F_{DB} = 0$ *Ans.*

Joint B

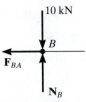

$\xrightarrow{+} \Sigma F_x = 0;$ $F_{BA} = 0$ *Ans.*

$+\uparrow \Sigma F_y = 0;$ $N_B - 10.0 = 0$ $N_B = 10.0 \text{ kN}$

F3–4. Joint D

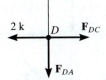

$\xrightarrow{+} \Sigma F_x = 0;$ $F_{DC} - 2 = 0$ $F_{DC} = 2.00 \text{ k (T)}$ *Ans.*

$+\uparrow \Sigma F_y = 0;$ $F_{DA} = 0$ *Ans.*

Joint C

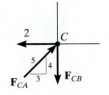

$\xrightarrow{+} \Sigma F_x = 0;$ $F_{CA}\left(\frac{3}{5}\right) - 2 = 0$

$F_{CA} = 3.333 \text{ k (C)} = 3.33 \text{ k (C)}$ *Ans.*

$+\uparrow \Sigma F_y = 0;$ $3.333\left(\frac{4}{5}\right) - F_{CB} = 0$

$F_{CB} = 2.667 \text{ k (T)} = 2.67 \text{ k (T)}$ *Ans.*

Joint A

$\xrightarrow{+} \Sigma F_x = 0;$ $F_{AB} - 3.333\left(\frac{3}{5}\right) = 0$ $F_{AB} = 2.00 \text{ k (T)}$ *Ans.*

$+\uparrow \Sigma F_y = 0;$ $N_A - 3.333\left(\frac{4}{5}\right) = 0$ $N_A = 2.667 \text{ k}$

F3–5. Joint D

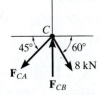

$\xrightarrow{+} \Sigma F_x = 0;$ $F_{DC} = 0$ *Ans.*

$+\uparrow \Sigma F_y = 0;$ $F_{DA} = 0$ *Ans.*

Joint C

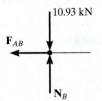

$\xrightarrow{+} \Sigma F_x = 0;$ $8 \cos 60° - F_{CA} \cos 45° = 0$

$F_{CA} = 5.657 \text{ kN (T)} = 5.66 \text{ kN (T)}$ *Ans.*

$+\uparrow \Sigma F_y = 0;$ $F_{CB} - 5.657 \sin 45° - 8 \sin 60° = 0$

$F_{CB} = 10.93 \text{ kN (C)} = 10.9 \text{ kN (C)}$ *Ans.*

Joint B

$\xrightarrow{+} \Sigma F_x = 0;$ $F_{AB} = 0$ *Ans.*

$+\uparrow \Sigma F_y = 0;$ $N_B = 10.93 \text{ kN}$

F3–6.   Entire truss

$\zeta + \Sigma M_A = 0$; $E_y(8) - 600(2) - 800(4) - 600(6) = 0$

$E_y = 1000$ N

Joint E

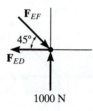

$+\uparrow \Sigma F_y = 0$; $1000 - F_{EF} \sin 45° = 0$

$F_{EF} = 1414.21$ N (C) $= 1.41$ kN (C)   *Ans.*

$\xrightarrow{+} \Sigma F_x = 0$; $1414.21 \cos 45° - F_{ED} = 0$

$F_{ED} = 1000$ N (T) $= 1.00$ kN (T)   *Ans.*

Joint F

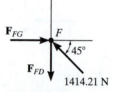

$\xrightarrow{+} \Sigma F_x = 0$; $F_{FG} - 1414.21 \cos 45° = 0$

$F_{FG} = 1000$ N (C) $= 1.00$ kN (C)   *Ans.*

$+\uparrow \Sigma F_y = 0$; $1414.21 \sin 45° - F_{FD} = 0$

$F_{FD} = 1000$ N (T) $= 1.00$ kN (T)   *Ans.*

Joint D

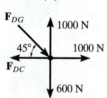

$+\uparrow \Sigma F_y = 0$; $1000 - 600 - F_{DG} \sin 45° = 0$

$F_{DG} = 565.69$ N (C) $= 566$ N (C)   *Ans.*

$\xrightarrow{+} \Sigma F_x = 0$; $1000 + 565.69 \cos 45° - F_{DC} = 0$

$F_{DC} = 1400$ N (T) $= 1.40$ kN (T)   *Ans.*

Joint C

$+\uparrow \Sigma F_y = 0$; $F_{CG} - 800 = 0$ $F_{CG} = 800$ N (T)   *Ans.*

Due to symmetry,

$F_{BC} = F_{DC} = 1.40$ kN (T) $F_{BG} = F_{DG} = 566$ N (C)

$F_{HG} = F_{FG} = 1.00$ kN (C)   *Ans.*

$F_{HB} = F_{FD} = 1.00$ kN (T) $F_{AH} = F_{EF} = 1.41$ kN (C)

$F_{AB} = F_{ED} = 1.00$ kN (T)   *Ans.*

F3–7.   For the entire truss

$\zeta + \Sigma M_E = 0$; $2(5) + 2(10) + 2(15) - A_y(20) = 0$

$A_y = 3.00$ k

$\xrightarrow{+} \Sigma F_x = 0$; $A_x = 0$

For the left segment

$+\uparrow \Sigma F_y = 0$; $3.00 - 2 - F_{BG} \sin 45° = 0$

$F_{BG} = 1.41$ k (C)   *Ans.*

$\zeta + \Sigma M_B = 0$; $F_{HG}(5) - 3(5) = 0$

$F_{HG} = 3.00$ k (C)   *Ans.*

$\zeta + \Sigma M_G = 0$; $F_{BC}(5) + 2(5) - 3.00(10) = 0$

$F_{BC} = 4.00$ k (T)   *Ans.*

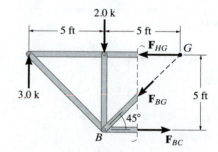

F3–8.   For the entire truss

$\zeta + \Sigma M_E = 0$; $600(16) + 600(12) + 600(8) + 600(4)$
$$- A_y(16) = 0$$

$A_y = 1500$ lb
$\overset{+}{\rightarrow} \Sigma F_x = 0$; $A_x = 0$

For the left segment

$\zeta + \Sigma M_C = 0$; $F_{HI}(3) + 600(4) + 600(8) - 1500(8) = 0$

$F_{HI} = 1600$ lb (C)

$\zeta + \Sigma M_I = 0$; $F_{BC}(3) + 600(4) - 1500(4) = 0$

$F_{BC} = 1200$ lb (T)   *Ans.*

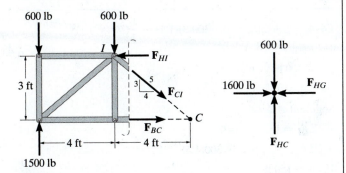

Joint *H*

$\overset{+}{\rightarrow} \Sigma F_x = 0$; $1600 - F_{HG} = 0$ $F_{HG} = 1600$ lb (C)   *Ans.*
$+\uparrow \Sigma F_y = 0$; $F_{HC} - 600 = 0$ $F_{HC} = 600$ lb (C)   *Ans.*

F3–9.   For the entire truss

$\zeta + \Sigma M_A = 0$; $N_C(4) - 8(2) - 6(2) = 0$ $N_C = 7.00$ kN

Consider the right segment

$+\uparrow \Sigma F_y = 0$; $7.00 - F_{BD} \sin 45° = 0$

$F_{BD} = 9.899$ kN (T) $= 9.90$ kN (T)   *Ans.*

$\zeta + \Sigma M_B = 0$; $7.00(2) - 6(2) - F_{ED}(2) = 0$

$F_{ED} = 1.00$ kN (C)   *Ans.*

$\zeta + \Sigma M_D = 0$; $0 - F_{BC}(2) = 0$ $F_{BC} = 0$ *Ans.*

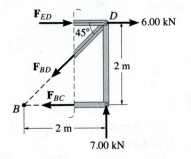

F3–10.   For the entire truss

$\zeta + \Sigma M_A = 0$; $N_E(32) - 400(8) - 400(16) - 400(24)$
$$- 400(32) = 0$$

$N_E = 1000$ lb

Consider the right segment

$\zeta + \Sigma M_E = 0$; $400(8) - F_{CF}\left(\frac{3}{5}\right)(16) = 0$

$F_{CF} = 333.33$ lb (C) $= 333$ lb (C)   *Ans.*

$\zeta + \Sigma M_C = 0$; $1000(16) - 400(16) - 400(8)$
$$- F_{GF}\left(\frac{3}{5}\right)(16) = 0$$

$F_{GF} = 666.67$ lb (C) $= 667$ lb (C)   *Ans.*

$\zeta + \Sigma M_F = 0$; $1000(8) - 400(8) - F_{CD}(6) = 0$

$F_{CD} = 800$ lb (T)   *Ans.*

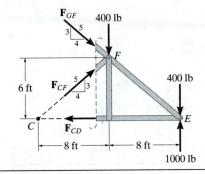

F3–11.   For the entire truss

$\zeta + \Sigma M_A = 0$; $N_D(6) - 2(6) - 4(3) = 0$ $N_D = 4.00$ kN

Consider the right segment

$+\uparrow \Sigma F_y = 0$; $4.00 - 2 - F_{FC} \sin 45° = 0$

$F_{FC} = 2.828$ kN (C) $= 2.83$ kN (C)   *Ans.*

$\zeta + \Sigma M_F = 0$; $4.00(3) - 2(3) - F_{BC}(1.5) = 0$

$F_{BC} = 4.00$ kN (T)   *Ans.*

$\zeta + \Sigma M_C = 0$; $4.00(1.5) - 2(1.5) - F_{FE}(1.5) = 0$

$F_{FE} = 2.00$ kN (C)   *Ans.*

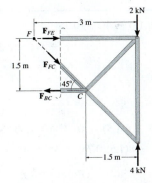

F3–12. For the entire truss

$\zeta + \Sigma M_A = 0;$ $N_E(16) - 500(4) - 500(8) - 500(12) = 0$

$N_E = 750$ lb

Consider the right segment

$\zeta + \Sigma M_F = 0;$ $750(4) - F_{CD}(3) = 0$

$F_{CD} = 1000$ lb (T) *Ans.*

$\zeta + \Sigma M_C = 0;$ $750(8) - 500(4) - F_{GF}\left(\dfrac{1}{\sqrt{17}}\right)(16) = 0$

$F_{GF} = 1030.78$ lb $= 1.03$ k (C) *Ans.*

$\zeta + \Sigma M_O = 0;$ $F_{CF}\left(\dfrac{3}{5}\right)(16) + 500(12) - 750(8) = 0$

$F_{CF} = 0$ *Ans.*

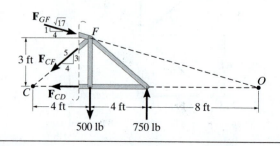

500 lb   750 lb

Chapter 4

F4–1. $\zeta + \Sigma M_A = 0;$ $B_y(2) + 20 - 10(4) = 0$

$B_y = 10.0$ kN

Segment CB

$\xrightarrow{+} \Sigma F_x = 0;$ $N_C = 0$ *Ans.*

$+\uparrow \Sigma F_y = 0;$ $V_C + 10 - 10 = 0$

$V_C = 0$ *Ans.*

$\zeta + \Sigma M_C = 0;$ $-M_C + 10(1) - 10(3) = 0$

$M_C = -20$ kN $\cdot$ m *Ans.*

F4–2. $\zeta + \Sigma M_A = 0;$ $B_y(3) - 4(1.5)(0.75)$

$- 8(1.5)(2.25) = 0$

$B_y = 10.5$ kN

Segment CB

$\xrightarrow{+} \Sigma F_x = 0;$ $N_C = 0$ *Ans.*

$+\uparrow \Sigma F_y = 0;$ $V_C + 10.5 - 8(1.5) = 0$ $V_C = 1.50$ kN *Ans.*

$\zeta + \Sigma M_C = 0;$ $10.5(1.5) - 8(1.5)(0.75) - M_C = 0$

$M_C = 6.75$ kN $\cdot$ m *Ans.*

F4–3. $\zeta + \Sigma M_B = 0;$ $\dfrac{1}{2}(6)(6)(3) - A_y(6) = 0$

$A_y = 9.00$ kN

$\xrightarrow{+} \Sigma F_x = 0;$ $A_x = 0$

Segment AC

$\xrightarrow{+} \Sigma F_x = 0;$ $N_C = 0$ *Ans.*

$+\uparrow \Sigma F_y = 0;$ $9.00 - \dfrac{1}{2}(3)(1.5) - V_C = 0$

$V_C = 6.75$ kN *Ans.*

$\zeta + \Sigma M_C = 0;$ $M_C + \dfrac{1}{2}(3)(1.5)(0.5) - 9.00(1.5) = 0$

$M_C = 12.4$ kN $\cdot$ m *Ans.*

F4–4. $\zeta + \Sigma M_B = 0;$ $300(3)(1.5) - \dfrac{1}{2}(300)(3)(1)$

$- A_y(3) = 0$

$A_y = 300$ lb

$\xrightarrow{+} \Sigma F_x = 0;$ $A_x = 0$

Segment AC

$\xrightarrow{+} \Sigma F_x = 0;$ $N_C = 0$ *Ans.*

$+\uparrow \Sigma F_y = 0;$ $300 - 300(1.5) - V_C = 0$

$V_C = -150$ lb *Ans.*

$\zeta \Sigma M_C = 0;$ $M_C + 300(1.5)(0.75) - 300(1.5) = 0$

$M_C = 112.5$ lb *Ans.*

F4–5. Reactions

$\zeta + \Sigma M_A = 0;$ $F_B \sin 45°(3) - 5(6)(3) = 0$ $F_B = 42.43$ kN

$\xrightarrow{+} \Sigma F_x = 0;$ $42.43 \cos 45° - A_x = 0$ $A_x = 30.0$ kN

$+\uparrow \Sigma F_y = 0;$ $42.43 \sin 45° - 5(6) - A_y = 0$ $A_y = 0$

Segment AC

$\xrightarrow{+} \Sigma F_x = 0;$ $N_C - 30.0 = 0$ $N_C = 30.0$ kN *Ans.*

$+\uparrow \Sigma F_y = 0;$ $-5(1.5) - V_C = 0$ $V_C = -7.50$ kN *Ans.*

$\zeta + \Sigma M_C = 0;$ $M_C + 5(1.5)(0.75) = 0$

$M_C = -5.625$ kN $\cdot$ m *Ans.*

F4–6. Reactions

$\zeta + \Sigma M_A = 0;$ $B_y(15) - 150(9)(10.5) - 600(6) - 800(3) = 0$

$B_y = 1345$ lb

Segment CB

$\xrightarrow{+} \Sigma F_x = 0;$ $N_C = 0$ *Ans.*

$+\uparrow \Sigma F_y = 0;$ $V_C + 1345 - 150(6) = 0$ $V_C = -445$ lb *Ans.*

$\zeta + \Sigma M_C = 0;$ $1345(6) - 150(6)(3) - M_C = 0$

$M_C = 5370$ lb $\cdot$ ft $= 5.37$ k $\cdot$ ft *Ans.*

F4–7.   Left segment

$$+\uparrow\Sigma F_y = 0; \quad -6 - \frac{1}{2}\left(\frac{18}{3}x\right)(x) - V = 0$$

$$V = \left\{-3x^2 - 6\right\} \text{ kN} \qquad\qquad Ans.$$

$$\zeta + \Sigma M_O = 0; \quad M + \frac{1}{2}\left(\frac{18}{3}x\right)(x)\left(\frac{x}{3}\right) + 6x = 0$$

$$M = \left\{-x^3 - 6x\right\} \text{ kN} \cdot \text{m} \qquad Ans.$$

F4–8.   Reaction

$$\zeta + \Sigma M_B = 0; \quad \frac{1}{2}(12)(6)(2) - A_y(6) = 0 \quad A_y = 12.0 \text{ kN}$$

Left segment

$$+\uparrow\Sigma F_y = 0; \quad 12.0 - \frac{1}{2}\left(\frac{12}{6}x\right)(x) - V = 0$$

$$V = \left\{12.0 - x^2\right\} \text{ kN} \qquad\qquad Ans.$$

$$\zeta + \Sigma M_O = 0; \quad M + \frac{1}{2}\left(\frac{12}{6}x\right)(x)\left(\frac{x}{3}\right) - 12.0x = 0$$

$$M = \left\{12.0x - \frac{1}{3}x^3\right\} \text{ kN} \cdot \text{m} \qquad Ans.$$

F4–9.   Reactions

$$\zeta + \Sigma M_A = 0; \quad B_y(8) - 8(4)(6) = 0 \quad B_y = 24.0 \text{ kN}$$

$$\zeta + \Sigma M_B = 0; \quad 8(4)(2) - A_y(8) = 0 \quad A_y = 8.00 \text{ kN}$$

$0 \le x < 4$ m left segment

$$+\uparrow\Sigma F_y = 0; \quad 8.00 - V = 0 \quad V = \{8\} \text{ kN} \qquad Ans.$$

$$\zeta + \Sigma M_O = 0; \quad M - 8.00x = 0 \quad M = \{8x\} \text{ kN} \cdot \text{m} \qquad Ans.$$

4 m $< x < 8$ m right segment

$$+\uparrow\Sigma F_y = 0; \quad V + 24.0 - 8(8 - x) = 0$$

$$V = \{40 - 8x\} \text{ kN} \qquad\qquad Ans.$$

$$\zeta + \Sigma M_O = 0; \quad 24.0(8 - x) - 8(8 - x)\left(\frac{8 - x}{2}\right) - M = 0$$

$$M = \left\{-4x^2 + 40x - 64\right\} \text{ kN} \cdot \text{m} \qquad Ans.$$

F4–10. $0 \le x < 2$ m

$$+\uparrow\Sigma F_y = 0; \quad V = 0 \qquad\qquad Ans.$$

$$\zeta + \Sigma M_O = 0; \quad M + 20 = 0 \quad M = -20 \text{ kN} \cdot \text{m} \qquad Ans.$$

2 m $< x \le 4$ m

$$+\uparrow\Sigma F_y = 0; \quad -5(x - 2) - V = 0$$

$$V = \{10 - 5x\} \text{ kN} \qquad\qquad Ans.$$

$$\zeta + \Sigma M_O = 0; \quad M + 5(x - 2)\left(\frac{x - 2}{2}\right) + 15 + 20 = 0$$

$$M = \left\{-\frac{5}{2}x^2 + 10x - 45\right\} \text{ kN} \cdot \text{m} \qquad Ans.$$

F4–11.   Reactions

$$+\uparrow\Sigma F_y = 0; \quad A_y - 5(2) - 15 = 0 \quad A_y = 25.0 \text{ kN}$$

$$\zeta + \Sigma M_A = 0; \quad M_A - 5(2)(1) - 15(4) = 0$$

$$M_A = 70.0 \text{ kN} \cdot \text{m}$$

$0 \le x < 2$ m left segment

$$+\uparrow\Sigma F_y = 0; \quad 25.0 - 5x - V = 0$$

$$V = \{25 - 5x\} \text{ kN} \qquad\qquad Ans.$$

$$\zeta + \Sigma M_O = 0; \quad M + 5x\left(\frac{x}{2}\right) + 70.0 - 25.0x = 0$$

$$M = \left\{-\frac{5}{2}x^2 + 25x - 70\right\} \text{ kN} \cdot \text{m} \qquad Ans.$$

2 m $< x \le 4$ m right segment

$$+\uparrow\Sigma F_y = 0; \quad V - 15 = 0 \quad V = 15 \text{ kN} \qquad Ans.$$

$$\zeta + \Sigma M_O = 0; \quad -M - 15(4 - x) = 0$$

$$M = \{15x - 60\} \text{ kN} \cdot \text{m} \qquad Ans.$$

F4–12.   Support reactions

$$\zeta + \Sigma M_A = 0; \quad B_y(24) - 2(12)(6) - 18(12) = 0$$

$$B_y = 15.0 \text{ k}$$

$$\zeta + \Sigma M_B = 0; \quad 18(12) + 2(12)(18) - A_y(24) = 0$$

$$A_y = 27.0 \text{ k}$$

$0 \le x < 12$ ft left segment

$$+\uparrow\Sigma F_y = 0; \quad 27.0 - 2x - V = 0$$

$$V = \{27 - 2x\} \text{ k} \qquad\qquad Ans.$$

$$\zeta + \Sigma M_O = 0; \quad M + 2x\left(\frac{x}{2}\right) - 27.0x = 0$$

$$M = \left\{-x^2 + 27x\right\} \text{ k} \cdot \text{ft} \qquad Ans.$$

12 ft $< x \le 24$ ft right segment

$$+\uparrow\Sigma F_y = 0; \quad V + 15.0 = 0 \quad V = \{-15 \text{ k}\} \qquad Ans.$$

$$\zeta + \Sigma M_O = 0; \quad 15.0(24 - x) - M = 0$$

$$M = \{-15x + 360\} \text{ k} \cdot \text{ft} \qquad Ans.$$

F4–13.

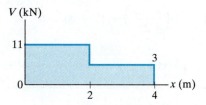

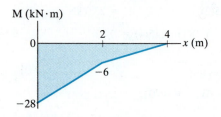

F4–14.

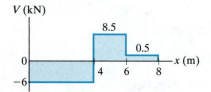

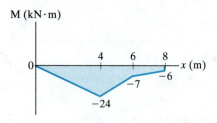

F4–15.

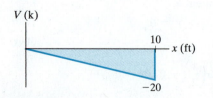

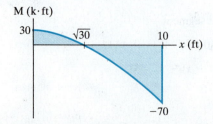

F4–16.

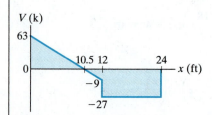

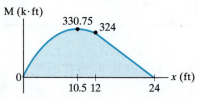

F4–17.

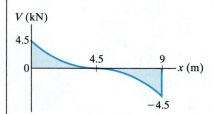

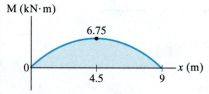

F4–18.

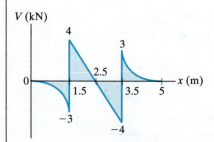

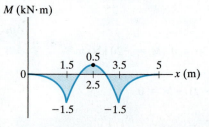

F4–19.

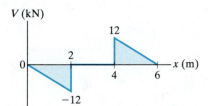

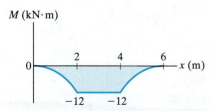

F4–20.

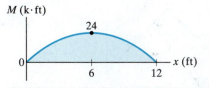

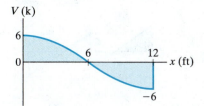

F4–21.

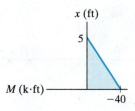

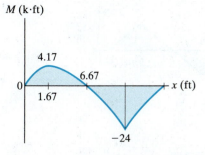

F4–22.

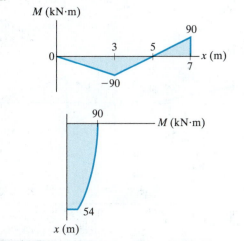

F4–23.

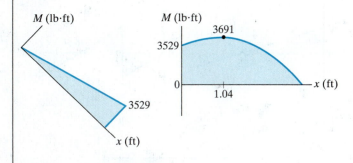

F4–24.

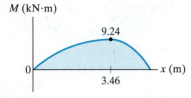

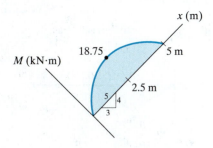

Chapter 6

F6–1.

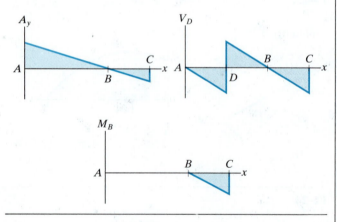

F6–2.

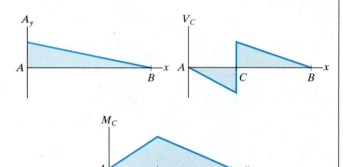

F6–3.

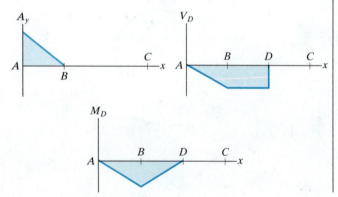

F6–4.

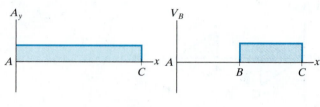

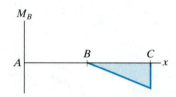

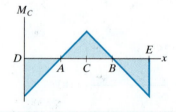

F6–5.

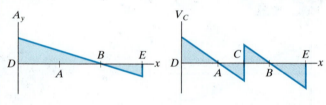

F6–6.

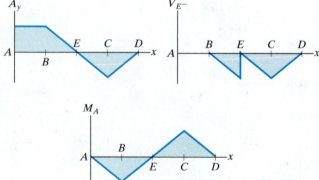

F6–7.

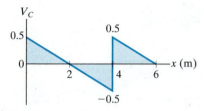

V_C

0.5

0.5

0

2   4   6 —x (m)

−0.5

M_C

1

0

2   4   6 —x (m)

−1

$$(M_C)_{max(+)} = 8(1) + \left[\frac{1}{2}(6-2)(1)\right](1.5)$$

$$+ \left[\frac{1}{2}(2)(-1)\right](2) + \left[\frac{1}{2}(6-2)(1)\right](2)$$

$$= 13.0 \text{ kN} \cdot \text{m} \qquad Ans.$$

$$(V_C)_{max(+)} = 8(0.5) + \left[\frac{1}{2}(2)(0.5)\right](1.5) + \left[\frac{1}{2}(6-4)(0.5)\right](1.5)$$

$$+ \left[\frac{1}{2}(2)(0.5)\right](2) + \left[\frac{1}{2}(4-2)(-0.5)\right](2) +$$

$$\left[\frac{1}{2}(6-4)(0.5)\right](2)$$

$$= 6.50 \text{ kN} \qquad Ans.$$

F6–8.

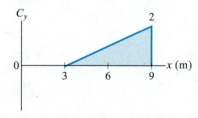

C_y

2

0

3   6   9 —x (m)

M_A

3

0

3   6   9 —x (m)

−3

a) $(C_y)_{max(+)} = 6(2) + \left[\frac{1}{2}(9-3)(2)\right](2)$

$$+ \left[\frac{1}{2}(9-3)(2)\right](4) = 48 \text{ kN} \qquad Ans.$$

b) $(M_A)_{max(-)} = 6(-3) + \left[\frac{1}{2}(6-0)(-3)\right](2)$

$$+ \left[\frac{1}{2}(6-0)(-3)\right](4) + \left[\frac{1}{2}(9-6)(3)\right](4)$$

$$= -54 \text{ kN} \cdot \text{m} \qquad Ans.$$

Chapter 8

F8–1.

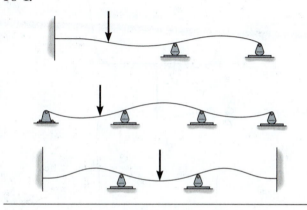

F8–2.

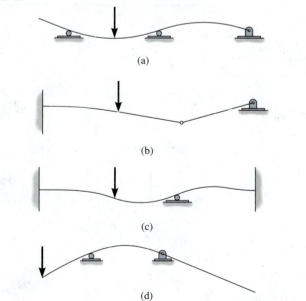

(a)

(b)

(c)

(d)

F8–3.

F8–4.

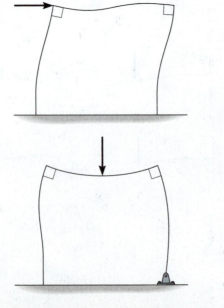

F8–5.

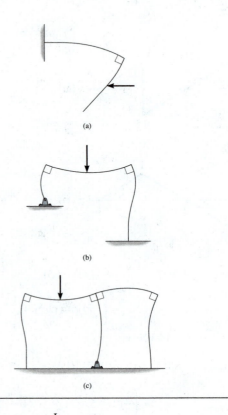

(a)

(b)

(c)

F8–6.   For $0 \leq x_1 < \dfrac{L}{2}$

$$M_1 = \frac{P}{2}x_1$$

$$EI\frac{d^2v_1}{dx_1^2} = \frac{P}{2}x_1$$

$$EI\frac{dv_1}{dx_1} = \frac{P}{4}x_1^2 + C_1 \tag{1}$$

$$EI\,v_1 = \frac{P}{12}x_1^3 + C_1x_1 + C_2 \tag{2}$$

For $\dfrac{L}{2} < x_2 \leq L$

$$M_2 = \frac{P}{2}(L - x_2) = \frac{PL}{2} - \frac{P}{2}x_2$$

$$EI\frac{d^2v_2}{dx_2^2} = \frac{PL}{2} - \frac{P}{2}x_2$$

$$EI\frac{dv_2}{dx_2} = \frac{PL}{2}x_2 - \frac{P}{4}x_2^2 + C_3 \tag{3}$$

$$EI\,v_2 = \frac{PL}{4}x_2^2 - \frac{P}{12}x_2^3 + C_3x_2 + C_4 \tag{4}$$

$v_1 = 0$ at $x_1 = 0$. From Eq (2), $C_2 = 0$

$\dfrac{dv_1}{dx_1} = 0$ at $x_1 = \dfrac{L}{2}$. From Eq (1), $C_1 = -\dfrac{PL^2}{16}$

$\dfrac{dv_2}{dx_2} = 0$ at $x_2 = \dfrac{L}{2}$. From Eq (3), $C_3 = -\dfrac{3PL^2}{16}$

$v_2 = 0$ at $x_2 = L$. From Eq (4), $C_4 = \dfrac{PL^3}{48}$

$v_1 = \dfrac{Px_1}{48EI}(4x_1^2 - 3L^2)$ *Ans.*

$v_2 = \dfrac{P}{48EI}(-4x_2^3 + 12Lx_2^2 - 9L^2x_2 + L^3)$ *Ans.*

F8–7. $M = Px - PL$

$EI\dfrac{d^2v}{dx^2} = Px - PL$

$EI\dfrac{dv}{dx} = \dfrac{P}{2}x^2 - PLx + C_1$ (1)

$EI\,v = \dfrac{P}{6}x^3 - \dfrac{PL}{2}x^2 + C_1x + C_2$ (2)

$\dfrac{dv}{dx} = 0$ at $x = 0$. From Eq (1), $C_1 = 0$

$v = 0$ at $x = 0$. From Eq (2), $C_2 = 0$

$v = \dfrac{Px}{6EI}(x^2 - 3Lx)$ *Ans.*

F8–8. $M = M_0 - \dfrac{M_0}{L}x$

$EI\dfrac{d^2v}{dx^2} = M_0 - \dfrac{M_0}{L}x$

$EI\dfrac{dv}{dx} = M_0x - \dfrac{M_0}{2L}x^2 + C_1$

$EI\,v = \dfrac{M_0}{2}x^2 - \dfrac{M_0}{6L}x^3 + C_1x + C_2$ (1)

$v = 0$ at $x = 0$. From Eq (1), $C_2 = 0$

$v = 0$ at $x = L$. From Eq (1), $C_1 = -\dfrac{M_0L}{3}$

$v = \dfrac{M_0}{6EIL}(-x^3 + 3Lx^2 - 2L^2x)$ *Ans.*

F8–9. For $0 \le x_1 < \dfrac{L}{2}$

$M = -\dfrac{M_0}{L}x_1$

$EI\dfrac{d^2v_1}{dx_1^2} = -\dfrac{M_0}{L}x_1$

$EI\dfrac{dv_1}{dx_1} = -\dfrac{M_0}{2L}x_1^2 + C_1$ (1)

$EIv_1 = -\dfrac{M_0}{6L}x_1^3 + C_1x_1 + C_2$ (2)

For $\dfrac{L}{2} < x_2 \le L$

$M = M_0 - \dfrac{M_0}{L}x_2$

$EI\dfrac{d^2v_2}{dy_2^2} = M_0 - \dfrac{M_0}{L}x_2$

$EI\dfrac{dv_2}{dx_2} = M_0x_2 - \dfrac{M_0}{2L}x_2^2 + C_3$ (3)

$EIv_2 = \dfrac{M_0}{2}x_2^2 - \dfrac{M_0}{6L}x_2^3 + C_3x_2 + C_4$ (4)

$v_1 = 0$ at $x_1 = 0$. From Eq (2), $C_2 = 0$

$v_2 = 0$ at $x_2 = L$. From Eq (4),

$0 = C_3L + C_4 + \dfrac{M_0L^2}{3}$ (5)

$\dfrac{dv_1}{dx_1} = \dfrac{dv_2}{dx_2}$ at $x_1 = x_2 = \dfrac{L}{2}$. From Eqs (1)

and (3), $C_1 - C_3 = \dfrac{M_0L}{2}$ (6)

$v_1 = v_2$ at $x_1 = x_2 = \dfrac{L}{2}$. From Eqs (2) and (4),

$C_1L - C_3L - 2C_4 = \dfrac{M_0L^2}{4}$ (7)

Solving Eqs (5), (6) and (7)

$C_4 = \dfrac{M_0L^2}{8}$ $C_3 = -\dfrac{11M_0L}{24}$ $C_1 = \dfrac{M_0L}{24}$

$v_1 = \dfrac{M_0}{24EIL}(-4x_1^3 + L^2x_1)$ *Ans.*

$v_2 = \dfrac{M_0}{24EIL}(-4x_2^3 + 12Lx_2^2 - 11L^2x_2 + 3L^3)$ *Ans.*

F8–10. $M = -\dfrac{w}{2}x^2 + wLx - \dfrac{wL^2}{2}$

$$EI\dfrac{d^2v}{dx^2} = -\dfrac{w}{2}x^2 + wLx - \dfrac{wL^2}{2}$$

$$EI\dfrac{dv}{dx} = -\dfrac{w}{6}x^3 + \dfrac{wL}{2}x^2 - \dfrac{wL^2}{2}x + C_1 \qquad (1)$$

$$EI\,v = -\dfrac{w}{24}x^4 + \dfrac{wL}{6}x^3 - \dfrac{wL^2}{4}x^2 + C_1 x + C_2 \qquad (2)$$

$\dfrac{dv}{dx} = 0$ at $x = 0$. From Eq (1) $C_1 = 0$

$v_1 = 0$ at $x = 0$. From Eq (2) $C_2 = 0$

$$v = \dfrac{w}{24EI}\left(-x^4 + 4Lx^3 - 6L^2x^2\right) \qquad \textit{Ans.}$$

F8–11. $M = -\dfrac{w_0}{6L}x^3$

$$EI\dfrac{d^2v}{dx^2} = -\dfrac{w_0}{6L}x^3$$

$$EI\dfrac{dv}{dx} = -\dfrac{w_0}{24L}x^4 + C_1 \qquad (1)$$

$$EI\,v = -\dfrac{w_0}{120L}x^5 + C_1 x + C_2 \qquad (2)$$

$\dfrac{dv}{dx} = 0$ at $x = L$. From Eq (1), $C_1 = \dfrac{w_0 L^3}{24}$

$v = 0$ at $x = L$. From Eq (2), $C_2 = -\dfrac{w_0 L^4}{30}$

$$v = \dfrac{w_0}{120EIL}\left(-x^5 + 5L^4x - 4L^5\right) \qquad \textit{Ans.}$$

F8–12.

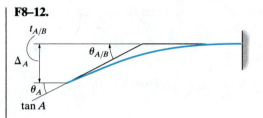

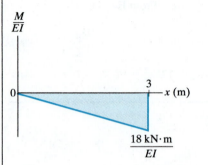

$$\theta_A = \left|\theta_{A/B}\right| = \left|\dfrac{1}{2}\left(\dfrac{-18\ \text{kN}\cdot\text{m}}{EI}\right)(3\ \text{m})\right| = \dfrac{27\ \text{kN}\cdot\text{m}^2}{EI} \qquad \nearrow \textit{Ans.}$$

$$\Delta_A = \left|t_{A/B}\right| = \left|\left[\dfrac{1}{2}\left(\dfrac{-18\ \text{kN}\cdot\text{m}}{EI}\right)(3\ \text{m})\right]\left[\dfrac{2}{3}(3\ \text{m})\right]\right|$$

$$= \dfrac{54\ \text{kN}\cdot\text{m}^3}{EI}\downarrow \qquad \textit{Ans.}$$

F8–13.

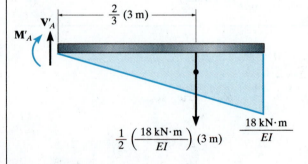

$$+\uparrow\Sigma F_y = 0;\quad V'_A - \dfrac{1}{2}\left(\dfrac{18\ \text{kN}\cdot\text{m}}{EI}\right)(3\ \text{m}) = 0$$

$$\theta_A = \dfrac{27\ \text{kN}\cdot\text{m}^2}{EI} \quad \nearrow \qquad \textit{Ans.}$$

$$\zeta +\Sigma M_A = 0;\quad -M'_A - \left[\dfrac{1}{2}\left(\dfrac{18\ \text{kN}\cdot\text{m}}{EI}\right)(3\ \text{m})\right]\left[\dfrac{2}{3}(3\ \text{m})\right] = 0$$

$$M'_A = \Delta_A = -\dfrac{54\ \text{kN}\cdot\text{m}^3}{EI} = \dfrac{54\ \text{kN}\cdot\text{m}^3}{EI}\downarrow \qquad \textit{Ans.}$$

F8–14.

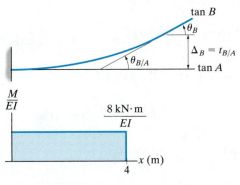

$$\theta_B = |\theta_{B/A}| = \left(\frac{8 \text{ kN} \cdot \text{m}}{EI}\right)(4 \text{ m}) = \frac{32 \text{ kN} \cdot \text{m}^2}{EI} \quad \text{⟋} \qquad Ans.$$

$$\Delta_B = |t_{B/A}| = \left[\left(\frac{8 \text{ kN} \cdot \text{m}}{EI}\right)(4 \text{ m})\right]\left[\frac{1}{2}(4 \text{ m})\right]$$

$$= \frac{64 \text{ kN} \cdot \text{m}^3}{EI}\uparrow \qquad Ans.$$

F8–15.

$$\left(\frac{8 \text{ kN} \cdot \text{m}}{EI}\right)(4 \text{ m})$$

$$+\uparrow\Sigma F_y = 0; \quad \left(\frac{8 \text{ kN} \cdot \text{m}}{EI}\right)(4 \text{ m}) - V'_B = 0$$

$$\theta_B = \frac{32 \text{ kN} \cdot \text{m}^2}{EI} \quad \text{⟋} \qquad Ans.$$

$$\zeta +\Sigma M_B = 0; \quad M'_B - \left[\left(\frac{8 \text{ kN} \cdot \text{m}}{EI}\right)(4 \text{ m})\right](2 \text{ m}) = 0$$

$$M'_B = \Delta_B = \frac{64 \text{ kN} \cdot \text{m}^3}{EI}\uparrow \qquad Ans.$$

F8–16.

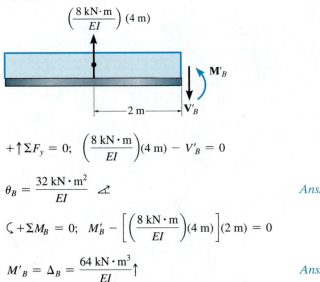

$$t_{B/A} = \left[\frac{1}{2}\left(\frac{5 \text{ kN} \cdot \text{m}}{EI}\right)(3 \text{ m})\right]\left[\frac{2}{3}(3 \text{ m})\right] = \frac{15 \text{ kN} \cdot \text{m}^3}{EI}$$

$$t_{C/A} = \left[\frac{1}{2}\left(\frac{2.5 \text{ kN} \cdot \text{m}}{EI}\right)(1.5 \text{ m})\right]\left[\frac{2}{3}(1.5 \text{ m})\right]$$

$$\quad + \left[\left(\frac{2.5 \text{ kN} \cdot \text{m}}{EI}\right)(1.5 \text{ m})\right]\left[\frac{1}{2}(1.5 \text{ m})\right]$$

$$= \frac{4.6875 \text{ kN} \cdot \text{m}^3}{EI}$$

$$\Delta' = \frac{1}{2}t_{B/A} = \frac{1}{2}\left(\frac{15 \text{ kN} \cdot \text{m}^3}{EI}\right) = \frac{7.5 \text{ kN} \cdot \text{m}^3}{EI}$$

$$\theta_A = \frac{|t_{B/A}|}{L_{AB}} = \frac{15 \text{ kN} \cdot \text{m}^3/EI}{3 \text{ m}} = \frac{5 \text{ kN} \cdot \text{m}^2}{EI} \quad \text{⟍} \qquad Ans.$$

$$\Delta_C = \Delta' - t_{C/A} = \frac{7.5 \text{ kN} \cdot \text{m}^3}{EI} - \frac{4.6875 \text{ kN} \cdot \text{m}^3}{EI}$$

$$= \frac{2.81 \text{ kN} \cdot \text{m}^3}{EI}\downarrow \qquad Ans.$$

F8–17.

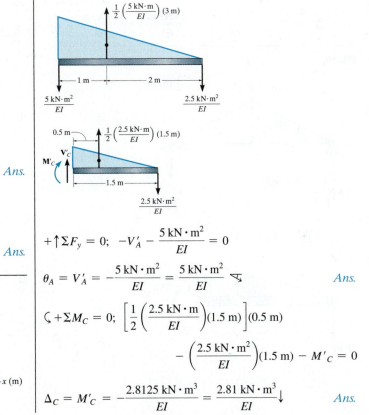

$$+\uparrow\Sigma F_y = 0; \quad -V'_A - \frac{5 \text{ kN} \cdot \text{m}^2}{EI} = 0$$

$$\theta_A = V'_A = -\frac{5 \text{ kN} \cdot \text{m}^2}{EI} = \frac{5 \text{ kN} \cdot \text{m}^2}{EI} \quad \text{⟍} \qquad Ans.$$

$$\zeta +\Sigma M_C = 0; \quad \left[\frac{1}{2}\left(\frac{2.5 \text{ kN} \cdot \text{m}}{EI}\right)(1.5 \text{ m})\right](0.5 \text{ m})$$

$$\quad - \left(\frac{2.5 \text{ kN} \cdot \text{m}^2}{EI}\right)(1.5 \text{ m}) - M'_C = 0$$

$$\Delta_C = M'_C = -\frac{2.8125 \text{ kN} \cdot \text{m}^3}{EI} = \frac{2.81 \text{ kN} \cdot \text{m}^3}{EI}\downarrow \qquad Ans.$$

F8–18.

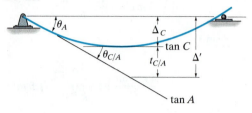

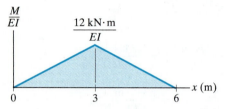

$$\theta_A = \theta_{C/A} = \frac{1}{2}\left(\frac{12\ \text{kN} \cdot \text{m}}{EI}\right)(3\ \text{m}) = \frac{18\ \text{kN} \cdot \text{m}^2}{EI} \ \ \ \ \text{Ans.}$$

$$t_{C/A} = \left[\frac{1}{2}\left(\frac{12\ \text{kN} \cdot \text{m}}{EI}\right)(3\ \text{m})\right]\left[\frac{1}{3}(3\ \text{m})\right] = \frac{18\ \text{kN} \cdot \text{m}^3}{EI}$$

$$\Delta' = \theta_A L_{AC} = \left(\frac{18\ \text{kN} \cdot \text{m}^2}{EI}\right)(3\ \text{m}) = \frac{54\ \text{kN} \cdot \text{m}^3}{EI}$$

$$\Delta_C = \Delta' - t_{C/A} = \frac{54\ \text{kN} \cdot \text{m}^3}{EI} - \frac{18\ \text{kN} \cdot \text{m}^3}{EI} = \frac{36\ \text{kN} \cdot \text{m}^3}{EI}\downarrow \ \textit{Ans.}$$

F8–19.

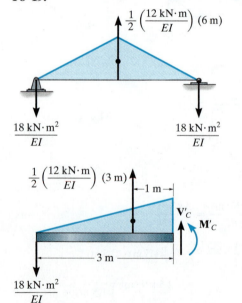

$$+\uparrow \Sigma F_y = 0; \quad -V'_A - \frac{18\ \text{kN} \cdot \text{m}^2}{EI} = 0$$

$$V'_A = \theta_A = -\frac{18\ \text{kN} \cdot \text{m}^2}{EI} = \frac{18\ \text{kN} \cdot \text{m}^2}{EI} \ \ \ \ \ \ \ \ \textit{Ans.}$$

$$\zeta + \Sigma M_C = 0; \quad M'_C + \left(\frac{18\ \text{kN} \cdot \text{m}^2}{EI}\right)(3\ \text{m})$$

$$- \left[\frac{1}{2}\left(\frac{12\ \text{kN} \cdot \text{m}}{EI}\right)(3\ \text{m})\right](1\ \text{m}) = 0$$

$$M'_C = \Delta_C = -\frac{36\ \text{kN} \cdot \text{m}^3}{EI} = \frac{36\ \text{kN} \cdot \text{m}^3}{EI}\downarrow \ \ \ \ \ \textit{Ans.}$$

F8–20.

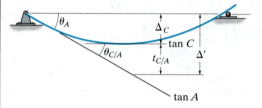

$$\theta_A = \theta_{C/A} = \frac{1}{2}\left(\frac{8\ \text{kN} \cdot \text{m}}{EI}\right)(2\ \text{m}) + \left(\frac{8\ \text{kN} \cdot \text{m}}{EI}\right)(2\ \text{m})$$

$$= \frac{24\ \text{kN} \cdot \text{m}^2}{EI} \ \ \ \ \ \ \ \ \ \ \ \ \ \ \textit{Ans.}$$

$$t_{C/A} = \left[\frac{1}{2}\left(\frac{8\ \text{kN} \cdot \text{m}}{EI}\right)(2\ \text{m})\right]\left[2\ \text{m} + \frac{1}{3}(2\ \text{m})\right]$$

$$+ \left[\left(\frac{8\ \text{kN} \cdot \text{m}}{EI}\right)(2\ \text{m})\right](1\ \text{m}) = \frac{37.33\ \text{kN} \cdot \text{m}^3}{EI}$$

$$\Delta' = \theta_A L_{AC} = \left(\frac{24\ \text{kN} \cdot \text{m}^2}{EI}\right)(4\ \text{m}) = \frac{96\ \text{kN} \cdot \text{m}^3}{EI}$$

$$\Delta_C = \Delta' - t_{C/A} = \frac{96\ \text{kN} \cdot \text{m}^2}{EI} - \frac{37.33\ \text{kN} \cdot \text{m}^3}{EI}$$

$$= \frac{58.7\ \text{kN} \cdot \text{m}^3}{EI}\downarrow \ \ \ \ \ \ \ \ \ \ \ \ \ \ \textit{Ans.}$$

F8–21.

$$\left(\frac{8 \text{ kN}\cdot\text{m}}{EI}\right)(4 \text{ m}) + \frac{1}{2}\left(\frac{8 \text{ kN}\cdot\text{m}}{EI}\right)(4 \text{ m}) = \frac{48 \text{ kN}\cdot\text{m}^2}{EI}$$

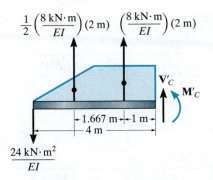

$\frac{24 \text{ kN}\cdot\text{m}^2}{EI}$ $\frac{24 \text{ kN}\cdot\text{m}^2}{EI}$

$\frac{1}{2}\left(\frac{8 \text{ kN}\cdot\text{m}}{EI}\right)(2 \text{ m})$ $\left(\frac{8 \text{ kN}\cdot\text{m}}{EI}\right)(2 \text{ m})$

$\frac{24 \text{ kN}\cdot\text{m}^2}{EI}$

$+\uparrow\Sigma F_y = 0; \quad -V'_A - \frac{24 \text{ kN}\cdot\text{m}^2}{EI} = 0$

$\theta_A = V'_A = \frac{24 \text{ kN}\cdot\text{m}^2}{EI}$ ⟻ *Ans.*

$\zeta + \Sigma M_C = 0; \quad M'_C + \left(\frac{24 \text{ kN}\cdot\text{m}^2}{EI}\right)(4 \text{ m})$

$$- \left[\frac{1}{2}\left(\frac{8 \text{ kN}\cdot\text{m}}{EI}\right)(2 \text{ m})\right](2.667 \text{ m})$$

$$- \left(\frac{8 \text{ kN}\cdot\text{m}}{EI}\right)(2 \text{ m})(1 \text{ m}) = 0$$

$\Delta_C = M'_C = \frac{58.7 \text{ kN}\cdot\text{m}^3}{EI}\downarrow$ *Ans.*

F8–22.

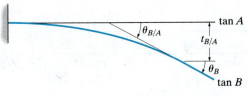

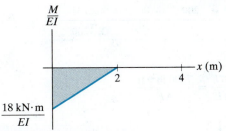

$\theta_B = |\theta_{B/A}| = \left|\frac{1}{2}\left(-\frac{18 \text{ kN}\cdot\text{m}}{EI}\right)(2 \text{ m})\right| = \frac{18 \text{ kN}\cdot\text{m}^2}{EI}$ ⟻ *Ans.*

$\Delta_B = |t_{B/A}| = \left|\left[\frac{1}{2}\left(-\frac{18 \text{ kN}\cdot\text{m}}{EI}\right)(2 \text{ m})\right]\left[2 \text{ m} + \frac{2}{3}(2 \text{ m})\right]\right|$

$= \frac{60 \text{ kN}\cdot\text{m}^3}{EI}\downarrow$

F8–23.

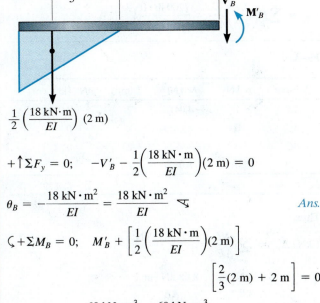

$\frac{1}{2}\left(\frac{18 \text{ kN}\cdot\text{m}}{EI}\right)(2 \text{ m})$

$+\uparrow\Sigma F_y = 0; \quad -V'_B - \frac{1}{2}\left(\frac{18 \text{ kN}\cdot\text{m}}{EI}\right)(2 \text{ m}) = 0$

$\theta_B = -\frac{18 \text{ kN}\cdot\text{m}^2}{EI} = \frac{18 \text{ kN}\cdot\text{m}^2}{EI}$ ⟻   *Ans.*

$\zeta + \Sigma M_B = 0; \quad M'_B + \left[\frac{1}{2}\left(\frac{18 \text{ kN}\cdot\text{m}}{EI}\right)(2 \text{ m})\right]$

$\left[\frac{2}{3}(2 \text{ m}) + 2 \text{ m}\right] = 0$

$M'_B = \Delta_B = -\frac{60 \text{ kN}\cdot\text{m}^3}{EI} = \frac{60 \text{ kN}\cdot\text{m}^3}{EI}\downarrow$ *Ans.*

Chapter 9

F9–1.

Member	n (lb)	N (lb)	L (ft)	nNL (lb$^2 \cdot$ft)
AB	-1.667	-250	10	4166.67
AC	1	150	6	900.00
BC	1.333	200	8	2133.33
				$\Sigma\ 7200$

Thus,

$$1\ \text{lb} \cdot \Delta_{B_v} = \sum \frac{nNL}{AE} = \frac{7200\ \text{lb}^2 \cdot \text{ft}}{AE}$$

$$\Delta_{B_v} = \frac{7200\ \text{lb} \cdot \text{ft}}{AE} \downarrow \qquad\qquad Ans.$$

F9–2.

Member	N	$\dfrac{\delta N}{\delta P}$	$N\ (P=150\ \text{lb})$	L (ft)	$N\left(\dfrac{\delta N}{\delta P}\right)L$ (lb·ft)
AB	$-1.667P$	-1.667	-250	10	4166.67
AC	P	1	150	6	900.00
BC	$1.333P$	1.333	200	8	2133.33
					$\Sigma\ 7200$

$$\Delta_{B_v} = \sum N\left(\frac{\delta N}{\delta P}\right)\frac{L}{AE} = \frac{7200\ \text{lb} \cdot \text{ft}}{AE} \downarrow \qquad Ans.$$

F9–3.

Member	n (kN)	N (kN)	L (m)	nNL (kN$^2 \cdot$m)
AB	1	-4.041	2	-8.0829
AC	0	8.0829	2	0
BC	0	-8.0829	2	0
CD	0	8.0829	1	0
				$\Sigma\ -8.0829$

Thus,

$$1\ \text{kN} \cdot \Delta_{A_h} = \sum \frac{nNL}{AE} = -\frac{8.0829\ \text{kN}^2 \cdot \text{m}}{AE}$$

$$\Delta_{A_h} = -\frac{8.08\ \text{kN} \cdot \text{m}}{AE} = \frac{8.08\ \text{kN} \cdot \text{m}}{AE} \rightarrow \qquad Ans.$$

F9–4.

Member	N (kN)	$\dfrac{\delta N}{\delta P}$	$N\ (P=0)$(kN)	L (m)	$N\left(\dfrac{\delta N}{\delta P}\right)L$ (kN·m)
AB	$P-4.041$	1	-4.041	2	-8.083
AC	8.083	0	8.083	2	0
BC	-8.083	0	-8.083	2	0
CD	8.083	0	8.083	1	0
					$\Sigma\ -8.083$

$$\Delta_{A_h} = \sum N\left(\frac{\delta N}{\delta P}\right)\frac{L}{AE} = -\frac{8.083\ \text{kN} \cdot \text{m}}{AE} = \frac{8.08\ \text{kN} \cdot \text{m}}{AE} \rightarrow \quad Ans.$$

F9–5.

Member	n (kN)	N (kN)	L (m)	nNL (kN$^2 \cdot$m)
AB	0	0	3	0
AC	1.414	8.485	$3\sqrt{2}$	50.91
BC	-1	-6	3	18.00
AD	0	-6	3	0
CD	-1	0	3	0
				$\Sigma\ 68.91$

$$1\ \text{kN} \cdot \Delta_{D_h} = \sum \frac{nNL}{AE} = \frac{68.91\ \text{kN}^2 \cdot \text{m}}{AE}$$

$$\Delta_{D_h} = \frac{68.9\ \text{kN} \cdot \text{m}}{AE} \rightarrow \qquad\qquad Ans.$$

F9–6.

Member	N (kN)	$\dfrac{\delta N}{\delta P}$	$N\ (P=0)$(kN)	L (m)	$N\left(\dfrac{\delta N}{\delta P}\right)L$ (kN·m)
AB	0	0	0	3	0
AC	$\sqrt{2}(P+6)$	$\sqrt{2}$	$6\sqrt{2}$	$3\sqrt{2}$	50.91
BC	$-(P+6)$	-1	-6	3	18.00
AD	-6	0	-6	3	0
CD	$-p$	-1	0	3	0
					$\Sigma\ 68.91$

$$\Delta_{D_h} = \sum N\left(\frac{\delta N}{\delta P}\right)\frac{L}{AE} = \frac{68.9\ \text{kN} \cdot \text{m}}{AE} \rightarrow \qquad Ans.$$

F9–7.

Member	n (kN)	N (kN)	L (m)	nNL (kN$^2 \cdot$m)
AB	0.375	18.75	3	21.09
BC	0.375	18.75	3	21.09
AD	−0.625	−31.25	5	97.66
CD	−0.625	−31.25	5	97.66
BD	0	50	4	0
				Σ 237.5

$$1 \text{ kN} \cdot \Delta_{D_v} = \sum \frac{nNL}{AE} = \frac{237.5 \text{ kN}^2 \cdot \text{m}}{AE}$$

$$\Delta_{D_v} = \frac{237.5 \text{ kN} \cdot \text{m}}{AE} \downarrow \qquad \qquad \textit{Ans.}$$

F9–8.

Member	N (kN)	$\dfrac{\delta N}{\delta P}$	$N (P=0)$ (kN)	L (m)	$N\left(\dfrac{\delta N}{\delta P}\right)L$ (kN·m)
AB	$\frac{3}{8}P + 18.75$	0.375	18.75	3	21.09
BC	$\frac{3}{8}P + 18.75$	0.375	18.75	3	21.09
AD	$-\left(\frac{5}{8}P + 31.25\right)$	−0.625	−31.25	5	97.66
CD	$-\left(\frac{5}{8}P + 31.25\right)$	−0.625	−31.25	5	97.66
BD	50	0	50	4	0
					Σ 237.5

$$\Delta_{D_v} = \sum N\left(\frac{\delta N}{\delta P}\right)\frac{L}{AE} = \frac{237.5 \text{ kN} \cdot \text{m}}{AE} \downarrow \qquad \qquad \textit{Ans.}$$

F9–9.

Member	n (kN)	N (kN)	L (m)	nNL (kN$^2 \cdot$m)
AB	0	−6	1.5	0
BC	0	−6	1.5	0
BD	1	0	2	0
CD	0	10	2.5	0
AD	−1.25	−10	2.5	31.25
DE	0.75	12	1.5	13.5
				Σ 44.75

$$1 \text{ kN} \cdot \Delta_{B_v} = \sum \frac{nNL}{AE} = \frac{44.75 \text{ kN}^2 \cdot \text{m}}{AE}, \quad \Delta_{B_v} = \frac{44.75 \text{ kN} \cdot \text{m}}{AE} \downarrow \qquad \textit{Ans.}$$

F9–10.

	N (kN)	$\dfrac{\delta N}{\delta P}$	$N\,(P=0)$(kN)	L (m)	$N\left(\dfrac{\delta N}{\delta P}\right)L$ (kN·m)
AB	−6	0	−6	1.5	0
BC	−6	0	−6	1.5	0
BD	*P*	1	0	2	0
CD	10	0	10	2.5	0
AD	−(1.25*P* + 10)	−1.25	−10	2.5	31.25
DE	0.75*P* + 12	0.75	12	1.5	13.5
					Σ 44.75

$$\Delta_{B_v} = \sum N\left(\frac{\delta N}{\delta p}\right)\frac{L}{AE} = \frac{44.75 \text{ kN} \cdot \text{m}}{AE}\downarrow \qquad \textit{Ans.}$$

F9–11.

Member	n (kN)	N (kN)	L (m)	nNL (kN²·m)
AB	0.5	50	2	50.00
DE	0.5	50	2	50.00
BC	0.5	50	2	50.00
CD	0.5	50	2	50.00
AH	−0.7071	−70.71	$2\sqrt{2}$	141.42
EF	−0.7071	−70.71	$2\sqrt{2}$	141.42
BH	0	30	2	0
DF	0	30	2	0
CH	0.7071	28.28	$2\sqrt{2}$	56.57
CF	0.7071	28.28	$2\sqrt{2}$	56.57
CG	0	0	2	0
GH	−1	−70	2	140.00
FG	−1	−70	2	140.00
				Σ 875.98

$$1 \text{ kN} \cdot \Delta_{C_v} = \sum \frac{nNL}{AE} = \frac{875.98 \text{ kN}^2 \cdot \text{m}}{AE}, \quad \Delta_{C_v} = \frac{876 \text{ kN} \cdot \text{m}}{AE}\downarrow \qquad \textit{Ans.}$$

F9–12.

Member	N (kN)	$\dfrac{\delta N}{\delta P}$	N ($P = 40$ kN)	L (m)	$N\left(\dfrac{\delta N}{\delta P}\right)L$ (kN·m)
AB	$0.5P + 30$	0.5	50	2	50.00
DE	$0.5P + 30$	0.5	50	2	50.00
BC	$0.5P + 30$	0.5	50	2	50.00
CD	$0.5P + 30$	0.5	50	2	50.00
AH	$-(0.7071P + 42.43)$	-0.7071	-70.71	$2\sqrt{2}$	141.42
EF	$-(0.7071P + 42.43)$	-0.7071	-70.71	$2\sqrt{2}$	141.42
BH	30	0	30	2	0
DF	30	0	30	2	0
CH	$0.7071P$	0.7071	28.28	$2\sqrt{2}$	56.57
CF	$0.7071P$	0.7071	28.28	$2\sqrt{2}$	56.57
CG	0	0	0	2	0
GH	$-(P + 30)$	-1	-70	2	140.00
FG	$-(P + 30)$	-1	-70	2	140.00
					Σ 875.98

$$\Delta_{C_v} = \sum N\left(\frac{\delta N}{\delta P}\right)\frac{L}{AE} = \frac{875.98 \text{ kN} \cdot \text{m}}{AE}$$

$$\Delta_{C_v} = \frac{876 \text{ kN} \cdot \text{m}}{AE} \downarrow \qquad\qquad Ans.$$

F9–13.   For the slope,

$$1 \text{ kN} \cdot \text{m} \cdot \theta_A = \int_0^2 \frac{m_\theta M}{EI}dx = \int_0^{3m} \frac{(-1)(-30x)}{EI}dx$$

$$= \frac{135 \text{ kN}^2 \cdot \text{m}^3}{EI}$$

$$\theta_A = \frac{135 \text{ kN} \cdot \text{m}^2}{EI} \quad \nearrow \qquad\qquad Ans.$$

For the displacement,

$$1 \text{ kN} \cdot \Delta_{A_v} = \int_0^L \frac{mM}{EI}dx = \int_0^{3m} \frac{(-x)(-30x)}{EI}dx = \frac{270 \text{ kN}^2 \cdot \text{m}^3}{EI}$$

$$\Delta_{A_v} = \frac{270 \text{ kN} \cdot \text{m}^3}{EI} \downarrow \qquad\qquad Ans.$$

F9–14.   For the slope, $M = -30x - M'$. Then, $\dfrac{\partial M}{\partial M'} = -1$.

Set $M' = 0$. Then, $M = (-30x)$ kN · m.

$$\theta_A = \int_0^L M\left(\frac{\partial M}{\partial M'}\right)\frac{dx}{EI} = \int_0^{3m} \frac{(-30x)(-1)dx}{EI}$$

$$= \frac{135 \text{ kN} \cdot \text{m}^2}{EI} \quad \nearrow \qquad\qquad Ans.$$

For the displacement, $M = -Px$. Then $\dfrac{\partial M}{\partial P} = -x$.

Set $P = 30$ kN. Then $M = (-30x)$ kN · m.

$$\Delta_{A_v} = \int_0^L M\left(\frac{\partial M}{\partial P}\right)\frac{dx}{EI} = \int_0^{3m} \frac{(-30x)(-x)dx}{EI} = \frac{270 \text{ kN} \cdot \text{m}^3}{EI} \downarrow$$

F9–15.   For the slope, $m_\theta = 1$ kN · m and $M = 4$ kN · m.

$$1 \text{ kN} \cdot \text{m} \cdot \theta_A = \int_0^L \frac{m_\theta M}{EI}dx = \int_0^{3m} \frac{(1)(4)dx}{EI} = \frac{12 \text{ kN}^2 \cdot \text{m}^3}{EI}$$

$$\theta_A = \frac{12 \text{ kN} \cdot \text{m}^2}{EI} \quad \nwarrow \qquad\qquad Ans.$$

For the displacement, $m = x$ kN · m and $M = 4$ kN · m.

$$1 \text{ kN} \cdot \Delta_{A_v} = \int_0^L \frac{mM}{EI}dx = \int_0^{3m} \frac{x(4)dx}{EI} = \frac{18 \text{ kN}^2 \cdot \text{m}^3}{EI}$$

$$\Delta_{A_v} = \frac{18 \text{ kN} \cdot \text{m}^3}{EI} \uparrow \qquad\qquad Ans.$$

F9–16. For the slope, $M = M'$. Then $\dfrac{\partial M}{\partial M} = 1$.

Set $M' = 4$ kN $\cdot$ m. Then $M = 4$ kN $\cdot$ m.

$$\theta_A = \int_0^L M\left(\frac{\partial M}{\partial M'}\right)\frac{dx}{EI} = \int_0^{3m} \frac{4(1)dx}{EI} = \frac{12 \text{ kN} \cdot \text{m}^2}{EI} \;\; \text{Ans.}$$

For the displacement, $M = (Px + 4)$ kN $\cdot$ m. Then $\dfrac{\partial M}{\partial P} = x$.

Set $P = 0$. Then $M = 4$ kN $\cdot$ m.

$$\Delta_{A_v} = \int_0^L M\left(\frac{\partial M}{\partial P}\right)\frac{dx}{EI} = \int_0^{3m} \frac{4(x)dx}{EI} = \frac{18 \text{ kN} \cdot \text{m}^3}{EI} \uparrow \;\; \text{Ans.}$$

F9–17. For the slope, $m_\theta = -1$ kN $\cdot$ m and $M = (-x^3)$ kN $\cdot$ m.

$$1 \text{ kN} \cdot \text{m} \cdot \theta_B = \int_0^L \frac{m_\theta M}{EI}dx = \int_0^{3m} \frac{(-1)(-x^3)}{EI}dx = \frac{20.25 \text{ kN}^2 \cdot \text{m}^3}{EI}$$

$$\theta_B = \frac{20.25 \text{ kN} \cdot \text{m}^2}{EI} \;\; \text{Ans.}$$

For the displacement, $m = (-x)$ kN $\cdot$ m and $M = (-x^3)$ kN $\cdot$ m.

$$1 \text{ kN} \cdot \Delta_{B_v} = \int_0^L \frac{mM}{EI}dx = \int_0^{3m} \frac{(-x)(-x^3)}{EI}dx$$

$$= \frac{48.6 \text{ kN}^2 \cdot \text{m}^3}{EI}$$

$$\Delta_{B_v} = \frac{48.6 \text{ kN} \cdot \text{m}^3}{EI} \downarrow \;\; \text{Ans.}$$

F9–18. For the slope, $M = -(M' + x^3)$ kN $\cdot$ m. Then $\dfrac{\partial M}{\partial M'} = -1$.

Set $M' = 0$. Then $M = (-x^3)$ kN $\cdot$ m.

$$\theta_B = \int_0^L M\left(\frac{\partial M}{\partial M'}\right)\frac{dx}{EI} = \int_0^{3m} \frac{(-x^3)(-1)dx}{EI}$$

$$= \frac{20.25 \text{ kN} \cdot \text{m}^2}{EI} \;\; \text{Ans.}$$

For the displacement, $M = -(Px + x^3)$ kN $\cdot$ m.

Then $\dfrac{\partial M}{\partial P} = -x$. Set $P = 0$, then $M = (-x^3)$ kN $\cdot$ m.

$$\Delta_{B_v} = \int_0^L M\left(\frac{\delta M}{\partial P}\right)\frac{dx}{EI} = \int_0^{3m} \frac{(-x^3)(-x)dx}{EI}$$

$$= \frac{48.6 \text{ kN} \cdot \text{m}^3}{EI} \downarrow \;\; \text{Ans.}$$

F9–19. For the slope, $m_\theta = (1 - 0.125x)$ kN $\cdot$ m and $M = (32x - 4x^2)$ kN $\cdot$ m.

$$1 \text{ kN} \cdot \text{m} \cdot \theta_A = \int_0^L \frac{m_\theta M}{EI}dx = \int_0^{8m} \frac{(1 - 0.125x)(32x - 4x^2)}{EI}dx$$

$$= \frac{170.67 \text{ kN}^2 \cdot \text{m}^3}{EI}$$

$$\theta_A = \frac{171 \text{ kN} \cdot \text{m}^2}{EI} \;\; \text{Ans.}$$

For the displacement, $m = (0.5x)$ kN $\cdot$ m and $M = (32x - 4x^2)$ kN $\cdot$ m.

$$1 \text{ kN} \cdot \Delta_{C_v} = \int \frac{mM}{EI}dx = 2\int_0^{4m} \frac{0.5x(32x - 4x^2)}{EI}dx$$

$$= \frac{426.67 \text{ kN}^2 \cdot \text{m}^3}{EI}$$

$$\Delta_{C_v} = \frac{427 \text{ kN} \cdot \text{m}^3}{EI} \downarrow \;\; \text{Ans.}$$

F9–20. For the slope, $M = M' - 0.125M'x + 32x - 4x^2$.

Then $\dfrac{\partial M}{\partial M'} = 1 - 0.125x$.

Set $M' = 0$, then $M = (32x - 4x^2)$ kN $\cdot$ m.

$$\theta_A = \int_0^L M\left(\frac{\partial M}{\partial M'}\right)\frac{dx}{EI} = \int_0^{8m} \frac{(32x - 4x^2)(1 - 0.125x)}{EI}dx$$

$$= \frac{170.67 \text{ kN} \cdot \text{m}^2}{EI} = \frac{171 \text{ kN} \cdot \text{m}^2}{EI} \;\; \text{Ans.}$$

For the displacement, $M = 0.5Px + 32x - 4x^2$. Then

$\dfrac{\partial M}{\partial P} = 0.5x$. Set $P = 0$, then $M = (32x - 4x^2)$ kN $\cdot$ m.

$$\Delta_{C_v} = \int M\left(\frac{\partial M}{\partial P}\right)\frac{dx}{EI} = 2\int_0^{4m} \frac{(32x - 4x^2)(0.5x)dx}{EI}$$

$$= \frac{426.67 \text{ kN} \cdot \text{m}^3}{EI} = \frac{427 \text{ kN} \cdot \text{m}^3}{EI} \downarrow \;\; \text{Ans.}$$

F9–21.   For the slope, $(m_\theta)_1 = 0$, $(m_\theta)_2 = -1$ kN$\cdot$m,

$M_1 = (-12x_1)$ kN$\cdot$m, and $M_2 = -12(x_2 + 2)$ kN$\cdot$m.

$$1 \text{ kN} \cdot \text{m} \cdot \theta_C = \int_0^L \frac{m_\theta M}{EI}\, dx = \int_0^{2\text{m}} \frac{0(-12x_1)}{EI}\, dx$$

$$+ \int_0^{2\text{m}} \frac{(-1)\left[-12(x_2 + 2)\right]}{EI}\, dx$$

$$1 \text{ kN} \cdot \text{m} \cdot \theta_C = \frac{72 \text{ kN}^2 \cdot \text{m}^3}{EI}$$

$$\theta_C = \frac{72 \text{ kN} \cdot \text{m}^2}{EI} \quad \measuredangle \qquad \qquad Ans.$$

For the displacement, $m_1 = 0$, $m_2 = -x_2$, $M_1 = (-12x_1)$ kN$\cdot$m,

and $M_2 = -12(x_2 + 2)$ kN$\cdot$m.

$$1 \text{ kN} \cdot \Delta_C = \int_0^L \frac{mM}{EI}\, dx = \int_0^{2\text{m}} \frac{0(-12x_1)}{EI}\, dx$$

$$+ \int_0^{2\text{m}} \frac{(-x_2)\left[-12(x_2 + 2)\right]}{EI}\, dx$$

$$1 \text{ kN} \cdot \Delta_{C_v} = \frac{80 \text{ kN}^2 \cdot \text{m}^3}{EI}$$

$$\Delta_{C_v} = \frac{80 \text{ kN} \cdot \text{m}^3}{EI} \downarrow \qquad \qquad Ans.$$

F9–22.   For the slope, $M_1 = (-12x_1)$ kN$\cdot$m, and

$M_2 = -12(x_2 + 2) - M'$.

Thus, $\dfrac{\partial M_1}{\partial M'} = 0$ and $\dfrac{\partial M_2}{\partial M'} = -1$. Set $M' = 0$, $M_2 = -12(x_2 + 2)$.

$$\theta_C = \int_0^L M\left(\frac{\partial M}{\partial M'}\right)\frac{dx}{EI} = \int_0^{2\text{m}} \frac{-12x_1(0)}{EI}\, dx +$$

$$\int_0^2 \frac{\left[-12(x_2 + 2)\right](-1)}{EI}\, dx$$

$$= \frac{72 \text{ kN} \cdot \text{m}}{EI} \quad \measuredangle \qquad \qquad Ans.$$

For the displacement, $M_1 = (-12x_1)$ kN$\cdot$m and

$M_2 = -12(x_2 + 2) - Px_2$.

Thus, $\dfrac{\partial M_1}{\partial P} = 0$ and $\dfrac{\partial M_2}{\partial P} = -x_2$. Set $P = 0$,

$M_2 = -12(x_2 + 2)$ kN$\cdot$m.

$$\Delta_C = \int_0^L M\left(\frac{\partial M}{\partial P}\right)\frac{dx}{EI} = \int_0^{2\text{m}} \frac{(-12x_1)(0)}{EI}\, dx +$$

$$\int_0^{2\text{m}} \frac{\left[-12(x_2 + 2)\right](-x_2)}{EI}\, dx$$

$$= \frac{80 \text{ kN} \cdot \text{m}^3}{EI} \downarrow$$

F9–23.

$M_1 = 0.5x_1$, $M_2 = 0.5x_2$, $M_1 = \left(24x_1 - \dfrac{1}{6}x_1^3\right)$ kN$\cdot$m

and $M_2 = \left(48x_2 - 6x_2^2 + \dfrac{1}{6}x_2^3\right)$ kN$\cdot$m.

$$1 \text{ kN} \cdot \Delta_{C_v} = \int_0^L \frac{mM}{EI}\, dx = \int_0^{6\text{m}} \frac{(0.5x_1)\left(24x_1 - \dfrac{1}{6}x_1^3\right)}{EI}\, dx_1 +$$

$$\int_0^{6\text{m}} \frac{(0.5x_2)\left(48x_2 - 6x_2^2 + \dfrac{1}{6}x_2^3\right)}{EI}\, dx_2$$

$$= \frac{1620 \text{ kN}^2 \cdot \text{m}^3}{EI}$$

$$\Delta_{C_v} = \frac{1620 \text{ kN} \cdot \text{m}^3}{EI} \downarrow \qquad \qquad Ans.$$

F9–24. $M_1 = 0.5Px_1 + 24x_1 - \dfrac{1}{6}x_1^3$,

$M_2 = 0.5Px_2 + 48x_2 - 6x_2^2 + \dfrac{1}{6}x_2^3$.

Then $\dfrac{\partial M_1}{\partial P} = 0.5x_1$, $\dfrac{\partial M_2}{\partial P} = 0.5x_2$.

Set $P = 0$, $M_1 = \left(24x_1 - \dfrac{1}{6}x_1^3\right)$ kN$\cdot$m and

$M_2 = \left(48x_2 - 6x_2^2 + \dfrac{1}{6}x_2^3\right)$ kN$\cdot$m

$$\Delta_{C_v} = \int_0^L M\left(\frac{\partial M}{\partial P}\right)\frac{dx}{EI} = \int_0^{6\text{m}} \frac{\left(24x_1 - \dfrac{1}{6}x_1^3\right)(0.5x_1)}{EI}\, dx_1$$

$$+ \int_0^{6\text{m}} \frac{\left(48x_2 - 6x_2^2 + \dfrac{1}{6}x_2^3\right)(0.5x_2)}{EI}\, dx_2$$

$$= \frac{1620 \text{ kN} \cdot \text{m}^3}{EI} \downarrow \qquad \qquad Ans.$$

Chapter 10

F10–1. Superposition

$$\Delta'_B = \frac{Px^2}{6EI}(3L - x) = \frac{40(2^2)}{6EI}[3(4) - 2] = \frac{266.67 \text{ kN} \cdot \text{m}^3}{EI} \downarrow$$

$$f_{BB} = \frac{(L/2)^3}{3EI} = \frac{L^3}{24EI} = \frac{4^3}{24EI} = \frac{2.667 \text{ m}^3}{EI} \uparrow$$

$$\Delta_B = \Delta'_B + B_y f_{BB}$$

$$(+\uparrow) \, 0 = -\frac{266.67 \text{ kN} \cdot \text{m}^3}{EI} + B_y\left(\frac{2.667 \text{ m}^3}{EI}\right)$$

$$B_y = 100 \text{ kN} \hspace{3cm} Ans.$$

Equilibrium

$$\xrightarrow{+} \Sigma F_x = 0; \quad A_x = 0 \hspace{3cm} Ans.$$

$$+\uparrow \Sigma F_y = 0; \quad 100 - 40 - A_y = 0 \quad A_y = 60 \text{ kN} \hspace{1cm} Ans.$$

$$\zeta + \Sigma M_A = 0; \quad 100(2) - 40(4) - M_A = 0$$

$$M_A = 40 \text{ kN} \cdot \text{m} \hspace{3cm} Ans.$$

F10–2. Superposition

$$\Delta'_B = \int_0^L \frac{mM}{EI}dx = \int_0^L \frac{(-x)\left(-\dfrac{w_0}{6L}x^3\right)}{EI}dx = \frac{w_0 L^4}{30 \, EI} \downarrow$$

$$f_{BB} = \int_0^L \frac{mm}{EI}dx = \int_0^L \frac{(-x)(-x)}{EI}dx = \frac{L^3}{3EI} \downarrow$$

$$\Delta_B = \Delta'_B + B_y f_{BB}$$

$$(+\downarrow) \, 0 = \frac{w_0 L^4}{30 \, EI} + B_y\left(\frac{L^3}{3EI}\right) \hspace{1cm} B_y = -\frac{w_0 L}{10} = \frac{w_0 L}{10} \uparrow \, Ans.$$

Equilibrium

$$\xrightarrow{+} \Sigma Fx = 0; \quad A_x = 0$$

$$+\uparrow \Sigma F_y = 0; \quad A_y - \frac{1}{2}w_0 L + \frac{w_0 L}{10} = 0 \quad A_y = \frac{2w_0 L}{5} \hspace{0.5cm} Ans.$$

$$\zeta + \Sigma M_A = 0; \quad M_A + \frac{w_0 L}{10}(L) - \left(\frac{1}{2}w_0 L\right)\left(\frac{L}{3}\right) = 0$$

$$M_A = \frac{w_0 L^2}{15} \hspace{3cm} Ans.$$

F10–3. Superposition

$$\Delta'_B = \frac{wL^4}{8EI} = \frac{10(6^4)}{8EI} = \frac{1620 \text{ kN} \cdot \text{m}^3}{EI} =$$

$$\frac{1620(10^3) \text{ N} \cdot \text{m}^3}{\left[200(10^9) \text{ N/m}^2\right]\left[300(10^{-6}) \text{ m}^4\right]} = 0.027 \text{ m} \downarrow$$

$$f_{BB} = \frac{L^3}{3EI} = \frac{6^3}{3EI} = \frac{72 \text{ m}^3}{EI} =$$

$$\frac{72 \text{ m}^3}{\left[200(10^9) \text{ N/m}^2\right]\left[300(10^{-6}) \text{ m}^4\right]} = 1.2(10^{-6}) \text{ m/N} \uparrow$$

$$\Delta_B = \Delta'_B + B_y f_{BB}$$

$$(+\downarrow) 5(10^{-3}) \text{ m} = 0.027 \text{ m} + B_y\left[-1.2(10^{-6}) \text{ m/N}\right]$$

$$B_y = 18.33(10^3) \text{ N} = 18.33 \text{ kN} = 18.3 \text{ kN} \hspace{1cm} Ans.$$

Equilibrium

$$\xrightarrow{+} \Sigma F_x = 0; \quad A_x = 0 \hspace{3cm} Ans.$$

$$+\uparrow \Sigma F_y = 0; \quad A_y + 18.33 - 60 = 0$$

$$A_y = 41.67 \text{ kN} = 41.7 \text{ kN} \hspace{3cm} Ans.$$

$$\zeta + \Sigma M_A = 0; \quad M_A + 18.33(6) - 60(3) = 0$$

$$M_A = 70.0 \text{ kN} \cdot \text{m} \hspace{3cm} Ans.$$

F10–4. Superposition

$$\Delta'_B = \frac{M_0 x}{6EIL_{AC}}(L_{AC}^2 - x^2) = \frac{M_0(L)}{6EI(2L)}\left[(2L)^2 - L^2\right] = \frac{M_0 L^2}{4EI} \downarrow$$

$$f_{BB} = \frac{L_{AC}^3}{48EI} = \frac{(2L)^3}{48EI} = \frac{L^3}{6EI} \uparrow$$

$$\Delta_B = \Delta'_B + B_y f_{BB}$$

$$(+\uparrow) 0 = -\frac{M_0 L^2}{4EI} + B_y\left(\frac{L^3}{6EI}\right) \quad B_y = \frac{3M_0}{2L} \hspace{1cm} Ans.$$

Equilibrium

$$\xrightarrow{+} \Sigma F_x = 0; \quad A_x = 0 \hspace{3cm} Ans.$$

$$\zeta + \Sigma M_A = 0; \quad -C_y(2L) + \frac{3M_0}{2L}(L) - M_0 = 0$$

$$C_y = \frac{M_0}{4L} \hspace{3cm} Ans.$$

$$+\uparrow \Sigma F_y = 0; \quad \frac{3M_0}{2L} - \frac{M_0}{4L} - A_y = 0 \quad A_y = \frac{5M_0}{4L} \hspace{1cm} Ans.$$

F10–5. Superposition

$$\Delta'_B = \frac{Pbx}{6EIL_{AC}}(L_{AC}^2 - b^2 - x^2) = \frac{50(2)(4)}{6EI(8)}(8^2 - 2^2 - 4^2)$$

$$= \frac{366.67 \text{ kN} \cdot \text{m}^3}{EI} \downarrow$$

$$f_{BB} = \frac{L_{AC}^3}{48EI} = \frac{8^3}{48EI} = \frac{10.667 \text{ m}^3}{EI} \uparrow$$

$$\Delta_B = \Delta'_B + B_y f_{BB}$$

$$(+\uparrow) \ 0 = -\frac{366.67 \text{ kN} \cdot \text{m}^3}{EI} + B_y\left(\frac{10.667 \text{ m}^3}{EI}\right)$$

$$B_y = 34.375 \text{ kN} = 34.4 \text{ kN} \qquad\qquad\qquad Ans.$$

Equilibrium

$$\zeta + \Sigma M_A = 0; \quad 34.375(4) - 50(2) - C_y(8) = 0$$

$$C_y = 4.6875 \text{ kN} = 4.69 \text{ kN} \qquad\qquad\qquad Ans.$$

$$+\uparrow \Sigma F_y = 0; \quad A_y + 34.375 - 50 - 4.6875 = 0$$

$$A_y = 20.3125 \text{ kN} = 20.3 \text{ kN} \qquad\qquad\qquad Ans.$$

$$\overset{+}{\to} \Sigma F_x = 0; \quad A_x = 0 \qquad\qquad\qquad Ans.$$

F10–6.
$$\Delta'_B = \frac{5wL_{AC}^4}{384EI} = \frac{5(10)(12^4)}{384EI} = \frac{2700 \text{ kN} \cdot \text{m}^3}{EI}$$

$$= \frac{2700(10^3) \text{ N} \cdot \text{m}^3}{[200(10^9) \text{ N/m}^2][300(10^{-6}) \text{ m}^4]} = 0.045 \text{ m} \downarrow$$

$$f_{BB} = \frac{L_{AC}^3}{48EI} = \frac{12^3}{48EI} = \frac{36 \text{ m}^3}{EI}$$

$$= \frac{36 \text{ m}^3}{[200(10^9) \text{ N/m}^2][300(10^{-6}) \text{ m}^4]} = 0.6(10^{-6}) \text{ m/N} \uparrow$$

$$\Delta_B = \Delta'_B + B_y f_{BB}$$

$$(+\downarrow) \ 5(10^{-3}) \text{ m} = 0.045 \text{ m} + B_y[-0.6(10^{-6}) \text{ m/N}]$$

$$B_y = 66.67(10^3) \text{ N} = 66.7 \text{ kN} \qquad\qquad\qquad Ans.$$

Equilibrium

$$\zeta + \Sigma M_A = 0; \quad C_y(12) + 66.67(6) - 120(6) = 0$$

$$C_y = 26.67 \text{ kN} = 26.7 \text{ kN} \qquad\qquad\qquad Ans.$$

$$+\uparrow \Sigma F_y = 0; \quad A_y + 26.67 + 66.67 - 120 = 0$$

$$A_y = 26.67 \text{ kN} = 26.7 \text{ kN} \qquad\qquad\qquad Ans.$$

$$\overset{+}{\to} \Sigma F_x = 0; \quad A_x = 0 \qquad\qquad\qquad Ans.$$

Answers to Selected Problems

Chapter 1

1–1.	48.3 k
1–2.	1008 lb/ft
1–3.	1.16 k/ft
1–5.	$DL = 638$ lb/ft; $LL = 225$ lb/ft
1–6.	173 kN
1–7.	633 lb/ft
1–9.	9.36 kN/m
1–10.	240 lb/ft
1–11.	$p_x = 4.12$ psf; $p_y = 7.14$ psf
1–13.	Internal: $p = \pm 199$ N/m^2;
	External: $p = -657$ N/m^2
1–14.	0.96 kN/m^2
1–15.	24 lb/ft^2
1–17.	-21.9 psf or -8.85 psf
1–18.	192 kN
1–19.	-9.96 psf or -18.6 psf
1–21.	0.576 kN/m^2
1–22.	1.15 kN/m^2
1–23.	0.816 kN/m^2
1–25.	External windward: -14.1 psf;
	External leeward: -6.83 psf;
	Internal: ± 3.85 psf

Chapter 2

2–1.	*BE*: 14.2 kN/m; *FED*: 35.6 kN at *E*
2–2.	*BE*: $w_{max} = 21.4$ kN/m;
	FED: $w_{max} = 10.7$ kN/m
	26.7 kN at *E*
2–3.	*BF*: 0.9 k/ft; *ABCDE*: 13.5 k at *B*, *C* and *D*
2–5.	*BF*: 0.675 k/ft; *ABCDE*: 6.75 k at *B*, *C* and *D*
2–6.	*BG*: 276 lb/ft; *ABCD*: 2484 lb at *B* and *C*
2–7.	*BG*: $w_{max} = 460$ lb/ft; *ABCD*: $w_{max} = 230$ lb/ft
2–9.	*BE*: $w_{max} = 4.05$ k/ft
	FED: $w_{max} = 2.025$ k/ft 15.2 k at *E*
2–10.	*BE*: 2.70 k/ft; *FED*: 24.3 k at *E*
2–11.	a. Statically determinate.
	b. Statically indeterminate to second degree.
	c. Statically determinate.
	d. Statically indeterminate to first degree.
	e. Statically indeterminate to first degree.
2–13.	a. Stable and statically indeterminate to first degree.
	b. Stable and statically determinate.
	c. Stable and statically indeterminate to first degree.

2–14.	a. Unstable.
	b. Stable and statically determinate.
	c. Stable and statically indeterminate to second degree.
2–15.	a. Unstable.
	b. Statically determinate.
	c. Statically determinate.
2–17.	a. Statically determinate.
	b. Statically determinate.
	c. Stable and statically determinate.
	d. Unstable.
2–18.	$A_y = 80.0$ k; $B_y = 100$ k; $B_x = 0$
2–19.	$F_B = 52.0$ k; $A_x = 26.0$ k; $A_y = 45$ k
2–21.	$A_y = 4.00$ kN; $M_B = 63.0$ kN·m; $B_y = 17.0$ kN; $B_x = 0$
2–22.	$A_x = 0$; $A_y = 480$ lb; $B_y = 620$ lb
2–23.	$N_C = 2.25$ kN; $B_y = 3.75$ kN; $A_x = 0$; $A_y = 21.75$ kN; $M_A = 38.25$ kN·m
2–25.	$C_y = 45.0$ kN; $C_x = 45.0$ kN; $A_x = 45.0$ kN; $A_y = 45.0$ kN
2–26.	$B_y = 5.12$ kN; $A_y = 14.7$ kN; $B_x = 20.0$ kN
2–27.	$N_E = 12.0$ kN; $N_C = 4.00$ kN; $A_x = 0$; $A_y = 10.0$ kN; $M_A = 30.0$ kN·m
2–29.	$B_y = 16$ k; $A_y = 10$ k; $C_y = 30$ k; $D_y = 12$ k
2–30.	$N_A = 13.0$ kN; $B_x = 0$; $B_y = 26.0$ kN; $M_B = 60.0$ kN·m
2–31.	$w_1 = \dfrac{2P}{L}$; $w_2 = \dfrac{4P}{L}$; For $P = 500$ lb, $L = 12$ ft: $w_1 = 83.3$ lb/ft; $w_2 = 167$ lb/ft
2–33.	$A_x = 12.0$ kN; $A_y = 16.0$ kN; $C_x = 12.0$ kN; $C_y = 16.0$ kN
2–34.	$N_B = 7.42$ kN; $A_x = 1.55$ kN; $A_y = 2.06$ kN;
2–35.	$F = 311$ kN; $A_x = 460$ kN; $A_y = 7.85$ kN
2–37.	$A_x = 875$ N; $A_y = 1.09$ kN; $C_x = 875$ N; $C_y = 1.09$ kN
2–38.	$T = 350$ lb; $A_y = 700$ lb; $A_x = 1.88$ k; $D_x = 1.70$ k; $D_y = 1.70$ k
2–39.	$A_y = 500$ lb; $A_x = 1067$ lb; $D_x = 1067$ lb; $D_y = 900$ lb
2–41.	$A_x = 6.96$ kN; $A_y = 34.7$ kN; $C_x = 15.0$ kN; $C_y = 37.3$ kN
2–42.	$C_x = 45.0$ kN; $D_x = 45.0$ kN; $C_y = 7.00$ kN; $A_y = 83.0$ kN; $A_x = 45.0$ kN; $D_y = 7.00$ kN
2–43.	$B_y = 30$ k; $D_x = 0$; $D_y = 30$ k; $C_y = 135$ k; $A_x = 0$; $A_y = 75$ k; $F_y = 135$ k; $E_x = 0$; $E_y = 75$ k
2–1P.	79.7 k

Chapter 3

3–1. **a.** Unstable.
b. Statically indeterminate to first degree.
c. Statically determinate.
d. Statically determinate.

3–2. **a.** Internally and externally stable.
Statically determinate.
b. Internally and externally stable.
Statically indeterminate to second degree.

3–3. **a.** Internally and externally stable.
Statically determinate.
b. Internally and externally stable.
Statically indeterminate to first degree.
c. Internally unstable.

3–5. $F_{ED} = 8.33$ kN (T); $F_{CD} = 6.67$ kN (C);
$F_{BC} = 6.67$ kN (C); $F_{CE} = 5$ kN (T);
$F_{GF} = 20$ kN (T); $F_{GA} = 15$ kN (T);
$F_{AF} = 18.0$ kN (C); $F_{AB} = 10.0$ kN (C);
$F_{BE} = 4.17$ kN (C); $F_{FB} = 7.50$ kN (T);
$F_{FE} = 12.5$ kN (T)

3–6. $F_{DC} = 40.4$ kN (C); $F_{DE} = 20.2$ kN (T);
$F_{CE} = 5.77$ kN (C); $F_{CB} = 17.3$ kN (C);
$F_{BA} = 28.9$ kN (C); $F_{BE} = 5.77$ kN (T)

3–7. $F_{DC} = 9.24$ kN (T); $F_{DE} = 4.62$ kN (C);
$F_{CE} = 9.24$ kN (C); $F_{CB} = 9.24$ kN (T);
$F_{BE} = 9.24$ kN (C); $F_{BA} = 9.24$ kN (T);
$F_{EA} = 4.62$ kN (C)

3–9. $F_{AF} = 3.33$ k (T); $F_{AB} = 2.67$ k (C);
$F_{BF} = 9.00$ k (C); $F_{BC} = 2.67$ k (C);
$F_{FC} = 5.00$ k (T); $F_{FE} = 1.33$ k (C);
$F_{CE} = 3.00$ k (C); $F_{CD} = 1.33$ k (T);
$F_{DE} = 1.67$ k (C)

3–10. $F_{GB} = F_{BF} = 0$; $F_{DE} = 13.8$ kN (C);
$F_{DC} = 11.1$ kN (T); $F_{EF} = 13.8$ kN (C);
$F_{EC} = 6.00$ kN (C); $F_{CF} = 5.37$ kN (T);
$F_{CB} = 5.07$ kN (T); $F_{FG} = 10.8$ kN (C);
$F_{GA} = 10.8$ kN (C); $F_{BA} = 5.07$ kN (T)

3–11. $F_{ED} = F_{DC} = 0$; $F_{EF} = F_{CG} = 0$;
$F_{DF} = F_{DG} = 4.24$ kN (C);
$F_{AF} = F_{GB} = 4.24$ kN (C); $F_{AB} = 3$ kN (T);
$F_{EA} = F_{CB} = 4$ kN (C)

3–13. $F_{AB} = 0$; $F_{AN} = 90.0$ k (C); $F_{NB} = 106$ k (T);
$F_{NM} = 75.0$ k (C); $F_{ML} = 75.0$ k (C);
$F_{MB} = 30.0$ k (C); $F_{BL} = 63.6$ k (C);
$F_{BC} = 120$ k (T); $F_{CD} = 120$ k (T); $F_{CL} = 0$;
$F_{LD} = 21.2$ k (T); $F_{LK} = 135$ k (C);
$F_{KD} = 30.0$ k (C); $F_{FG} = 0$; $F_{HI} = 75.0$ k (C);
$F_{IF} = 30.0$ k (C); $F_{EF} = 120$ k (T); $F_{JE} = 0$;
$F_{KJ} = 135$ k (C)

3–14. $F_{ED} = 6.71$ k (T); $F_{EF} = 6.00$ k (C);
$F_{FD} = 6.00$ k (T); $F_{FG} = 6.00$ k (C);
$F_{DG} = 6.71$ k (C); $F_{DC} = 13.4$ k (T);

$F_{GC} = 9.00$ k (T); $F_{GA} = 12.0$ k (C);
$F_{CA} = 8.49$ k (C); $F_{CB} = 20.1$ k (T)

3–15. $F_{GF} = F_{EF} = 2.25$ kN (C);
$F_{GB} = F_{EC} = 3.75$ kN (T); $F_{AB} = F_{DC} = 0$;
$F_{AG} = F_{DE} = 6.00$ kN (C);
$F_{BF} = F_{CF} = 3.75$ kN (C); $F_{CB} = 4.50$ kN (T)

3–17. $F_{DC} = 10.0$ kN (C); $F_{DE} = 8.00$ kN (C);
$F_{CE} = 34.2$ kN (T); $F_{CB} = 24.0$ kN (C);
$F_{EA} = 24.0$ kN (T); $F_{EB} = 12.0$ kN (C);
$F_{BA} = 30.0$ kN (C)

3–18. $F_{BD} = 7.00$ kN (T); $F_{BC} = 1.00$ kN (T);
$F_{ED} = 7.00$ kN (C)

3–19. $F_{CD} = 2.00$ k (T); $F_{JN} = 2.50$ k (T); $F_{JK} = 4.03$ k (C);
Members *KN, NL, MB, BL, CL, IO, OH, GE, EH,
HD* are zero force members.

3–21. $F_{FG} = 5.33$ kN (C); $F_{GD} = 3.33$ kN (C);
$F_{CD} = 10.0$ kN (T); $F_{GA} = 8.00$ kN (C)

3–22. $F_{HB} = 0$; $F_{AB} = 3.00$ kN (T); $F_{HC} = 5.41$ kN (T);
$F_{HG} = 6.71$ kN (C)

3–23. $F_{GC} = 3.00$ kN (T); $F_{BC} = 3.00$ kN (T);
$F_{GF} = 6.71$ kN (C); $F_{HC} = 5.41$ kN (T);

3–25. $F_{JI} = 117.5$ kN (C); $F_{DE} = 97.5$ kN (T);
$F_{JD} = 61.7$ kN (C)

3–26. $F_{HI} = 16.0$ k (T); $F_{ID} = 8.94$ k (T); $F_{DC} = 24.7$ k (C);

3–27. $F_{GF} = 1.80$ k (C); $F_{FB} = 693$ lb (T);
$F_{BC} = 1.21$ k (T)

3–29. $F_{IH} = 6.00$ kN (T); $F_{ID} = 4.24$ kN (T);
$F_{CD} = 10.1$ kN (C)

3–30. $F_{JI} = 9.00$ kN (T); $F_{IC} = 6.00$ kN (C);
$F_{CD} = 10.1$ kN (C)

3–31. $F_{BC} = 9.75$ kN (T); $F_{HC} = 1.06$ kN (C);
$F_{GH} = 10.1$ kN (C)

3–33. $F_{AB} = 0$; $F_{AG} = 1.50$ k (C); $F_{GB} = 0.707$ k (T);
$F_{GL} = 0.500$ k (C); $F_{GI} = 0.707$ k (C);
$F_{LI} = 0.707$ k (T); $F_{LK} = 0.500$ k (C);
$F_{IK} = 0.707$ k (C); $F_{IF} = 0.707$ k (T);
$F_{BF} = 2.12$ k (T); $F_{BC} = 1.00$ k (C);
$F_{FC} = 0.707$ k (T); $F_{FH} = 2.12$ k (T);
$F_{KH} = 0.707$ k (T); $F_{KJ} = 1.50$ k (C);
$F_{JH} = 2.12$ k (T); $F_{CD} = 0$; $F_{DE} = 0.500$ k (C);
$F_{CE} = 0.707$ k (C); $F_{HE} = 0.707$ k (T);
$F_{JE} = 1.50$ k (C)

3–34. $F_{AD} = 0$; $F_{AF} = 4.00$ kN (C); $F_{FD} = 8.94$ kN (T);
$F_{FE} = 11.3$ kN (C); $F_{BC} = 4.00$ kN (C);
$F_{CE} = 8.94$ kN (T); $F_{BE} = 0$; $F_{CD} = 11.3$ kN (C);
$F_{ED} = 16.0$ kN (C)

3–35. $F_{AB} = 3.90$ kN (C); $F_{BC} = 2.50$ kN (C);
$F_{CD} = 2.50$ kN (C);
$F_{BG} = F_{GF} = F_{GH} = F_{CH} = F_{HE} = 0$

3–37. $F_{DE} = F_{CD} = F_{AD} = F_{CE} = F_{AC} = 0$;
$F_{AE} = 671$ lb (C); $F_{BE} = 900$ lb (T); $F_{BC} = 0$;
$F_{AB} = 600$ lb (C)

3–38. $F_{BC} = F_{BD} = 1.34$ kN (C); $F_{AB} = 2.4$ kN (C);
$F_{AG} = F_{AE} = 1.01$ kN (T); $F_{BG} = 1.80$ kN (T);
$F_{BE} = 1.80$ kN (T)

3–39. $F_{FE} = 0; F_{ED} = 0$

3–41. $C_z = 0; A_z = 667$ lb; $B_z = 667$ lb; $C_y = 300$ lb;
$A_y = 200$ lb; $B_x = 0; F_{AD} = 786$ lb (T);
$F_{AC} = 391$ lb (C); $F_{AB} = 167$ lb (T);
$F_{BD} = 731$ lb (C); $F_{BC} = 250$ lb (T); $F_{DC} = 0$

3–42. $F_{AB} = 287$ lb (C); $F_{FE} = 962$ lb (C);
$F_{BD} = 962$ lb (T)

3–43. $F_{AE} = 0; F_{AF} = 577$ lb (T); $F_{FD} = 770$ lb (T)

3–1P. $F_{HG} = 7.00$ k (C); $F_{GF} = 7.00$ k (C);
$F_{GC} = 4.20$ k (T); $F_{DE} = F_{AB} = 8.40$ k (T);
$F_{DC} = F_{BC} = 8.40$ k (T); $F_{EF} = F_{AH} = 10.5$ k (C);
$F_{BH} = F_{DF} = 0; F_{HC} = F_{FC} = 3.50$ k (C)

Chapter 4

4–1. $N_C = 0; V_C = 0.667$ kN; $M_C = 0.667$ kN $\cdot$ m;
$N_D = 0; V_D = -5.33$ kN; $M_D = -9.33$ kN $\cdot$ m

4–2. $N_C = 0; V_C = 0; M_C = -24.0$ kN $\cdot$ m

4–3. $N_A = 0; V_A = 450$ lb; $M_A = -1.125$ kip $\cdot$ ft;
$N_B = 0; V_B = 850$ lb; $M_B = -6.325$ kip $\cdot$ ft;
$V_C = 0; N_C = -1.20$ kip; $M_C = -8.125$ kip $\cdot$ ft

4–5. $w = 100$ N/m

4–6. $N_C = 0; V_C = -0.75$ kN; $M_C = -0.375$ kN $\cdot$ m;
$N_D = 0; V_D = 1.25$ kN; $M_D = 1.875$ kN $\cdot$ m

4–7. $N_C = 0; V_C = 0; M_C = 48.0$ kN $\cdot$ m

4–9. $N_C = 0; V_C = 10.5$ kN; $M_C = 40.5$ kN $\cdot$ m

4–10. $N_B = 0; V_B = -600$ lb; $M_B = -600$ lb $\cdot$ ft;
$N_C = 0; V_C = 2000$ lb; $M_C = 9600$ lb $\cdot$ ft

4–11. $N_C = 0; V_C = 250$ lb; $M_C = 1500$ lb $\cdot$ ft

4–13. $M = (-9.9x + 1600)$ N $\cdot$ m

4–14. For $0 \le x < a, V = \dfrac{M_O}{L}; M = \dfrac{M_O}{L}x;$

For $a < x \le L, V = \dfrac{M_O}{L}; M = -\dfrac{M_O}{L}(L - x)$

4–15. $V = 1.00$ kN; $M = (x + 28)$ kN $\cdot$ m

4–17. For $0 \le x < 1$ m, $V = -4$ kN; $M = \{-4x\}$ kN $\cdot$ m;
For 1 m $< x < 2$ m, $V = \{-12\}$ kN $\cdot$ m;
$M = \{-12x + 8\}$ kN $\cdot$ m; For 2 m $< x \le 3$ m,
$V = \{-20\}$ kN; $M = \{-20x + 24\}$ kN $\cdot$ m

4–18. For $0 \le x < 6$ ft, $V = \{30.0 - 2x\}$ k;
$M = \{-x^2 + 30.0x - 216\}$ k $\cdot$ ft;
For 6 ft $< x \le 10$ ft, $V = 8.00$ k;
$M = \{8.00x - 120\}$ k $\cdot$ ft

4–19. For $0 \le x < 4$ ft, $V = -250$ lb;
$M = \{-250x\}$ lb $\cdot$ ft; $V = \{1050 - 150x\}$ lb;
For 4 ft $< x < 10$ ft,
$M = \{-75x^2 + 1050x - 4000\}$ lb $\cdot$ ft
For 10 ft $< x \le 14$ ft, $V = 250$ lb;
$M = \{250x - 3500\}$ lb $\cdot$ ft

4–21. $V = -\dfrac{wx^2}{60} - P; M = -\dfrac{wx^3}{180} - Px$

4–22. For $0 \le x \; 10$ ft, $V = \{1350 - 200x\}$ lb;
$M = \{1350x - 100x^2\}$ lb $\cdot$ ft;
For 10 ft $< x \le 20$ ft, $V = -650$ lb;
$M = \{-650x + 10,000\}$ lb $\cdot$ ft

4–23. $V_{max} = -386$ lb; $M_{max} = -2400$ lb $\cdot$ ft

4–25. $V_{max} = -4.89$ kN; $M_{max} = -20$ kN $\cdot$ m

4–26. $V_{max} = -143$ kN; $M_{max} = 372$ kN $\cdot$ m

4–27. $V_{max} = -3040$ lb; $M_{max} = 11.6$ k $\cdot$ ft

4–29. $V_{max} = -30.0$ kN; $M_{max} = -50.0$ kN $\cdot$ m

4–30. $V_{max} = 15.5$ kN; $M_{max} = 30.0$ kN $\cdot$ m

4–31. $V_{max} = \pm 10.0$ kN; $M_{max} = -7.50$ kN $\cdot$ m

4–33. $V_{max} = \pm 1800$ lb; $M_{max} = -3600$ lb $\cdot$ ft

4–34. $V_{max} = \pm 1200$ lb; $M_{max} = 6400$ lb $\cdot$ ft

4–35. $V_{max} = -375$ lb; $M_{max} = -675$ lb $\cdot$ ft

4–37. $V_{max} = 24.5$ kN; $M_{max} = 34.5$ kN $\cdot$ m

4–38. $V_{max} = 83.0$ kN; $M_{max} = -180$ kN $\cdot$ m

4–39. $V_{max} = 2.8$ k; $M_{max} = -30.4$ k $\cdot$ ft

4–41. $V_{max} = -13.75$ kN; $M_{max} = 23.6$ kN $\cdot$ m

4–42. $V_{max} = 20.0$ k; $M_{max} = -144$ k $\cdot$ ft

4–43. $V_{max} = 16.0$ kN; $M_{max} = 24.0$ kN $\cdot$ m

4–45. $V_{max} = 16.0$ kN; $M_{max} = 26.7$ kN $\cdot$ m

4–46. $V_{max} = 11.7$ k; $M_{max} = 34.0$ k $\cdot$ ft

4–47. $V_{max} = 4.10$ kN; $M_{max} = 18.8$ kN $\cdot$ m

4–49. $V_{max} = \pm 9.00$ kN; $M_{max} = 16.0$ kN $\cdot$ m

4–1P. Front girder: $M_{max} = 14.9$ k $\cdot$ ft;
Side girder: $M_{max} = 37.3$ k $\cdot$ ft

4–2P. $V_{max} = -2475$ lb; $M_{max} = 5389$ lb $\cdot$ ft

Chapter 5

5–1. $T_{BC} = 1.60$ kN; $T_{CD} = 3.72$ kN; $T_{AB} = 2.99$ kN;
$y_D = 2.10$ m

5–2. $P = 7.14$ k

5–3. $P_2 = 10.3$ kN; $P_1 = 17.1$ kN

5–5. $P = 89.4$ lb

5–6. $T_{AB} = 67.2$ lb; $T_{BC} = 34.4$ lb; $T_{CD} = 97.5$ lb;
$L_T = 22.0$ ft

5–7. $y_D = 0.644$ m; $P_1 = 6.58$ kN

5–9. $T_{min} = 400$ kN; $T_{max} = 431$ kN

5–10. $T_{min} = 7.50$ k; $T_{max} = 10.6$ k

5–11. $T_{max} = 14.4$ k; $T_{min} = 13.0$ k

5–13. $T_{max} = 10.9$ k

5–14. $F_{min} = 100$ k; $F_{max} = 117$ k; $T = 10$ k

5–15. $M_{max} = 6.25$ k $\cdot$ ft

5–17. $T_{min} = 6.25$ MN; $T_{max} = 6.93$ MN

5–18. $T_D = T_E = 11.7$ k; $T_F = 9.75$ k

5–19. $F_F = 7.0$ kN; $F_E = F_D = 8.75$ kN; $T = 1.31$ kN

5–21. $B_y = 1.43$ kN; $B_x = 4.64$ kN; $A_x = 4.64$ kN;
$A_y = 8.57$ kN; $C_x = 4.64$ kN; $C_y = 6.43$ kN

5–22. $F_A = 19.9$ kN; $F_B = 11.9$ kN; $F_C = 10.4$ kN

5–23. $M_D = 10.8$ kN·m

5–25. $B_x = 46.7$ k; $B_y = 5.00$ k; $A_x = 46.7$ k; $A_y = 95.0$ k; $C_x = 46.7$ k; $C_y = 85$ k

5–26. $h_1 = 43.75$ ft; $h_2 = 75.0$ ft; $h_3 = 93.75$ ft

5–27. $B_x = 2.72$ k; $B_y = 0.216$ k; $A_x = 2.72$ k; $A_y = 3.78$ k; $C_x = 0.276$ k; $C_y = 0.216$ k

5–29. $F_{AC} = 0.850$ k; $N_C = 2.35$ k; $A_x = 0.200$ k; $A_y = 0.850$ k

Chapter 6

6–15. $(M_C)_{max} = 142$ kN·m; $(V_C)_{max} = 20$ kN

6–17. $(B_y)_{max(+)} = 12.4$ k; $(M_B)_{max(-)} = -37.5$ k·ft

6–18. $(M_C)_{max(+)} = 112.5$ k·ft; $(B_y)_{max(+)} = 24.75$ k

6–19. $(A_y)_{max(+)} = 70.1$ k; $(M_C)_{max(+)} = 151$ k·ft; $(V_{A^+})_{max(+)} = 40.1$ k

6–21. At point B. $(M_D)_{max} = -4$ k·ft

6–22. $(M_E)_{max(+)} = 89.6$ kN·m; $(V_E)_{max(+)} = 24.6$ kN

6–23. $(M_E)_{max(-)} = -13.5$ kN·m; $(V_E)_{max(+)} = 5.25$ kN

6–26. $(V_{BC})_{max(+)} = 7.15$ kN; $(M_G)_{max(-)} = -9.81$ kN·m

6–27. $(M_E)_{max(+)} = 30.0$ kN·m

6–30. $(V_{AB})_{max} = 2.73$ k; $(M_D)_{max} = 61.25$ k·ft

6–31. $(V_{BC})_{max} = 109$ kN; $(M_C)_{max} = 196$ kN·m

6–33. $(V_{AB})_{max(-)} = -3.80$ k; $(M_B)_{max(-)} = -15.2$ k·ft

6–34. $(V_{CD})_{max(+)} = 17.6$ kN; $(M_C)_{max(+)} = 164$ kN·m

6–35. -6 k

6–37. $(V_{BC})_{max(-)} = -8.21$ kN; $(M_B)_{max(+)} = 12.3$ kN·m

6–59. 5.79 kN (T)

6–61. 7.54 k (T)

6–62. 44.1 kN·m

6–63. -8.75 k·ft

6–65. 6.92 k

6–66. 20.0 kN·m

6–67. 16.8 k·ft

6–69. $F_{GF} = 0$

6–70. $(V_C)_{max} = 2.62$ k; $(M_C)_{max} = 52.5$ k·ft

6–71. 67.8 kN·m

6–73. 8.04 kip·ft

6–74. $V_{max} = +12$ kN; $M_{max} = -46.8$ kN·m

6–75. $V_{max} = 17.8$ k; $M_{max} = 71.1$ k·ft

6–77. 67.5 kN

6–78. 164 kN·m

6–79. 13.75 kN

6–81. 114 k·ft

6–82. 12.5 k

6–83. 90.1 k·ft

6–1P. $M_{max} = 30.9$ kN·m; $T_{max} = 169$ kN

6–2P. Case a: $F_{max(C)} = 12.3$ kN; $F_{max(T)} = 8.71$ kN
Case b: $F_{max(C)} = 12.3$ kN; $F_{max(T)} = 11.6$ kN
Choose Case a.

Chapter 7

7–1. $F_{BF} = 1.67$ kN (T); $F_{AE} = 1.67$ kN (C); $F_{EF} = 1.33$ kN (C); $F_{AB} = 1.33$ kN (T); $F_{BD} = 1.67$ kN (T); $F_{CE} = 1.67$ kN (C); $F_{DE} = 1.33$ kN (C); $F_{BC} = 1.33$ kN (T); $F_{AF} = 7.00$ kN (C); $F_{BE} = 2.00$ kN (C); $F_{CD} = 9.00$ kN (C)

7–2. $F_{AE} = 0$; $F_{BF} = 3.33$ kN (T); $F_{EF} = 2.67$ kN (C); $F_{AB} = 0$; $F_{CE} = 0$; $F_{BD} = 3.33$ kN (T); $F_{DE} = 2.67$ kN (C); $F_{BC} = 0$; $F_{AF} = 8.00$ kN (C); $F_{BE} = 4.00$ kN (C); $F_{CD} = 10.0$ kN (C)

7–3. $F_{AB} = F_{DE} = 1.875$ kN (C); $F_{BC} = F_{DC} = 1.875$ kN (C); $F_{JI} = F_{GF} = 1.875$ kN (T); $F_{IH} = F_{HG} = 1.875$ kN (T); $F_{JB} = F_{FD} = 3.125$ kN (C); $F_{AI} = F_{GE} = 3.125$ kN (T); $F_{IC} = F_{GC} = 3.125$ kN (T); $F_{BH} = F_{HD} = 3.125$ kN (C); $F_{JA} = F_{EF} = 2.50$ kN (C); $F_{IB} = F_{DG} = 5$ kN (C); $F_{HC} = 5$ kN (C)

7–5. $F_{BH} = 23.3$ kN (T); $F_{AG} = 23.3$ kN (C); $F_{GH} = 16.5$ kN (C); $F_{AB} = 7.50$ kN (T); $F_{CG} = 2.12$ kN (T); $F_{BF} = 2.12$ kN (C); $F_{BC} = 25.5$ kN (T); $F_{FG} = 34.5$ kN (C); $F_{CE} = 19.1$ kN (T); $F_{DF} = 19.1$ kN (C); $F_{CD} = 13.5$ kN (T); $F_{EF} = 22.5$ kN (C); $F_{AH} = 31.5$ kN (C); $F_{BG} = 15.0$ kN (C); $F_{CF} = 15.0$ kN (C); $F_{DE} = 28.5$ kN (C)

7–6. $F_{AG} = 0$; $F_{BH} = 46.7$ kN (T); $F_{AB} = 9.00$ kN (C); $F_{GH} = 33.0$ kN (C); $F_{BF} = 0$; $F_{CG} = 4.24$ kN (T); $F_{FG} = 36.0$ kN (C); $F_{BC} = 24.0$ kN (T); $F_{DF} = 0$; $F_{CE} = 38.2$ kN (T); $F_{CD} = 0$; $F_{EF} = 36.0$ kN (C); $F_{AH} = 48.0$ kN (C); $F_{BG} = 33.0$ kN (C); $F_{CF} = 30.0$ kN (C); $F_{DE} = 42.0$ kN (C)

7–7. $F_{FB} = 250$ lb (T); $F_{AE} = 250$ lb (C); $F_{FE} = 200$ lb (C); $F_{AB} = 200$ lb (T); $F_{AF} = 350$ lb (C); $F_{BD} = 250$ lb (T); $F_{CE} = 250$ lb (C); $F_{DE} = 200$ lb (C); $F_{BC} = 200$ lb (T); $F_{CD} = 450$ lb (C); $F_{BE} = 300$ lb (C)

7–9. $F_{CF} = 2.48$ k (T); $F_{DG} = 2.48$ k (C); $F_{FG} = 3.25$ k (C); $F_{CD} = 3.25$ k (T); $F_{BG} = 3.89$ k (T); $F_{AC} = 3.89$ k (C); $F_{BC} = 7.75$ k (T); $F_{AG} = 7.75$ k (C); $F_{EF} = 2.12$ k (C); $F_{DE} = 1.50$ k (T); $F_{DF} = 0.250$ k (C); $F_{CG} = 1.00$ k (C); $F_{AB} = 2.75$ k (T)

7–10. $F_{DG} = F_{AC} = 0$; $F_{CF} = 4.95$ k (T); $F_{CD} = 1.50$ k (T); $F_{FG} = 5.00$ k (C); $F_{BG} = 7.78$ k (T); $F_{BC} = 5.00$ k (T);

$F_{AG} = 10.5$ k (C); $F_{EF} = 2.12$ k (C);
$F_{DE} = 1.50$ k (T); $F_{DF} = 2.00$ k (C);
$F_{CG} = 5.50$ k (C); $F_{AB} = 0$

7–11. $F_{CE} = 6.67$ kN (C); $F_{DF} = 6.67$ kN (T);
$F_{CD} = 5.33$ kN (C); $F_{EF} = 5.33$ kN (T);
$F_{BF} = 15.0$ kN (C); $F_{AC} = 15.0$ kN (T);
$F_{BC} = 22.7$ kN (C); $F_{AF} = 22.7$ kN (T);
$F_{DE} = 4.00$ kN (C); $F_{CF} = 5.00$ kN (C);
$F_{AB} = 9.00$ kN (T)

7–13. $M_E = 4.86$ kN·m; $M_F = 8.64$ kN·m

7–14. $M_F = 10.8$ k·ft; $M_D = 6.075$ k·ft

7–17. $M_{HG} = 2.025$ kN·m; $M_{JI} = 2.025$ kN·m;
$M_{JK} = 3.60$ kN·m

7–18. $M_F = 5.40$ k·ft; $M_E = 8.10$ k·ft

7–19.   Pinned: $A_x = 6.00$ kN; $A_y = 18.0$ kN;
$B_x = 6.00$ kN; $B_y = 18.0$ kN
Fixed: $A_x = 6.00$ kN; $A_y = 9.00$ kN;
$M_A = 18.0$ kN·m; $B_x = 6.00$ kN;
$B_y = 9.00$ kN; $M_B = 18.0$ kN·m

7–21. $F_{EH} = F_{FI} = 0$; $F_{EG} = 10.5$ k (C); $F_{CD} = 0$;
$F_{GH} = F_{CH} = 9.375$ k (T); $F_{FG} = 4.50$ k (T);
$F_{DI} = F_{GI} = 9.375$ k (C)

7–22. $F_{EH} = F_{FI} = 0$; $F_{EG} = 8.25$ k (C); $F_{CD} = 0$;
$F_{GH} = F_{CH} = 6.56$ k (T); $F_{FG} = 2.25$ k (T);
$F_{DI} = F_{GI} = 6.56$ k (C)

7–23. $A_x = 1.50$ k; $A_y = 1.875$ k; $M_A = 9.00$ k·ft;
$B_x = 1.50$ k; $B_y = 1.875$ k; $M_B = 9.00$ k·ft;
$F_{DG} = 3.125$ k (C); $F_{CD} = 2.00$ k (T);
$F_{FG} = 1.00$ k (C); $F_{DF} = 3.125$ k (T);
$F_{DE} = 3.00$ k (C)

7–25. $F_{EG} = 27.5$ kN (T); $F_{EF} = 24.0$ kN (C);
$F_{CG} = 4.00$ kN (C); $F_{CE} = 27.5$ kN (C);
$F_{DE} = 20.0$ kN (T)

7–26. $F_{EG} = 15.0$ kN (T); $F_{CG} = 4.00$ kN (C);
$F_{EF} = 14.0$ kN (C); $F_{CE} = 15.0$ kN (C);
$F_{DE} = 10.0$ kN (T)

7–27. $A_x = B_x = 11.0$ kN; $A_y = B_y = 32.7$ kN;
$F_{CG} = 40.8$ kN (T); $F_{FG} = 54.0$ kN (C);
$F_{CD} = 18.5$ kN (T); $F_{GD} = 40.8$ kN (C);
$F_{GH} = 5.00$ kN (C); $F_{DH} = 40.8$ kN (T);
$F_{DE} = 30.5$ kN (C); $F_{HE} = 40.8$ kN (C);
$F_{HI} = 44.0$ kN (T)

7–29. $A_x = 2.00$ k; $A_y = 1.875$ k; $B_x = 2.00$ k;
$B_y = 1.875$ k; $F_{GK} = 3.125$ k (C); $F_{GF} = 0$;
$F_{JK} = 0.500$ k (T)

7–30. $F_{FG} = 0$; $F_{EH} = 0.500$ k (T); $F_{FH} = 3.125$ k (C)

7–31. $F_{FG} = 0$; $F_{EH} = 0.500$ k (C); $F_{FH} = 1.875$ k (C)

7–33. $F_{HG} = 2.52$ kN (C); $F_{KL} = 1.86$ kN (T);
$F_{HL} = 2.99$ kN (C)

7–1P. $F = 79.7$ k
It is not reasonable to assume the members are
pin connected, since such a framework is unstable.

Chapter 8

8–1. $\theta_A = \dfrac{Pa(a - L)}{2EI}$; $v_1 = \dfrac{Px_1}{6EI}\left[x_1^2 + 3a(a - L) \right]$;

$v_2 = \dfrac{Pa}{6EI}\left[3x_2(x_2 - L) + a^2 \right]$;

$v_{max} = \dfrac{Pa}{24EI}(4a^2 - 3L^2)$

8–2. $\theta_A = \dfrac{3}{8}\dfrac{PL^2}{EI}$; $v_C = \dfrac{PL^3}{6EI}$

8–3. $v_B = \dfrac{11PL^3}{48EI}$

8–5. $\theta_B = \dfrac{wa^3}{6EI}$; $v_1 = \dfrac{wx_1^2}{24EI}(-x_1^2 + 4ax_1 - 6a^2)$;

$v_3 = \dfrac{wa^3}{24EI}(4x_3 + a - 4L)$;

$v_B = \dfrac{wa^3}{24EI}(a - 4L)$

8–6. $v = \dfrac{M_0 x}{2EI}(x - L)$

$\theta_A = \dfrac{M_0 L}{2EI}$

$v_{max} = \dfrac{M_0 L^2}{8EI}$

8–7. $v_1 = \dfrac{Pb}{6EIL}\left[x_1^3 - (L^2 - b^2)x_1 \right]$

$v_2 = \dfrac{Pa}{6EIL}\left[3x_2^2 L - x_2^3 - (2L^2 + a^2)x_2 + a^2 L \right]$

8–9. $\theta_B = \dfrac{7wa^3}{6EI}$; $v_1 = \dfrac{wax_1}{12EI}(2x_1^2 - 9ax_1)$;

$v_C = \dfrac{7wa^4}{12EI}$

$v_3 = \dfrac{w}{24EI}(-x_3^4 + 8ax_3^3 - 24a^2x_3^2 + 4a^3x_3 - a^4)$

8–10. $\theta_B = 0.00268$ rad

$\Delta_{max} = 0.322$ in

8–11. $\theta_B = 0.00268$ rad

$\Delta_{max} = 0.322$ in

8–13. $\Delta_C = \dfrac{18.6 \text{ kN·m}^3}{EI}$

$\theta_A = \dfrac{11.8 \text{ kN·m}^2}{EI}$

8–14. $0.153 L$

8–15. $0.153 L$

8–17. $\dfrac{L}{3}$

8–18. $\theta_D = \dfrac{45 \text{ kN·m}^2}{EI}$

$\Delta_C = \dfrac{1620 \text{ kN·m}^3}{EI}$

8–19. $\Delta_C = \dfrac{1620 \text{ kN} \cdot \text{m}^3}{EI}$

$\theta_D = \dfrac{45 \text{ kN} \cdot \text{m}^2}{EI}$

8–21. $\theta_C = 0.00171 \text{ rad}$

$\Delta_C = 3.86 \text{ mm}$

8–22. $\theta_C = \dfrac{4}{3} \dfrac{M_0 L}{EI}$

$\Delta_C = \dfrac{5 M_0 L^2}{6EI}$

8–23. $\theta_C = \dfrac{4 M_0 L}{3EI}; \Delta_C = \dfrac{5 M_0 L^2}{6EI}$

8–25. $\theta_B = 0.00322 \text{ rad}$

$\Delta_{\max} = 0.579 \text{ in.}$

8–26. $\theta_C = \dfrac{Pa^2}{8EI}$

$\Delta_C = \dfrac{9 Pa^3}{4EI}$

8–27. $\theta_C = \dfrac{Pa^2}{8EI}$

$\Delta_C = \dfrac{9 Pa^3}{4EI}$

8–29. $\theta_C = \dfrac{M_0 a}{EI}$

$\Delta_C = \dfrac{M_0 a^2}{2EI}$

8–30. $\Delta_B = \dfrac{224 \text{ kN} \cdot \text{m}^3}{EI}$

8–31. $\theta_A = \dfrac{7 Pa^2}{6EI}$

$\Delta_D = \dfrac{Pa^3}{4EI}$

8–33. $\theta_B = \dfrac{5 Pa^2}{12EI}$

$\Delta_C = \dfrac{3 Pa^3}{4EI}$

8–34. $\Delta_B = \dfrac{M_0 L^2}{16EI}, \theta_A = \dfrac{M_0 L}{3EI}$

8–35. $\Delta_B = \dfrac{M_0 L^2}{16EI}, \theta_A = \dfrac{M_0 L}{3EI}$

8–37. $\theta_B = \dfrac{24 \text{ k} \cdot \text{ft}^2}{EI}$

$\Delta_C = \dfrac{252 \text{ k} \cdot \text{ft}^3}{EI}$

8–38. $\theta_D = \dfrac{180 \text{ kN} \cdot \text{m}^2}{EI}$

$\Delta_C = \dfrac{405 \text{ kN} \cdot \text{m}^3}{EI}$

Chapter 9

9–1. 5.15 mm

9–2. 5.15 mm

9–3. 6.23 mm

9–5. 3.79 mm

9–6. 3.79 mm

9–7. 0.0579 in

9–9. 0.156 in.

9–10. 0.156 in.

9–11. 6.95 mm

9–13. $\dfrac{809 \text{ k} \cdot \text{ft}}{AE}$

9–14. $\dfrac{809 \text{ k} \cdot \text{ft}}{AE}$

9–15. 0.0341 in.

9–17. 0.507 in.

9–18. 1.12 in.

9–19. $\Delta_C = \dfrac{PL^3}{48 \, EI}$

$\theta_B = \dfrac{PL^2}{16 \, EI}$

9–21. $\theta_C = \dfrac{3.375 \text{ kN} \cdot \text{m}^2}{EI}$

$\Delta_C = \dfrac{118 \text{ kN} \cdot \text{m}^3}{EI}$

9–22. $\theta_C = \dfrac{3.375 \text{ kN} \cdot \text{m}^2}{EI}$

$\Delta_C = \dfrac{118 \text{ kN} \cdot \text{m}^3}{EI}$

9–23. $\Delta_D = \dfrac{12{,}519 \text{ k} \cdot \text{ft}^3}{EI}$

9–25. $\Delta_C = 3.86 \text{ mm}$

$\theta_C = 0.00171 \text{ rad}$

9–26. $\theta_C = 0.00391 \text{ rad}$

$\Delta_C = 0.425 \text{ in}$

9–27. $\theta_C = 0.00391 \text{ rad}$

$\Delta_C = 0.425 \text{ in}$

9–29. $\theta_C = 0.00670$

$\Delta_C = 0.282 \text{ in.}$

9–30. $\theta_C = \dfrac{49.5 \text{ kN} \cdot \text{m}^2}{EI}$

$\Delta_C = \dfrac{140 \text{ kN} \cdot \text{m}^3}{EI}$

9–31. $\theta_C = \dfrac{49.5 \text{ kN} \cdot \text{m}^2}{EI}$

$\Delta_C = \dfrac{140 \text{ kN} \cdot \text{m}^3}{EI}$

9–33. $\Delta_C = \dfrac{1037 \text{ kN} \cdot \text{m}^3}{EI}$

9–34. $\theta_B = 0.00448 \text{ rad}$
$\Delta_B = 0.455 \text{ in}$

9–35. $\theta_B = 0.00448 \text{ rad}$
$\Delta_B = 0.455 \text{ in}$

9–37. $\dfrac{w_0 L^4}{120 EI}$

9–38. $\dfrac{w_0 L^4}{120 EI}$

9–39. 0.122 in.

9–41. $4.05(10^{-2}) \text{ rad}$

9–42. $\dfrac{1397 \text{ k} \cdot \text{ft}^3}{EI}$

9–43. $\dfrac{1397 \text{ k} \cdot \text{ft}^3}{EI}$

9–45. 0.124 in

9–46. $\dfrac{962 \text{ kN} \cdot \text{m}^3}{EI}$

9–47. $\dfrac{337.5 \text{ kN} \cdot \text{m}^2}{EI}$

9–49. 0.0433 in

9–50. 0.0433 in

9–51. 1.91 in.

9–53. 2.33 in.

9–54. 2.33 in.

9–55. $\dfrac{211 \text{ kN} \cdot \text{m}^3}{EI}$

9–57. $\theta_A = \dfrac{16.7 \text{ kN} \cdot \text{m}^2}{EI}$;
$\Delta_{B_v} = \dfrac{18.1 \text{ kN} \cdot \text{m}^3}{EI}$

9–58. $\theta_A = \dfrac{16.7 \text{ kN} \cdot \text{m}^2}{EI}$;
$\Delta_{B_v} = \dfrac{18.1 \text{ kN} \cdot \text{m}^3}{EI}$

9–59. 1.70 in.

9–61. $\dfrac{667 \text{ k} \cdot \text{ft}^3}{EI}$

9–62. $\dfrac{667 \text{ k} \cdot \text{ft}^3}{EI}$

Chapter 10

10–1. $B_y = 600 \text{ lb}; A_x = 0; A_y = 2400 \text{ lb};$
$M_A = 4800 \text{ lb} \cdot \text{ft}$

10–2. $C_x = 0; B_y = 30.75 \text{ kip}; A_y = 2.625 \text{ kip};$
$C_y = 14.6 \text{ kip}$

10–3. $B_y = \dfrac{3wL}{8}; M_A = \dfrac{wL^2}{8}; A_y = \dfrac{5wL}{8}; A_x = 0$

10–5. $A_x = 0; A_y = 75 \text{ lb}; B_y = 75 \text{ lb}; M_A = 200 \text{ lb} \cdot \text{ft}$

10–6. $A_x = 0; A_y = C_y = 17.1 \text{ k}; B_y = 37.7 \text{ k}$

10–7. $0.414 L$

10–10. $B_y = 1.25 \text{ k}; A_x = 0; A_y = 1.25 \text{ k};$
$M_A = 7.50 \text{ k} \cdot \text{ft}$

10–11. $B_y = 8.74 \text{ k}; M_{max} = 52.4 \text{ k} \cdot \text{ft}$

10–13. $C_x = 3.75 \text{ k}$
$A_x = 21.75 \text{ k}; C_y = 42.4 \text{ k};$
$A_y = 29.6 \text{ k}$

10–14. $C_y = 18.75 \text{ kN}$
$A_x = 12.0 \text{ kN}; A_y = 0.750 \text{ kN};$
$M_A = 5.25 \text{ kN} \cdot \text{m}$

10–15. $A_y = 0.9375 \text{ k}; C_x = 10.0 \text{ k}; C_y = 0.975 \text{ k};$
$M_C = 48.75 \text{ k} \cdot \text{ft}$

10–17. $A_y = 14.8 \text{ kN}; C_x = 0; C_y = 17.2 \text{ kN};$
$M_C = 4.92 \text{ kN} \cdot \text{m}$

10–18. $M_A = 55.4 \text{ kN} \cdot \text{m}; A_x = 24 \text{ kN}; C_y = 2.08 \text{ kN};$
$A_y = 2.08 \text{ kN}; M_B = 16.6 \text{ kN} \cdot \text{m}$

10–19. $A_y = 18.2 \text{ k}$
$D_x = 10.0 \text{ k}; D_y = 11.8 \text{ k}; M_D = 67.5 \text{ k} \cdot \text{ft}$

10–21. $D_x = 5.41 \text{ k}$
$A_x = 2.59 \text{ k}; D_y = 4.65 \text{ k}; A_y = 4.65 \text{ k}$

10–22. $A_x = 31.4 \text{ kN}; A_y = 11.4 \text{ kN}; B_x = 48.6 \text{ kN};$
$B_y = 11.4 \text{ kN}; M_C = 68.6 \text{ kN} \cdot \text{m}$

10–23. $B_x = 1.53 \text{ kN}$
$A_x = 1.53 \text{ kN}; B_y = 7.50 \text{ kN}; A_y = 15.0 \text{ kN}$

10–25. $F_{CD} = 2.83 \text{ kN (C)}; F_{AD} = 2.83 \text{ kN (T)}; F_{BD} = 0$

10–26. $F_{CB} = 3.06 \text{ k (C)}; F_{AC} = 0.823 \text{ k (C)};$
$F_{DC} = 6.58 \text{ k (T)}; F_{DB} = 5.10 \text{ k (T)};$
$F_{AB} = 10.1 \text{ k (C)}; F_{DA} = 4.94 \text{ k (T)}$

10–27. 7.91 kN (C)

10–29. 2.73 kN (T)

10–30. 2.13 kN (T)

10–31. 3.70 kN (C)

10–33. 1.19 k (T)

10–34. $F_{sp} = \dfrac{3wL}{10}; B_y = \dfrac{7wL}{5}$

10–35. 3.02 k

10–37. $F_{CD} = 32.4 \text{ k (C)}$
$F_{CA} = F_{CB} = 42.1 \text{ k (T)}$

10–38. $M_{max} = 45.8 \text{ k} \cdot \text{ft}$

10–39. $D_x = 0; D_x = \dfrac{P}{2}; M_D = \dfrac{PL}{6}$

10–41. $A_y = \dfrac{P}{2}; A_x = 0; M_A = \dfrac{PL}{2}; B_y = \dfrac{P}{2};$
$B_x = 0; M_B = \dfrac{PL}{2}$

Chapter 11

11–1. $M_{BC} = \frac{2}{7}M_0; M_{AB} = -\frac{1}{7}M_0;$

$M_{BA} = -\frac{2}{7}M_0$

11–2. $M_{max} = \frac{7}{40}PL$

11–3. $M_{CD} = -38.4 \text{ k} \cdot \text{ft}; M_{BA} = -9.60 \text{ k} \cdot \text{ft};$
$M_{BC} = 9.60 \text{ k} \cdot \text{ft}; M_{CB} = 38.4 \text{ k} \cdot \text{ft}$

11–5. $M_{AB} = 5 \text{ kN} \cdot \text{m}; M_{BA} = 10 \text{ kN} \cdot \text{m};$
$M_{BC} = -10 \text{ kN} \cdot \text{m}; M_{CB} = 25 \text{ kN} \cdot \text{m}$

11–6. $A_y = 61.8 \text{ k}; D_y = 34.3 \text{ k}; B_y = 144 \text{ k};$
$C_y = 136 \text{ k}$

11–7. $M_{AB} = 2.25 \text{ kN} \cdot \text{m}; M_{BA} = 4.50 \text{ kN} \cdot \text{m};$
$M_{BC} = -4.50 \text{ kN} \cdot \text{m}$

11–9. $A_y = 10.7 \text{ k}; B_y = 14.4 \text{ k};$
$C_y = -1.07 \text{ k}$

11–10. $M_{AB} = -2680 \text{ k} \cdot \text{ft}; M_{BA} = -1720 \text{ k} \cdot \text{ft};$
$M_{BC} = 1720 \text{ k} \cdot \text{ft}; M_{CB} = 0$

11–11. $M_{AB} = -24.5 \text{ k} \cdot \text{ft}; M_{BA} = -0.923 \text{ k} \cdot \text{ft};$
$M_{BC} = 0.923 \text{ k} \cdot \text{ft}; M_{CB} = 27.2 \text{ k} \cdot \text{ft};$
$M_{CD} = -27.2 \text{ k} \cdot \text{ft}$

11–13. $M_{BA} = 42.9 \text{ k} \cdot \text{ft}; M_{BD} = -20.7 \text{ k} \cdot \text{ft};$
$M_{DB} = 64.0 \text{ k} \cdot \text{ft}; M_{BC} = -22.2 \text{ k} \cdot \text{ft};$
$M_{CB} = -11.1 \text{ k} \cdot \text{ft}$

11–14. $A_x = 2.95 \text{ k}; A_y = 4.10 \text{ k}; C_x = 4.55 \text{ k};$
$C_y = 0.902 \text{ k}$

11–15. $M_{AB} = 3.93 \text{ k} \cdot \text{ft}; M_{BA} = 7.85 \text{ k} \cdot \text{ft};$
$M_{BC} = -7.85 \text{ k} \cdot \text{ft}; M_{CB} = 7.85 \text{ k} \cdot \text{ft};$
$M_{CD} = -7.85 \text{ k} \cdot \text{ft}; M_{DC} = -3.93 \text{ k} \cdot \text{ft}$

11–17. $M_{AB} = -1.50 \text{ kN} \cdot \text{m}; M_{CB} = 1.50 \text{ kN} \cdot \text{m};$
$M_{BA} = 6.00 \text{ kN} \cdot \text{m}; M_{BC} = 3.00 \text{ kN} \cdot \text{m}$

11–18. $M_{AB} = -330 \text{ kN} \cdot \text{m}; M_{CB} = 400 \text{ kN} \cdot \text{m};$
$M_{DB} = 0$

11–19. $M_{BC} = -2.90 \text{ k} \cdot \text{ft}; M_{BA} = 2.90 \text{ k} \cdot \text{ft}$

11–21. $M_{AB} = -10.4 \text{ kN} \cdot \text{m}; M_{BA} = -6.26 \text{ kN} \cdot \text{m};$
$M_{BC} = 6.26 \text{ kN} \cdot \text{m}; M_{DC} = -7.30 \text{ kN} \cdot \text{m}$

11–22. $M_{AB} = -4.64 \text{ k} \cdot \text{ft}; M_{BA} = -110 \text{ k} \cdot \text{ft};$
$M_{BC} = 110 \text{ k} \cdot \text{ft}; M_{CB} = 155 \text{ k} \cdot \text{ft};$
$M_{CD} = -155 \text{ k} \cdot \text{ft}; M_{DC} = -243 \text{ k} \cdot \text{ft}$

11–23. $M_{AB} = -19.4 \text{ k} \cdot \text{ft}; M_{BA} = -15.0 \text{ k} \cdot \text{ft};$
$M_{BC} = 15.0 \text{ k} \cdot \text{ft}; M_{CB} = 20.1 \text{ k} \cdot \text{ft};$
$M_{CD} = -20.1 \text{ k} \cdot \text{ft}; M_{DC} = -36.9 \text{ k} \cdot \text{ft}$

11–1P. $M_{max} = 14.0 \text{ k} \cdot \text{ft}$

Chapter 12

12–1. $M_{AB} = -34.8 \text{ k} \cdot \text{ft}; M_{BA} = 45.6 \text{ k} \cdot \text{ft};$
$M_{BC} = -45.6 \text{ k} \cdot \text{ft}; M_{CB} = 67.2 \text{ k} \cdot \text{ft}$

12–2. $M_{AB} = 6.67 \text{ k} \cdot \text{ft}; M_{BA} = 13.3 \text{ k} \cdot \text{ft};$
$M_{BC} = -13.3 \text{ k} \cdot \text{ft}; M_{CB} = 23.3 \text{ k} \cdot \text{ft}$

12–3. $M_{AB} = -167 \text{ k} \cdot \text{ft}; M_{BA} = 66.7 \text{ k} \cdot \text{ft};$
$M_{BC} = -66.7 \text{ k} \cdot \text{ft}; M_{CB} = -33.3 \text{ k} \cdot \text{ft}$

12–5. $M_{AB} = -11.6 \text{ k} \cdot \text{ft}; M_{BA} = 12.8 \text{ k} \cdot \text{ft};$
$M_{BC} = -12.8 \text{ k} \cdot \text{ft}; M_{CB} = 13.8 \text{ k} \cdot \text{ft}$

12–6. $M_{BA} = 16.0 \text{ kN} \cdot \text{m}; M_{BC} = -16.0 \text{ kN} \cdot \text{m};$
$M_{CB} = 16.0 \text{ kN} \cdot \text{m}; M_{CD} = -16.0 \text{ kN} \cdot \text{m}$

12–7. $M_{AB} = 44.4 \text{ k} \cdot \text{ft}; M_{BA} = 88.8 \text{ k} \cdot \text{ft};$
$M_{BC} = -88.8 \text{ k} \cdot \text{ft}; M_{CB} = 72.4 \text{ k} \cdot \text{ft};$
$M_{CD} = -72.4 \text{ k} \cdot \text{ft}; M_{DC} = -36.2 \text{ k} \cdot \text{ft}$

12–9. $A_y = 6 \text{ kN}; D_y = 6 \text{ kN}$

12–10. $M_{AD} = 48 \text{ k} \cdot \text{ft}; M_{AB} = -48 \text{ k} \cdot \text{ft};$
$M_{BA} = -24 \text{ k} \cdot \text{ft}; M_{BC} = 24 \text{ k} \cdot \text{ft};$
$M_{CB} = 48 \text{ k} \cdot \text{ft}; M_{CE} = -48 \text{ k} \cdot \text{ft}$

12–11. $M_{AB} = -30 \text{ kN} \cdot \text{m}; M_{BA} = 15 \text{ kN} \cdot \text{m};$
$M_{BC} = -15 \text{ kN} \cdot \text{m}$

12–13. $M_{BA} = 19.6 \text{ kN} \cdot \text{m}; M_{BC} = -19.6 \text{ kN} \cdot \text{m}$

12–14. $M_{AB} = 0; M_{BA} = 76.2 \text{ k} \cdot \text{ft}; M_{BD} = 21.8 \text{ k} \cdot \text{ft};$
$M_{BC} = -98.0 \text{ k} \cdot \text{ft}; M_{CB} = 89.4 \text{ k} \cdot \text{ft};$
$M_{CE} = -89.4 \text{ k} \cdot \text{ft}; M_{EC} = 0; M_{DB} = 0$

12–15. $A_x = 29.3 \text{ k}; A_y = 96.0 \text{ k}; M_A = 146 \text{ k} \cdot \text{ft};$
$D_x = 29.3 \text{ k}; D_y = 96.0 \text{ k}; M_D = 146 \text{ k} \cdot \text{ft}$

12–17. $M_{AB} = 0; M_{BA} = 181 \text{ k} \cdot \text{ft}; M_{BD} = -77.3 \text{ k} \cdot \text{ft};$
$M_{BC} = -103 \text{ k} \cdot \text{ft}; M_{CB} = 38.3 \text{ k} \cdot \text{ft};$
$M_{DB} = 0$

12–18. $M_{BA} = 19.9 \text{ k} \cdot \text{ft}; M_{BC} = -19.9 \text{ k} \cdot \text{ft};$
$M_{CB} = 22.4 \text{ k} \cdot \text{ft}; M_{CD} = -6.77 \text{ k} \cdot \text{ft};$
$M_{CE} = -15.6 \text{ k} \cdot \text{ft}$

12–19. $M_B = -14.9 \text{ k} \cdot \text{ft}$

12–21. $M_{BA} = 168 \text{ k} \cdot \text{ft}; M_{BD} = -168 \text{ k} \cdot \text{ft};$
$M_{DB} = -47.8 \text{ k} \cdot \text{ft}; M_{DE} = 47.8 \text{ k} \cdot \text{ft};$
$M_{AB} = M_{DC} = 0$

12–22. $M_{AB} = -24.8 \text{ k} \cdot \text{ft}; M_{BA} = 26.1 \text{ k} \cdot \text{ft};$
$M_{BC} = -26.1 \text{ k} \cdot \text{ft}; M_{CB} = 50.7 \text{ k} \cdot \text{ft};$
$M_{CD} = -50.7 \text{ k} \cdot \text{ft}; M_{DC} = -40.7 \text{ k} \cdot \text{ft}$

12–23. $M_{AB} = -94.5 \text{ k} \cdot \text{ft}; M_{BA} = -2.19 \text{ k} \cdot \text{ft};$
$M_{BC} = 2.19 \text{ k} \cdot \text{ft}; M_{CB} = 116 \text{ k} \cdot \text{ft};$
$M_{CD} = -116 \text{ k} \cdot \text{ft}; M_{DC} = -118 \text{ k} \cdot \text{ft}$

12–25. $A_x = 2.44 \text{ k}; A_y = 6.50 \text{ k}; D_x = 2.44 \text{ k};$
$D_y = 3.50 \text{ k}$

12–26. $M_{DA} = 14.2 \text{ k} \cdot \text{ft}; M_{DC} = -14.2 \text{ k} \cdot \text{ft};$
$M_{CD} = -7.54 \text{ k} \cdot \text{ft}; M_{CB} = 7.54 \text{ k} \cdot \text{ft}$

Chapter 13

13–1. $M_{AB} = -348 \text{ k} \cdot \text{ft}; M_{BA} = 301 \text{ k} \cdot \text{ft};$
$M_{BC} = -301 \text{ k} \cdot \text{ft}; M_{CB} = 348 \text{ k} \cdot \text{ft}$

13–2. $M_{AB} = -348 \text{ k} \cdot \text{ft}; M_{BA} = 301 \text{ k} \cdot \text{ft};$
$M_{BC} = -301 \text{ k} \cdot \text{ft}; M_{CB} = 348 \text{ k} \cdot \text{ft}$

13–3. $M_{AC} = 37.6 \, \text{k} \cdot \text{ft}; M_{CA} = 75.1 \, \text{k} \cdot \text{ft};$
$M_{CB} = -75.1 \, \text{k} \cdot \text{ft}; M_{BC} = 369 \, \text{k} \cdot \text{ft}$

13–5. $M_{AB} = 0; M_{BA} = -42.3 \, \text{k} \cdot \text{ft};$
$M_{BF} = -44.7 \, \text{k} \cdot \text{ft}; M_{BC} = 87.0 \, \text{k} \cdot \text{ft};$
$M_{CB} = 185 \, \text{k} \cdot \text{ft}; M_{CE} = 213 \, \text{k} \cdot \text{ft};$
$M_{CD} = -398 \, \text{k} \cdot \text{ft}; M_{EC} = 106 \, \text{k} \cdot \text{ft};$
$M_{FB} = -22.4 \, \text{k} \cdot \text{ft}; M_{DC} = 0$

13–6. $M_{AB} = 0; M_{BA} = -42.3 \, \text{k} \cdot \text{ft};$
$M_{BF} = -44.7 \, \text{k} \cdot \text{ft}; M_{BC} = 87.0 \, \text{k} \cdot \text{ft};$
$M_{CB} = 185 \, \text{k} \cdot \text{ft}; M_{CE} = 213 \, \text{k} \cdot \text{ft};$
$M_{CD} = -398 \, \text{k} \cdot \text{ft}; M_{EC} = 106 \, \text{k} \cdot \text{ft};$
$M_{FB} = -22.4 \, \text{k} \cdot \text{ft}; M_{DC} = 0$

13–7. $M_{AB} = 1.75 \, \text{k} \cdot \text{ft}; M_{BA} = 3.51 \, \text{k} \cdot \text{ft};$
$M_{BC} = -3.51 \, \text{k} \cdot \text{ft}; M_{CB} = 3.51 \, \text{k} \cdot \text{ft};$
$M_{CD} = -3.51 \, \text{k} \cdot \text{ft}; M_{DC} = -1.75 \, \text{k} \cdot \text{ft}$

13–9. $M_{BA} = 29.0 \, \text{k} \cdot \text{ft}; M_{BC} = -29.0 \, \text{k} \cdot \text{ft};$
$M_{CB} = 29.0 \, \text{k} \cdot \text{ft}; M_{CD} = -29.0 \, \text{k} \cdot \text{ft}$

13–10. $M_{BA} = 29.0 \, \text{k} \cdot \text{ft}; M_{BC} = -29.0 \, \text{k} \cdot \text{ft}$
$M_{CB} = 29.0 \, \text{k} \cdot \text{ft}; M_{CD} = -29.0 \, \text{k} \cdot \text{ft}$

13–11. $M_{CD} = M_{BA} = 180 \, \text{k} \cdot \text{ft}; M_{CF} = M_{BE} = 94.6 \, \text{k} \cdot \text{ft}$
$M_{CB} = M_{BC} = -274 \, \text{k} \cdot \text{ft}; M_{FC} = M_{EB} = 47.3 \, \text{k} \cdot \text{ft}$

Chapter 14

14–1.

$$\mathbf{K} = AE \begin{bmatrix} 8.0194\text{E}{-}02 & 3.8058\text{E}{-}03 & -4.419\text{E}{-}02 & 4.419\text{E}{-}02 & 0 & 0 & -0.036 & -0.048 \\ 3.8058\text{E}{-}03 & 0.2332 & 4.419\text{E}{-}02 & -4.419\text{E}{-}02 & 0 & -0.125 & -0.048 & -0.064 \\ -4.419\text{E}{-}02 & 4.419\text{E}{-}02 & 4.419\text{E}{-}02 & -4.419\text{E}{-}02 & 0 & 0 & 0 & 0 \\ 4.419\text{E}{-}02 & -4.419\text{E}{-}02 & -4.419\text{E}{-}02 & 4.419\text{E}{-}02 & 0 & 0 & 0 & 0 \\ 0 & 0 & 0 & 0 & 0 & 0 & 0 & 0 \\ 0 & -0.125 & 0 & 0 & 0 & 0.125 & 0 & 0 \\ -0.036 & -0.048 & 0 & 0 & 0 & 0 & 0.036 & 0.048 \\ -0.048 & -0.064 & 0 & 0 & 0 & 0 & 0.048 & 0.064 \end{bmatrix}$$

14–2. $q_1 = 4.52 \, \text{k}$
$q_2 = 1.20 \, \text{k}$
$q_3 = -3.00 \, \text{k}$

14–3.

$$\mathbf{K} = 10^6 \begin{bmatrix} 0.15 & 0 & 0 & 0 & -0.075 & 0 & -0.075 & 0 \\ 0 & 0.1 & 0 & -0.1 & 0 & 0 & 0 & 0 \\ 0 & 0 & 0.0768 & 0 & -0.0384 & -0.0288 & -0.0384 & 0.0288 \\ 0 & -0.1 & 0 & 0.1432 & -0.0288 & -0.0216 & -0.0288 & -0.0216 \\ -0.075 & 0 & -0.0384 & -0.0288 & 0.1134 & 0.0288 & 0 & 0 \\ 0 & 0 & -0.0288 & -0.0216 & 0.028 & 0.0216 & 0 & 0 \\ -0.075 & 0 & -0.0384 & 0.0288 & 0 & 0 & 0.1134 & -0.0288 \\ 0 & 0 & 0.0288 & -0.0216 & 0 & 0 & -0.0288 & 0.0216 \end{bmatrix}$$

14–5.

$$\mathbf{K} = AE \begin{bmatrix} 0.03536 & -0.03536 & 0 & 0 & -0.03536 & 0.03536 & 0 & 0 & 0 & 0 \\ -0.03536 & 0.13536 & 0 & -0.10 & 0.03536 & -0.03536 & 0 & 0 & 0 & 0 \\ 0 & 0 & 0.10 & 0 & -0.10 & 0 & 0 & 0 & 0 & 0 \\ 0 & -0.10 & 0 & 0.20 & 0 & 0 & 0 & 0 & 0 & -0.10 \\ -0.03536 & 0.03536 & -0.10 & 0 & 0.17071 & 0 & 0 & 0 & -0.03536 & -0.03536 \\ 0.03536 & -0.03536 & 0 & 0 & 0 & 0.17071 & 0 & -10 & -0.03536 & -0.03536 \\ 0 & 0 & 0 & 0 & 0 & 0 & 0 & 0 & 0 & 0 \\ 0 & 0 & 0 & 0 & 0 & -0.10 & 0 & -0.10 & 0 & 0 \\ 0 & 0 & 0 & 0 & -0.03536 & -0.03536 & 0 & 0 & 0.03536 & 0.03536 \\ 0 & 0 & 0 & -0.10 & -0.03536 & -0.03536 & 0 & 0 & 0.03536 & 0.03536 \end{bmatrix}$$

14–6. $D_1 = \dfrac{933}{AE}$

$q_1 = 11.3 \text{ k (C)}$

14–7.

$$\mathbf{K} = AE \begin{bmatrix} 0.8536 & 0.3536 & -0.3536 & -0.3536 & 0 & 0 & -0.5 & 0 \\ 0.3536 & 0.3536 & -0.3536 & -0.3536 & 0 & 0 & 0 & 0 \\ -0.3536 & -0.3536 & 1.0606 & 0.3536 & -0.3536 & -0.3536 & -0.3536 & 0.3536 \\ -0.3536 & -0.3536 & 0.3536 & 1.0606 & -0.3536 & -0.3536 & 0.3536 & -0.3536 \\ 0 & 0 & -0.3536 & -0.3536 & 0.8536 & 0.3536 & 0 & -0.5 \\ 0 & 0 & -0.3536 & -0.3536 & 0.3536 & 0.3536 & 0 & 0 \\ -0.5 & 0 & -0.3536 & 0.3536 & 0 & 0 & 0.8536 & -0.3536 \\ 0 & 0 & 0.3536 & -0.3536 & -0.5 & 0 & -0.3536 & 0.8536 \end{bmatrix}$$

14–9.

$$\mathbf{K} = 10^3 \begin{bmatrix} 113.4 & 28.8 & -38.4 & -28.8 & 0 & 0 & -75 & 0 \\ 28.8 & 121.6 & -28.8 & -21.6 & 0 & 100 & 0 & 0 \\ -38.4 & -28.8 & 113.4 & 28.8 & -75 & 0 & 0 & 0 \\ -28.8 & -21.6 & 28.8 & 121.6 & 0 & 0 & 0 & -100 \\ 0 & 0 & -75 & 0 & 113.4 & -28.8 & -38.4 & 28.8 \\ 0 & -100 & 0 & 0 & -28.8 & 121.6 & 28.8 & -21.6 \\ -75 & 0 & 0 & 0 & -38.4 & 28.8 & 113.4 & -28.8 \\ 0 & 0 & 0 & -100 & 28.8 & -21.6 & -28.8 & 121.6 \end{bmatrix}$$

14–10. $q_6 = 729 \text{ N}$

14–11.

$$\mathbf{K} = AE \begin{array}{c} \begin{array}{cccccc} 1 & 2 & 3 & 4 & 5 & 6 \end{array} \\ \begin{bmatrix} 0.40533 & 0.096 & 0.01697 & -0.11879 & -0.33333 & 0 \\ 0.096 & 0.128 & 0.02263 & -0.15839 & 0 & 0 \\ 0.01697 & 0.02263 & 0.129 & -0.153 & 0 & 0.17678 \\ -0.11879 & -0.15839 & -0.153 & 0.321 & 0 & -0.17678 \\ -0.33333 & 0 & 0 & 0 & 0.33333 & 0 \\ 0 & 0 & 0.17678 & -0.17678 & 0 & 0.25 \end{bmatrix} \begin{array}{c} 1 \\ 2 \\ 3 \\ 4 \\ 5 \\ 6 \end{array} \end{array}$$

Chapter 15

15–1. $q_6 = 22.5 \text{ kN} \cdot \text{m}; q_2 = -11.25 \text{ kN} \cdot \text{m}$

15–2. $Q_4 = 4.125 \text{ kN}; Q_5 = 15.75 \text{ kN};$
$Q_6 = 4.125 \text{ kN}$

15–3. $F_1 = F_3 = 25.5 \text{ k}; F_2 = 21.0 \text{ k}$

15–5. $M_2 = M_3 = 30.0 \text{ k} \cdot \text{ft}$

15–6. $M_3 = 2.25 \text{ kN} \cdot \text{m}; M_1 = 4.50 \text{ kN} \cdot \text{m}$

15–7. $M_2 = 18.5 \text{ kN} \cdot \text{m}; M_3 = 20.4 \text{ kN} \cdot \text{m}$

15–9. $Q_2 = -\dfrac{wL^2}{6}; Q_3 = wL; Q_4 = -\dfrac{wL^2}{3}$

15–10. $Q_3 = 122 \text{ k} \cdot \text{ft}; Q_5 = 230 \text{ k} \cdot \text{ft}$

15–11. $Q_3 = 131 \text{ k} \cdot \text{ft}; Q_5 = 236 \text{ k} \cdot \text{ft}$

Chapter 16

16–1.

$$\mathbf{K} = \begin{bmatrix}
851250 & 0 & 22500 & 22500 & -11250 & 0 & -840000 & 0 & 0 \\
0 & 1055760 & -14400 & 0 & 0 & -1050000 & 0 & -5760 & -14400 \\
22500 & -14400 & 108000 & 30000 & -22500 & 0 & 0 & 14400 & 24000 \\
22500 & 0 & 30000 & 60000 & -22500 & 0 & 0 & 0 & 0 \\
-11250 & 0 & -22500 & -22500 & 11250 & 0 & 0 & 0 & 0 \\
0 & -1050000 & 0 & 0 & 0 & 1050000 & 0 & 0 & 0 \\
-840000 & 0 & 0 & 0 & 0 & 0 & 840000 & 0 & 0 \\
0 & -5760 & 14400 & 0 & 0 & 0 & 0 & 5760 & 14400 \\
0 & -14400 & 24000 & 0 & 0 & 0 & 0 & 14400 & 48000
\end{bmatrix}$$

16–2. $Q_5 = -36.3$ kN
$Q_6 = -46.4$ kN
$Q_7 = 36.3$ kN
$Q_8 = 46.4$ kN
$Q_9 = 77.1$ kN·m

16–3.

$$\mathbf{K} = \begin{bmatrix}
2469.11 & 0 & 3776.04 & -2416.67 & 0 & 0 & -52.44 & 0 & 3776.04 & 0 & 0 & 0 \\
0 & 1782.11 & 8496.09 & 0 & -177.00 & 8496.09 & 0 & -1611.11 & 0 & 0 & 0 & 0 \\
3776.04 & 8496.09 & 906250 & 0 & -8496.09 & 271875 & -3776.04 & 0 & 181250 & 0 & 0 & 0 \\
-2416.67 & 0 & 0 & 2469.11 & 0 & 3776.04 & 0 & 0 & 0 & -52.44 & 0 & 3776.04 \\
0 & -177.00 & -8496.09 & 0 & 1788.11 & -8496.09 & 0 & 0 & 0 & 0 & -1611.11 & 0 \\
0 & 8496.09 & 271875 & 3776.04 & -8496.09 & 906250 & 0 & 0 & 0 & -3776.04 & 0 & 181250 \\
-52.44 & 0 & -3776.04 & 0 & 0 & 0 & 52.44 & 0 & -3776.04 & 0 & 0 & 0 \\
0 & -1611.11 & 0 & 0 & 0 & 0 & 0 & 1611.11 & 0 & 0 & 0 & 0 \\
3776.04 & 0 & 181250 & 0 & 0 & 0 & -3776.04 & 0 & 362500 & 0 & 0 & 0 \\
0 & 0 & 0 & -52.44 & 0 & -3776.04 & 0 & 0 & 0 & 52.44 & 0 & -3776.04 \\
0 & 0 & 0 & 0 & -1611.11 & 0 & 0 & 0 & 0 & 0 & 1611.11 & 0 \\
0 & 0 & 0 & 3776.04 & 0 & 181250 & 0 & 0 & 0 & -3776.04 & 0 & 362500
\end{bmatrix}$$

16–5.

$$\mathbf{K} = \begin{bmatrix}
6123.25 & 0 & 5873.84 & 0 & 5873.84 & -6041.67 & 0 & -81.58 & 0 \\
0 & 6123.25 & 5873.84 & 5873.84 & 0 & 0 & -81.58 & 0 & -6041.67 \\
5873.84 & 5873.84 & 1127777.78 & 281944.44 & 281944.44 & 0 & -5873.84 & -5873.84 & 0 \\
0 & 5871.84 & 281944.44 & 561888.89 & 0 & 0 & -5873.84 & 0 & 0 \\
5873.84 & 0 & 281944.44 & 0 & 563888.89 & 0 & 0 & -5873.84 & 0 \\
-6041.67 & 0 & 0 & 0 & 0 & 6041.67 & 0 & 0 & 0 \\
0 & -81.58 & -5873.84 & -5873.84 & 0 & 0 & 81.53 & 0 & 0 \\
-81.58 & 0 & -5873.84 & 0 & -5873.84 & 0 & 0 & 81.85 & 0 \\
0 & -6041.67 & 0 & 0 & 0 & 0 & 0 & 0 & 6041.67
\end{bmatrix}$$

16–6. $Q_7 = 4.00$ k
$Q_8 = 0$
$Q_9 = 4.00$ k

16–7.

$$\mathbf{K} = \begin{bmatrix} 4833.33 & 0 & 0 & -4833.33 & 0 & 0 & 0 & 0 & 0 \\ 0 & 130.90 & 7854.17 & 0 & -130.90 & 7854.17 & 0 & 0 & 0 \\ 0 & 7854.17 & 628333.33 & 0 & -7854.17 & 314166.67 & 0 & 0 & 0 \\ -4833.33 & 0 & 0 & 4909.08 & 0 & 5454.28 & -75.75 & 0 & 5454.28 \\ 0 & -130.90 & -7854.17 & 0 & 4158.68 & -7854.17 & 0 & -4027.78 & 0 \\ 0 & 7854.17 & 314166.67 & 5454.28 & -7854.17 & 1151944.44 & -5454.28 & 0 & 261805.55 \\ 0 & 0 & 0 & -75.75 & 0 & -5454.28 & 75.75 & 0 & -5454.28 \\ 0 & 0 & 0 & 0 & -4027.78 & 0 & 0 & 4027.78 & 0 \\ 0 & 0 & 0 & 5454.28 & 0 & 261805.55 & -5454.28 & 0 & 523611.11 \end{bmatrix}$$

16–9.

$$\mathbf{K} = \begin{bmatrix} 2249.89 & 0 & 11328.13 & 11328.13 & 0 & -236.00 & 0 & -2013.89 & 0 \\ 0 & 3090.76 & -5034.72 & 0 & -5034.72 & 0 & -3020.83 & 0 & -69.93 \\ 11328.13 & -5034.72 & 1208333.33 & 362500 & 241666.67 & -11328.13 & 0 & 0 & 5034.72 \\ 11328.13 & 0 & 362500 & 725000 & 0 & -11328.13 & 0 & 0 & 0 \\ 0 & -5034.72 & 241666.67 & 0 & 483333.33 & 0 & 0 & 0 & 5034.72 \\ -236.00 & 0 & -11328.13 & -11328.13 & 0 & 236.00 & 0 & 0 & 0 \\ 0 & -3020.83 & 0 & 0 & 0 & 0 & 3020.83 & 0 & 0 \\ -2013.89 & 0 & 0 & 0 & 0 & 0 & 0 & 2013.89 & 0 \\ 0 & -69.93 & 5034.72 & 0 & 5034.72 & 0 & 0 & 0 & 69.93 \end{bmatrix}$$

16–10. For member 1:
$q_{Nx'} = 0.260$ k; $q_{Ny'} = 1.03$ k; $q_{Nz'} = 0$; $q_{Fx'} = -0.260$ k; $q_{Fy'} = 1.37$ k; $q_{Fz'} = -2.08$ k $\cdot$ ft

For member 2:
$q_{Nx'} = 1.37$ k; $q_{Ny'} = 0.260$ k; $q_{Nz'} = 2.08$ k $\cdot$ ft; $q_{Fx'} = -1.37$ k; $q_{Fy'} = -0.260$ k; $q_{Fz'} = 0$

16–11.

$$\mathbf{k}_1 = \begin{bmatrix} 81.581 & 0 & -5873.843 & -81.581 & 0 & -5873.843 \\ 0 & 6041.667 & 0 & 0 & -6041.667 & 0 \\ -5873.843 & 0 & 563888.88 & 5873.843 & 0 & 281944.44 \\ -81.581 & 0 & 5873.843 & 81.581 & 0 & 5873.843 \\ 0 & -6041.667 & 0 & 0 & 6041.667 & 0 \\ -5873.843 & 0 & 281944.44 & 5873.843 & 0 & 563888.88 \end{bmatrix}$$

$$\mathbf{k}_2 = \begin{bmatrix} 6041.667 & 0 & 0 & -6041.667 & 0 & 0 \\ 0 & 81.581 & 5873.843 & 0 & -81.581 & 5873.843 \\ 0 & 5873.843 & 563888.88 & 0 & -5873.843 & 281944.44 \\ -6041.667 & 0 & 0 & 6041.667 & 0 & 0 \\ 0 & -81.581 & -5873.843 & 0 & 81.581 & -5873.843 \\ 0 & 5873.843 & 281944.44 & 0 & -5873.843 & 563888.88 \end{bmatrix}$$

16–13.

$$
\mathbf{K} = \begin{bmatrix}
2249.892 & 0 & 11328.125 & 11328.125 & 0 & -236 & 0 & -2013.89 & 0 \\
0 & 3090.76 & -5034.722 & 0 & -5034.722 & 0 & -3020.833 & 0 & -69.927 \\
11328.125 & -5034.722 & 1208.33(10^2) & 362500 & 241666.67 & -11328.125 & 0 & 0 & 5034.722 \\
11328.125 & 0 & 362500 & 725000 & 0 & -11328.125 & 0 & 0 & 0 \\
0 & -5034.722 & 241666.67 & 0 & 483333.33 & 0 & 0 & 0 & 5034.722 \\
-236 & 0 & -11328.125 & -11328.125 & 0 & 236 & 0 & 0 & 0 \\
0 & -3020.833 & 0 & 0 & 0 & 0 & 3020.833 & 0 & 0 \\
-2013.89 & 0 & 0 & 0 & 0 & 0 & 0 & 2013.89 & 0 \\
0 & -69.927 & 5034.722 & 0 & 5034.722 & 0 & 0 & 0 & 69.927
\end{bmatrix}
$$

16–14. $D_4 = -0.0680(10^{-3})$ rad; $D_5 = -0.271(10^{-3})$ rad; $Q_6 = -0.818$ k; $Q_7 = 3.05$ k; $Q_8 = 0.818$ k; $Q_9 = 1.95$ k

16–15. $Q_9 = -1.11$ k; $Q_{10} = 2.50$ k; $Q_{11} = -2.89$ k; $Q_{12} = -2.50$ k

16–17. $Q_6 = 8.50$ kN; $Q_7 = 52.6$ kN; $Q_8 = 56.5$ kN; $Q_9 = 3.43$ kN

Index

Geometric Properties of Areas

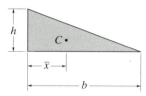

Triangle

$$A = \frac{1}{2}bh$$

$$\bar{x} = \frac{1}{3}b$$

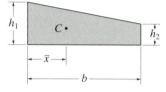

Trapezoid

$$A = \frac{1}{2}b(h_1 + h_2)$$

$$\bar{x} = \frac{b(2h_2 + h_1)}{3(h_1 + h_2)}$$

Semi Parabola

$$A = \frac{2}{3}bh$$

$$\bar{x} = \frac{3}{8}b$$

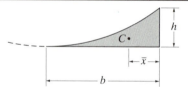

Parabolic spandrel

$$A = \frac{1}{3}bh$$

$$\bar{x} = \frac{1}{4}b$$

Semi-segment of nth degree curve

$$A = bh\left(\frac{n}{n+1}\right)$$

$$\bar{x} = \frac{b(n+1)}{2(n+2)}$$

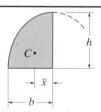

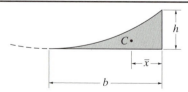

Spandrel of nth degree curve

$$A = bh\left(\frac{1}{n+1}\right)$$

$$\bar{x} = \frac{b}{n+2}$$